SPECTROSCOPIC ASTROPHYSICS

Otto Struve

SPECTROSCOPIC ASTROPHYSICS

An Assessment of the Contributions of
Otto Struve

Edited by

G. H. HERBIG

Lick Observatory
University of California

University of California Press
Berkeley, Los Angeles, London   1970

UNIVERSITY OF CALIFORNIA PRESS
BERKELEY AND LOS ANGELES, CALIFORNIA

UNIVERSITY OF CALIFORNIA PRESS, LTD.
LONDON, ENGLAND

COPYRIGHT © 1970, BY
THE REGENTS OF THE UNIVERSITY OF CALIFORNIA

STANDARD BOOK NUMBER 520-01410-3

LIBRARY OF CONGRESS CATALOG CARD NUMBER: 69-15939
PRINTED IN THE UNITED STATES OF AMERICA

. . . a man who had most convincingly demonstrated the strength
of his natural sagacity, and was in the very highest degree
worthy of admiration in that respect. For by native insight . . .
he was beyond other men, with the briefest deliberation, both
a shrewd judge of the immediate present and wise in forecasting
what would happen in the most distant future. Moreover, he had
the ability to expound to others the enterprises he had in
hand, and on those in which he had not yet essayed he could
without fail pass competent judgement. . . .

PREFACE

Otto Struve died in Berkeley on April 6, 1963. About a year later, the American Astronomical Society appointed a small committee (J. L. Greenstein, G. H. Herbig, W. A. Hiltner, *Chairman*) to study a number of suggestions that had been made regarding a proper memorial to this man who had done so much to shape modern astronomy, largely through skilled and imaginative use of the spectrograph. A memorial volume of a special kind was finally decided upon: a volume in which direct reprints of ten of the best of Struve's own papers, one from each of the fields to which he made major contributions, would be followed by commentaries on the present status of those same areas. The former colleagues of Struve's who were invited to write these commentaries, on subjects in which they themselves were active, lent their devoted cooperation to the endeavor.

This volume is the fruit of that effort. It appears six years after Struve's death, a time in which new problems and totally new phenomena have arisen in astronomy. Yet perhaps a certain advantage has come of this delay, in that as transients decay and the frontier moves on, a better judgment can be made of Struve and the significance of his work. The specialized assessments are here, for one to study in detail. But one can sense the weight of it all simply by considering how much poorer modern astronomy would be, had it not been for Otto Struve.

A few editorial remarks: no attempt has been made to homogenize the commentaries, or even to impose a consistent notation. At the risk of a whiff of editorial untidiness, perhaps some feeling for the atmosphere of opinion that pervades these areas has thereby been preserved. In the papers of Struve that are reproduced here, some passages have been omitted that no longer seem relevant, as well as large masses of tabular data. We are indebted to the editors of *Annales d'Astrophysique*, *Astrophysical Journal*, and the *Journal of the Washington Academy of Sciences* for permission to reproduce articles from those sources. Finally, this volume could never have appeared without the financial assistance of the National Science Foundation, and without the material and moral support of the University of California Press, to whose Messrs. Lloyd Lyman and Joel Walters this enterprise owes a great deal.

The epigraph used on page v is quoted from Thucydides, *History of the Peloponnesian War* (tr. by C. F. Smith, Harvard University Press, 1919–1923).

George Herbig

Santa Cruz, California
May 14, 1969

CONTENTS

1. INTRODUCTION: A PERSONAL AND SCIENTIFIC APPRECIATION OF OTTO STRUVE

G. H. Herbig

Lick Observatory
University of California

This volume is a tribute to a great astronomer, Otto Struve. It is offered by those of his friends and colleagues whose names appear in the Table of Contents, together with a number of his former associates who have supported this publication in other ways. All give testimony here, each in his own fashion, to the influence that Otto Struve has had over us, our work, and our science.

Struve's career has already been related in detail by astronomers who knew him well. Attention is called particularly to the biographical accounts by T. G. Cowling (*Biographic Memoirs of Fellows of The Royal Society*, **10**, 283, 1964) and by A. Unsöld (*Mitteilungen der Astronomische Gesellschaft 1963*, p. 5, 1964). As an astronomical internationalist, Struve made major contributions through his support of foreign scientists in the United States, and his postwar efforts for international understanding through the International Astronomical Union. As a scientific administrator he was very successful, notably in the elevation of the Yerkes Observatory to a position of eminence in the 1930's and in the organization of the McDonald Observatory later in that decade.

However, this volume honors Struve's accomplishments not as an organizer or as a public figure of great prestige, but as a scientist. Like ripples on a pond, the impact of his contributions has spread far and in many directions, and corks are still bobbing in direct response to his work. It is not unprecedented for a scientist to work in several distinct fields during his career, but it is unusual for such a person to dominate almost every area that he essayed. Struve did just that: the Table of Contents lists the main topics in which, for the time he gave them his full attention, he was a recognized leader.

It is interesting to ponder upon the question: why or how did Struve do so supremely well? Cowling has made the point that many of the fields or ideas to which Struve's name is now firmly attached were not opened or conceived by him. This is of course no criticism: an assiduous burrower in the literature can usually find a precursor for any major discovery. The greater credit properly goes to the one who nurtured and developed and pressed the idea upon the often-reluctant scientific community, not to the source of a relatively casual suggestion. Likewise, Struve usually did not enjoy the advantage of overwhelming instrumental power. He never worked with the largest telescope in the world or with the most sophisticated instrumentation. He often showed an impatience with instrumental elaboration for its own sake, and his reputation largely rests on the energetic and skillful use of conventional equipment. In fact, it is remarkable what scientific results Struve could produce with old instruments or from spectrograms of indifferent quality. His philosophy seemed to be that telescopes and instrumentation are an essential link between the observable universe and the inquiring mind of the astronomer, but that they deserved for their own sake no more attention or devotion or elegance than does any other means of communication, such as a copper wire or a telephone.

This is not to say that Struve was uninterested in the technological side of astronomy: the part that he played in the construction of the 82-inch McDonald reflector and its superb Cassegrain spectrograph is proof enough of that. On a smaller

1

scale, the story of the conception of the Yerkes and McDonald nebular spectrographs illustrates his readiness to seize upon an astronomically applicable opportunity. Late on a cloudy night in 1937, the two observers at the 40-inch refractor, L. G. Henyey and J. L. Greenstein, were in the Yerkes library with Struve. He brought up the question of the fastest spectrograph that could theoretically be built. Before the night was over, they had produced the tentative design of a device having a wooden slit at the upper end of the tube of the 40-inch, diaphragms along the tube, and at the bottom, the two quartz prisms and an $f/1$ Schmidt camera that were destined for the McDonald Observatory Cassegrain spectrograph. Two nights later the instrument was ready. That first night was cloudy, but nevertheless Greenstein pointed the new spectrograph at the clouds through the open slit of the dome, and obtained an excellent auroral spectrum. Subsequently, under a clear sky, the instrument was used by Struve and C. T. Elvey and by Henyey and Greenstein, and was so successful that it was replaced by an 150-foot equivalent on a hillside in Texas, the spectrograph with which Struve discovered the galactic H II regions.

In trying to analyze the reason for Struve's success as an astronomer, perhaps one who (like the writer) did not know him too long or too intimately has a certain advantage. I did not know him well until 1948 when, although it was not apparent at the time, most of his great achievements were behind him, as were the years of strong scientific direction and policy-making that had given rise to so many fascinating stories. I found Otto Struve to be a very tired, but a kind and exceedingly generous man. He was an unfailing source of three commodities that at a certain stage in one's career are more precious than gold: support, encouragement, appreciation. Much of Struve's words and actions were due, I think, to the fact that he believed astronomy to be a terribly important matter, so important that it was entirely justifiable for one to wrap himself in it to the exclusion of almost everything else. With Struve, astronomy was no eight-to-five, Monday-through-Friday occupation, any more than was the act of breathing. It was his life, his *raison d'être*. There is a story at Yerkes that, following an evening staff meeting over which Struve presided, the faculty went home to bed while Struve went up to the 40-inch dome to observe for the rest of the night. To a young man, this energy and all-encompassing devotion that Struve poured into astronomy was an inspiration. Struve gave the impression that the sky was filled with marvelous and important things, free for joyful harvesting by anyone with perception and the opportunity. It was unforgivable in his eyes for anyone to fall short of full commitment: his harshest judgments descended upon those who through sloth, distraction, or weakness of will failed to take full advantage of their scientific opportunities. I found this a stern philosophy on bitter nights when sixty-mile-an-hour winds roared around the open dome, and snow blew down on the spectrograph, and the seeing disc exploded to the point of invisibility. Perhaps the basic reason for Struve's success is that he asked no more of anyone—on cold nights or otherwise—than he was prepared to give himself, and that he lived according to his own guideline: if astronomy is worth doing at all, it is worth everything that the individual astronomer can bring to it.

A corollary of his philosophy was: if something is worth doing, it is thereby worth publishing. There are few, if any, references in the literature to "unpublished work by O. Struve." His bibliography (contained in the obituary article by Unsöld) lists some 444 titled scientific papers, plus 186 semipopular articles in *Sky and Telescope* and *Popular Astronomy*, and many reports, reviews, and minor notes. Not every one of these is an imperishable document, but a great many are first-rate, and most of them exerted a perceptible influence upon the astronomy of the time. The fertility of Struve's mind was sometimes beyond his own physical ability to put words on paper. Much of his book *Stellar Evolution* was dictated into a recorder during an observing session at the McDonald Observatory in 1948–49. Probably not many realize that the subtitle of that volume (*An Exploration from the Observatory*) was almost literally true: some of it he recorded at the Cassegrain focus of the 82-inch. Struve once remarked that "if one did not publish, science would get along without him." He was impatient with those who, in search of perfection or for other reasons, did not justify their scientific existence by publication.

Struve's work pervades such a vast sector of modern astronomy that it is difficult to appreciate the real magnitude of his contribution. The chapter titles of this volume identify most of the subjects in which he made a major mark. It is not easy to imagine the shape that these topics would have today, if it were not for him. But it is obvious that our picture of stellar astronomy and astrophysics would be a much more limited one. In a sense, this volume is an attempt to place Struve's larger contributions in their proper perspective, to show how essential they were then, and are now. In each of his major fields of activity, his best paper is reproduced here, directly from the pages of the original journal. All of these papers are still very much worth reading, although one dates from 1929. Struve's own paper is then followed by a review of that subject as it stands today, written in every case by one of his former associates or colleagues who has specialized in that area. Very probably, it is this kind of a memorial that Struve himself would have preferred. And probably it is only through a publication of this kind that an intelligent appreciation can be gained of the scientific legacy left to us by Otto Struve.

Si monumentum requiris, perlege.

2. SPECTRAL CLASSIFICATION

O. STRUVE: The Problem of Classifying Stellar Spectra. *Astrophysical Journal,* **78,** 73, 1933.

W. W. MORGAN: Struve's Approach to Spectral Classification.

THE PROBLEM OF CLASSIFYING STELLAR SPECTRA

By O. STRUVE

ABSTRACT

The subdivisions of the Draper classification, intended for statistical investigations with small-dispersion spectrograms, require greater refinement for detailed studies of stellar atmospheres. The first part of this paper deals with the limitations imposed upon such a process of refining by the use of conflicting spectroscopic criteria.

A method is suggested by which discrepancies between criteria may be used to throw light upon physical parameters, other than temperature and pressure, which define the appearance of the spectrum. This method is applied to the B-type stars.

Peculiarities in the intensities of He I, O II, Mg II, etc., are pointed out and discussed.

I

The spectral classification used by Miss Cannon in the *Draper Catalogue* of the Harvard College Observatory represents an empirical system designed to arrange the majority of stellar spectra into a linear sequence. The fact that this was at all possible proves that the more conspicuous differences in the intensities of stellar absorption lines are traceable to one physical parameter, and this we now know to be the temperature of the atmosphere of the star. The Draper classification was adopted by the International Solar Union in 1910, and again, with minor additions, by the International Astronomical Union in 1922. The most important of these additions deals with the effect of pressure or absolute magnitude. A letter "g" is prefixed to the designation of the class if the spectrum exhibits giant properties, and a letter "d" is used to designate dwarfs.

7

There are other characteristics for which special symbols have been recommended: "e" designates presence of emission lines, "k" stands for interstellar lines; the letters "n" and "s" are probably rough measures of axial rotation.

There remains the question whether a normal stellar spectrum, devoid of emission lines and not suggestive of rapid rotation, is fully defined by the two physical parameters, temperature and pressure. There is no a priori reason to suppose that such should be the case. For example, we do not know whether the relative abundance of different elements is constant for any given values of temperature and pressure. Nor have we any right to assume that the effective depth of a stellar atmosphere does not vary from star to star, unless we are willing to depend upon the treacherous results derived theoretically from star models chosen because they agree, in general, with observation.

If it should be true that the spectrum of a star is fully defined by two physical parameters which can be separated both observationally and theoretically, there would be no reason why the present classification could not be greatly refined by the introduction of a larger number of subdivisions. It should be remembered that the Harvard classification depends upon objective-prism spectrograms, the majority of which have a small linear dispersion. Numerous attempts have been made within recent years to increase the number of spectral subdivisions, and in some cases workers have tried to classify stellar spectra to one-hundredth of a spectral class. The results have been somewhat disappointing: spectral classes assigned at different observatories often differ by several subdivisions, and even at the same observatory contradictory results are sometimes obtained if different criteria are used. Table I is taken from a summary by R. H. Curtiss.[1] It is obvious that the differences in results between observatories greatly exceed possible errors in estimating line intensities. Thus, a difference of seven-tenths of a spectral class between Mount Wilson and Victoria cannot be due to errors in the estimates. It is clearly due to the fact that the criteria were not the same.

[1] *Handbuch der Astrophysik*, **5**, Part I, 100, 1932.

In order to make progress in the study of stellar spectra it is desirable to investigate a number of questions:

1. If two stars of equal atmospheric temperature and pressure are given, are their spectra necessarily identical? We assume here

TABLE I

Star	Harv. Vis. Mag.	Mt. W. Abs. Mag.	Vict. Abs. Mag.	Harv. Class	Mt. W. Est.	Mt. W. Meas.	Vict. Class	Mich. Class
β Cr. B.	3.7	−0.3	−0.4 / −0.2	Fop	F2	F4	F2	F3.2
ν Her.	4.5	−0.2	+1.2 / +1.5	F0	F2p	F3	F3	F3.5
20 C. Vn.	4.7	−1.4	+0.7 / +1.0	F0	F4	F3	F3	F4.1
ε Aur.	3.4	−2.0	−2.9 / −2.7	F5p	F5p	F9	F2	F3.4
β Dra.	3.0	−3.5	−2.1 / −1.5	G0	G0p	F8	F9	F9.0
29 Mon.	4.4	−4.2	−2.1 / −2.5	G0	G3p	F9	G0	F8.5
ε Leo.	3.1	−0.9	......	G0p	G0	G0	......	G0.1
ε U. Mi.	4.4	−1.4	......	G5	G2	G2	......	G2.6
o U. Ma.	4.5	−0.8	......	G0	G2	G2	......	G2.5
η Her.	3.6	+0.9	+1.1 / +1.0	K0	G5	G5	G6	G5.6
ε Vir.	3.0	0.0	+1.4 / +1.4	K0	G6	G6	G6	G5.0
β Her.	2.8	−1.0	+0.7 / +1.0	K0	G5	G6	G5	G5.1
ε Boo.	2.7	−0.9	+0.5 / +1.0	K0	G8	G7	G6	G6.3

that the stars show no rotational broadening, and are not complicated by emission lines, etc.

2. If the answer to the preceding question should be negative, we shall have proved that pressure and temperature are insufficient to characterize the spectra. One or more additional physical param-

eters would be needed, and an effort should be made to discover these.

3. It is possible that among stars of certain classes the two principal parameters are sufficient, while in other classes they are not.

4. The intensity of an absorption line depends upon the number of absorbing atoms, and this in turn depends upon (a) effective thickness of absorbing layer, (b) density of atoms in the lower energy state of the line, (c) absorption coefficient of the given line. The first factor is believed to be dominated by the general opacity of the stellar atmosphere. The second represents the combined effect of temperature, pressure, and abundance. The third involves the probability that the line will be absorbed.

It seems impossible to predict the theoretical intensity of a stellar absorption line. In his recent work Milne has obtained widely differing results depending upon assumptions which might be considered equally plausible. Under these circumstances it is desirable to proceed by purely observational methods, in order not to be biased by theoretical results which still rest on uncertain foundations. By doing so we shall be able to provide a satisfactory basis for future theoretical investigations.

In the next section I shall discuss more particularly the classification of the B-type stars. The A's have been studied at the Yerkes Observatory by W. W. Morgan, and the F's are now under investigation by J. A. Hynek.

II

The spectral classification of stars of classes O, B, A, and F has remained in a somewhat unsatisfactory state because of the great variety in the character (or in the width) of the absorption lines. Shajn, Elvey, and I have shown that these differences may be satisfactorily interpreted as an effect of rapid axial rotation. A statistical study of rotational velocities has been made by Miss Westgate.[2] Since it is technically difficult, if not impossible, to compare a star having wide lines with one having narrow lines, I have limited this

[2] *Astrophysical Journal*, **77**, 141, 1933; **78**, 46, 1933.

discussion to spectra having narrow lines. Whether or not by doing so we obtain a physically homogeneous group is not certain. Narrow lines indicate one of two things: either the velocity of rotation is small or the axis of the star nearly coincides with the line of sight. We do not yet know whether the spectra of these two groups of stars are identical.

I have shown elsewhere how we may divide the B stars into three groups:[3] giants, intermediate stars, and dwarfs. I have now selected only the stars for which good spectrograms on Eastman Process emulsion were available, and have divided them into two groups: giants and dwarfs. The criteria were: (a) absence or presence of hydrogen wings, (b) intensity of forbidden helium lines, (c) intensity of the interstellar calcium line K. Criteria (a) and (b) are directly related to pressure, so that our division is essentially independent of any other physical parameters.

Table II gives the stars, 30 dwarfs and 17 giants. The details have been taken from Schlesinger's *B.S. Catalogue*. Plates III, IV, V, and VI contain reproductions of some of the spectra. In each of the two groups I have arranged the stars, as nearly as was possible, in order of decreasing temperature.

An inspection of the plates reveals many interesting details which are either entirely new or which, though known for some time, have not been carefully investigated and explained. We shall discuss these below.

1. In the later types there is a wide gap between giants and dwarfs. Does a similar gap exist in the B stars, or do they merely show a measurable amount of dispersion in atmospheric pressure? The fact that in an earlier paper[3] I subdivided the stars into giants, average stars, and dwarfs might indicate that the latter is true. However, I have now found, with somewhat better material, that it is possible in almost every case to assign a star to either the giants or the dwarfs. There are few intermediate cases; possibly β Canis Majoris, φ Herculis, η Leonis, and 4 Lacertae should be called intermediate. I am now inclined to think, on spectroscopic evidence alone,

[3] *Ibid.*, **74**, 225, 1931.

TABLE II

Name	No. in Plates	α	δ	Mag.	Spec.	Proper Motion α	Proper Motion δ	Par.	Rad. Vel. in Km	Remarks
					Giants					
1. 39 λ Ori	22	5^{h}29^{m}38^s	+9° 52'	{3.66 / 5.56}	Oe5 / Oe5	+.004	−.011	+.006	{+35 / Var.}	4'', 45°, fixed; $v_0 = +35$ km
2. 19 Cep	23	22 2 4	+61 48	5.17	Oe5	+.013	−.003	.003	Var.	$v_0 = -13$ km
3. 53 κ Ori	24	5 43 1	+9 42	2.20	Bo	+.007	−.005	.013	+20	
4. 47 ρ Leo	……	10 27 33	+9 49	3.85	Bop	+.004	−.004	.012	+41	
5. 15 κ Cas	25	0 27 19	+62 23	4.24	Bo	+.012	−.001	.002	Var.	$v_0 = -7$ km
6. 2 β C. Ma	26	6 18 18	−17 54	1.99	B1	−.003	.000	.012	Var.	$v_0 = +33$ km
7. 44 ζ Per	27	3 47 51	+31 35	2.91	B1	+.015	+.018	.006	+20	9.3M, 13'', 208°, c.p.m.
8. 21 ε C. Ma	28	6 54 42	−28 50	1.63	B1	+.005	.000	.012	+28	9M, 8'', 160°, fixed
9. 9 Cep	29	21 35 14	+61 38	4.87	B2p	+.008	.000	.003	−15	
10. 55 Cyg	30	20 45 32	+45 45	4.89	B2	+.006	−.006	.006	Var.	
11. 25 χ Aur	31	5 26 13	+32 7	4.88	B1	+.008	−.016	.004	Var.	655 days, $v_0 = +4$ km
12. 24 o³ C. Ma	32	6 58 51	−23 41	3.12	B5p	+.003	−.005	.007	Var.	24.3 days, $v_0 = +49$ km
13. 67 Oph	33	17 55 38	+2 56	3.92	B5p	+.004	−.014	.006	−3	8M, 55'', 142°, fixed
14. 19 β Ori	34	5 9 44	−8 19	0.34	B8p	+.005	−.002	.006	Var.	7M, 9'', 201°, fixed
15. 2 Hev Cam	35	3 20 58	+59 36	4.42	B9p	+.004	−.002	.009	−5	9.0M, 2''.4, 160°
16. 67 σ Cyg	36	21 13 29	+38 59	4.28	Aop	+.001	−.008	−.007	Var.	11.0 days, $v_0 = -2$ km
17. 50 α Cyg	37	20 38 1	+44 55	1.33	A2p	+.004	−.002	−.005	Var.	Deneb, $v_0 = -4$ km

TABLE II—*Continued*

NAME	No. IN PLATES	α	δ	MAG.	SPEC.	PROPER MOTION α	PROPER MOTION δ	PAR.	RAD. VEL. IN KM	REMARKS
										Dwarfs
1. 10 Lac.	1	22h 34m 46s	+38° 32′	4.91	Oe5	+.003	−.011	....	Var.	$v_0 = -9$ km
2. 36 ν Ori.	2	5 27 6	−7 23	4.64	B3	+.007	−.012	+.004	+18	8.4 years, $v_0 = +32$ km
3. 37 φ Ori.	....	5 29 20	+9 25	4.53	B0	+.002	−.008	.003	Var.	
4. 23 τ Sco.	3	16 29 39	−28 1	2.91	B0	−.009	−.036	.007	0	
5. 6 λ Lep.	....	5 14 58	−13 17	4.29	B1	+.004	+.003	.005	+20	
6. 8 β Cep.	4	21 27 22	+70 7	3.32	B1	+.013	+.003	.008	Var.	8M, 14″, 250°, fixed 0.19 days
7. 48 ν Eri.	....	4 31 19	−3 33	4.12	B2	+.003	−.003	.005	Var.	$v_0 = +11$ km
8. 82 δ Cet.	5	2 34 21	−0 6	4.04	B2	+.013	+.001	.010	Var.	Algenib
9. 88 γ Peg.	6	0 8 5	+14 38	2.87	B2	+.003	−.015	.005	+5	$v_0 = -14$ km
10. 102 Her.	7	18 4 29	+20 48	4.32	B3	+.003	−.017	.005	Var.	Vel. perhaps var.
11. 17 ζ Cas.	8	0 31 24	+53 21	3.72	B3	+.023	−.011	.005	+3	0.29 days
12. 42 θ Oph.	....	17 15 52	−24 54	3.37	B3	+.001	−.031	.009	Var.	8M, 28″, 82°, fixed
13. 20 η Lyr.	9	19 10 21	+38 58	3.79	B3	−.004	−.004	.004	Var.	Vel. possibly var.
14. 85 ι Her.	10	17 36 38	+46 4	3.79	B3	−.004	−.002	.007	−18	
15. 45 ε Cas.	11	1 47 12	+63 11	3.44	B3	+.039	−.019	.013	−7	8.5M, 36″, 173°, c.p.m.
16. 29 π And.	12	0 31 32	+33 10	4.44	B3	+.023	−.011	.008	Var.	
17. 2 ε Del.	13	20 28 26	+10 58	3.98	B5	+.012	−.027	.009	−19	$v_0 = -19$ km
18. 22 τ Her.	14	16 16 44	+46 33	3.91	B5	−.008	+.031	.011	−14	14M, 7″, 147°, c.p.m.
19. 20 c Tau.	15	3 39 52	+24 4	4.02	B5	+.031	−.046	.013	+8	Maia
20. 22 ζ Dra.	16	17 8 30	+65 50	3.22	B5	−.010	+.021	.026	−14	
21. 112 β Tau.	17	5 19 58	+28 31	1.78	B8	+.034	+.177	.035	+10	
22. 4 γ Crv.	18	12 10 40	−16 59	2.78	B8	−.158	+.013	.021	Var.	
23. 4 Lac.	....	22 20 28	+48 58	4.64	B8p	−.005	−.013	.007	−26	
24. 11 α Dra.	19	14 1 41	+64 51	3.64	Aop	−.049	+.017	.023	Var.	51.4 days, $v_0 = -15$ km
25. α Lyr.	21	18 33 33	+38 41	0.14	Ao	+.209	+.278	.123	−14	Vega
26. α And.	....	0 3 13	+28 32	2.15	Aop	+.143	+.163	.040	Var.	96.7 days, $v_0 = -11$ km
27. 11 φ Her.	20	16 5 37	+45 12	4.26	B9p	−.018	−.024	.015	−16	
28. τ⁹ Eri.	....	3 55 40	−24 18	4.69	Aop	+.012	+.003	.012	Var.	
29. 30 η Leo.	....	10 1 53	+17 15	3.58	Aop	+.002	−.010	....	+2	$v_0 = +24$ km
30. 13 Mon.	....	6 27 30	+7 24	4.50	Aop	+.006	−.010	.007	+12	

that there are two fairly distinct groups of stars having different atmospheric pressures.[4] The ratio

$$\frac{\text{dwarfs}}{\text{giants}} = \frac{30}{17} \approx \frac{2}{1}$$

is somewhat affected by selection: I have included wider lines among the giants than among the dwarfs. It is also possible that slow rotation is more frequent among the giants than among the dwarfs. The real ratio may therefore easily be 3/1 or even 4/1. This result would agree well with that derived statistically by G. Strömberg.[5]

2. The arrangement of the stars within each group is according to decreasing temperature. In general, it is possible to assign to each star a definite place in the sequence. The Harvard spectral classes appear on the right-hand margin of the plates. They show some scatter, owing to the difficulty of classifying stars on small-scale objective-prism plates, but on the whole they agree with my arrangement. It is obvious, however, that a much finer classification than that of the *Draper Catalogue* is possible. Thus, among the dwarfs, Harvard class B3 includes spectra which are widely different in the intensities of the *Si* III lines, which range from fairly strong in 102 Herculis to complete invisibility in 29 π Andromedae. A similar variation is shown by the *O* II lines 4415, 4417. The rapid decrease in intensity of *Si* IV 4116 from B1 to B2 among the dwarfs is noteworthy. Among the giants *He* II 4542 shows greatly differing intensities in Oe5. The unidentified line at 4420 shows considerable variations in intensity for the three B1 giants; it is exceedingly faint in ϵ Canis Majoris. These examples prove that the definition now obtainable on single-prism spectrograms is sufficient to classify the B stars to approximately one-twentieth of the interval from O9 to A0.

3. The preceding result raises another question: Is the refinement of spectral classification limited merely by the definition of the

[4] Miss E. T. R. Williams has shown (*ibid.*, **75**, 386, 1932) that my giants, intermediate stars, and dwarfs have markedly different absolute magnitudes. I am inclined to think, however, that the "intermediate" stars are, with a few exceptions, a mixture of giants and dwarfs which could not be properly classified on the plates then available. Among the later B's and the A's this division is not so pronounced.

[5] *Ibid.*, **74**, 342, 1931.

plates and by our inability to estimate or measure small differences in line intensity? The answer is negative. If care is taken in making the estimates, it becomes apparent that our two groups of stars, the giants and the dwarfs, do not represent linear sequences. Consider, for example, the dwarf star 20 c Tauri (Maia). The helium lines 4472 and 4388 are distinctly too weak for the place assigned to this star in our scheme. Moving it downward to a place between γ Corvi and α Draconis would remedy this, but it would leave Mg II 4481 in the wrong place. This line is weaker in 20 c Tauri than in several stars preceding it in our sequence, which would suggest that it should be moved upward among the B3 stars. There is a discrepancy in the classification, depending upon which of two criteria is used.

This discrepancy is not caused by small differences in atmospheric pressure, because the wings of the hydrogen lines definitely place Maia among the dwarfs. While both He I and Mg II are too weak for the place assigned to this star, Fe II 4549 and 4233 are rather strong, while C II 4267 is slightly too weak, and Si II 4128 and 4130 are about normal.

Neither temperature nor pressure explains these facts. We are therefore forced to admit the existence of other physical parameters which control the appearance of a spectrum. Three simple hypotheses suggest themselves: (a) the effective depth of a stellar atmosphere may not be fully defined by pressure and temperature, as has been assumed in all earlier work; (b) the relative abundance of different elements may not be the same in all stars; (c) the tendency toward the formation of emission lines, which is pronounced in the brighter members of the Pleiades, may affect the appearance of the absorption lines.

Hypothesis (a) is ruled out because He I and Mg II would demand a small depth of atmosphere, while Fe II would require normal depth.

Hypothesis (b) may be considered in the light of Russell's recent theoretical results on the opacity of stellar atmospheres.[6] Great abundance of hydrogen in any given star would increase the opacity and reduce the intensities of all other lines. In Maia this is contrary to observation.[7] There remains the possibility that He and Mg are

[6] *Publications of the American Astronomical Society,* **7,** 99, 1932.

[7] However, Russell has shown that the effect may be traced in certain stars.

less abundant than usual, while the abundance of *Fe* II is normal or even greater than normal. The hydrogen lines seem to be normal.

Hypothesis (*c*) is improbable; a tendency to form emission lines might be expected to cut down the intensity of *H* and of *He*. If this were so, 20 c Tauri would have to be moved to the hotter stars. But this is contrary to the strength of the *Fe* II lines.

Whether such discrepancies are frequent is difficult to say. I have the impression that they are more pronounced among the later subdivisions of class B, and the work of Morgan[8] would indicate that they are even more pronounced among the A stars. His suggested division into two parallel spectral branches can, of course, be interpreted in a manner similar to that used in this section.

We may conclude that pressure and temperature are not sufficient to describe a stellar absorption-line spectrum, and the investigation of small discrepancies similar to those observed in 20 c Tauri promises to throw light upon new physical parameters such as abundance of elements, etc.

4. The behavior of *He*-triplets and *He*-singlets has been discussed in my paper on B stars.[9] It is now possible to confirm and expand these results. Comparing triplets 4472 and 4026 with singlets 4388 and 4009, we notice that with increasing temperature the singlets fade out more rapidly in the giants than in the dwarfs. For the cooler stars there is little if any difference in the ratio triplet/singlet between giants and dwarfs. Maximum intensity occurs at about B2 for both the triplets and the singlets.

If the Harvard classes are retained as corrected by my sequences, there is a null effect in the intensities of *He* I for the hottest stars. At and near maximum (B2) and past maximum, the *He* I lines are systematically stronger in the giants. It should be noted, however, that we have no zero point for our scale of temperatures; a Bo giant may not have the same temperature as a Bo dwarf.

5. Morgan[10] has already pointed out that *Mg* II 4481 shows a null effect between about O9 and B5. My more complete material confirms this for the earlier classes, O9–B2. At B5 there seems to be a slight increase for the giants.

[8] *Astrophysical Journal*, **77**, 330, 1933.

[9] *Ibid.*, **74**, 248, 1931.

[10] *Ibid.*, **77**, 291, 1933.

6. *O* II behaves similarly. For the hottest stars there is no difference between giants and dwarfs, while for B0, B1, and B2 the strengthening is appreciable.

7. Silicon is especially interesting. *Si* II shows a null effect at about B2, but is markedly enhanced in the cooler giants. *Si* III has a null effect at O9, but is enhanced in the giants from B0 on. *Si* IV is enhanced in all giants observed by me. The result of these effects is that in several giants, notably in ε Canis Majoris, all three stages of ionization are visible simultaneously. I have found only one dwarf, γ Pegasi, where this is true, but even in this star *Si* IV is at the very limit of visibility. The result is interesting in connection with the theoretical work of Milne. It is somewhat disconcerting, however, that in the last paper on this subject Milne and Chandrasekhar[11] do not predict definite null effects on the hotter sides of the maxima.

8. The effect of absolute magnitude upon intensity for other elements may be estimated directly from the illustrations.

9. Ionized oxygen provides a remarkably interesting result. The lines of lower excitation potential come to a maximum at appreciably cooler temperatures than do lines of higher excitation potential. Consider, for example, the line 4303.8. Its lower excitation potential is 28.7 v. On the other hand, lines 4317 and 4320 have a potential of 22.9 v. All these belong to the quartet system. It is apparent from the illustration that λ 4304 comes to an earlier maximum than do λ 4317 and λ 4319. The same tendency may be noticed in other *O* II lines.

The *O* II lines near *H*γ, 4346, 4347, 4349, and 4351, are somewhat affected by the wings of *H*γ. Thus λ 4349 fades out much earlier than do 4415 and 4417, although their potentials are similar. This is probably caused by the greater opacity in cooler stars, resulting from the wing of *H*γ.

10. The peculiar difference, noted elsewhere,[12] between the intensities of the quartet lines of *O* II, 4346 and 4349 and those of the doublet lines 4347 and 4351, is clearly shown in the reproductions. The quartet lines are relatively stronger in most giants than in dwarfs.

[11] *Monthly Notices of the Royal Astronomical Society,* **92,** 150, 1932.

[12] Struve and Morgan, *Proceedings of the National Academy of Sciences,* **18,** 590, 1932.

(β Can. Maj. seems to be an exception; this star is, as I remarked before, not a typical giant.)

It is surprising, however, that this difference does not affect all quartets and doublets. Notice, for example, the two pairs of lines

$$4317 \; y^4P - e^4P^o \qquad 4415 \; z^2P - e^2D^o \; ,$$
$$4320 \; y^4P - e^4P^o \qquad 4417 \; z^2P - e^2D^o \; .$$

The first belongs to the quartets, the second to the doublets. Yet there is no appreciable difference in the relative intensities for giants and for dwarfs.

It seems to me more probable that this difference is of another origin. Notice the pair of faint O II lines at 4312 and 4313 which are clearly visible in the dwarf τ Scorpii. Then consider that these lines are practically invisible in all of the giants, in spite of the fact that the strong lines 4317 and 4320 are enormously strengthened in the giants. Does this not indicate that strong lines are much more strengthened in giants than are weak lines? If so, the phenomenon is probably similar to that which I observed in 17 Leporis,[13] where the multiplet intensities of the metallic lines are abnormal.

The idea is at least worth considering. I have measured many more faint O II lines in τ Scorpii than in any other star, in spite of the fact that typical giants like ζ Persei contain several stronger lines of O II than does τ Scorpii. Accurate measurements with the microphotometer would be valuable in this connection.

III

The great variety of interesting phenomena which come to light in a detailed study of stellar spectra makes it imperative that new and more refined methods than are at present available be used in stellar spectroscopy. That the B stars are not unique is shown by the work of Morgan on A stars. He found that the latter can be arranged satisfactorily into two parallel branches of stars which do not appreciably differ in absolute magnitude. He excluded the c stars, and was therefore dealing principally with stars of average luminosity. This result proves, I think conclusively, that two parameters are insufficient to describe the spectrum of an early type star. If this is

[13] *Op. cit.*, **76**, 85, 1932; J. A. Hynek, *ibid.*, **78**, 54, 1933.

the case, the present system of classification cannot be relied upon for accurate work, because the Harvard subdivisions are assigned as a compromise between several slightly conflicting criteria.

By making such a compromise we discard information which promises to throw light upon physical parameters not yet identified. To retain these data we must, for precise work, replace the Harvard subdivisions with numerical estimates of line intensities.

In the case of the B stars the principal criterion ought to be the intensity of *He* I 4472. This would put every star, without ambiguity, in its proper place. Thus, Maia would fall into the later classes, between γ Corvi and α Draconis. The intensities of other lines would then be somewhat conflicting—more conflicting, in fact, than they now are. Thus, the intensities of *Mg* II 4481 would not all be alike, even though *He* I 4471 would be of similar intensity. This would merely indicate that more than one physical parameter is effective. In order to separate the parameters we should have to observe more lines. By observing two lines, one of which is sensitive to changes of pressure while the other is not, we might eliminate pressure from other, unknown, factors.

This is essentially the procedure which I have used in my discussion of the B stars in section II, although I have not carried it through in arranging the table and the reproductions.

I wish to make it clear that I do not advise modifications or refinements of the Draper classification in spectroscopic work intended for statistical purposes. The Draper classification remains unsurpassed for investigations involving large numbers of stars.

YERKES OBSERVATORY
June 14, 1933

NOTE ADDED JULY 2

Dr. Morgan has called my attention to the fact that the stars 4 Lacertae, η Leonis, and 13 Monocerotis, which are included in my list of dwarfs, have been regarded as rather luminous stars by other investigators (Morgan, *Astrophysical Journal*, **77**, 294, 1933; E. Öpik and M. Olmsted, *Harvard College Observatory Circular*, No. 380, 1933). The explanation of this discrepancy is that in dividing the spectra into two groups I have tried to follow consistently the same criterion: thus, the hydrogen lines in 13 Monocerotis resemble those of γ Pegasi in the relative importance of the wings to the total absorptions of the lines; on the other hand, α Cygni or β Orionis are almost entirely devoid of wings in the hydrogen lines, resembling such stars as ζ Persei or α Aurigae. The three stars mentioned above have relatively weak hydrogen lines and it is possible that the wings in them are relatively less pronounced than in other dwarfs, such as α Lyrae, but this will have to be investigated by means of accurate spectrophotometric measures. It is probable that my group of dwarfs includes among the later classes stars of somewhat different luminosity, although all of them are presumably less luminous than the stars which I have called giants. In order to avoid confusion, it would probably have been better not to use the definitions "giants" and "dwarfs" in my tables, but to regard the division as one based upon a purely observational criterion. This would also have been more consistent with the suggestions outlined in the last section. I might mention, however, that other methods of classifying early type stars into giants and dwarfs are no more satisfactory: thus the criterion of total intensity of hydrogen lines has given excellent results in statistical discussions. It is clear, however, that this method does not differentiate between real giants in which the hydrogen lines are free of Stark broadening, and therefore of small total intensity, and dwarfs in which the abundance of hydrogen is low. It would be of great interest to investigate this matter further, and to determine whether individual departures from the average abundance of hydrogen are actually measurable.

The peculiar differences in the intensity gradients of the O II absorption lines discussed on page 84 seem to be of great interest, especially in view of the observations by Morgan of lines of other elements (p. 158 of this issue). The phenomenon seems to be much more important than we had believed at first, and there is a wide field for accurate spectrophotometric investigations on line intensities. Dr. Elvey has undertaken such a study for certain early type stars. I am wondering whether the peculiar intensities of the He triplets and singlets, reported on page 82, are not in part different aspects of the same phenomenon. Since the triplets are systematically stronger than the corresponding singlet lines, part of the observational results could be interpreted as differences in the intensity gradients of the He lines. The observational results are still too incomplete to enable us to advance an explanation of this phenomenon. Attention might be called, however, to the fact that at least in two "abnormal" stars, 17 Leporis and ϵ Aurigae, the reversing layers are almost certainly enormously extensive, their thickness being roughly of the same order as the diameters of the stars. It is possible that the theoretical expressions of line contours derived for thin reversing layers are not applicable to such extensive stellar atmospheres.

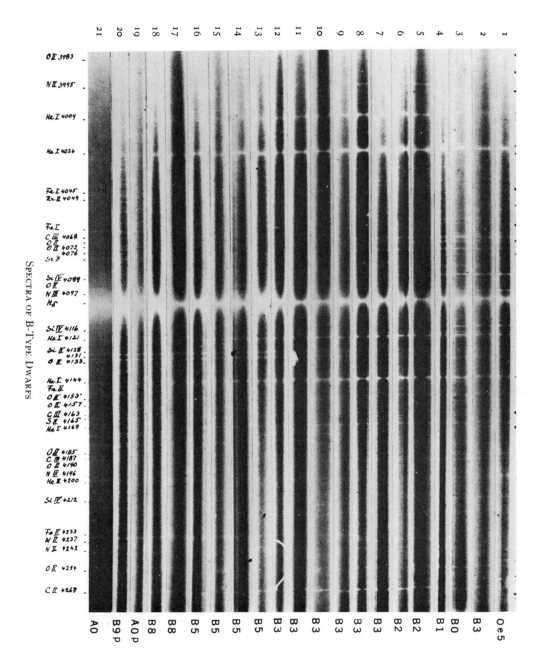

SPECTRA OF B-TYPE DWARFS

21

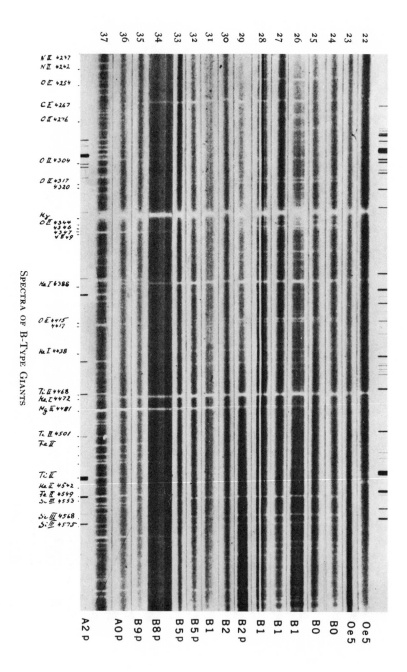

Spectra of B-Type Giants

22

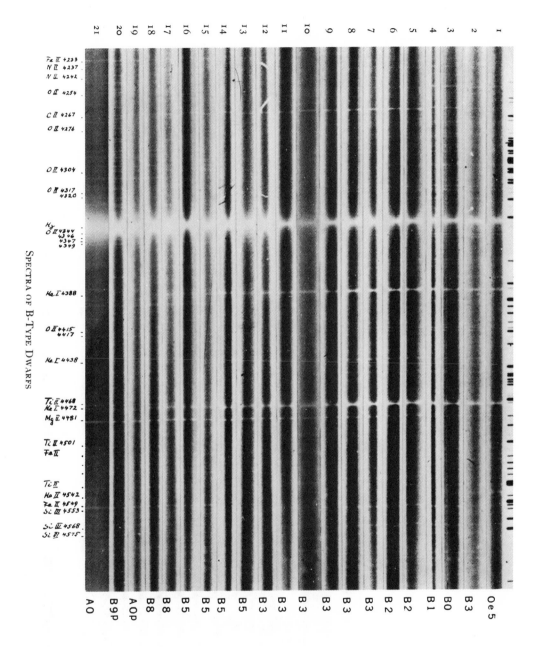

SPECTRA OF B-TYPE DWARFS

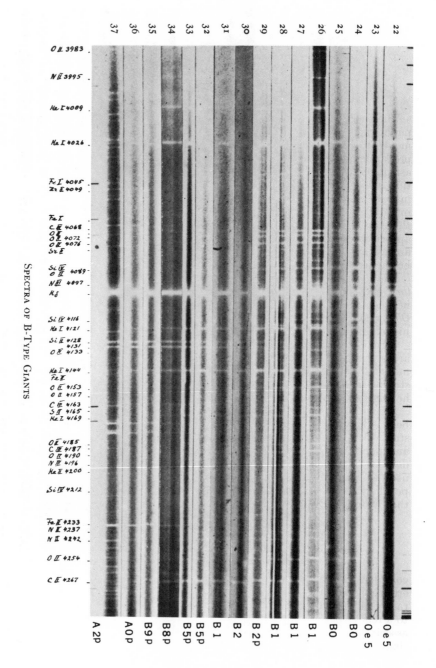

SPECTRA OF B-TYPE GIANTS

24

STRUVE'S APPROACH TO SPECTRAL CLASSIFICATION

W. W. MORGAN

Yerkes Observatory
University of Chicago

I

The path from the first inward stirrings of an idea to its formulation, solution, and publication differs widely from individual to individual. At one extreme are those who move an idea through the mind for months or years before reaching a formulation preparatory to the serious attack on the problem. In the observational field of astronomy, there are some individuals who soak up great numbers of bits of information, which then undergo a subconscious digestive process; the result of this process may be apparent only after a long period of time in the sudden emergence of an integrated picture formed from the earlier disassociated bits. At the other extreme, there are creative scientists who find it impossible to retain new information for an appreciable length of time; the impression here is that, for these individuals, the bits are too hot to hold on to—literally—and whatever picture-building is possible must be carried out immediately. Struve was of this latter type.

His sheer productivity, as shown in his *Contributions from the McDonald Observatory*, and elsewhere, was miraculous. In addition to his administrative responsibilities, he was author—or joint author—of more than fifty *Contributions from the McDonald Observatory* in the period 1941–1949. In the amount and intensity of effort shown in his published record, Struve is in a class by himself among American astronomers.

Struve was a plower-up of fields; his exploratory work brought to light great new areas ripe for investigation. His discoveries were of great importance; however, the measure of his impact on twentieth-century astronomy is more in the realm of a stirring-up, a communication-by-example, of the excitement and enthusiasm which he felt and which is so well illustrated in his published record.

Struve's individual papers, such as "The Problem of Classifying Stellar Spectra," bear evidence of the intensity with which he worked. His most important efforts start with the germ of a new idea, for which evidence is collected as rapidly as possible and published with the greatest dispatch. Such a procedure on occasion precludes detailed treatment; the time that would have been required for elaboration Struve used to tackle new problems. For each scientist, there is a personal path that is peculiarly right for himself; and Struve's proper path was without question the one that he followed. In following it, he achieved greatness.

II

In the years immediately preceding the publication of this paper there had been considerable activity on the spectra of the peculiar A-type stars at Yerkes; the general result was that the frequency of such stars known in the solar neighborhood is rather high. However, the peculiarities in the case of the A-type spectra were confined to a relatively small number of elements; and the most spectacular phenomena were associated with the "rare-earth stars," whose spectra contained strong lines of singly ionized europium which frequently varied in intensity. Lines of Sr II, Ca II,

Cr II and other elements also varied in certain stars. The general discriminant for the peculiar A stars as a class was their obvious "strangeness," which set them off clearly from normal A stars.

In this situation Struve turned his attention toward the classification of the B-type stars. The observational feature that made this step promising was the demonstration by Struve that stellar spectra of the brighter B stars could be obtained on the fine-grain, high-contrast Eastman Process emulsion with sufficient dispersion to show large numbers of lines.

Now the spectra of the brighter stars of classes B0–B5 differ from those of classes B9–A5 in an important respect: the obviously strange silicon-strontium-europium A stars have no clear parallel in the stars of earlier type; there certainly are peculiar spectra among the latter, but they do not have the pronounced strangeness of the Ap stars. And so Struve's investigation of the reclassification of the B stars was a more difficult venture; it consisted basically in pointing out what he considered to be discrepancies in the intensities of lines of the "ordinary" elements.

He first divided the spectra of the B stars into two categories according to the relative intensity of the wings of the Balmer lines of hydrogen; he labeled these two groups "giants" and "dwarfs," and reached the conclusion that there was probably no intermediate category. He then reclassified the members of each group into spectral types with as high accuracy as possible.

Then came the crucial step for his conclusions: a careful examination of spectra classified at similar types within each of the two sequences showed that the spectra were not identical; there were considerable differences in the intensities of the various lines used for classification. That is, the spectral types arrived at were compromises between discrepant criteria. For example, the star 20 Tau (Maia), which Struve placed among his dwarfs because of the appearance of the wings of the hydrogen lines, gave highly discrepant types from some lines of He I and Mg II; the intensity of He I $\lambda 4387$ and $\lambda 4471$ indicated a type of B8, while the intensity of Mg II $\lambda 4481$ suggested a type near B3. Other examples of such ambiguities were noted by Struve, who reached the conclusion from his data that additional physical factors to temperature and "atmospheric pressure" were indicated.

A short time later, Russell, C. P. Gaposchkin, and Menzel (1935) published a discussion of Struve's paper. They criticized Struve's reasoning concerning the physical conditions under which the spectra of two stars should be identical. They considered that the Henry Draper classification continued to be highly satisfactory, and that any reclassification should be deferred. Among their recommendations was one that "a precise classification must involve knowledge obtained with high dispersion." The paper of Russell, Gaposchkin, and Menzel is still of interest.

The importance of Struve's paper on spectral classification lies in his conclusion that the brighter B stars cannot be classified unambiguously in his two-sequence array of "giants" and "dwarfs." As far as it goes, this conclusion is valid today.

With the introduction later of a complete two-dimensional classification for the B stars (a process which is only reaching a completely satisfactory solution in 1968), most of the discrepancies pointed out by Struve disappear.

But an interesting phenomenon has appeared very recently: as most of Struve's discrepancies have been resolved in a more complex classification, new discrepancies have come to light which cannot be explained by introducing finer structure into a two-dimensional classification scheme. And so, Struve's discovery was the forerunner of the present fantastic complexity of the classification of the B stars.

REFERENCE

Russell, H. N., Gaposchkin, C. P., Menzel, D. H. 1935, *Ap. J.*, **81**, 107.

3. CURVES OF GROWTH

O. STRUVE and C. T. ELVEY: The Intensities of Stellar Absorption Lines. *Astrophysical Journal*, **79**, 409, 1934.

JESSE L. GREENSTEIN: The Curve of Growth Method.

THE INTENSITIES OF STELLAR ABSORPTION LINES

By OTTO STRUVE and C. T. ELVEY

ABSTRACT

The intensities of stellar absorption lines have been measured in the spectra of α Canis Majoris, α Lyrae, α Cygni, α Persei, ε Aurigae and 17 Leporis. With the help of theoretical multiplet intensities, curves have been constructed, for each star separately, showing the relation between total absorption and number of atoms. These curves differ considerably. All parts of the curves lie between the limits $A = $ Const. and $A \propto N$.

An application of the theory of Schütz based upon the simultaneous effects of radiation damping and thermal Doppler broadening shows agreement with the observations in α Canis Majoris, α Lyrae, and α Cygni, provided, however, the damping constant is assumed to be ten times that given by the classical theory.

For the remaining three stars the hypothesis is advanced that in their atmospheres turbulent motions far exceed the motions caused by thermal agitation. The theory of Schütz is enlarged to include the case of turbulent motions, and excellent agreement with the observations is found.

The most probable turbulent velocity is found to be 67 km/sec. in 17 Leporis, 20 km/sec. in ε Aurigae, and 7 km/sec. in α Persei.

The interpretation of many other observational data in the light of the turbulence hypothesis is briefly discussed.

I. THE GRADIENT EFFECT

Several years ago Unsöld showed[1] that the theory of radiation damping could be applied to the wide wings of strong stellar absorption lines. The total absorptions of such lines increase in proportion to the square roots of the numbers of absorbing atoms. Essentially the same result has also been derived by J. Q. Stewart,[2] and applied by him, by Fairley[3] and by Korff,[4] to the sodium absorption lines in the sun and in the laboratory. Miss Payne[5] found from measurements of Harvard spectrograms that the square-root law is actually obeyed, over a wide range of spectral classes, in the case of the calcium lines H and K. For weak solar lines Minnaert[6] and Woolley[7] found departures from the square-root law which have since been explained by Minnaert and Mulders[6] on the basis of theoretical

---

[1] *Zeitschrift für Physik*, **44**, 793, 1927.

[2] *Astrophysical Journal*, **59**, 30, 1924.

[3] *Ibid.*, **67**, 114, 1928.   [4] *Ibid.*, **76**, 124, 1932.

[5] *Harvard Observatory Circular*, No. 334, 2, 1928; *Proceedings of the National Academy of Sciences*, **14**, 404, 1928.

[6] *Zeitschrift für Astrophysik*, **1**, 192, 1930; **2**, 165, 1931.

[7] *Annals of the Solar Physics Observatory, Cambridge*, **3**, Part II, 1933.

29

developments by Schütz.[8] The latter showed that if the atoms of an absorbing gas are not quiescent but are subject to thermal agitation, the absorption coefficient as first derived by Voigt[9] gives a satisfactory explanation of the observed departures from the square-root law. For very faint lines the contours are essentially defined by the thermal Doppler effect. Radiation damping may be neglected when the number of atoms is small. Accordingly, the absorption coefficient is proportional to the number of atoms having any given velocity v, and the total intensity of the line increases proportionally to N. However, as the line becomes stronger, its central part becomes more and more saturated, causing a departure from the relation $A \propto N$, where A designates the total absorption. Were it not for radiation damping, A would become more and more independent of N, approaching a constant for large values of N. The effect of radiation damping prevents A from becoming a constant by producing wide wings which again permit A to increase with N. The ultimate result is again $A \propto \sqrt{N}$, as in the case of pure radiation damping.[10] A summary of the theoretical work on this problem has recently been published by Weisskopf.[11] Laboratory verifications are difficult and rest primarily upon measurements by Minkowski,[12] by Schütz,[13] and by Korff.[4]

The first application of Schütz's theory to astronomical spectra was made by Unsöld[14] in a discussion of the intensities of interstellar calcium lines. An even more convincing confirmation of the theory has been given by Minnaert and Mulders[6] for the faint lines in the sun's spectrum. Pannekoek,[15] and Minnaert and Slob,[16] have elaborated the theory of Schütz and made it directly applicable to stellar

[8] *Zeitschrift für Astrophysik*, **1**, 300, 1930; *Zeitschrift für Physik*, **64**, 682, 1930.

[9] *Sitzungsberichte der Akademie, München*, p. 602, 1912.

[10] This is correct only for values of N which are of practical significance. The formula for the absorption coefficient derived from the theory of radiation damping also demands $A \propto N$, for very small values of N (see Ladenburg and Reiche, *Annalen der Physik*, **42**, 181, 1913).

[11] *Physikalische Zeitschrift*, **34**, 1, 1933; *Observatory*, **56**, 302, 1933.

[12] *Zeitschrift für Physik*, **36**, 839, 1926.   [13] *Loc. cit.*

[14] Unsöld, Struve, and Elvey, *Zeitschrift für Astrophysik*, **1**, 314, 1930.

[15] *Monthly Notices*, **91**, 139, 1930.

[16] *Proceedings of the Amsterdam Academy*, **34**, 542, 1931.

spectra. Pannekoek[17] has published measurements of line intensities in the spectrum of a Cygni and has interpreted these in the light of the theory as derived by Minnaert and Slob.

Total absorptions of certain multiplet lines have been measured by Shajn,[18] although the greater part of his work dealt with the central intensities. His conclusion was that real departures from the square-root law exist.

In 1930 the writers published a series of measurements[19] of the total absorptions of the Si III triplet λλ 4552, 4567, and 4574. The investigation was made principally to find whether the relative intensities in stars having nebulous lines are the same as those observed in stars having narrow lines. This was found to be the case, as had been predicted by the rotational hypothesis of broad and diffuse stellar absorption lines. It was also found that, on the average, the relative intensities were proportional to the square roots of the theoretical multiplet intensities derived from the rules of Burger and Dorgelo. There was found to be no marked difference between giants such as ζ Persei and dwarfs such as γ Pegasi.

In 1932 Struve[20] found a remarkable difference in the relative intensities of different members of multiplets in the stars 17 Leporis and ε Aurigae: while the strongest lines of Fe II and Ti II were slightly stronger in 17 Leporis than in ε Aurigae, the weaker lines were practically absent in the former though they were fairly strong in the latter. This phenomenon has been confirmed, from microphotometric measurements, by Hynek.[21]

New observational evidence of differences in the gradients of intensity of stellar absorption lines was announced by Struve and Morgan.[22] They found that the relative intensities of the members of the Fe I multiplet $a^3F - y^3F^o$ were not the same in several stars of classes A and F, and it was suspected that rotationally broadened lines had a tendency to possess smaller gradients. However, differ-

[17] Ibid., p. 755.   [18] Monthly Notices, **91**, 675, 1931.

[19] Astrophysical Journal, **72**, 267, 1930.

[20] Ibid., **76**, 85, 1932; Pl. XI of this article contains a good illustration of the gradient effect. See also ibid., **78**, Pl. VII, 1933.

[21] Ibid., **78**, 54, 1933.

[22] Proceedings of the National Academy of Sciences, **18**, 590, 1932.

ences in the gradient could also be detected between stars having perfectly narrow lines.

The same investigators also found peculiar differences between the intensities of several O II lines in B stars. This subject was more fully investigated by Struve,[23] who found complete analogy between the phenomenon noted in 17 Leporis and ϵ Aurigae and the behavior of the O II lines. In a general way, all the stronger O II lines are greatly enhanced in the giants, while moderately faint lines appear to be of about equal intensity in giants and in dwarfs; the faintest O II lines are visible only in the dwarfs, such as τ Scorpii, and are not seen in the giants, such as β Canis Majoris.

It is true that the excitation potentials of the O II lines which are observed in stellar spectra are not all the same; they vary approximately between 23 and 31 volts. The intensities of the O II lines should, therefore, exhibit changes depending upon the degree of ionization, and such an effect has in fact been found by Struve.[24] However, a careful analysis of the lines, by multiplets, proves that the effect of gradient cannot be explained by differences in excitation. Consider, for example, the two multiplets shown in the accompanying tabulation. From the comparison of the intensities in τ

λ	Int. Lab.	Int. τ Sco (Dwarf)	Int. ϵ CMa (Giant)
	$e^4D^o - z^4F$		
4070	4+6	8	12
4072	8	6	10
4076	10	7	11
4079	4	3	3
4085	3	2	2
4093	5	2	1
	$e^4P^o - x^4P$		
4129	2	1	Blend
4133	6	5	3
4153	7	5	6
4156	3	3	1

[23] Astrophysical Journal, 78, 73, 1933.　　　[24] Op. cit., p. 83.

Scorpii and in ε Canis Majoris it appears reasonably certain that the strong lines are enhanced in the giant, while the weak lines are enhanced in the dwarf. This explains an interesting phenomenon repeatedly noticed in measuring the wave-lengths of stellar absorption lines: contrary to expectation, the faintest lines of any given element are not found in those stars in which the principal lines are strongest. One might, for example, be tempted to measure a star like ε Canis Majoris or β Canis Majoris in the hope of obtaining the faintest O II lines. In reality it is in the dwarf τ Scorpii that the greatest number of O II lines have been measured.[25]

The discovery of the gradient effect for O II made it desirable to reinvestigate the intensities of the Si III lines in B stars. Elvey's results[26] confirmed the earlier measurements,[19] but he found, in addition, that the gradient was not quite the same in all stars. Departures from the square-root law were generally in the direction of smaller gradients. The difference between giants and dwarfs is much less conspicuous than is the case for O II, but there is a slight tendency in the same direction. Thus we have:

For six giants: $\lambda 4552/\lambda 4567/\lambda 4574 = 41.3/35.2/23.5$
For six dwarfs: $\lambda 4552/\lambda 4567/\lambda 4574 = 40.6/33.6/25.8$

In examining the spectra of stars for the occurrence of the gradient effect, Morgan[27] found that ε Aurigae, the comparison star used by Struve and by Hynek in studying the gradient effect in 17 Leporis, was itself abnormal. A comparison with α Persei and 41 Cygni indicated that the gradient in ε Aurigae is much steeper than in the other two stars. ε Aurigae is a supergiant, while α Persei and 41 Cygni are normal giants of class F. It would be natural to expect an even smaller gradient in the dwarfs, but this is apparently not the case. Visual examination reveals no clear-cut difference in gradient between α Persei and a normal dwarf of the same spectral class.

On the other hand, Morgan[28] found striking differences in the intensity gradients of the Fe I lines in A-type stars. Thus, the peculiar star 73 Draconis shows a very small gradient, while in the normal A star ε Serpentis the gradient is larger. In the case of Fe I the intensities of the lines are correlated with excitation potentials, the

[25] For criteria of absolute magnitude in B stars we refer to earlier papers: O. Struve, *Astrophysical Journal*, **78**, 73, 1933, etc.

[26] *Ibid.*, p. 219.   [27] *Ibid.*, p. 158.   [28] *Ibid.*, **77**, 77, 1933.

fainter lines usually having slightly higher excitation potentials. It is improbable, however, that the differences noticed by Morgan are caused by normal excitation. Further work on the relative intensities of lines in A and F stars is now being carried out by Morgan and by Hynek, respectively.

The recognition of the gradient effect in many stars of different spectral classes makes it appear probable that other phenomena are directly or indirectly caused by it. Thus, the peculiar differences, noted by Struve,[29] of the singlet and triplet lines of *He* I are at least in part manifestations of the gradient effect. It so happens that the singlet lines are usually fainter than the corresponding triplet lines, consequently the conditions are suitable for the gradient effect. Whether all of the triplet-singlet differences can be accounted for in this manner has not yet been investigated.

An interesting manifestation of the gradient effect in the *He* I lines has recently been found by Morgan.[30] It appears that the variations of the intensities of the *He* I lines discovered by him in the spectroscopic binary μ Sagittarii[31] and by Struve in the spectroscopic variable 27 Canis Majoris[32] are simply changes in the intensity gradients. A similar effect is also present in the spectrum of β Lyrae.[30]

A first attempt to measure the differences in gradient of the lines of *Fe* II and *Ti* II has recently been made by Elvey.[33] His results confirmed the visual observations, but completely failed to give any clue to the cause of the variations. There was no definite correlation with either atomic number or spectral type, or even absolute magnitude of the stars.

The absence of a correlation with absolute magnitude was especially perplexing, because such a correlation had been rather definitely suspected for *O* II in the B stars. It had also been shown in the comparison of 17 Leporis, ϵ Aurigae, α Persei, and 41 Cygni. On the other hand, for *Si* III the absolute-magnitude effect was, if present at all, much smaller than were individual differences between stars of approximately the same luminosity.

The phenomenon is a complicated one, and the present investiga-

[29] *Ibid.*, **74**, 225, 1931; **78**, 73, 1933.

[30] Unpublished.

[31] *Astrophysical Journal*, **75**, 407, 1932.

[32] *Ibid.*, **73**, 301, 1931.

[33] *Ibid.*, **79**, 263, 1934.

tion is an attempt to explain it on the basis of the working hypothesis proposed in section IV.

II. THE OBSERVATIONS

A closer examination of the intensities of the lines of selected multiplets of *Fe* II and *Ti* II published by Elvey[33] indicated that the lines of both elements could be combined into one set showing a correlation between the total absorptions and the theoretical intensities. For three of the stars—a Persei, a Canis Minoris, and 41 Cygni—there was very little change in the total absorption, with a large change in the theoretical intensities, the total absorption varying with about the ninth root of the theoretical intensities. On the other hand, in a Cygni and a Lyrae the total absorption varied with the square roots of the theoretical intensities, and in 17 Leporis approximately with the theoretical intensities.

It is difficult to find a sufficient range in the intensities of the lines within a single multiplet to make satisfactory tests of the relationships between the theoretical and the observed intensities of absorption lines. The foregoing data showed, as have other data, that different multiplets can be combined to form relationships between the absorption and the number of atoms. Such combinations, covering a large range in intensities, tend to smooth over the irregularities exhibited by individual multiplets such as have been found by Woolley[7] and attributed by him to "interlocking." The foregoing data also indicate the course to be taken for further accumulation of observational material. Since the lines that have been observed are among the more intense lines in the spectrum of the particular star, and since there are such great differences among the various stars in the relationship between total absorption and theoretical intensity, it is desirable to obtain as much observational material in a given spectrum as is possible.

To accomplish this we have used the Bruce spectrograph with a dispersion of three prisms, the scale being 10 A/mm at λ 4500, to obtain spectrograms of some of the brighter stars on plates of high contrast. The emulsion used is Eastman O IV, and the developer was D-11. Because of the high contrast of the plates, it was necessary to make use of several spectrograms of each star, using different exposures, in order to have satisfactory densities in the various parts of the spectrum for the accurate determination of the intensities.

Tracings of the spectrograms were made with the recording microphotometer, the magnification being about 30. Each part of the spectrum was run through the microphotometer a second time with a displacement of the plate by half a turn of the screw, to eliminate the periodic errors due to one turn of the screw. The central intensity of a line was measured in the usual manner and the total intensity was obtained by measuring the base of the line on the tracing and computing the area, assuming the line to have a triangular contour. The procedure is justified, since previous work[33] has shown no appreciable systematic differences, for lines of the intensities used in this investigation, between results obtained in this manner and those obtained by integrating the entire contour. The resulting total intensities are expressed in angstrom units of complete absorption of the continuous spectrum.

Practically all the measurements of intensity have been confined to those parts of the spectrum for which the tracings fell on the straight-line portion of the reduction-curve for the plate. The intensities in different parts of the spectrum were measured on different spectrograms, but there was sufficient overlapping to insure against systematic errors from plate to plate.

The resulting intensities for the five stars, α Canis Majoris, α Lyrae, α Cygni, α Persei, and ϵ Aurigae, are listed in Table I. The total absorptions in Angstrom units have been expressed as log A. The measured wave-lengths and the identifications are partly by Morgan.[30] The tabulated wave-length is that measured in α Cygni, whenever available; when not, the wave-length in α Lyrae is given. In the identifications, only the most important contributor to the line has been listed, except where there is some uncertainty or where the blend is from two lines of about equal strength. In a few cases a line has been measured as single on the spectrogram, while on the tracing it was possible to sketch the shapes of two lines from the asymmetry of the contour and obtain a measure of the total absorption. In such cases the table gives a single wave-length for the blend, but separate intensities for the components. For example, the line λ 4227.01 is composed of the two lines Ca I 4226.73 and Fe I 4227.45, each of which has been measured.

TABLE I

λ	LOG A					IDENTIFICATION	
	α CMa	α Lyr	α Cyg	α Per	7 ε Aur		
4215.60...	−0.93	−1.35	−1.28			Sr II 15.52	$5^2S - 5^2P^o$
4217.....	1.58					Fe I 17.56	$z^5F^o - 9$
4219.34...	1.47	1.92				Fe I 19.36	.....
4222.24...	1.51	.....				Fe I 22.23	$z^7D^o - e^7D$
4227.01... {	1.24	1.12	1.37			Ca I 26.73	$4^1S - 4^1P^o$
	1.11	1.41	1.42			Fe I 27.45	$z^5F^o - 46$
4233.05...	0.80	0.91	0.25			Fe II 33.16	$b^4P - z^4D^o$
						(Y II 35.70)	$a^3D - z^3P^o$
4235.62...	1.17	1.62	2.05			Fe I 35.95	$z^7D^o - e^7D$
4239.01...	1.43	1.40	2.05			Fe I 38.83	$z^5F^o - 59$
4242.32...	1.05	1.48	0.76			Cr II 42.35	$a^4F - z^4D^o$
4244.62...	.....	.....	2.00			Ni II 44.80	$a^2G - z^4F^o$
4245.....	1.62					Fe I 45.26	$b^3P - 20^o$
4246.89...			0.82			Sc II 46.83	$a^1D - z^1D^o$
4247.34...	1.44					Fe I 47.44	$z^5F^o - 28$
4250.19... {	1.22	2.22	1.85			Fe I 50.13	$z^7D^o - e^7D$
	1.21	1.35	1.58			Fe I 50.79	$a^3F - z^3G^o$
4252.60...	1.42	1.88	1.39			Cr II 52.63	$a^4F - z^4D^o$
4254.53...	1.15	.....	1.52			Cr I 54.34	$a^7S - z^7P^o$
4256.18...			1.85			Fe I 56.21	$z^5F^o - 19$
4258.21...	1.37		0.90			Fe II 58.16	$b^4P - z^4F^o$
4259.13...			1.92				
4260.31...	1.07	1.42	1.43			Fe I 60.49	$z^7D^o - e^7D$
4261.81...	1.05	1.55	0.85			Cr II 61.91	$a^4F - z^4D^o$
4263.96...			1.58				
4267.90...			1.58			Fe I 67.83	.....
4269.26...	1.36		1.32			Cr II 69.30	$a^4F - z^4D^o$
4270.66...			1.85				
4271.70... {	1.28	1.60}	1.24			Fe I 71.17	$z^7D^o - e^7D$
	1.12	1.38}				Fe I 71.76	$a^3F - z^3G^o$
4273.32...	1.29	1.85	0.88			Fe II 73.31	$b^4P - z^4D^o$
						Cr I 74.80	$a^7S - z^7P^o$
4275.48... {	1.34					Cr II 75.55	$a^4F - z^4D^o$
	1.24	1.89	1.00			Fe II 78.13	.....
4278.21...	1.55		1.13			Fe I 82.41	$a^5P - z^5S^o$
4282.50...	1.28					Cr II 84.20	$a^4F - z^4D^o$
4284.35...	1.25		1.06			Zr II 86.51	.....
4286.39...	.....		1.72			Ti II 87.88	$a^2D - z^2D^o$
4287.93...	1.31		1.17			Cr I 89.72	$a^7S - z^7P^o$
4290.27... {	1.51					Ti II 90.23	$a^4P - z^4D^o$
	1.10	1.04	0.66			Ti II 94.10	$a^2D - z^2D^o$
4294.12...	1.12	1.14	0.77			Fe I 94.13	$a^3F - z^5G^o$
4296.54...	1.12		0.62			Fe II 96.56	$b^4P - z^4F^o$
						Fe I 99.25	$z^7D^o - e^7D$
4299.97... {	1.35	1.77	.....			Ti II 00.05	$a^4P - z^4D^o$
	0.92	1.04	0.52			Ti II 01.93	$a^4P - z^4D^o$
4301.91...	1.17	1.47	0.96			Fe II 03.18	$b^4P - z^4D^o$
4303.28...	1.02	1.21	0.56			Sr II 05.46	$5^2P^o - 6^2S$
						Fe I 05.46	$3M - z^3S^o$
4305.67...	1.57	1.85	.....			Ti II 07.89	$a^4P - z^4D$
4307.81...	−0.90	−1.00	−0.79			Fe I 07.91	$a^3F - z^3G^o$

TABLE I—*Continued*

λ	α CMa	α Lyr	α Cyg	α Per	γ ε Aur		Identification	
4309.66...	−1.54	......	......	......	......	Y II	09.63	a³D−z³P°
4312.79...	1.08	−1.24	−0.96	......	......	Ti II	12.88	a⁴P−z⁴D°
4314.18...	1.32	1.38	0.77	......	......	Sc II	14.09	a³F−z³D°
4314.87...	1.00	1.40	0.97			{ Ti II	14.98	a⁴P−z⁴D°
						Fe I	15.09	a⁵P−z⁵S°
4316.87...	......	......	1.55			Ti II	16.80	b²P−z²P°
4320.85...	1.54	......	1.96			{ Sc II	20.73	a³F−z³D°
						Ti II	20.95	a⁴P−z⁴D°
4325.52... {			1.44			Sc II	25.00	a³F−z³D°
	0.84	1.20	0.96			Fe I	25.77	a³F−z³G°
4330.55...	......	......	1.22			{ Ti II	30.25	b²P−z²P°
						Ti II	30.71	a⁴P−z⁴D°
4337.95...			0.98			Ti II	37.92	a²D−z²D°
4343.10...			2.05			Fe I	43.27	
4344.21...			1.24		−0.26	{ Ti II	44.31	a²D−z²D°
						Cr I	44.51	a⁵D−z⁵P°
4347.83...				−1.05		Fe I	47.85	z⁵P°−f⁵D
4351.80...	0.67	1.13	0.28	0.67	0.07	Fe II	51.77	b⁴P−z⁴D°
4352.75...				0.84		Fe I	52.74	a⁵P−z⁵S°
4354.23...			1.52	0.87	1.19	Sc II	54.60	a³F−z³F°
4358.44...	1.57					{ Y II	58.74	a³D−z³P°
						Fe I	58.51	b³G−18°
4359.69...	......			0.90	1.13	Zr I	59.74	b⁴P−z⁴D°
4362.02...	1.48	1.46	1.48	......	1.03	Ni II	62.10	a²G−z⁴F°
4367.84...	1.24	1.17	0.96	0.59	0.40	{ Fe I	67.58	b³G−z³H°
						Ti II	67.67	b²F−y²G°
4369.39...	1.33	1.80	1.01	0.75	0.57	{ Fe II	69.41	b⁴P−z⁴F°
						Fe I	69.78	z³F°−f³D
4371.11...	......	......		0.77	1.16	Zr II	70.95	b⁴P−z⁴D°
4373.6....				1.02		Fe I	73.57	{ b³F−w⁵D°
								b³G−x³F°
4374.67...	1.05	1.64	1.07		0.02	{ Sc II	74.46	a³F−z³F°
						Y II	74.96	a¹D−z¹D°
4379.86...					0.97	Zr II	79.78	a²H−z²G°
4383.48...	0.84	1.03	0.96		0.18	Fe I	83.55	a³F−z⁵G°
4384.33...	1.07	1.49	0.77			{ ☉ II	84.32	......
						Mg II	84.64	4²P°−5²D
4385.40...	0.97	1.33	0.58		0.85	Fe II	85.39	b⁴P−z⁴D°
4386.83...	1.36	1.80	1.24	0.68	0.49	Ti II	86.84	b²F−y²G°
4387.96...	......	......	1.41			{ Fe I	87.90	4M−z³S°
						He I	87.93	2P°−5D
4390.65...	0.86	1.34	0.97	1.06		{ Mg II	90.59	4²P°−5²D
						Fe I	90.96	b³G−z³H°
4391.0....	......			0.72	0.44	Ti II	91.02	b⁴P−z⁴D°
4393.86...	1.31	1.70	1.24			Ti II	94.06	a²P−z⁴D°
4395.01...	0.85	1.14	0.53	0.52		Ti II	95.04	a²D−z²F°
4395.84...				0.66		Ti II	95.85	b⁴P−z⁴D°
4398.08...					0.67	{ Y II	98.03	a³D−z³P°
						Ti II	98.31	b⁴P−z⁴D°
4399.77...	−1.17	−1.35	0.85	......	......	Ti II	99.77	a²P−z⁴D°
4400......	......	......	−1.51	−0.49	−0.08	Sc II	00.38	a³F−z³F°

TABLE I—*Continued*

λ	LOG A					IDENTIFICATION		
	α CMa	α Lyr	α Cyg	α Per	γ ε Aur			
4403.28	...	...	.:..	−1.00	−0.18	⊙ I	03.19	...
						Zr II	03.33	b⁴P−z⁴D°
4404.76	−0.88	−1.30	−1.35	...	0.42	Fe I	04.75	a³F−z⁵G°
4407.66	...	...	...	...	0.92	Ti II	07.68	a²P−z⁴D°
						Fe I	07.72	a⁵P−x⁵D°
4408.46	...	...	...	0.71	...	Fe I	08.42	a⁵P−x⁵D°
4409.32	...	...	...	0.61	0.63	Ti II	09.24	b⁴P−z⁴D°
						Ti II	09.54	b⁴P−z⁴D°
4411.12	1.36	...	1.31	0.72	0.48	Ti II	11.08	c²D−x²F°
4411.9	...	...	...	0.91	1.05	Ti II	11.94	b⁴P−z⁴D°
4414.47	...	...	...	0.92	...	Zr II	14.52	b⁴P−z⁴D°
4415.37	1.01	...	1.28	0.41	0.19	Fe I	15.13	a³F−z⁵G°
						Sc II	15.55	a³F−z³F°
4416.85	1.03	1.38	0.47	0.58	...	Fe II	16.81	b⁴P−z⁴D°
4417.70	1.09	1.39	0.96	0.94	...	Ti II	17.71	a⁴P−z²D°
4420.63	...	...	...	0.92	...	Sc II	20.66	a⁵P−z³F°
4421.98	...	...	1.39	...	0.35	Ti II	21.95	b²P−y²D°
4425.45	...	...	...	0.84	1.48	Ca I	25.43	4³P°−4³D
4427.97	...	...	1.46	...	1.00	Ti II	27.90	b⁴P−z⁴D°
						Mg II	28.00	4²P°−6²S
4430.61	...	...	...	0.82	...	Fe I	30.62	a⁵P−x⁵D°
4431.37	...	...	...	1.02	0.87	Sc II	31.35	a³F−z³F°
4433.96	...	...	1.32	...	...	Fe I	33.81	z⁵P°−56
						Mg II	33.99	4²P°−6²S
4434.80	...	...	...	0.52	0.91	Ca I	34.95	4³P°−4³D
						Fe I	35.15	a⁵D−z⁷F°
4435.66	...	...	...	0.62	...	Ca I	35.67	4³P°−4³D
4436.33	...	...	...	1.30	...	Mn I	36.36	a⁴D−z⁴D°
4438.40	...	...	...	1.57	...	Fe I	38.36	z⁵P°−f⁵D
4440.44	...	...	...	0.94	...	Zr II	40.44	b⁴P−z⁴D°
4441.70	...	...	...	...	0.47	Ti II	41.72	...
4443.84	...	...	0.55	...	0.03	Ti II	43.80	a²D−z²F°
4446.99	...	...	...	1.02	...	Fe I	46.85	z⁵P°−f⁵D
						Fe I	47.14	
4447.76	...	...	...	...	1.16	Fe I	47.73	a⁵P−x⁵D°
4450.44	...	...	0.94	0.39	0.11	Ti II	50.49	a²D−z²F
4451.60	...	...	1.27	0.83	...	Mn I	51.58	a⁴D−z⁴D°
4454.47	...	...	...	1.09	...	Fe I	54.39	b³P−x³D°
4454.90	...	...	...	0.27	0.63	Ca I	54.77	4³P°−4³D
						Zr II	54.80	a²F−z⁴G°
4455.30	...	...	1.19	...	...	Ti I	55.32	b³F−v³F°
						Mn I	55.32	z⁶P°−e⁶D
4455.89	...	...	...	0.65	...	Ca I	55.88	4³P°−4³D
4456.33	...	...	...	...	0.91	Ca I	56.61	4³P°−4³D
						Ti II	56.62	c²D−x²F°
4459.20	...	...	...	0.54	0.94	Fe I	59.13	a⁵P−x⁵D°
4461.57	1.36	1.64	1.06	...	0.69	Zr II	61.21	a²G−z²F°
						Fe I	61.66	a⁵D−z⁷F°
4464.45	1.62	1.58	1.09	...	0.34	Ti II	64.47	a⁴P−z²D°
4466.46	1.57	1.64	1.60	0.57	0.94	Fe I	66.56	b³P−x³D°
4468.49	−0.97	−1.03	−0.54	−0.35	−0.48	Ti II	68.49	a²G−z²F°

TABLE I—*Continued*

λ	LOG A					IDENTIFICATION		
	α CMa	α Lyr	α Cyg	α Per	γ ε Aur			
4470.88...	−1.44	......	......	......	−0.48	*Ti* II	70.88	a⁴P−z²D°
4471.46...	1.43	......	−1.52	......	......	*He* I *He* I	71.48 71.69	2³P°−4³D
4472.94...	1.39	......	0.98	−0.58	0.62	*Fe* II	72.93	b⁴F−z⁴F°
4475.90...	1.39	−1.92	0.82	0.58	1.11	*Fe* I	76.02	b³P−x³D°
4479.52...	......	......	......	1.15		*Fe* I	79.61	z⁵P°−f⁵D
4481.25...	0.26	0.29	......	......	0.08	*Mg* II *Mg* II	81.13 81.33	3²D−4²F° 3²D−4²F°
4482.25...	......	......	......	0.53		*Fe* I *Fe* I	82.18 82.26	a⁵D−z⁷F° a⁵P−x⁵D°
4484.81...	......	......	......	0.88		*Fe* I	84.24	z⁵P°−f⁵D
4488.36...	1.39	1.89	1.00	0.61	......	*Ti* II	88.32	c²D−x²F°
4489.19...	1.17	1.48	0.62	0.47	......	*Fe* II	89.21	b⁴F−z⁴F°
4491.40...	1.10	1.38	0.51	0.48	0.32	*Fe* II	91.41	b⁴F−z⁴F°
4493.55...	......	......	......	......	1.00	*Ti* II	93.52	a²D−z⁴F°
4494.53...	......	......	......	0.55	0.95	*Fe* I	94.57	a⁵P−x⁵D°
4496.79...	......	......	......	......	0.88	*Zr* II	96.96	a⁴F−z⁴G°
4499.04...	......	......	......	1.19		*Mn* I	98.90	a⁴D−z⁴D°
4501.28...	1.08	1.10	0.52	0.35	0.06	*Ti* II	01.27	a²G−z²F°
4502.25...	......	......	......	1.33		*Mn* I	02.22	a⁴D−z⁴D°
4508.29...	0.97	1.10	0.30	0.46	0.14	*Fe* II	08.29	b⁴F−z⁴D°
4512.79...	......	......	......	1.36		*Ti* I	12.73	a⁵F−y⁵F°
4515.37...	1.05	1.17	0.37	0.49	0.30	*Fe* II	15.34	b⁴F−z⁴F°
4517.64...	2.15					*Fe* I	17.53	4M−y³P°
4518.36...	......	......	......	0.83		⊙ II	18.34	....
4520.21...	1.08	1.28	0.40	0.49	0.27	*Fe* II	20.24	b⁴F−z⁴F°
4422.65...	1.03	1.26	0.33	0.33	0.19	*Fe* II	22.64	b⁴F−z⁴D°
4524.99...	......	......	1.70	......	1.08	*Ti* II *Fe* I	24.72 25.15	b⁴P−z²D° z⁵P°−17
4526.35...	......	......	......	1.60		*Fe* I	26.57	
4528.30...	1.36	......	......	0.54	0.79	*Fe* I	28.62	a⁵P−x⁵D°
4529.52...	......	......	......	0.54	0.59	*Ti* II	29.46	a²H−z²G°
4534.04...	0.90	1.05	0.44	0.24	0.10	*Ti* II	33.97	a²P−z²D°
4541.53...	1.31	1.48	0.58	0.54	0.39	*Fe* II	41.52	b⁴F−z⁴D°
4544.06...	......	......	......	0.60	1.00	*Ti* II	44.01	b⁴P−z²D°
4544.92...	......	......	1.51	......	0.94	*Ti* II	45.14	a²G−z⁴F°
4549.53...	0.54	0.71	0.14	......	0.07	*Fe* II *Ti* II	49.48 49.64	b⁴F−z⁴D° a²H−z²G°
4552.27...	......	......	1.57	0.67	1.02	*Ti* II *Fe* I	52.29 52.55	 a²G−z⁴F°
4554.02...	......	......	......	0.45		*Ba* II	54.04	6²S−6²P°
4554.96...	1.29	1.82	0.83	......	......	*Cr* II	54.99	b⁴F−z⁴F°
4556.00..	1.13	1.28	0.38	0.37	0.15	*Fe* II *Fe* I	55.90 56.13	b⁴F−z⁴F° b³G−x⁵G°
4558.66...	0.99	1.28	0.36	0.53	0.32	*Cr* II	58.66	b⁴F−z⁴D°
4563.79...	1.11	1.24	0.50	0.42	0.10	*Ti* II	63.77	a²P−z²D°
4565.77...	1.66	......	1.13	......	0.86	*Fe* I	65.68	z⁵D°−e⁵F
4568.33...	......	......	......	1.12		*Ti* II	68.30	b⁴P−z²D°
4571.97...	1.03	1.17	0.50	0.26	0.16	*Ti* II	71.98	a²H−z²G°
4576.36...	−1.37	−1.96	−0.57	0.58	−0.39	*Fe* II	76.31	b⁴F−z⁴D°
4578.56...	......	......	......	−1.01	......	*Ca* I	78.57	3³D−4³F°

TABLE I—*Continued*

λ	LOG A					IDENTIFICATION		
	α CMa	α Lyr	α Cyg	α Per	7 ε Aur			
4579.90...	−1.27	....	−0.88	......	−0.63			..F..
4581.42...				−0.84		{ *Ca* I	81.41	$3^3D-4^3F^o$
						{ *Fe* I	81.53	$z^5D^o-e^3F$
4582.82...	1.33	−1.82	0.55	0.45		*Fe* II	82.84	$b^4F-z^4F^o$
4583.89...	0.88	1.08	0.24	0.27		*Fe* II	83.84	$b^4F-z^4D^o$
4585.92...				0.85		*Ca* I	85.87	$3^3D-4^3F^o$
4588.19...	1.15	1.62	0.38	0.52	0.32	*Cr* II	88.21	$b^4F-z^4D^o$
4589.85...	1.54	1.96	0.89	0.48	0.47	*Ti* II	89.96	$a^2P-z^2D^o$
4592.03...	1.27	1.92	0.72	0.44	0.64	*Cr* II	92.05	$b^4F-z^4D^o$
4593.92...			1.92					
4595.90...	1.64		0.97		1.04	*Fe* II	95.70	$b^4F-z^4D^o$
4598.22...			1.42	0.85		*Fe* I	98.14	$z^5D^o-e^5F$
4600.11...					1.22	{ *V* II	00.20	
						{ *Ni* I	00.36	$z^5G^o-e^5F$
4601.33...					1.28	*Fe* II	01.38	$a^6S-z^4D^o$
4605.....	1.74				1.46	*Ni* I	04.99	$z^5G^o-e^5F$
4607.63...				0.91		*Fe* I	07.67	$z^5D^o-e^5F$
4611.29...	1.68					*Fe* I	11.29	z^5P^o-17
4616.69...	1.17	....	0.80	0.42	0.66	*Cr* II	16.67	$b^4F-z^4D^o$
4618.85...	1.12	1.82	0.59	0.48	0.44	*Cr* II	18.82	$b^4F-z^4D^o$
4620.53...	1.34		0.78	0.44	0.61	*Fe* II	20.52	$b^4F-z^4D^o$
4625.02...				0.89		*Fe* I	25.06	$z^5D^o-e^5F$
4626.07...				0.99		*Cr* I	26.18	$a^5D-y^5P^o$
4629.34...	1.08	1.64	0.33	0.37	0.23	*Fe* II	29.33	$b^4F-z^4F^o$
4634.05...	1.12	−1.64	0.57	0.56	0.47	*Cr* II	34.12	$b^4F-z^4D^o$
4635.42...	1.32	....	1.01		0.87	*Fe* II	35.35	
4637.50...				0.76		*Fe* I	37.52	$z^5D^o-e^5F$
4638.....					1.23	*Fe* I	38.02	z^5P^o-31
4646.19...				0.70	1.24	*Cr* I	46.17	$a^5D-y^5P^o$
4647.41...				0.71		*Fe* I	47.44	$b^3G-y^3G^o$
4648.74...				0.74	1.00	*Ni* I	48.66	$z^5G^o-e^5F$
4651.29...				1.17		*Cr* I	51.30	$a^5D-y^5P^o$
4652.19...				1.15	1.62	*Cr* I	52.17	$a^5D-y^5P^o$
4654.56...			......	0.64		{ *Fe* I	54.50	$a^3F-y^5F^o$
						{ *Fe* I	54.64	$z^5D^o-e^5F$
4657.04...	1.26	....	0.83	......	0.49	*Fe* II	56.98	$a^6S-z^4D^o$
4661.94...				1.05		*Fe* I	61.98	$b^3G-y^3G^o$
4663.30...	1.21		0.92		0.79	*Fe* II	63.71	$a^6S-z^4F^o$
4664.81...				1.22		*Cr* I	64.81	$z^7F^o-g^7D$
4666.76...	1.26		0.70	0.41	0.40	*Fe* II	66.75	$b^4F-z^4F^o$
4668.14...				0.61		*Fe* I	68.15	$z^5D^o-e^5F$
4670.16...	−1.55		−0.99		0.39	*Sc* II	70.40	$b^1D-z^1F^o$
4682.69...					1.55	*Y* II	82.32	$a^1D-z^3P^o$
4686.13...				1.42		*Ni* I	86.21	$z^5G^o-e^5F$
4691.44...				0.91		*Fe* I	91.42	$b^3G-y^3G^o$
4698.69...					1.01	{ *Sc* II	98.30	$a^3F-z^1F^o$
						{ *Cr* II	98.74	$a^4P-z^4F^o$
4702.97...					1.03	*Mg* I	03.00	$3P^o-5D$
4707.31...				0.73		*Fe* I	07.29	$z^5D^o-e^5F$
4708.75...				0.60	−0.72	*Ti* II	08.66	$a^2P-z^2F^o$
4710.17...				−1.12		*Fe* I	10.28	$b^3G-y^3G^o$

TABLE I—*Continued*

λ	LOG A					IDENTIFICATION	
	α CMa	α Lyr	α Cyg	α Per	γ ε Aur		
4714.41				−0.80	−1.51	Ni I	14.42	z⁵G°−e⁵F
4715.81					1.37	Ni I	15.76	z⁵G°−e⁵F
4718.41					1.29	Cr I	18.45	z⁷F°−g⁷D
4719.8					1.13	Ti II	19.51	b⁴P−z²F°
4731.32			−0.78		0.44	Fe II	31.49	a⁶S−z⁴D°
4736.89				0.60	1.08	Fe I	36.79	z⁵D°−e⁵F
4756.47				0.90		Ni I	56.53	z⁵G°−e⁵F
4764.23					0.57	Ni I	63.95	z³F°−f³F
4770.26					1.64	C I	70.00	3s³P°−4p³P
4780.07			0.94	−0.61	0.53	Ti II	79.99	b²P−z²S°
4783.17					1.42	Mn I	83.43	z⁸P°−e⁸S
4786					1.42	Y II	86.58	a³F−z³D°
4798.57					0.90	Ti II	98.52	a²D−z⁴G°
4805			0.82		0.42	Ti II	05.11	b²P−z²S°
4812.11			1.15		0.77	Cr II	12.37	a⁴F−z⁴F°
4823.92			0.55		0.30	Cr II	24.13	a⁴F−z⁴F°
4836.38			1.06		0.97	Cr II	36.23	a⁴F−z⁴F°
4848.27			0.64		−0.51	Cr II	48.27	a⁴F−z⁴F°
4854.8			−1.49			Y II	54.89	a³F−z³D°

Pannekoek[17] has measured the total absorptions of a number of lines of *Fe* II, *Ti* II, and *Cr* II in α Cygni, and a comparison between our measures and his shows a systematic difference of about 15 per cent, in the sense that our total absorptions are smaller than his (Fig. 1). Such a systematic error may result from differences in judgment as to the position of the continuous spectrum across a line and also in measuring the point where the contour of the line joins the continuous spectrum. For the narrow lines of the metals the latter is probably the source of the differences.

A comparison of the total absorptions with theory requires a knowledge of the emission intensities of the lines, since they are proportional to the numbers of atoms producing the lines. The laboratory intensities are unsatisfactory, largely owing to a lack of accurate photometric data. For a large portion of the lines the theoretical intensities are available from the formulae of Russell and of others, but these are only relative intensities within a given multiplet. We have made use of an investigation of the flash spectrum by Pannekoek and Minnaert[34] in which they have made a comparison of the intensities

[34] Results of observations of the total solar eclipse of June 29, 1927; photometry of the flash spectrum, Amsterdam, 1928.

observed in the flash spectrum with the theoretical intensities. They found linear relationships between the logarithms of the flash intensities and the logarithms of the theoretical intensities. These results were obtained by first plotting as ordinates the logarithms of the flash intensities and as abscissae the logarithms of the theoretical intensities for individual multiplets of an atom, and by then shifting the curves with respect to one multiplet along the abscissa until all points fell on a smooth curve, in this case a straight line. The amount of the shift of each multiplet relative to one chosen for zero point gives a correction to be applied to the logarithms of the intensities in that multiplet. Pannekoek and Minnaert do not give the corrections for all of the multiplets, so we have determined them from their data. These cor-

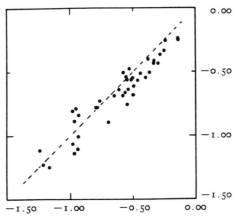

FIG. 1.—Comparison of Yerkes measures (ordinates) of log A for α Cygni with those of Pannekoek (abscissae).

rections are given in Table II and are in the form of the logarithms which must be added to the logarithms of the theoretical intensities given in their paper.

This gives us a homogeneous set of theoretical intensities for each atom, which we can use for all the stars. To find the relationship between the total absorptions for a given star and the theoretical intensities of all of the lines, we have plotted as ordinate log A (total absorption) and as abscissa log E (theoretical intensity) for each atom. Since we have no information on the relative abundance of the atoms forming the lines in question, we must adjust empirically the curves for the various elements in exactly the same manner as for the multiplets in the flash spectrum. In all cases we have taken the Fe II lines as the basis for establishing the zero point. The plots for the various atoms were shifted until the best smooth curve was formed. The resulting curves are shown in Figures 2–6. Figure 7 is a curve for the star 17 Leporis constructed from data published by

Hynek.[21] There is some uncertainty about this curve since Hynek did not make accurate measures of the fainter lines. The five faintest lines plotted on the diagram are listed by him as "traces." In the discussion of this spectrum Hynek estimates the intensities of the lines which he called "traces" as 0.1 on his scale. Converting these into

TABLE II

Fe I:		Fe II:	
a³F −z³G°.........	−1.00	b⁴P −z⁴D°.........	−0.55
a⁵P −x⁵D°.........	−0.30	b⁴F −z⁴F°.........	−2.00
a⁵P −z⁵S°.........	+0.70	b⁴F −z⁴D°.	−1.60
z⁷D°−e⁷D	−1.40	b⁴P −z⁴F°.........	+0.20
z⁵D°−e⁵F..........	−1.00		
Ca I:		Ti II:	
4³P°−4³D	0.00	a²D −z²F°.........	+0.47
4³P°−p³P	+0.20	a²D −z²D°.........	+0.29
3³D −4³F°........	+0.10	a²G −z²F°.........	+0.10
		a²P −z²D°.........	+0.53
		a²H −z²G°.........	−0.38
		a⁴P −z⁴D°.........	−0.74
		a²P −z⁴D°........	(−0.63)
		b⁴P −z⁴D°.........	−2.25
		b²F −y²G°.........	−1.86
Mn I:		Sc II:	
a⁴D −z⁴F°.........	−0.60	a³F −z³F°.........	−0.80
a⁴D −z⁴D°.........	0.00	a³F −z³D°.........	0.00
a⁴D −y⁴P°.........	+0.80		
Cr I:		Cr II:	
a⁷S −z⁷P°.........	+2.20	b⁴F −z⁴D°.........	0.00
a⁵S −y⁵P°.........	+0.70	a⁴F −z⁴F°.........	−0.70
a⁵D −y⁵P°.........	0.00	a⁴F −z⁴D°.........	−0.90
a⁵D −z⁵F°.........	−0.40		

our scale of total absorptions, we obtain 0.05 A. We have compared these five lines in the spectrum of 17 Leporis with faint lines in other spectra which we have measured, and we feel confident that the adopted estimates of intensity are not greatly in error.

III. SUMMARY OF OBSERVATIONAL DATA

From a comparison of the observational curves given in section II (Figs. 2–7) the following conclusions are derived:

1. The intensity gradient varies from star to star.

2. It is approximately constant in some stars. On the scales used in the figures the slope of the straight line is approximately $45°$ for a Cygni, $65°$ for 17 Leporis, $40°$ for Sirius, and $45°$ for Vega.

3. In other stars the gradient is variable. Thus, in a Persei and in ϵ Aurigae it is large for small values of N and A and decreases for large values of N and A. A slight amount of curvature in the opposite direction is suspected in the case of Sirius.

4. The gradient varies between extreme values of somewhat more than $0°$ and $65°$, but it never exceeds the latter value. For the scales used the latter corresponds to $A \propto N$.

5. The gradient effect is not primarily a function of atomic number, excitation potential, or multiplet character, but depends, for each star, upon the total absorptions of the lines investigated.

6. We have made no attempt to study possible interlocking effects which may be present. Nor have we made a minute comparison of different elements in each star. We assert, merely, that the gradient effect is more pronounced than other effects. The work of Woolley and, more recently, that of Thackeray[35] indicate that there are real differences between multiplets, which are independent of the total absorptions of the lines considered. From a preliminary inspection of the data we are inclined to agree with them, and a more complete investigation of this phenomenon will be made at the Yerkes Observatory. In this paper we are concerned merely with the gradient effect obtained after smoothing over the results derived from many multiplets.

IV. THEORETICAL DISCUSSION

The theoretical work of Schütz, as elaborated by Pannekoek and by Minnaert and Slob, provides a good starting-point for the discussion. According to the latter,[16] thermal Doppler effect dominates the line contours for small numbers of atoms. For stars of temperature $T = 10,000°$ and for an element of atomic weight 50 (intermediate between Fe [56] and Ti [48]), the curve flattens out for lines the total absorptions of which are in the neighborhood of $A = 0.1$ A.

An inspection of our observed curves proves conclusively that this

---

[35] *Observatory*, **57**, 10, 1934; *Monthly Notices*, **94**, 99, 1933.

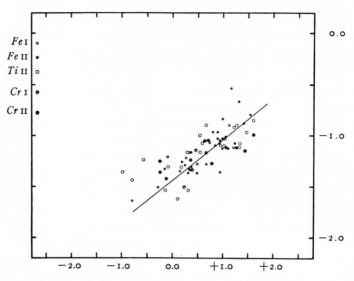

Fig. 2.—The variation of log A (total absorption) with log E (theoretical intensity) for lines in the spectrum of α Canis Majoris.

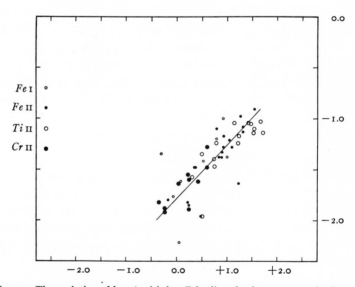

Fig. 3.—The variation of log A with log E for lines in the spectrum of α Lyrae

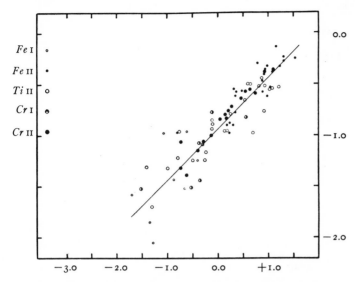

FIG. 4.—The variation of log A with log E for lines in the spectrum of α Cygni

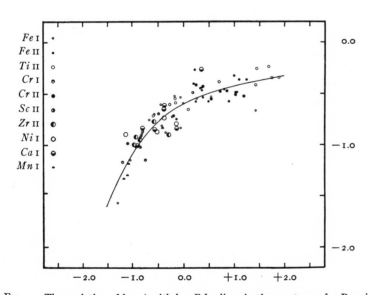

FIG. 5.—The variation of log A with log E for lines in the spectrum of α Persei

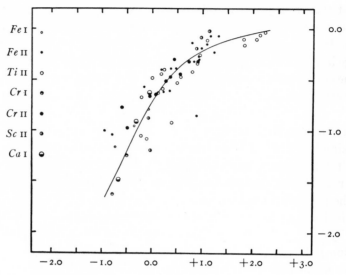

Fe I
Fe II
Ti II
Cr I
Cr II
Sc II
Ca I

FIG. 6.—The variation of log A with log E for lines in the spectrum of ϵ Aurigae

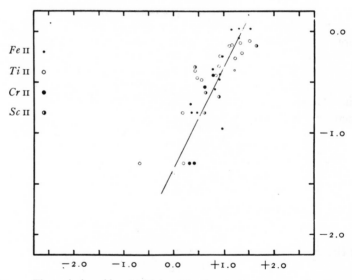

Fe II
Ti II
Cr II
Sc II

FIG. 7.—The variation of log A with log E for lines in the spectrum of 17 Leporis

is not true in all stars. In particular, 17 Leporis, ϵ Aurigae, and α Persei show a decided departure from the curve of Minnaert and Slob. In 17 Leporis the entire curve fits the relation $A \propto N$, over a range in A from 0.05 to 1.0 A. In ϵ Aurigae and in α Persei the shape of the curve resembles that of Minnaert and Slob, but the flat part of the curve is reached at about 1.0 A in ϵ Aurigae and at 0.4 A in α Persei.

In order to explain these departures we propose the following working hypothesis: The atmospheres of the stars are agitated not only by thermal motion of the individual atoms, but also by currents of a macroscopic character, which we shall refer to as "turbulence."

The possibility that turbulence is responsible for the gradient effect had already been tentatively considered by Struve and Morgan[22] and by Elvey,[33] but it is only with the help of the measures given in section II that a test of the hypothesis can be made.

We shall assume that the distribution of turbulent velocities is defined by Maxwell's law, and that consequently the absorption coefficient is [36]

$$\sigma = \frac{\sqrt{\pi}\, \epsilon^2 \lambda_0^2 Nf}{mc^2 b}\, e^{-\left(\frac{\Delta\lambda}{b}\right)^2},\qquad (1)$$

where

$$b = \lambda_0\, \frac{v_0}{c}$$

is the Doppler shift corresponding to the most probable turbulent velocity v_0. We have

$\epsilon = $ Charge of the electron
$m = $ Mass of the electron
$\lambda_0 = $ Wave-length of the center of the line
$N = $ Number of atoms per cubic centimeter
$f = $ Oscillator strength
$c = $ Velocity of light

For the case of pure thermal agitation

$$b_0 = \frac{\lambda_0}{c}\sqrt{\frac{2RT}{M}},\qquad (2)$$

[36] A. Unsöld, *Astrophysical Journal*, **69**, 214, 1929.

where

$$R = \text{Boltzmann constant}$$
$$M = \text{Atomic weight}$$
$$T = \text{Absolute temperature.}$$

If thermal agitation and turbulence are both effective, b in formula (1) must be replaced by β, where

$$\beta = \sqrt{b^2 + b_0^2} \ .$$

For convenience we shall next assume that the reversing layer has a thickness H and that it absorbs exponentially:

$$I = I_0 e^{-\sigma H} \ . \tag{3}$$

This formula is not strictly correct, but the use of expressions more plausible theoretically, such as the Schuster-Schwarzschild approximation or the formula of Eddington,[37] does not appreciably alter the results, as was shown by Minnaert and Mulders[38] for the case of thermal agitation in the sun.[39] The total absorption of the line is

$$A = \int_{-\infty}^{+\infty} (1 - I) d\lambda \ .$$

The integral may be written in the form

$$A = 2 \int_0^{+\infty} \left(1 - e^{-a e^{-\frac{x^2}{b^2}}}\right) dx \ ,$$

where

$$a = \frac{\sqrt{\pi}\, \epsilon^2 \lambda_0^2 N f H}{mc^2 b} 10^8 = 10^{-13} \frac{NfH}{b} \ .$$

Introducing

$$\frac{x}{b} = y \ ; \qquad dx = b\,dy \ ,$$

[37] *Monthly Notices*, **89**, 620, 1929.

[38] *Zeitschrift für Astrophysik*, **2**, 171, 1931.

[39] In a private discussion Dr. D. H. Menzel has expressed the opinion that the use of eq. (3) may not be justified under the abnormal conditions prevailing in turbulent stellar atmospheres. This point should be further investigated.

we obtain

$$A = 2b \int_0^\infty \left(1 - e^{-ae^{-y^2}}\right) dy \, . \tag{4}$$

Expanding the exponential into a series, we obtain the following expression, which is convenient for small values of a:

$$A = 2\sqrt{\pi}\, ba \left\{ 1 - \frac{a}{2!\sqrt{2}} + \frac{a^2}{3!\sqrt{3}} - \cdots \right\} \, . \tag{5}$$

For large values of a this series is inconvenient, although it always converges. Professor Walter Bartky has derived for us the following asymptotic expansion[40] for large values of a:

$$A = 2b \left\{ \sqrt{\log_e a} + \frac{0.2886}{\sqrt{\log_e a}} - \frac{0.1335}{(\sqrt{\log_e a})^3} + \frac{0.0070}{(\sqrt{\log_e a})^5} + \cdots \right\} \, . \tag{6}$$

With the help of these two expansions we compute the theoretical values of A corresponding to different sets of NfH and b. The limiting straight lines have already been given by Unsöld.[14] They are

$$A = 1.8 \times 10^{-13} N \quad \text{(pure Doppler effect)} \tag{7}$$

and

$$A = 6.5 \times 10^{-9} \sqrt{N} \quad \text{(classical radiation damping)} \, . \tag{8}$$

Equation (8) has been found to be correct for ultimate lines, such as H and K of calcium. There is some doubt whether the constant is the same for subordinate lines. Indeed, Weisskopf and Wigner,[41] and also Oppenheimer,[42] have shown that the damping constant in the exact classical expression for the atomic absorption coefficient

$$\sigma = \frac{\gamma_{cl}}{(\tfrac{1}{2}\gamma_{cl})^2 + 4\pi^2(\nu - \nu_0)^2} \, ; \qquad \gamma_{cl} = \frac{8\pi^2 \epsilon^2 \nu_0^2}{3mc^3}$$

[40] Unpublished.

[41] *Zeitschrift für Physik*, **63**, 54, 1930; **65**, 18, 1931.

[42] *Physical Review*, **35**, 461, 1930. For a discussion of the absorption coefficient see D. H. Menzel, *Lick Observatory Publications*, **17**, 227, 1931.

is to be replaced by the following expression:

$$\gamma_{cl} = \frac{1}{T_n} + \frac{1}{T_m} = \sum A_n + \sum A_m ,$$

where T_n and T_m are the mean lifetimes of the atom in the upper and in the lower quantum states, and A_n and A_m are the corresponding transition probabilities. Since these quantities cannot, in general, be computed, the quantum-theoretical damping constant must be regarded as unknown. Minnaert and Mulders[6] found by computation that for certain lines the true damping constant could exceed the classical constant by a factor of 3 to 4.5. Actual measurements gave for the sun

$$\frac{\gamma_{obs}}{\gamma_{cl}} = 9 .$$

Pannekoek[17] found for α Cygni a value of

$$\frac{\gamma_{obs}}{\gamma_{cl}} = 19 ,$$

but since this result rests entirely upon the small amount of curvature shown in his Figure 3 (p. 762), which is not confirmed by our measurements (Fig. 4), we are inclined to consider it as relatively uncertain.

There remains the question whether the loss of energy through radiation is the only source of damping. The collisional damping of Lorentz which results from sudden changes in the phases of the oscillators may contribute an appreciable amount to the damping constant. Pannekoek[17] discarded collisional damping because his value of γ_{obs}/γ_{cl} for α Cygni was larger than Minnaert's for the sun. Since α Cygni is a supergiant, while the sun is a dwarf, the opposite should have been observed if collisional damping were effective. But if, for the reasons stated above, the curvature in Pannekoek's Figure 3 is spurious, this result no longer holds.

We shall assume that the damping constant is unknown, and write

$$\sigma_{obs} = C \sigma_{cl\ damp} .$$

Then for either the pure exponential absorption formula, or for the Schuster-Schwarzschild expression,[43]

$$A_{\text{damp}} = \text{Const. } \sqrt{C} \ .$$

Accordingly, an observed displacement of the limiting curve (8) along the axis of log A by an amount Δ log A should lead to a determination of the damping constant:

$$\log \frac{\gamma_{\text{obs}}}{\gamma_{\text{cl}}} \approx \log C = 2\Delta \log A \ .$$

Figure 8 shows the integrand of formula (3) for different values of the parameter a. This illustrates the contours of lines broadened by

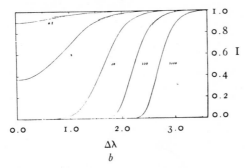

Fɪɢ. 8.—Theoretical contours of lines for different values of a

turbulence. Figure 9 shows the theoretical curves $A = \varphi \ (NfH)$ for $b = 1$, $b = 0.1$, and $b = 0.01$. The dotted line for $b = 0.03$ corresponds to a temperature of roughly 10,000° K. The limiting straight line on the right side of the diagram corresponds to the classical theory of radiation damping. Parallel to it run two other straight lines corresponding to $C = 4$ and $C = 10$, respectively. Figure 10 shows comparisons between theoretical and observed curves. For completeness we have included the average curve derived by Minnaert and Mulders for the sun. We see that the agreement is excellent if we assume that the three stars 17 Leporis, ϵ Aurigae, and a Persei have turbulent atmospheres, while in the sun, a Cygni, Vega, and Sirius, turbulence is not appreciable.

[43] O. Struve, *Monthly Notices*, **89**, 584, 1929.

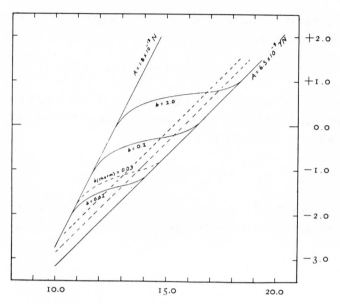

Fig. 9.—Theoretical relations between log A and log N

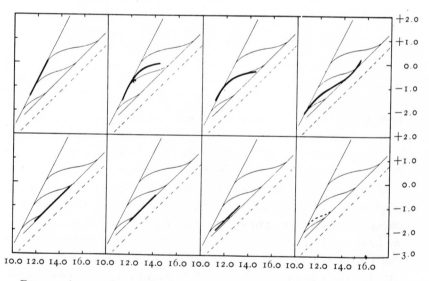

Fig. 10.—A comparison of the observed variations of log A with log E and of the theoretical variations. Upper row: 17 Lep., ϵ Aur., α Per., and Sun. Lower row: α Cyg., α CMa., α Lyr., and the theoretical curve for a temperature of 10,000° K. The theoretical curves are for a damping constant of ten times the classical value.

Estimating for each of the stars the value of b, we derive their average turbulent velocities. We can also compute the theoretical line widths from

$$e^{-ae^{-\frac{x^2}{b^2}}} = 0.5.$$

For the last three stars a similar computation has been made on the basis of pure radiation damping:

$$e^{-C\frac{2\pi e^4\lambda_0^2 NfH}{3m^2c^4x^2}} = 0.5.$$

The values of NfH used in the computations were taken from the curves of Figure 10 and refer to the strongest lines measured in each

TABLE III

STAR	b	v_0 TURB. VEL. OR THERM. VEL.	FICTITIOUS TEMPERATURE $(M = 50)$	LINE WIDTH Computed Turbu-lence	Rad. Damping	Observed
17 Lep............	1.0 A	67 km/sec.	3×10^7	1.2	....	1.5
ε Aur............	0.3	20	2×10^6	1.1	....	1.0
α Per............	0.1	7	3×10^5	0.5	....	0.6
α Cyg............	(0.03)	(2)	(10^4)	0.2	0.5	0.8
α CMa............	(0.03)	(2)	(10^4)	0.1	.1	(0.5)
α Lyr............	(0.03)	(2)	(10^4)	0.1	0.1	(0.4)
Sun............	(0.03)	(2)	(10^4)	....	....	....

star. The results given in Table III are surprisingly good for 17 Leporis, ε Aurigae, and α Persei. In all three stars an appreciable amount of broadening, produced by turbulence, is actually observed. In α Cygni the measured widths exceed the computed ones, although the character of the lines does not suggest rotation. For α Canis Majoris and α Lyrae the resolving power of the spectrograph is insufficient. The "fictitious temperature" is derived from (2) and merely shows what the temperature would have to be in order to produce the required values of b.

The curves for α Cygni, α Canis Majoris, and α Lyrae agree remarkably well with the law $A \propto \sqrt{N}$, and there is no evidence of turbulence in these stars. Their curves have been shifted parallel to themselves, toward smaller values of log NHf, so as to obtain the best possible fit with the theory of Schütz. This is obtained for $C = 10$. A larger value of C is irreconcilable with the curves for α Persei and for the sun. A smaller value is inconsistent with the theory of Schütz. Even for $C = 10$, the lower parts of the observed curves $A = \varphi(NHf)$ fail to flatten out, as might be expected from the thermal curve shown in the last section of Figure 10. This might be construed to mean that, unless Schütz's theory is in error, thermal agitation is less pronounced than might be expected from the effective temperatures of these stars. While this may be true, it should be remembered that the lines are faint and difficult to measure, and that the highly unsatisfactory procedure of fitting each multiplet and each element into the best smooth curve may easily introduce errors into the curves. Because of this inherent uncertainty we have intentionally drawn straight lines for all stars except ε Aurigae and α Persei, in which the curvature was too large to be ignored.

We have assumed here that C is constant for all stars. In reality, it probably varies from multiplet to multiplet, producing a scatter in our observations. It may also vary from star to star, especially if collisional damping is not negligible. However, as a first approximation, our result $C = 10$ may be considered an average value[44] for the multiplets listed in Table II.

V. A POSSIBLE OBJECTION

The measured intensities for 17 Leporis, ε Aurigae, and α Persei agree remarkably well with the proposed theory, However, in 17 Leporis a visual inspection of the three lines of Fe I, λλ 4045, 4063, 4071, which are very faint, suggests that for these members of the $a^3F - y^3F^\circ$ multiplet the gradient is smaller than would be expected if $A \propto N$. In fact, all three lines are faintly visible, and the difference between λ 4045 and λ 4071 is small. Ordinary eye-estimates are deceptive, and it is possible that accurate measurements would show that the theory is obeyed. Unfortunately we are unable, with our

---

[44] Since the curve for the sun by Minnaert and Mulders has been used in our Fig. 10, this value is not quite independent of their result: $C = 9$.

present equipment, to obtain high-contrast spectra of sufficient dispersion for accurate measurement.

Nevertheless it is wise to consider the possibility that for these lines the gradient is smaller than is required by the law $A \propto N$. It is our opinion that even if this should be the case, it would not invalidate the theory. Our reasons are:

1. As explained in section III, individual multiplets may have peculiar gradients.

2. The steep gradient for all measurable lines of 17 Leporis is well established.

3. The almost complete absence of very faint lines shows that the theory is essentially correct.

4. In the star 17 Leporis turbulence is highly probable, as has been shown by Struve.[20] The absorption lines for which the gradient is large result from several shells which expand with different velocities, so that the combined effect is that of turbulence. The great width of the lines and the steep gradient are thus immediate consequences of the general behavior of the star.

5. It is not impossible that a theoretical explanation could be found for the occurrence of small gradients in the faintest lines.

VI. SUMMARY

With certain reservations (more completely explained in the preceding section), the proposed hypothesis of turbulent motions which in some stars tend to raise the flat portions of the curves $A = \varphi(NfH)$, provides a satisfactory extension of the theory of Schütz. The latter appears fairly well established from the work of Minnaert and his associates, on the sun, and of Dunham, on stars.[45] Our results for stars in which turbulence is inappreciable agree with this theory fairly well if the observed damping constant is taken to be approximately ten times that derived from the classical electron theory.

In pursuing farther the hypothesis of turbulent motions we conclude that B stars in which the gradients of the O II lines are large have turbulent atmospheres. Since such stars give curved lines for

[45] *Publications of the American Astronomical Society*, **7**, 215, 1933; W. S. Adams, "Summary of the Year's Work at Mount Wilson," *Publications of the Astronomical Society of the Pacific*, **45**, 279, 1933.

the relation $A = \varphi(NfH)$, it is quite possible that the gradients for the relatively weak O II lines will be large, while those for the Si III lines will be relatively small.

In fact, there is no definite answer to the question: What is the effect upon the gradient if we pass from a quiescent atmosphere to a turbulent one? If NfH remains constant, the gradient will either increase or decrease, depending upon whether we are nearer to the Doppler-effect branch or to the radiation-damping branch. But, if NfH changes, as is likely to be the case if the quiescent atmosphere is that of a dwarf while the turbulent atmosphere belongs to a giant, no answer can be given unless the actual change in NfH is known.

In several earlier papers Struve,[46] and Struve and Dunham,[47] concluded that in several B-type stars the lines were so narrow as to exclude the possibility of any appreciable turbulence. In fact, thermal Doppler effect was found to be practically sufficient to explain the observed line widths in such stars as τ Scorpii and γ Pegasi. This agrees well with our new results; these stars are typical dwarfs and their gradients suggest absence of turbulence.

On the other hand, the giants, such as ϵ Canis Majoris, have slightly broadened lines. In *some* of these stars axial rotation may be present, but the fact that *all* stars having large gradients have slightly broadened lines is highly suggestive. It is improbable that the axes of all these stars are oriented in such a way as to produce widening of the lines. The hypothesis of turbulence is much more satisfactory.

It should be remembered in this connection that rotational broadening does not change the gradient.[48] No observations have as yet been made of the gradients in stars showing large rotational velocities.

The idea of turbulence in the outer atmospheres of the stars is not new. Unsöld,[36] and after him Keenan,[49] have explained various phenomena in the solar chromosphere on the basis of turbulence. Even the aspect of the chromosphere with its prominences suggests turbulence in at least the outermost layers of the sun's atmosphere.

[46] *Proceedings of the National Academy of Sciences*, **18**, 585, 1932.

[47] *Astrophysical Journal*, **77**, 321, 1933.

[48] Struve and Elvey, *ibid.*, **72**, 267, 1930; *Monthly Notices*, **91**, 664, 1931.

[49] *Astrophysical Journal*, **75**, 277, 1932.

An appreciable effect of turbulence upon stellar absorption lines has been suspected by Menzel[50] in Cepheid variables, and by Adams and St. John[51] in several bright stars whose spectra revealed differences in the radial velocities obtained from enhanced lines and from arc lines. There is no obvious correlation between the results of Adams and St. John and those reported in this paper, but the methods used are so different that this is not surprising.

Finally, attention may be directed to a series of investigations by P. Guthnick,[52] who has attempted to explain by turbulence a number of peculiar phenomena in the radial velocities of such stars as α Lyrae and ε Ursae Majoris.

Whether all manifestations of the gradient effect may be explained on the basis of the enlarged theory of Schütz is not certain. For example, it is quite possible that there are also real differences between He I singlets and triplets, in addition to those reduced to the gradient effect. But many interesting results already suggest themselves. Among these are the following:

1. There is a tendency for very luminous stars to have larger turbulent velocities than are observed in less luminous stars.

2. There is, however, no *close* correlation between luminosity and turbulence, because some of the most notable supergiants show no turbulence.

3. The greatest amount of turbulence yet recorded is in the star 17 Leporis, which is peculiar in many other respects.

4. If turbulence is responsible for the variation of gradient in μ Sagittarii and in 27 Canis Majoris, we must conclude that the turbulent motions are not constant in these stars, or that NfH varies in such a manner as to imitate a change of gradient. Either explanation may be reasonable. In μ Sagittarii the largest gradient is observed a short time after periastron passage.

5. The extremely small gradient in the A-type star 73 Draconis is perplexing. There is no obvious reason why this star should have a turbulent atmosphere.

[50] *Publications of the American Astronomical Society*, 6, 370, 1930.

[51] *Astrophysical Journal*, 60, 43, 1924.

[52] *Sitzungsberichte der Preussischen Akademie der Wissenschaften: Phys.-math. Klasse*, 29, Part II, Berlin, 1931.

6. It would be interesting to test the hypothesis of turbulence in novae, where the absorption lines may be broadened by uneven velocities of expansion of the shell.

7. Unfortunately none of the three turbulent stars investigated here is suitable for a study of their H and K lines. Theoretically, these lines, in at least the cooler stars, should lie on the right-hand branch of the curve, where $A \propto \sqrt{NfH}$, but blending with $H\epsilon$ makes measurements for Ca H uncertain.

8. The influence of the gradient effect upon the interpretation of absolute-magnitude effects is obvious and must be carefully investigated.

9. An investigation of separate multiplets and of the relative intensities of different multiplets of one element should be of interest.

10. We are inclined to think that the turbulence effect is most frequently observed in the hotter stars. It seems to be common in the B's and much rarer in the A's and the F's. In our observational work, 17 Leporis and ϵ Aurigae were not picked at random. Had we relied on a random selection of bright A and F stars, a Persei would have been the only one to show the gradient effect clearly.

YERKES OBSERVATORY
January 1934

THE CURVE OF GROWTH METHOD

JESSE L. GREENSTEIN

Mount Wilson and Palomar Observatories
Carnegie Institution of Washington
California Institute of Technology

I. INTRODUCTION

One of the pleasures of re-reading and reviewing a classic is its unexpected richness. Much of what we take for granted as primitive often turns out to be highly sophisticated. Since the preceding paper by Otto Struve and C. T. Elvey "The Intensities of Stellar Absorption Lines" appeared in 1934 an enormous number of papers on that subject have been published, and great progress has been made. Very elaborate criticisms, and elaborate re-justifications, of the curve-of-growth method have occupied excellent astrophysicists. In addition, the observers have valiantly plunged ahead with partially closed eyes, attempting to obtain answers to problems. It is certain that many of the large group of problems noted by Struve and Elvey are still among the central ones of current astrophysics.

I started my graduate study the year this paper appeared; my first actual contact with stellar spectroscopic problems was in 1939 when Otto Struve suggested casually that a peculiar supergiant, v Sgr (which I observed in 1939) and the second brightest star in the sky, Canopus (α Car, observed in 1941) should be excellent objects to study with the new 82-inch spectrographs. Since my own earlier work concerned interstellar matter, this revolutionary suggestion, in fact, brought me into a new world, which I have enjoyed ever since. It was typical of Struve and his attitude toward young scientists that he was quite willing to suggest a radical reorientation of my work, and that he gave me two of the finest stellar spectroscopic problems which eventually required much 82-inch observing time.

There are several important main themes in this 1934 paper (hereinafter referred to as SE). (1) The theory of the curve of growth (abbreviated COG) had been perfected in about the three preceding years. (2) There were some misconceptions carried over by the observers from the earlier period when the strong-line, $N^{1/2}$ law, was universal. (3) There were also carry-overs of theoretical developments, *e.g.*, "interlocking," which now seem irrelevant. (4) The two-parameter family of COG's connects theoretical relative numbers of atoms producing a set of lines and the observed equivalent widths in absorption. Needed, therefore, are good relative theoretical intensities and good measures. Both were lacking, and often still are. (5) The height of the nearly flat saturated portion of the COG depends on a total Doppler parameter β (SE, p. 430). This velocity distribution is assumed to be gaussian because thermal velocities are, and because (in SE, Eq. (1)) the "turbulent" velocities are, for convenience only, so taken. In fact this form is unfortunate. (6) The line-transfer problem can often be simplified (SE, pp. 430, 431), providing useful asymptotic forms of the COG. (7) The value of the radiation- or collision-damping constant, Γ, is based on either quantum mechanics or laboratory observation (SE, p. 432), and is often not known. (8) The available observations indicated that stars had a wide range of the parameter β (SE, Figure 10) (which we now call $\Delta\lambda_D$) and that the data seldom sufficed to give the damping constant (now called $a = \Gamma/\Delta\lambda_D$). (We expect a to vary from multiplet to multiplet, even from line to line.) (9) SE felt that the interaction between COG effects dependent on turbulence, and those dependent on damping, *e.g.*, for

61

He I, could explain anomalous line-intensity ratios, such as the singlet-triplet anomaly. SE, Figure 9 shows the schematic basis for this suggestion which has in fact not been worked out for many cases. (10) The paper establishes that the turbulence is large in a shell star (17 Lep), a very bright supergiant (ϵ Aur) and a moderately bright one (α Per). But it fails to distinguish between the very bright supergiant α Cyg and the main-sequence A-type stars α CMa and α Lyr. SE suggest the $N^{1/2}$ law as valid for all three stars. None of them, in fact, have strong enough lines to reach the damping portion of the COG.

II. CURVE OF GROWTH THEORY

Before reviewing some of these points in detail, let us display the current formalism and techniques for COG study. In the elementary method (coarse or grob analysis) it is assumed that a line is formed as if it were in a single layer of constant gas and electron pressure, excitation and ionization temperature, whose thickness is determined by the opacity. The normal relation between the boundary-value of the temperature, T_0, and the effective temperature, T_{eff}, is for a gray atmosphere $T_0/T_{\text{eff}} = 1/1.233$. Very approximately, lines are formed in regions of small optical depth, τ, with temperature near T_0, so that the excitation temperature, T_{exc}, in a Boltzmann formula, roughly describing the population of relatively low excited atomic states is

$$T_{\text{exc}} \approx T_0 \approx 0.81 \ T_{\text{eff}}. \tag{1}$$

Radiative transfer theory gives a limb-darkening coefficient which determines the maximum absorption at the center of a spectral line of infinite optical depth, if produced in "absorption" (rather than "scattering"). The depth and strength of a line, i.e., its profile, will depend on $h\nu/kT_0$, on the opacity relative to its mean value $k_\nu/\bar{k}$, on the hydrogen to metal ratio A, and on the actual concentration of atoms in the particular state producing the line. For general reviews, see chapters in Greenstein (1960) and in the text by Aller (1963).

The darkening parameter affects the COG, and it is necessary to have its mean value

$$\frac{B_0}{B_1} = \frac{8}{3} \frac{k_\nu}{\bar{k}} \frac{kT_0}{h\nu} (1 - e^{-h\nu/kT_0}) \tag{2}$$

and to allow for the change with ν, by second-order corrections. Such curves of growth are given conveniently by Wrubel (1949, 1950), and their use is described in detail by Aller (1963, pp. 377–387). For non-gray atmospheres, in case $k_\nu/\bar{k} \ll 1$, Eq. (1) can be seriously incorrect, as is Eq. (2) which refers only to the extreme boundary layers.

The shape of the line, in all COG analyses, is assumed to be that given by a Voigt profile, the result of folding the Doppler distribution of velocities (SE, Eq. (1)) into the atomic line absorption coefficient (SE, below Eq. (8)). Tables for resolving this Fourier transform are given by Harris (1948) and Elste (1953). Let N_{irs} be the effective number of atoms in a given state per gram of stellar material. The line absorption coefficient l_ν is obtained from

$$l_\nu = l_{\nu_0} \frac{a}{\pi} \int_{-\infty}^{+\infty} \frac{e^{-y^2} \, dy}{a^2 + (y - u)^2} , \tag{3a}$$

where the fictitious (i.e., zero damping) absorption at the center of the line is

$$l_{\nu_0} = \frac{\pi e^2}{mc} N_{irs} f_{st} \frac{1}{\pi^{1/2} \Delta\nu_D} , \tag{3b}$$

with f_{st} the absorption f-value for the transition s to t, and u the distance from the line center in units of the total Doppler width. The total velocity parameter (SE β)

is defined as V, and

$$u = \frac{\nu - \nu_0}{\Delta \nu_D}, \qquad a = \frac{\Gamma}{4\pi\Delta\nu_D}, \qquad \Delta\nu_D = \frac{V\nu_0}{c}. \tag{3c}$$

The relation between the Einstein spontaneous transition probability, A_{ts}, and f_{st} is

$$f_{st} = \frac{g_t A_{ts}}{g_s} \frac{1}{3\gamma_{cl}} = \frac{g_t A_{ts}}{g_s} \frac{mc^3}{8\pi^2 e^2 \nu^2}. \tag{3d}$$

Introducing the ratio of line to continuous absorption coefficient, $\eta_\nu = l_\nu/k_\nu$, we can write in terms of the Voigt functions $H(a, u)$,

$$\eta_\lambda = \eta_{\lambda_0} H(a, u) = \frac{\pi^{1/2} e^2}{mc^2} \frac{N_{irs} f_{st} \lambda^2}{\Delta\lambda_D k_\lambda} = \frac{\pi e^2}{mc} \frac{N_{irs} f_{st} \lambda}{\pi^{1/2} V k_\lambda}. \tag{4a}$$

We have converted to wavelength units, for practical reasons. If the atomic weight is A, and only thermal motions exist

$$\Delta\lambda_D = \frac{\lambda}{c}\left(\frac{2kT}{Am_H}\right)^{1/2} = \frac{\lambda V_{\text{Th}}}{c}. \tag{4b}$$

The introduction of "turbulence," if Gaussian, results according to SE in

$$V^2 = V_{\text{Th}}{}^2 + V_{\text{Tu}}{}^2. \tag{4c}$$

TABLE 1
M. E. COG FOR PURE ABSORPTION
(Wrubel 1949); $B_0/B_1 = \frac{1}{3}$
The Table gives log $W/\Delta\lambda_D$

log η_0	log a		
	−1	−2	−3
−2.0	−1.968	−1.969	−1.969
−1.6	−1.572	−1.572	−1.573
−1.2	−1.180	−1.181	−1.182
−0.8	−0.800	−0.804	−0.805
−0.4	−0.444	−0.453	−0.455
0.0	−0.131	−0.151	−0.154
+0.4	+0.124	+0.089	+0.084
+0.8	+0.325	+0.267	+0.260
+1.2	+0.490	+0.400	+0.387
+1.6	+0.641	+0.503	+0 481
+2.0	+0.796	+0.593	+0.554
+2.4	+0.964	+0.681	+0.614
+2.8	+1.145	+0.780	+0.668
+3.2	+1.335	+0.908	+0.724
+3.6	+1.531	+1.070	+0.791
+4.0	+1.728	+1.250	+0.883
+4.4	+1.927	+1.434	+1.008

The $H(a, u)$ functions are essentially Gaussian in shape, e^{-u^2}, up to 3 or 4 times $\Delta\lambda_D$, and switch over to a damping shape, u^{-2}, beyond this. An illustrative example of the change in profile of a given line as computed for increasing numbers of atoms is shown in Aller (1963), Figure 8–7, together with COG obtained by integrating the profiles to give the equivalent width. The curve of growth for a typical value of the darkening, for solar temperatures $B_0/B_1 = \frac{1}{3}$, is shown here in Figure 1 and Table 1. It is based on Wrubel's (1949) computations for the Milne-Eddington model ($\eta \neq f(\tau)$ and pure absorption). Note the linear portion reaching up to a point on the vertical

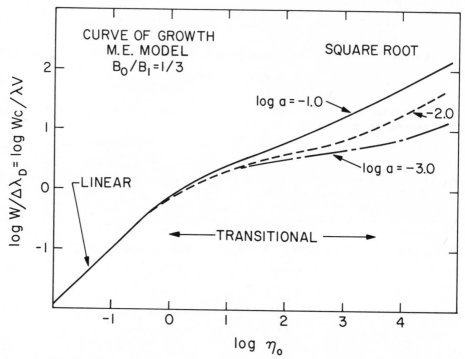

FIG. 1.—Theoretical curves of growth for the Milne-Eddington model.

axis, $W/\Delta\lambda_D \approx 1$, when $\eta_0 \approx 1$, on the horizontal axis. The equivalent width equals the Doppler width when the optical depth in the center of the line is about unity; from then on we have a nearly flat "transitional" section, on which $W/\Delta\lambda_D \approx \eta_0^{1/4}$; then we encounter the $\eta_0^{1/2}$ damping portion of the curve, starting at various η_0 which move to the left as a, $i.e.$, Γ, becomes larger. The ordinate can be written generally as $\log W/\Delta\lambda_D$, but this is also $\log Wc/\lambda V$, so in practice we plot $\log W/\lambda$ as the vertical coordinate. Similarly, for a given element, over a moderate range of λ, we can neglect the variation of k_λ and use as horizontal coordinate, instead of eq. (4a),

$$\log \eta_0 = \log (N_{irs} f_{st} \lambda) + \log \frac{\pi^{1/2} e^2}{mc} - \log V k_\lambda. \tag{5a}$$

The last two terms are constants for the star. The expression contains in the first term the population of the given state, which for an element i, in the rth state of ionization, can be written as

$$N_{irs} = N_i \frac{N_{ir}}{\Sigma N_{ir}} \frac{N_{irs}}{\Sigma N_{irs}} = \frac{g_s 10^{-\theta \mathrm{exc}(EP_s)} I_{ir} \zeta_i}{b_{ir}(T) \bar{A} m_H}. \tag{5b}$$

This is an abbreviated statement that atoms of type i, which occur in ratio to hydrogen ζ_i (by number), are distributed over various stages of ionization r, according to ionization equilibria which give a fraction in the required level of ionization, I_{ir}. Within this ionization level, the Boltzmann factor for the excitation potential EP_s, gives the population of state s, dependent on $b(\tau)$, the partition function (see Greenstein,

1960, Chap. 5, Table 8) but mainly on $\theta_{\text{exc}} EP_s$. The number of atoms per gram is obtained from the mean atomic weight, $\bar{A}$, of the gas mixture and the mass, m_H, of the hydrogen atom. A convenient way of writing the abscissa is then

$$\log g_s f_{st} \lambda - \theta_{\text{exc}} EP_s - \log V k_\lambda + \log I_{ir}/b_{ir}(T) + \log \zeta_i + \text{constant.} \quad (5c)$$

The combination of observed $\log W/\lambda$ and laboratory (or theoretical) $\log gf\lambda - \theta_{\text{exc}} EP$ is the quantity now plotted. The actual curve can be moved in both coordinates to fit a standard theoretical curve, with trials of different $\log a$ values. The vertical shift essentially gives V. The horizontal fit give the abundance ζ_i. The constant is $\log \pi^{1/2}e^2/mc\bar{A}\, m_H$. Once V is determined from the vertical shift, a gives Γ, the damping; V also affects the horizontal shift since a Doppler-broadened line has a smaller value of η_0. If the shifts can be determined for lines of very different EP, the value of θ_{exc} is determined. If it can be determined for two stages of ionization of the same element, in principle $N_{i,r+1}/N_{ir}$ gives the locus along which the electron pressure P_e and the ionization temperature θ_{ion} must lie. If enough pairs of elements with differing ionization potential exist, the θ_{ion} and P_e are separately determined. Otherwise a reasonable estimate of the ratio of $\theta_{\text{ion}}/\theta_{\text{exc}}$ must be made. Current technique depends most heavily on θ_{eff} as determined from the color temperature and Balmer jump of the star, as evidenced by its continuous spectrum. The latter is measured best by spectrum scans or six-color photoelectric observations of unreddened stars. Three-color data, plus estimates of space reddening are also sometimes usable. Continuum colors must be corrected for line blanketing. Various other subterfuges can be used; for example, if the stars have normal hydrogen/metal ratio, the strength of neutral lines is sensitive to and determines the temperature.

For purposes of this review, where turbulence, not abundance, is the major output of a curve of growth, it is necessary only to have the roughest estimates of temperature, since the thermal motion is

$$V_{\text{Th}} = 14(T_4/A)^{1/2} \text{ km/sec}; \quad (6)$$

here T_4 is the temperature in units of $10,000°\text{K}$, and A the atomic weight. For Fe, at solar $T_4 = 0.6$, $V_{\text{Th}} = 1.4$ km/sec, or $\Delta\lambda_D = 22$ mÅ at 5000 Å. In very few stars can we determine the actual value of $\Delta\lambda_D$ for different values of A, to check whether line broadening fits Eq. (6). Pressure or Stark broadening also affects H and He. For heavier elements, even a small V_{Tu} will mask the $A^{-1/2}$ dependence of V_{Th}. In fact, in stars with large turbulence, there are systematic differences of V_{Tu} from element to element, presumably dependent on height of formation, so that V_{Th} cannot be accurately determined. In practice, to determine the location of the flat part of the curve, a sufficiently large range must be covered by measurements to include weak lines on the linear portion of the COG. This may not be possible when V_{Tu} is large.

III. THE CURVES OF GROWTH OF STRUVE AND ELVEY

It is quite interesting to see what the basis is for their first claims for "turbulence" in stellar atmospheres. The observations by SE are based on 10 Å/mm prism spectrograms obtained with high-contrast emulsions. The data for 17 Lep are from Hynek, who used 30 Å/mm; the spectra were taken at "normal" phases of the variable spectrum, when the lines were not double. The tables give equivalent width measures down to 15 mÅ, far below the limit to which any satisfactory measures can now be made at comparable dispersion. There were no tables of theoretical or measured line intensities available. In fact, SE used chromospheric emission-line intensities plus the slender information provided by relative intensities within single multiplets (so that no θ_{exc} were determined). Have their actual COG's any value?

Many of these stars have been reanalyzed several times with modern techniques. But it is highly interesting to see the old observational data themselves reanalyzed with modern relative intensities. The largest amount of cohesive data, of fair quality, is in the 3300 gf values given by Corliss and Bozman (1962). In addition, extensive data for ionized metals are given in a recent publication by Warner (1967). I chose to use Fe II and Ti II, which cover a large range of stellar intensities. I have prepared COG's for reasonable θ_{exc} for three A-type stars, α CMa, α Lyr and the supergiant α Cyg; in addition, of course the shell of 17 Lep is worth such restudy. The results are shown in Figures 2–5, and Figure 6 gives measures by Searle (1958) for

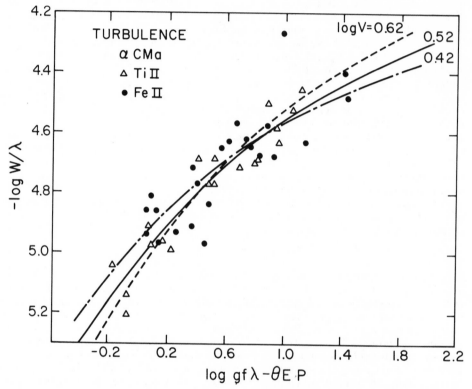

Fig. 2.—Struve and Elvey line strengths combined with modern laboratory gf-values, for α CMa. The fits show the probable extreme values of turbulence.

the shell of 1 Del, one of the few recent sets of measures of metallic-line shells. The fit to a theoretical COG is shown in Figure 2 for α CMa, for the two most extreme values of the vertical shift determining the velocity parameter, for which V ranges from 2.5 to 4.2 km/sec (where $V_{Th} = 1.8$ km/sec). In α Lyr, Figure 3, which shows very bad scatter and little curvature, $V = 2.1$ to 8.3 km/sec. In α Cyg, Figure 4, the solid curve that fits the observations is *not* a theoretical COG, and the deviations are so large that a wide range of V fits, approximately, from 12 to 34 km/sec. In Figure 5 for 17 Lep, there is essentially no curvature in the COG, so that we are on the linear part of the COG up to $\log W/\lambda = -3.6$. This requires that $V = 80$ km/sec; a $V < 50$ km/sec is definitely incompatible. Finally, in Figure 6, for 1 Del, the curve drawn is a theoretical COG for $V = 14$ km/sec; an allowable range for V beyond 10

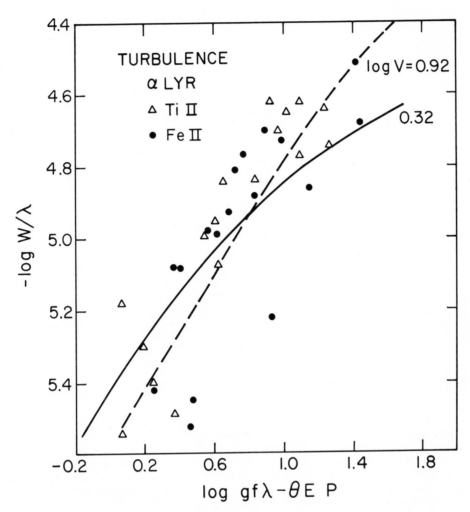

FIG. 3.—Same as Fig. 2, for α Lyr.

to 25 km/sec seems implausible. Thus our reanalysis of very old data shows, in essential features, that SE did in fact discover a meaningful effect, with the total V increasing significantly from the main-sequence stars through the supergiant to the shell star 17 Lep. They were also extremely fortunate in their choice, since 17 Lep has a truly exceptional COG. But their observational and theoretical errors were so serious that if such widely different stars had not been chosen, they would not have been able to discover the existence of turbulence! For example, they believed that most of the stars satisfied their expected theoretical relation $W \propto N^{1/2}$, whereas, in fact, none do, nor can this relation ever hold theoretically, for such weak lines. The $N^{1/2}$ law begins to hold only for lines for which $W \approx 10 \Delta\lambda_D$, i.e., $W > 0.5$ Å, and no lines of that strength are found in the SE data.

None of this criticism diminishes the remarkable, forward-looking nature of the

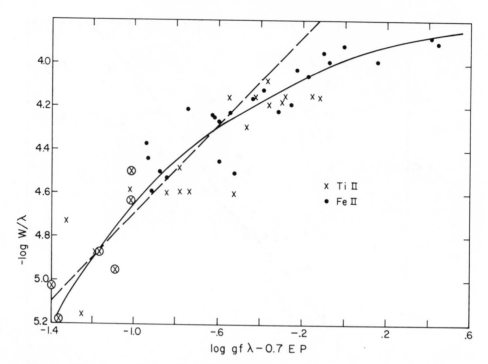

Fig. 4.—The Struve and Elvey line strengths for α Cyg, with modern line strengths. There is no good fit to a curve of growth. The solid line is not a theoretical curve. While certainly indicating that high turbulence is present, different portions of the curve give different values of log V.

work. Of the main themes touched upon, most are still subjects for current work. There is no doubt of the truth of their statement that "The atmospheres of the stars are agitated not only by thermal motion of the individual atoms, but also by currents of a macroscopic character, which we shall refer to as 'turbulence.'" The difference between microturbulence and macroturbulence was not specifically expressed, but SE noted that "in α Cyg the measured line widths exceeded the computed ones, although the character of the lines does not suggest rotation." The shell of 17 Lep was also already known to consist of several regions streaming with different velocities, so that even the difference between macroturbulence and line formation in a moving atmosphere with velocity gradient (a largely unsolved problem) is suggested. The now fairly well-established correlation of both microturbulence and macroturbulence with luminosity does not appear in SE's data, but is at least suggested by it.

IV. PRESENT TECHNIQUES FOR COG ANALYSIS

The enormous improvement in emulsions, stellar spectrographs, and telescopes has provided an order-of-magnitude increase in resolution and accuracy. Lines down to 5 mÅ are measured in sixth-magnitude stars; laboratory and theoretical line strengths are numerous; for example, rough gf values for important lines of the ions of the lighter elements have been computed, on the hydrogenic approximation, by Griem (1964). In particular, the technique of differential curves of growth has popularized the use of solar or stellar transition probabilities for lines not yet observable in laboratory sources. The best source of metallic-line strengths for main-sequence F–K stars,

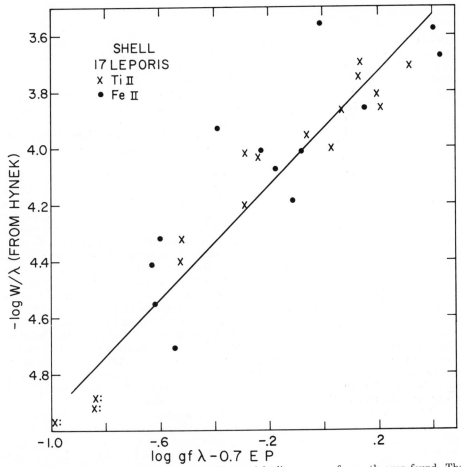

FIG. 5.—The shell of 17 Lep gave the only straight-line curve of growth ever found. The turbulence is very high.

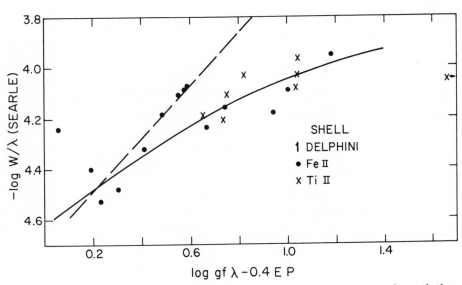

FIG. 6.—Searle's data for the shell of 1 Del, with a much more normal curve of growth than 17 Lep.

therefore, remains the sun, with Mrs. Sitterly's new solar identifications (Moore, Minnaert, and Houtgast 1966) and the prospect of the availability, soon, of the enormously expanded measures at Utrecht. In studies of A-type stars, the work by Greenstein (1948), and in K giants, that by Bonsack (largely unpublished) and the Cayrels (1963) on ϵ Vir yield convenient lists of stellar η_0's. In ϵ Vir, about 1400 lines were measured and tabulated; since in a K giant, lines of Rowland intensity -3 in the sun can be strong, this star provides an excellent source of about 1000 neutral and 200 ionized metallic line strengths. These are on a relative scale, dependent on the actual abundances in the star, just as are the solar line strengths. However, ϵ Vir has essentially normal abundances.

The availability of line strengths for various states of ionization for important elements such as C, N, O, Ne, Si, is much less complete. The higher levels of such atoms as N III, Mg II, Si II, Si III, Si IV, are nearly hydrogenic in character, can be computed from a Coulomb-field approximation, and improved quantum-mechanical calculations are often available. For the many lines of O II, O III, Ne II, the spectra and computations are more complex. The correlation between Griem's hydrogenic line strengths and the NBS values is surprisingly poor for the lighter elements, and for the low levels of the alkalis and alkali earths. It is clear that the lines of the partially ionized elements that dominate B-type spectra come from too highly excited levels for laboratory measurement, and will require complex quantum-mechanical computation. Plasmas and shock tubes work up to 15,000°K, and have already yielded some information on the first ions of O, Ne, A. A very complete bibliography of measured and computed line strengths is given by Glennon and Wiese (1966). Common-sense must be used in judging the reliability of a given transition probability; where it is possible to assume that the star should have nearly the same composition as the sun, the solar values provide an excellent guide. In the Glennon and Wiese bibliography, Table 1 lists several sources of line strengths, which often provide, by inter-comparison, an idea of the reliability of the adopted values.

Collisional damping constants, for strong lines, are much less commonly available. It can sometimes be assumed that collisional excitation cross-sections, and atomic transition probabilities, are somewhat correlated. For example, conservation theorems require that maximum cross-sections be proportional to $2J + 1$, i.e., to the statistical weight of the lower level, just as are multiplet intensities. Collisional strengths of intersystem transitions are also smaller than of those satisfying the $\Delta S = 0$ selection rule. But for simple LS-coupling transitions in the higher excited levels, quadratic Stark effect may appear, and cause, for example, such unexplained anomalies as the odd-even difference suggested for Fe I. Griem (1964) tabulates, in full detail, Stark-broadening parameters for certain strong lines of neutral and ionized elements, giving both line-width and shift caused by electron impact. However, in F–K-type stars, most impacts are with neutral hydrogen, for which only scattered laboratory data exist. Ch'en and Takeo (1957) give an excellent review of the complexities of line widths and shifts for neutral atom broadening, including the remarkable and unexplained violet shifts produced by helium. In addition, much useful theory, and data, is contained in Bates (1962). The line-broadening problem is somewhat less serious for electronic Stark-effect on the more highly ionized, light elements found in O- and B-type stars, and of course, these lines are much weaker and fewer in number, so that detailed quantum-mechanical computations can be carried out using a Coulomb-field approximation.

In addition, we now have a highly developed theory of Stark broadening for H, He II, and some He I lines. There now exists a good beginning in the program of providing, by computation and laboratory work, the atomic data required for the stellar curves of growth. The importance of line spectroscopy in plasma physics, and the growth of "laboratory astrophysics" promise further advances.

What then, is novel on the astronomical side? Measurements of equivalent widths by photographic techniques have greatly improved, but high accuracy is still unattainable. An I.A.U. subcommittee has studied the systematic differences between various observatories (Wright, Lee, Jacobson, and Greenstein 1964), with rather depressing results. Very elaborate precautions are required to avoid systematic errors of 10 percent. In crowded spectra of types later than F5, location of the continuum is almost impossible; later than G8 only the yellow and red spectral regions are sufficiently uncrowded. Lower dispersion than 10 Å/mm seems nearly useless (except for the hydrogen lines), and introduces systematic errors. Photoelectric scanning is useful in B and A stars, and eliminates at least errors of calibration, although not in the scattered light. There has been no extensive experience yet on the usefulness of image tubes for spectrophotometry of lines.

V. THEORETICAL ADVANCES IN COG ANALYSIS

It is clear that the primitive COG method must be revised, with the availability of large computing machines and the detailed model atmospheres they provide. The rapid growth of sophistication in this field has resulted in the incorporation of such effects as non-greyness, convection, and the effects of blanketing (statistical picket-fence model, or individual strong lines). Comparisons of models by several independent investigators (Gingerich et $al.$ 1965) shows excellent agreement; models for B and A stars are given by Mihalas (1965) and Strom and Avrett (1965). Some of the problems now concerning theoreticians are discussed in the reports of the Harvard-Smithsonian Conferences on Stellar Atmospheres (ed. Avrett, Gingerich, and Whitney 1964, 1965). The continuous absorption coefficient, necessary for such models, and also for the simpler method of coarse COG analysis has been elaborately tabulated by Bode (1965), including molecules, ions, and heavier elements from H to Ni. Once a grid of models is available, the best procedure, at present, seems to be to fit the continuum, including the Balmer jump, as measured by photoelectric scanning (Oke 1965). (This method is not very useful if the star is strongly reddened.) Temperatures are given by the scans, which determine very accurately the departure of the stellar flux from a black body. The models represent these scans, over favorable ranges of T, with considerable accuracy and also determine the surface gravity g, for $\theta_{eff} > 0.5$. (The hydrogen/helium and hydrogen/metal ratios prove to have only a small effect for hot stars.) For $\theta_{eff} \leq 0.5$, the scans and models determine only T. But the Stark-broadened hydrogen-line profiles are sensitive to g for $\theta \leq 0.5$, so that a combination of color, Balmer jump and $H\gamma$ profile gives both T and g for $0.15 < \theta_{eff} < 0.75$. Over this important range of types (B2 to G0) the models show very large departures from the grey-body $T(\tau)$. In fact, T_0/T_{eff} drops far below the grey-body value (0.81); at $\theta_{eff} = 0.12$, it is 0.75, suddenly decreases at $\theta_{eff} = 0.2$, is 0.70 at $\theta_{eff} = 0.35$ and only approaches 0.80 again at $\theta_{eff} = 0.70$. But these cooler models do not include the effect of metallic-line blanketing, which will also depress T_0. The expected $T(\tau)$ then, for all but the very hottest stars, will approach very low values at the boundary. Therefore, a deep temperature minimum must exist in stars of solar-temperature; hotter stars probably do not have magnetically or convectively heated chromospheres and a low temperature zone must exist at their boundary. Rotation may provide energy for their chromospheres. The importance of these effects for COG methods is based on the following fact. We may evaluate the contribution to the absorption in the line produced by an atom at optical depth τ in the continuum; we can use weighting functions $G(\tau)$, or the actual source function. In either case, while the continuum may be imagined as produced at $\tau_0 \approx 0.2$ to 0.4 (dependent on the temperature), the line must be formed near $\bar{\tau} \approx \tau_0/1 + \bar{\eta}_0$. Here $\bar{\eta}_0$ will vary from line to line, because of the depth-dependence of η_v, i.e., the stratification of the line-producing atoms, l_v, with respect to k_v, the continuum-producing atoms. Therefore, even without stratification, weak and strong lines are formed at different $\bar{\tau}$. In addition, the absorption in a line is roughly proportional to $(dB_v(T)/d\tau)_{\bar{\tau}}$, i.e., the temperature gradient which measures the limb-darkening coefficient. Strong lines may be formed near $\bar{\tau} = 0$ in the continuum, in a region of steep temperature gradient, which strengthens them relative to weak lines. Of course, stratification of ions with respect to neutral atoms further complicates this effect, as well as the variation of turbulence with height.

VI. IS "TURBULENCE" THE RIGHT WORD?

Let us now consider the peculiar importance of astronomical "turbulence." With our knowledge of the solar atmosphere, we no longer can believe that turbulence is a simple micro-scale eddy motion, superposed on the thermal motions. We know that in supergiants and shell stars, large-scale stream motions exist on a scale both large and small compared to the "depth of the reversing layer." We have evidence that turbulence is a function of height, generally increasing upwards (e.g., in ζ Aur). The sun has hot and cool, ascending or descending columns, with motion and temperature somewhat correlated. Solar spectra of good spatial resolution have doppler-displaced zigzag absorption lines, varying in intensity from granule to intergranular space. In addition, line centers must suffer in varying degree from chromospheric emission, probably roughly correlated with the underlying granulation. How big are these effects in a convective F- to K-type supergiant?

In reality, therefore, in stars of large turbulence we are faced with the problem of the composite curve of growth of lines formed in a moving atmosphere, not stratified in parallel planes, merging into an envelope which resembles a "field of prominences" which both absorb and emit. The problems of deviations from local-thermodynamic-equilibrium are enormous in this case. Exact results cannot be looked for, at present, in COG analysis of stars or shells with large macroturbulence, (indicated by broadened absorption lines) because the microturbulence concept is not meaningful when applied to an inhomogeneous atmosphere. When the macroturbulence is subsonic, and the microturbulence small, i.e., in luminosity classes II or fainter, the present differential curve of growth method is probably reliable.

Excellent insight into the problems of supergiants is given by study of eclipsing systems, such as ζ Aur, 31 Cyg, and so on. The review article by Wilson (1960) describes results from chromospheric absorption lines produced in the cool star as it eclipses the smaller, hotter companion. The curves of growth show microturbulent velocities of 7 to 13 km/sec, increasing nearly linearly with height, in the range from 5 to 50×10^6 km height. The apparent mean density of material (subject to correction depending on how much of the volume is actually filled or how inhomogeneous the gas streams are) can be described by a "scale height" which also increases with height. Neglecting magnetic forces, the scale height in an isothermal atmosphere should be constant in a continuous gas where pressure forces dominate. If the density distribution, however, is determined by dynamic forces (streams, mass exchange, or supersonic "turbulence") the scale height is not kT/gm_H but rather $V_{Tu}^2/2g$. In these supergiants, the observed radial velocities of absorption lines vary with height and from eclipse to eclipse. Propagation velocities and amplitudes of compressional waves in such an envelope should increase outwards, as the density decreases, into the transonic region. Then shock-wave formation should cause rapid energy loss and the formation of an ionized, high-temperature corona. There is little doubt that far-ultraviolet spectra of such envelopes will resemble those of symbiotic stars, i.e., will show highly ionized forbidden lines.

A review of the status of determinations of turbulent velocity by COG methods is given by Huang and Struve (1960). A homogeneous study of the dependence of V_{Tu} on type and luminosity was made by Bonsack (1959), in 46 stars of types G8 to M0. He finds few velocities higher than 4 km/sec, and only small differences between stars of luminosity classes Ib to III; however, V_{Tu} in luminosity class V is definitely lower. Apparently V_{Tu} decreases as T decreases. The observed total line widths (his quantity $\Delta\lambda'$) increase with luminosity; the line widths, caused by macroturbulence and/or rotation are about 5 km/sec for dwarfs and reach 17 km/sec for class Ib. The latter value is supersonic for late K- or M-type supergiants; some rotation might persist (caused by evolution from very rapidly rotating B stars). If it is all macroturbulence, it is likely to be a regional motion, or streaming, rather than part of the "hierarchy of eddies" required by the aerodynamic concept of turbulence. The important effect of the interaction between macroturbulence and microturbulence has been stated many years ago (Unsöld and Struve 1949; Huang 1950; Huang and Struve 1952). It is my personal feeling that the subject is worth new study; the theoretical basis has been better developed (*e.g.*, Sobolev 1960, 1963) and computing-machine techniques make the computations feasible. In addition, late-type cool stars with extended atmospheres are sources of stellar winds, and possibly have large enough mass losses to be significant in stellar evolution. Does a B star evolved into a K supergiant lose enough mass to become a white dwarf, or must it implode to a neutron star while exploding superficially as a supernova?

VII. THE REFINEMENT OF THE DIFFERENTIAL CURVE-OF-GROWTH METHOD

Neglecting stars of very large turbulence, it has been known for a long time that the line widths, and the microturbulence derived from COG techniques, both depended on excitation potential and on state of ionization (Wright 1947, 1950). Even the sun shows some residual effects of this type, so that values of solar log η_0 are erroneous if read off a single COG. However, both this effect and stratification of ions and neutral atoms vary similarly in stars of similar temperature and gravity, even if the turbulence differs slightly, so long as it is clearly subsonic. Therefore, the differential COG method I have used (Greenstein 1948) and which many of my collaborators have used, tends to be one in which many errors largely cancel, and relative abundances, star to sun, can be quite reliable.

Rapid computers make it possible to refine this technique. Cayrel and Jugaku (1963) assumed that for all stars with $1.4 \geqslant \theta_{eff} \geqslant 0.7$ the same non-grey temperature dependence was valid, and chose to base themselves on a $T(\tau)$ relation given by observations of the sun. Differences in line-blanketing were neglected even though different values of A (hydrogen/metal ratio) were used. Forty models, not necessarily obeying the flux-constancy condition, were computed, giving k_λ, P_e, T at each τ. Thus the level of LTE excitation and ionization was obtained, and therefore l_v. From $l_v/k_v = \eta_v$, the distribution of the absorbers in depth is given, and by integration through the atmosphere, a factor Γ is computed. These Γ are given at four wavelengths, for two stages of ionization, at six different excitation potentials (0 to 5 ev), for Na, Mg, Al, Si, Ca, Ti, Cr, Fe. The computations are carried through for the weak-line case, *i.e.*, the unsaturated lines. Then, for such lines, if we compute θ_{exc} from the tabulated values by Cayrel and Jugaku, we will find different values for different excitation or ionization potentials and wavelength regions. For example, Figure 7 shows their results for the dependence of θ_{exc} on θ_{eff} for neutral and ionized iron, in the yellow and blue. The quantity Γ is defined by Cayrel and Jugaku as

$$\frac{W}{\lambda_{\text{weak}}} = \Gamma\left(\frac{N_{el}}{N_H}\right)gf = \Gamma\zeta_i gf. \tag{7}$$

If the line is not weak, define

$$X = \Gamma\left(\frac{N_{\text{el}}}{N_H}\right)gf = \Gamma\zeta_i gf, \tag{8}$$

and read W/λ, as ordinate, from the actual curve of growth, using X as the abscissa. The nonlinear relation between W/λ and X is the saturation effect, measured by the COG. The relation between $\log X$ and $\log \eta_0$ is

$$\log \bar{\eta}_0 = \log X + \log \frac{c}{\pi^{1/2}V} - \log \int_0^\infty G_\lambda(\tau_\lambda)\,d\tau_\lambda. \tag{9}$$

Here $G_\lambda(\tau_\lambda)$ is the weighting function for weak lines (often taken as approximately $K_3(\tau_\lambda)$) and the mean value, $\bar{\eta}_0$, is obtained from the weight functions as

$$\bar{\eta}_0 = \frac{\int_0^\infty \eta(\tau_\lambda)G_\lambda(\tau_\lambda)\,d\tau_\lambda}{\int_0^\infty G_\lambda(\tau_\lambda)\,d\tau_\lambda}. \tag{10}$$

If we start first with an empirical curve of growth and a rough θ_{exc}, for a single-layer atmosphere, we can determine V. Then we make correction, for individual elements in two stages of ionization, to obtain independent values of ζ_i for ions and neutrals by the above techniques. Section VIII of the Cayrel and Jugaku paper gives the detailed technique, as well as results for the sun.

Pagel (1964) has developed his own modification of older methods, and applied the above ideas (Pagel 1965) to a reanalysis of an important red giant of low metal abundance, HD 122563. There is little doubt that even with small turbulence and neglecting non-LTE effects, the COG of stars is not a unique relation. At least two values of θ_{exc} must be used, separately for ionized and neutral elements, which tends to reduce the differences of V_{Tu} reported. It is also dangerous to use solar η_0's which are derived from one part of the COG, if they appear on a different portion in an other star (because of large changes of either θ_{eff} or A). Differential COG analyses should span only moderate ranges of θ, if they are to be reliable; i.e., a giant should be used as a source of stellar η_0's if another peculiar giant is to be analyzed. Figure 7 shows that θ_{exc} for ionized metals is expected to be smaller than θ_{eff} in cool stars. For neutral atoms, however, $\theta_{\text{exc}} > \theta_{\text{eff}}$, a serious difference when we attempt to derive ionization temperatures empirically from coarse analyses. This effect was noted many years ago and explored by Koelbloed (1953) in K giants and dwarfs. He found that $\Delta\theta_{\text{exc}} \approx 0.17$ between Fe atoms and ions, and also that θ_{exc} would decrease with excitation potential (Adams-Russell phenomenon). There are obviously many approximations in the Cayrel-Jugaku technique, and the curves given in Figure 7 even change with metal abundance. The very low θ_{exc} observed for A stars is not predicted, since near $\theta_{\text{eff}} = 0.7$ they find only small differences; possibly this is connected with the change of the $T(\tau)$ relation caused by the hydrogen-line blanketing, or additional continua arising from the metals.

To show some of the results of modern COG analysis with data of high accuracy, I present some results by Koelbloed (1967) on my Palomar spectra of two interesting late G-type stars of very large space motion (300–400 km/sec). The stars are HD 2665 (a bright giant) and HD 6755 (subgiant); spectra of 9 Å/mm in the blue (3 plates) and 14 Å/mm in the red (2 plates) were accurately measurable because of their uncrowded nature. Figure 8 shows the COG for Fe I in HD 2665 using NBS laboratory gf-values. The fitted theoretical COG has only 0.6 km/sec turbulence, quite characteristic of older stars. The scatter from the curve was definitely less for most elements with the solar η_0 values from Cowley, and these have been used in the final

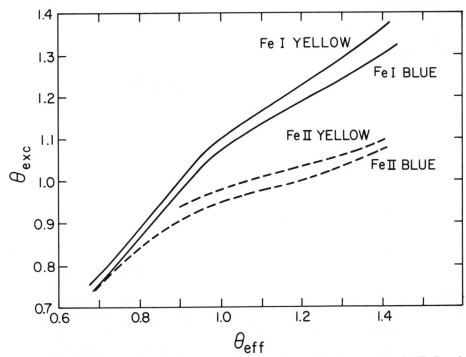

FIG. 7.—The theoretical relation between excitation and effective temperature for Fe I and Fe II (Cayrel and Jugaku).

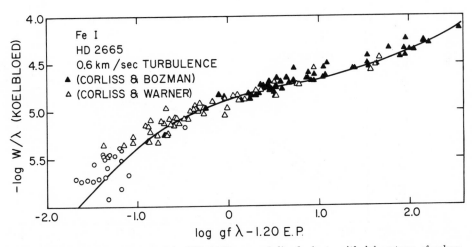

FIG. 8.—Palomar data for Fe I in HD 2665, a weak-lined giant, with laboratory gf-values. Open circles indicate uncertain stellar measurements. Turbulence from the fitted curve of growth is 0.6 km/sec.

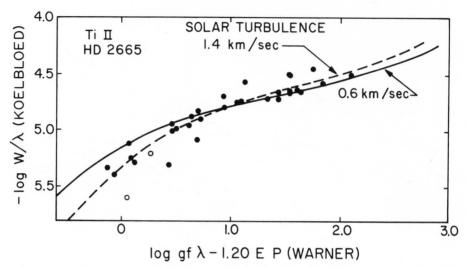

FIG. 9.—The laboratory *gf*-values for Ti II in HD 2665 suggest a higher turbulent velocity than did Fe I in Figure 8. The scatter is larger and the fit worse.

analysis. Figure 9 displays the result from the Ti II lines, with Warner's *gf*-values. The COG with 0.6 km/sec turbulence does not fit, and it could be imagined that $V_{\mathrm{Tu}} = 1.4$ km/sec would be preferred. But Figure 10, where the solar η_0 for Ti II are used, shows smaller scatter, and a good fit to the same COG as for Fe I, with $V_{\mathrm{Tu}} = 0.6$ km/sec. The errors are probably in the laboratory *gf*-values. To display the determination of damping constants, in Figure 11 we have shifted the Cr I solar COG, corrected for the difference of turbulence, so that the weak-line portions all coincide. Then the curves fan out for strong lines, with that for the bright giant HD 2665 lying below that for the subgiant HD 6755, and both below the sun. In

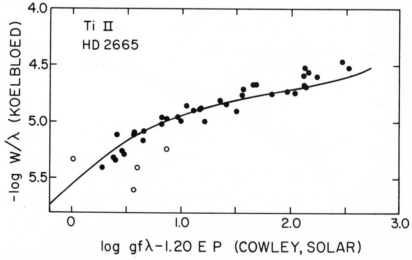

FIG. 10.—With solar *gf*-values from Cowley, the scatter for the same HD 2665 data as in Figure 9 is smaller and the same low turbulent velocity fits.

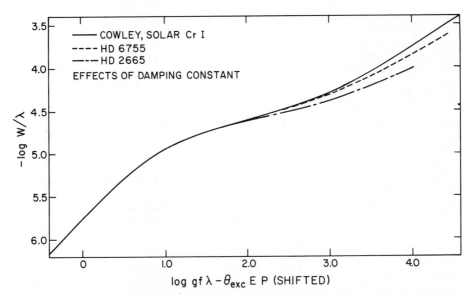

Fig. 11.—The curves of growth of Cr I for the sun and two weak-line stars are shifted to coincide on the straight-line portion, and the difference in turbulence eliminated. Note the lower damping constant for the most luminous star, HD 2665.

spite of the metal deficiency, the gas pressure is lower in these stars than in the sun, because of their lower surface gravity (higher luminosity).

With such curves, the horizontal shifts required to bring the COG of individual elements to coincide with the fitted theoretical curve gave the ratios of numbers of atoms. A conventional single-layer analysis was carried through; the elements in two stages of ionization determined the level of ionization. Varying θ_{ion} over a small range yielded the best (θ, P_e) consistent with θ_{exc}. In Table 2, column 3, the logarithmic

TABLE 2

HD 6755

CURVE OF GROWTH ANALYSIS ACCURACY ATTAINABLE AND COMPARISON
WITH MODEL ATMOSPHERES

Metal-Poor G Subgiant; Cowley Solar gf-Values. Parameters: $\theta_{exc} = 1.10$, log $P_e = -0.44$: Coarse Analysis; $\theta_{eff} = 1.00$, log $g = +2.40$: Model with Solar $T(\tau)$

Ion	[N] Model	[N] C. of G.	Model C. of G.	Ion	[N] Model	[N] C. of G.	Model C. of G.
Al I	−1.21	−1.11	−0.10	Na I	−1.51	−1.40	−0.11
Ba II	1.10	0.95	0.15	Ni I	1.32	1.20	0.12
Cr I	1.30	1.18	0.12	Sc II	1.11	0.96	0.15
Cr II	1.26	1.10	0.16	Sr II	1.33	1.18	0.15
Ca I	1.03	0.92	0.11	Ti I	1.08	0.96	0.12
Ce II	1.63	1.47	0.16	Ti II	1.11	0.96	0.15
Co I	1.15	1.02	0.13	V I	1.39	1.27	0.12
Fe I	1.26	1.11	0.15	V II	1.33	1.18	0.15
Fe II	1.26	1.13	0.13	Y II	1.34	1.20	0.14
La II	1.21	1.06	0.15	Zn I	1.46	1.31	0.15
Mg I	1.38	1.24	0.14	Zr II	−1.29	−1.14	−0.15
Mn I	−1.23	−1.10	−0.13				
				Mean	−1.27	−1.14	−0.13
				Neutral	−1.28	−1.15	0.13
				Ionized	−1.27	−1.12	−0.15

deficiencies in individual metal abundances, as compared with the sun, [N], are given for this coarse analysis. Then an improved method, like that suggested by Cayrel and Jugaku, and by Pagel, was applied to the same COG's. Rough models with solar $T(\tau)$ had been computed by Conti for relevant T and g. With these, and a conventional source-function LTE approach, the line strengths were computed and the shifts between these theoretical COG's and the observed ones re-determined. The elemental deficiencies in column (2) of Table 2, ("model") run parallel to those obtained by coarse analysis. The difference averages -0.14 in the logarithm, *i.e.*, the model predicts a 40 percent greater metal deficiency. There is no difference between the results from neutral and ionized metals. (The type of criticism that Pagel made of an older analysis of the very metal-poor star HD 122563 is not valid here.) Thus, the use of a mixture of coarse and fine analysis techniques, based on curves of growth, can give very good results, with errors presumably less than 0.14 in the logarithm of the metal to hydrogen ratio. The mean metal deficiencies were $[-1.2]$ for HD 6755 and $[-1.6]$ for HD 2665, with uncertainties not much greater than 0.2 in absolute value, and 0.1 in relative abundances from one metal to another.

VIII. DECAY OF TURBULENCE

In investigations of the G and K subdwarfs HD 140283 and HD 103095 the turbulence was found to be essentially zero. In the very metal-poor giants and subgiants this seems also to be true, although these stars should all have convective envelopes. The weakness of Ca II emission in population II stars, the lack of flare phenomena, the absence of rotation and the rarity of close binaries suggest that the initial angular momentum of the oldest stars was small. A weak-lined B star in the globular cluster M 13 has very sharp lines. If current speculations about non-uniform distribution of internal angular momentum are valid, such a star should rotate at the surface if it is rotating in the interior. The various mechanisms that provide energy for a chromosphere, *e.g.*, granulation, large-scale convection, mass-streaming, are all weak in halo population II stars. (Whether magnetic fields are systematically smaller is less well known.) Nevertheless, many current speculations in stellar evolution past the red-giant stage require significant mass loss, so that secondary phases of input of turbulent energy might be found.

In the young population I stars, many phenomena suggest a decay of the turbulent energy input with time. For example, the Ca II emission lines have decreasing strength in old clusters or systems (omitting close binaries). The sun's emission is very weak. There is a close correlation of K-line width and luminosity, possibly loosely connected with macroturbulence. There is some suggestion of lower microturbulence in older disk stars. The Li I line is strongest in young stars, and may be weaker in old, because of deep convection leading to thermonuclear destruction, or dilution. Because of the importance of large rotation in driving material, it is not simple to separate rotation from convection as a source of the energy of the disturbed envelopes of young stars. The large turbulence of the B and A-type supergiants require explanation, since convection should not occur there. There are many fascinating phenomena just now being studied, whose origin lies in the 1934 paper by Struve and Elvey.

This research was supported in part by the U.S. Air Force under contract AF 49(638)-1323, monitored by the Air Force Office of Scientific Research of the Office of Aerospace Research.

REFERENCES

Aller, L. H. 1963, *Astrophysics, The Atmospheres of the Sun and Stars* (New York: Ronald Press), Chaps. 5–8.
Avrett, E. H., Gingerich, O. J., and Whitney, C. A. (ed.) 1964, (Vol. I) and 1965 (Vol. II) *Proc. Harv.-Smithsonian Conf. on Stellar Atmospheres* (Cambridge), Smithsonian Spec. Reports No. 167, No. 174.

Bates, D. R. 1962, (ed.) *Atomic and Molecular Processes* (New York: Academic Press).
Bode, G. 1965, *Die Kontinuierliche Absorption von Sternatmosphären* (Kiel: Inst. f. Theor. Phy. und Sternwarte).
Bonsack, W. K. 1959, *Ap. J.*, **130**, 843.
Cayrel, G., and Cayrel, R. 1963, *Ap. J.*, **137**, 431.
Cayrel, R., and Jugaku, J. 1963, *Ann. d'Ap.*, **26**, 495.
Corliss, C. H., and Bozman, W. R. 1962, *Nat. Bur. Stand. Mono.*, No. 53.
Cowley, C. R. and Cowley, A. P. 1964, *Ap. J.*, **140**, 713.
Ch'en, S. Y., and Takeo, M. 1957, *Rev. Mod. Phys.*, **29**, 20.
Elste, G. 1953, *Zs. f. Ap.*, **33**, 39.
Glennon, B. M., and Wiese, W. L. 1966, *Bibliography on Atomic Transitional Probabilities*, N.B.S. Misc. Publ. 278.
Gingerich, O. J., Strom, S. E., Mihalas, D., and Matsushima, S. 1965, *Ap. J.*, **141**, 316.
Greenstein, J. L. 1948, *Ap. J.*, **107**, 151.
———. 1960, (ed.) *Stellar Atmospheres* (Chicago: University of Chicago Press), Chap. 3, Basic Theory of Line Formation; Chap. 4, 5, Quantitative Analysis and Interpretation of Normal Stellar Spectra.
Griem, H. R. 1964, *Plasma Spectroscopy* (New York: McGraw-Hill).
Harris, D. L. III 1948, *Ap. J.*, **108**, 112.
Huang, S.-S. 1950, *Ap. J.*, **112**, 399.
Huang, S.-S., and Struve, O. 1952, *Ap. J.*, **116**, 410.
———. 1960, *Stellar Atmospheres* ed. Greenstein, J. L. (Chicago: University of Chicago Press), Chap. 8.
Koelbloed, D. 1953, *Publ. Astr. Inst. Univ. of Amsterdam*, No. **10**.
———. 1967, *Ap. J.*, **149**, 299.
Mihalas, D. 1965, *Ap. J. Suppl.*, **9**, No. 92.
Moore, C. E., Minnaert, M. G. J., and Houtgast, J. 1966, *Nat. Bur. Stand. Mono.*, No. 61.
Oke, J. B. 1965, *Ann. Rev. Astronomy and Astrophysics* (Palo Alto: Ann. Rev.).
Pagel, B. E. J. 1964, *Roy. Obs. Bull.*, No. 87.
———. 1965, *Roy. Obs. Bull.*, No. 104.
Searle, L. 1958, *Ap. J.*, **128**, 61.
Sobolev, V. V. 1960, *Moving Envelopes of Stars*, tr. S. I. Gaposchkin (Cambridge, Mass.: Harvard Univ. Press).
———. 1963, *A Treatise on Radiative Transfer*, tr. S. I. Gaposchkin (Princeton, N.J.: Van Nostrand).
Strom, S. E., and Avrett, E. H. 1965, *Ap. J. Suppl.*, **12**, No. 103.
Struve, O., and Elvey, C. T. 1934, *Ap. J.*, **79**, 409.
Unsöld, A., and Struve, O. 1949, *Ap. J.*, **110**, 455.
Warner, B. 1967, *Mem. Roy. Astr. Soc.*, **70**, pt. 5.
Wilson, O. C. 1960, *Stellar Atmospheres*, ed. Greenstein, J. L. (Chicago: University of Chicago Press), Chap. 11.
Wright, K. O. 1947, *J. Roy. Ast. Soc. Canada*, **41**, 49.
———. 1950, *Pub. Dom. Ap. Obs.*, **8**, No. 9.
Wright, K. O., Lee, E. K., Jacobson, T. V., and Greenstein, J. L. 1964, *Pub. Dom. Ap. Obs. Victoria*, **12**, No. 7.
Wrubel, M. H. 1949, *Ap. J.*, **109**, 66.
———. 1950, *Ap. J.*, **111**, 157.

4. HYDROGEN LINES IN NORMAL STELLAR SPECTRA

O. STRUVE: The Stark Effect in Stellar Spectra. *Astrophysical Journal*, **69**, 173, 1929.

E. BÖHM-VITENSE: Stellar Analysis by Means of the Hydrogen Lines.

THE STARK EFFECT IN STELLAR SPECTRA

By OTTO STRUVE

ABSTRACT

The great widths of the hydrogen lines in stellar spectra are probably due in part to *mol-electric Stark broadening.* An estimate of the average *intensity* of the *intramolecular fields* gives 10^3–10^4 volts per centimeter. *Such fields,* if really present, *must affect* also the *atoms of elements other than H.*

Measurements of line-widths in B- and *A*-type stars show that narrow and wide lines occur in all spectral subdivisions. The narrow lines are usually strong in the center, and the wide lines are shallow. The total energy absorbed does not vary much within any one subdivision. This shows that the *mechanism of broadening is not identical* with the one due to *abundance* which has given good results in the later types. It is suggested that the *broadening in early types* is due in part to *Stark effect,* to *axial rotation,* and to *abundance.*

The *general appearance* of the lines *agrees, qualitatively,* with the requirement of the *theory* that the *H lines* should be *broadest,* that *He* and the other light elements should also be *widened appreciably,* but that the *widening of* the lines of the *heavier elements should be small.*

Since the intensity of the intramolecular fields depends upon the partial electron pressure, a *relation* was found connecting *line-width* and *absolute magnitude.* The stars with narrow lines should be more luminous than those with wide lines. This agrees with the well-known luminosity effect discovered by Adams and Joy.

The presence of *electric fields* of the order of 10^4 volts per centimeter should stimulate the appearance of *new lines forbidden* by the *selection principle* of the quantum theory. It is suggested that the *strongest forbidden line of He* belonging to the "near-diffuse" series (λ 4470) may be *identical* with a *line of unknown origin* at λ 4470 which is present in many B stars and reaches maximum intensity at about the same point of the spectral sequence as do the normal *He* lines.

The *He* lines are *known* to broaden unsymmetrically in the fluctuating Stark effect. The *expected shifts* have been *estimated* for the more prominent lines and were found to be in *good agreement* with determinations of wave-length by Albrecht, based upon measures of Frost and Adams.

The *Balmer lines should broaden* in the Stark effect in varying proportions. There should be a *marked increase in width* from $H\beta$ to $H\epsilon$. The measures of Elvey show that an increase exists, but its amount is smaller than might be expected. It is suggested that the *theoretical increase* in line-width is partially *balanced* by a *superposed mol-collisional broadening* depending upon the *motion* of the *charged particles.*

The *evidence* so far available *is regarded as favorable* to the idea that the *broadening in early spectral types* is partly due to the *interaction of the various atoms and free electrons* in the reversing layers of the stars.

83

I. INTRODUCTION

It was suggested by J. Stark[1] in 1906 that the electric fields of neighboring ions or free electrons might be expected to produce appreciable perturbations in the electronic vibrations of a radiating atom. The discovery, in 1913, of the resolution of spectral lines into a number of components when subjected to a strong electric field fully supported this view. The amount of resolution was found to be very nearly proportional to the intensity of the field, which in turn depends upon the strength of the charge and its distance from the radiating atom. In a gas containing many free electrons and positive ions the distances of the charged particles are distributed at random; consequently the components produced by the individual atoms will not coincide but will blend into a broadened spectral line. The amount and character of this broadening must be closely related to the original Stark effect in a constant field. If, for example, a line splits up into two components, both shifted toward the red, the integrated line must also be broadened toward the red. Stark[2] remarked, however, that in addition to this blended resolution into components there should be superimposed a symmetrical broadening due to the relative motion of the charged particle with respect to the radiating atom. The total effect should thus consist of two factors: the mol-electric broadening and the mol-collisional broadening, the mechanical action of the moving field of force being considered as a collision.

The ideas of Stark have been tested by many investigators in the laboratory, and the results were invariably in good agreement with the hypothesis. It is now generally believed that the Stark effect is responsible for at least a large part of the observed increase in width with increase of pressure; and in the presence of many charged particles this type of broadening becomes far more prominent than that due to the thermal agitation of the atoms, or to the ordinary mechanical collisions between them. It seems that under ordinary laboratory conditions—as, e.g., in a capillary discharge

[1] *Verhandlungen der Deutschen Physikalischen Gesellschaft*, **8**, 109, 1906; *Annalen der Physik*, **21**, 422, 1906.

[2] J. Stark, *Handbuch der Experimental-Physik*, **21**, 399, 1927; *Jahrbuch der Radiologie und Elektronik*, **12**, 349, 1915.

tube—the observed line-widths are fully accounted for[1] by the blended resolution into components, i.e., by that part of the effect which Stark called "mol-electric."

It is obvious that this type of broadening depends exclusively upon the concentration of charged particles in the given gas. It will be sponsored by strong ionization and high density. The first factor is present in the stars, but the pressure in the reversing layers is so low that it is not at once evident whether the Stark effect will play any important part in the broadening of stellar absorption lines.

Probably the first to consider the Stark broadening with respect to stellar spectra was R. de R. Atkinson.[2] He pointed out that, according to Debye,[3] the mean molecular field in a gas where there are 10^{-6} gr. per cubic centimeter of ions is about 10^4 volts per centimeter and that this field would tend to broaden the spectral emission lines of the photospheric atoms sufficiently to explain the continuous spectrum in giant stars having average densities of 10^{-6} gr. per cubic centimeter.

It was suggested by E. O. Hulburt[4] that the great width of the H lines in early spectral types might be due to Stark broadening. His numerical computations gave, it is true, predicted values which are much smaller than the observed ones, but he believed that the discrepancy could be removed by assuming a greater concentration of free electrons, resulting, perhaps, from the high state of ionization of the heavier elements in the hotter stars.

H. N. Russell and J. Q. Stewart[5] derived for the $H\beta$ line broadened by the Stark effect the expression:

$$W = (2.3 \times 10^5)(p'/T)^{2/3} , \qquad (1)$$

where p' is the pressure of the free electrons in atmospheres. Assuming for the sun $T = 5000°$, $p' = 10^{-5}$, they find $W = 0.4$ A. The

[1] E. O. Hulburt, *Physical Review*, **22**, 24, 1923; Mlle M. Hanot, *Journal de Physique et le Radium*, **9**, 156, 1928.

[2] *Monthly Notices of the Royal Astronomical Society*, **82**, 396, 1922.

[3] *Physikalische Zeitschrift*, **20**, 160, 1919.

[4] *Astrophysical Journal*, **59**, 177, 1924; *Journal of the Franklin Institute*, **201**, 777, 1926.

[5] *Astrophysical Journal*, **59**, 204, 1924.

observed value is about 0.6 A. Assuming for α Lyrae $T = 10,000°$
and $p' = 10^{-4}$, they find $W = 1.1$ A, the observed value being 20 A.
The agreement for the sun is satisfactory, but for α Lyrae the discrepancy is considerable, although not hopeless. An uncertainty in
p' corresponding to a factor of 10 would not be unreasonable, but,
what is more important, the total intensities of the H lines in α
Lyrae is greater than in the sun; consequently the lines will appear
relatively broader in α Lyrae because of the greater abundance of the
atoms capable of absorbing them.

A very recent discussion of the Stark broadening for the H lines
is by M. Vasnecov.[1] He obtains for the *Halbwertsbreite*, or "half-
width," δ, which corresponds to a point on the line-contour where
$I = \dfrac{I_0}{2}$:

$$\delta = (3.45 \times 10^5) f \cdot \lambda^2_0 \cdot (\chi \cdot p \cdot T^{-1})^{2/3} . \tag{2}$$

χ is the concentration of the ionized atoms, p is the pressure, and f
is the mean value for the given line of Sommerfeld's[2] expression depending upon the quantum numbers:

$$f = \{(n_2 - n_1)(n_1 + n_2 + n_3) - (k_2 - k_1)(k_1 + k_2 + k_3)\} . \tag{3}$$

The actual width, according to Vasnecov, is the value $\Delta = 2 (\lambda - \lambda_0)$,
corresponding to $I = 0.1 I_0$, and for the H lines

$$\Delta = 12\delta .$$

Assuming for the sun $T = 6000°$ and $P = 10^{-5}$, and for αLyrae $T = 10,000°$ and $P = 10^{-4}$, Vasnecov's tables give:

$$\Delta \text{ (sun)} = 1 \text{ A} ,$$
$$\Delta \text{ (α Lyrae)} = 42 \text{ A} .$$

These results are in good agreement with the observations.

Recently the whole subject has been reviewed by A. S. Eddington.[3] The evidence must be regarded as favorable to the Stark effect. He believes that even if the fluctuating Stark effect should not

[1] *Vestnik Kral. Ces. Spol. Nauk*, **2**, 1927.

[2] *Atomic Structure and Spectral Lines* (Eng. ed.), p. 286, 1923.

[3] *Der innere Aufbau der Sterne*, p. 443, 1928.

prove to be sufficient to explain the observed line-widths, it is likely that the broadening would be increased through the rapid motion of the charged particles. This idea, similar to that of Stark, attributes part of the broadening to the velocity of the particles. We must keep this possibility in view, but shall first proceed according to the point of view of Hulburt, and consider only the ordinary mol-electric broadening.

If the great width of the H lines in early type stars is really due to mol-electric broadening, we are led to assume that the average intramolecular field in the reversing layers is quite strong—of the order of 10^3–10^4 volts per centimeter—and that the number of atoms subjected to fields of 10^5 volts per centimeter is still sufficiently great to produce a measurable amount of absorption. Fields of such strength would affect also the atoms of elements other than hydrogen, and it is our purpose to test the expected manifestations of such fields.

II. GENERAL NATURE OF STARK EFFECT

We shall summarize briefly the various effects characteristic of the Stark broadening of spectral lines.[1]

a) The amount of broadening should be roughly proportional to the degree of resolution in a constant field, although because of the motion of the particles there may be superposed a symmetrical spreading-out, similar in character to the mol-collisional broadening.

b) Since the resolution into components is frequently not symmetrical, the broadening should in such cases be accompanied by a pressure shift. The H lines remain always symmetrical,[2] but beginning with He such shifts are very common. They are complicated by blending with new lines which occasionally appear very close to the parent-lines.

c) Under the stimulus of the electric field new lines appear in the spectrum which are normally forbidden by the selection principle.

[1] For a detailed description of the Stark effect see: Stark, *Handbuch der Experimental-Physik*, **21**, 399, 1927; W. Steubing, *Handbuch der physikalischen Optik*, **2**, 683, 1927; G. S. Fulcher, *Astrophysical Journal*, **41**, 359, 1915.

[2] It should be added, for completeness, that "in hydrogen, as well as in helium, the line-groups presenting a symmetrical appearance as a whole are in reality composed of highly asymmetric individual patterns" (J. Stuart Foster, *Proceedings of the Royal Society*, A, **117**, 137, 1927).

Lines violating the rule of azimuthal-quantum number, $\Delta k = \pm 1$, frequently appear, and in strong fields some of them may become as strong as the original lines.[1]

d) The Stark effect is most pronounced for elements of low atomic number and decreases for the heavier elements. A rough idea of the relative susceptibility of various elements can be gathered from Table I, which gives an estimate of the maximum displacement observed in the stronger lines of the region of wave-lengths covered by stellar spectrographs. The values given refer to a field of about

TABLE I

Element	Atom. Wt.	Max. Displ. in A	Element	Atom. Wt.	Max. Displ. in A
H.	1	10	Mg.	12	1
He.	2	5	Al.	13	0
He+.	2	2	A.	18	0.1
Li.	3	3	Ca.	20	1
N.	7	4	Fe.	26	0.5
N+.	7	0	Co.	27	1
O.	8	8	Ni.	28	1
O+.	8	0	Cu.	29	0.5
Ne.	10	3	Ag.	47	1
Na.	11	2	Au.	79	0.2

30,000 volts per centimeter. This is only a schematic representation, since individual lines of the same element may behave very differently.

e) The effect is most pronounced in the diffuse series of any given element. Next follows the principal series, then the sharp series, and finally the various intercombination series.

f) The effect is smallest for the lowest members of any given series and increases in size as the series limit is approached.

g) For the H lines the amount of resolution is proportional to the intensity of the field, F, up to about 10^5 volts per centimeter. For stronger fields a quadratic term becomes noticeable. In other elements the quadratic term sets in at much lower values of F, but for moderate fields the simplifying assumption

$$W \infty F \qquad (4)$$

[1] W. Grotrian, Struktur der Materie, 7, Part I, 119, 1928.

remains a good approximation. The broadest observed H lines are about 40–60 A wide, corresponding to maximum fields of something more than 10^5 volts per centimeter. It is probable that the average fields in the reversing layers of the stars never exceed 10^5 volts per centimeter or about 330 c.g.s. electrostatic units.

h) The theory of the Stark effect has been worked out by A. Sommerfeld[1] and by P. Epstein[2] for hydrogen and similar atomic models, such as He^+. The distance of any one component from its normal position is

$$\Delta\lambda = \frac{3h\lambda_0^2 \cdot F}{8\pi^2 mZc} \cdot f ,$$

(5)

where f is given by (3), m is the mass of the electron, Z is the nuclear charge, h is Planck's constant, and c is the velocity of light. The average value of f to be used to compute the effect of broadening has been given by Vasnecov:

$$f_{H\alpha} = 4; f_{H\beta} = 10; f_{H\gamma} = 18; f_{H\delta} = 28 .$$

i) Formula (5) predicts that He^+, where $Z = 2e$, should be broadened less than H. This has actually been observed. From analogy it seems probable that, in general, ionized atoms will show less Stark effect than neutral atoms. This has indeed been verified in several cases (e.g., N^+ and O^+). However, T. R. Merton[3] has pointed out that occasionally He^+ λ 4686 is the broadest line visible in the He spectrum, and has given a satisfactory explanation for this phenomenon.

III. LINE-WIDTHS IN STELLAR SPECTRA

Turning our attention to the interpretation of the observed line-widths in stellar spectra, we note that almost certainly a number of entirely different mechanisms are at work. Recent theoretical work by J. Q. Stewart[4] and by Unsöld,[5] based upon the classical electron theory, has shown that the width of a line must increase as the square

[1] *Atomic Structure and Spectral Lines*, p. 286, 1923.

[2] *Annalen der Physik*, **51**, 184, 1916.

[3] *Proceedings of the Royal Society*, A, **95**, 30, 1919.

[4] *Astrophysical Journal*, **59**, 30, 1924.

[5] *Zeitschrift für Physik*, **44**, 793, 1927.

root of the total number of absorbing atoms in the line of sight. Unsöld has shown that this mechanism of broadening satisfactorily explains the contours of many lines in the solar spectrum. Miss C. H. Payne[1] has applied Unsöld's formula to a large number of stars and has found substantial agreement in the later spectral types. The H lines, however, proved to be abnormal;[2] they were very much too shallow in the center and too widely spread—a phenomenon which becomes particularly striking in the A, B, and O stars.

According to an investigation by H. N. Russell and Miss C. E. Moore,[3] most of the winged lines in the solar spectrum bear out the argument of Stewart and of Unsöld that "the main cause of widening is shown to be almost entirely the abundance in the solar atmosphere of atoms which are in a condition to absorb the line in question," and "widening in the laboratory shows little correlation with widening in the Sun, and it is probably due to quite different causes." On the other hand, they state that "the widening of the H lines in the Sun . . . . may be largely due to the influence of neighboring molecules, since H is unusually sensitive to the Stark effect. It is also, of course, peculiarly sensitive to Doppler widening."

While most of the solar lines, including H and K of Ca^+, are widened principally by a mechanism that we may call "abundance-broadening," the H lines suggest that another type of mechanism is also present.

IV. LINE-WIDTHS IN STARS OF EARLY TYPE

The peculiarity noted in the H lines of the solar spectrum is quite general in stars of early spectral type. I have recently measured and estimated the line-widths (W) and the central intensities (I_0) in a large number of B- and A-type stars. The results for He λ 4472 and Mg^+ λ 4481 prove that the dispersion in W as well as in I_0 is considerable within a single spectral subdivision. A summary for λ 4472 in B stars is given in Table II. The central intensities and the line-widths are expressed in arbitrary units, but a calibration of each scale is given at the end of the table.

[1] *Proceedings of the National Academy of Sciences*, **14**, 399, 1928; *Harvard College Observatory Circular*, No. 334, 1928.

[2] Unsöld (*Zeitschrift für Physik*, **46**, 778, 1928) has also shown that the H lines in the solar spectrum cannot be represented satisfactorily by his formula. The discrepancy is least for $H\alpha$, and increases for the higher members of the Balmer series.

[3] *Astrophysical Journal*, **63**, 1, 1926.

TABLE II

W	No.	I_0	W	No.	I_0	W	No.	I_0
Type O			**Type B3**			**Type B8**		
3.3	3	9.0	1	6	10.3	1	2	4.5
5.3	3	7.7	2	10	8.6	2	7	4.6
9.0	3	3.3	3	12	8.8	3	10	4.7
			4	9	8.0	4	20	2.6
Type Bo			5	8	7.8	5	13	3.1
			6	10	6.8	6	5	1.8
2.4	5	8.6	7	5	5.6	7	4	2.0
4.5	6	7.3	8	9	6.2	8.7	3	2.3
7.3	3	6.1	9	8	5.8			
10.0	2	3.0	10	14	2.7	**Type B9**		
			15	1	1.0			
Type B1						1	2	4.5
			Type B5			2	5	2.8
1.6	5	10.3				3	11	2.8
3.3	3	10.0	1	2	7.5	4	4	2.5
5.7	4	6.2	2	5	6.2	5	6	2.6
10.0	2	3.0	3	9	4.2	6	2	2.5
			4	12	4.2	7	1	1.
Type B2			5	11	3.8	8	1	1.
			6	5	5.0	9	1	1.
1.0	4	9.5	7	8	2.8	10	1	1.
2.8	4	10.1	9.2	6	3.2			
4.2	4	7.8						
7.0	4	6.2						
9.0	4	5.2						

CALIBRATION OF LINE-WIDTHS

Arb. Un.	A	Arb. Un.	A
1	1.2	6	2.3
2	1.4	7	2.5
3	1.7	8	2.8
4	1.9	9	3.0
5	2.1	10	3.2

CALIBRATION OF INTENSITIES

Arb. Un.	Per Cent Absorption	Arb. Un.	Per Cent Absorption
1	9	6	42
2	17	7	47
3	24	8	52
4	31	9	56
5	37	10	60

An inspection of the table immediately shows that wide lines are invariably weak in the center, and narrow lines are invariably strong. This in itself suggests a mechanism other than "abundance-broadening."

Let us assume that the intensity of any point on the line-contour is given by an equation of the form

$$I = I_0 e^{-h^2(\lambda - \lambda_0)^2} . \tag{6}$$

The measured width W corresponds to such a value of $(\lambda-\lambda_0)$ that at this distance from the center of the line the value of I is the smallest perceptible difference in intensity on the background of the continuous spectrum. Putting this equal to I', we have

$$I' = I_0 e^{\frac{-h^2 W^2}{4}} ,$$

whence

$$W = \frac{2}{h} \sqrt{\log_e I_0 - \log_e I'} . \tag{7}$$

Thus W is a function not only of h but also of the ratio I_0/I'. If $I' = I_0$, the width is necessarily zero, irrespective of h. For large values of I_0/I', the factor under the root is of little importance, and then $W \infty \, 1/h$, as is frequently assumed to be the case. But for faint lines the corrective term becomes of considerable importance.

The numerical value of I' depends upon the contrast factor of the photographic plate. Under normal conditions the faintest definitely visible line corresponds to $I_0 = 9$ per cent absorption. The width of such a line may be assumed to be equal to its half-width, giving $I' = 4$ per cent.

The total energy absorbed in a line is, of course,

$$E = I_0 \int_{-\infty}^{+\infty} e^{-h^2(\lambda - \lambda_0)^2} \, d\lambda = \frac{I_0}{h} \int_{-\infty}^{+\infty} e^{-x^2} dx = \frac{I_0 \sqrt{\pi}}{h} = 1.77 \frac{I_0}{h} .$$

Substituting h from (7), we get

$$E = \frac{0.88 I_0 W}{\sqrt{\log_e I_0 - \log_e I'}} . \tag{8}$$

I_0 and I' are here expressed in terms of the intensity of the continuous spectrum. If this is known from the law of radiation, E can be expressed in absolute units.

A computation of E by means of formula (8), from the data of Table II, shows that it changes very little with W. For type B3, where the number of stars is large, there appears to be a slight increase in E with W, but the reality of this change may be doubted in view of the uncertainty of the calibrations of my scales of estimates. It seems that the law governing the broadening in these stars requires approximately:

$$E = \text{Const} .\tag{9}$$

Unsöld's mechanism does not help us here. His formula gives for any line $I_o = 100$ per cent absorption, and as a rough approximation

$$E \backsim W .\tag{10}$$

It is possible that the observed slight increase of E with W may be due to a superimposed "abundance-broadening," but in view of the uncertainty of my observations this result remains doubtful.

There are a number of broadening mechanisms that approximately satisfy the condition $E = \text{Const}$. Probably the most powerful ones are: (1) axial rotation (or pulsation in Cepheids, expansion in novae); (2) Stark effect.[1] The first mechanism has been discussed in a paper by Shajn and the writer.[2] It is probably present in a measurable degree in many stars, and may attain very large proportions in certain spectroscopic binaries (V Puppis, μ^1 Scorpii, W Ursae Majoris). However, the fact, mentioned first by H. C. Vogel,[3] that the H lines are frequently very much broader than any other lines, makes it practically certain that axial rotation is not the only mechanism responsible for (9). Accordingly, it becomes necessary to assume that the dispersion in W is due to at least three causes:

$$\sigma_{obs} = \sqrt{\sigma^2_{rot} + \sigma^2_{Stark} + \sigma^2_{abun}}\tag{11}$$

[1] It is not certain that the total energy absorbed in a line will remain undisturbed in the Stark effect. Indeed, Stark suggests that the intensity of the diffuse series decreases in favor of the new near-diffuse and almost-sharp series. The question has not been investigated sufficiently to permit definite conclusions. Our assumption that E remains approximately constant is based upon the fact that while the diffuse line λ 4472 gets fainter, the near-diffuse line, λ 4470, gets stronger so that the blend will probably change but little. The energy lost in the almost-sharp series is probably negligible.

[2] *Monthly Notices of the Royal Astronomical Society*, **89**, 222, 1929.

[3] *Astronomische Nachrichten*, **90**, 71, 1877.

The rotational dispersion can be separated, statistically, in short-period spectroscopic binaries. The dispersion due to abundance can be best studied in later spectral types and in lines not greatly influenced by Stark effect. The third mechanism must be investigated by comparing it with the known properties of the Stark effect.

V. ELEMENTS OF DIFFERENT ATOMIC WEIGHT

According to section II(d), we may expect that the broadest lines will be those of H, closely followed by the lighter elements. The heavier elements should always show fairly narrow lines. Barring all cases of widening due to abundance, this agrees qualitatively with the observations. Next to those of H, the broadest lines are doubtless those of He, followed by those of the other light elements. The metallic lines are usually much narrower. In this connection it is of interest to remark that Ca^+ λ 3933 is often much narrower than He λ 4472, even though their central intensities are nearly the same. As a matter of fact, these narrow stellar Ca^+ lines have frequently been confused with the sharp, detached (Hartmann) Ca^+ lines, but in spectral subdivisions B8 and B9 measurements of radial velocity prove them to be of ordinary stellar origin. Direct measurements also show that Mg^+ λ 4481 is, as a rule, considerably narrower than He λ 4472.[1]

A quantitative analysis is difficult in view of the very different central intensities. The H lines are strong, those of He are frequently faint. The average width of He λ 4472 in B stars is about 2.5 A, that of $H\beta$ roughly 15 A. The ratio $W_H/W_{He}=6$, considerably more than would be expected from the Stark effect alone. Indeed, the maximum resolution of λ 4472 in a field of 10^5 volts per centimeter is about 20 A, while that of $H\beta$ is about 40 A. The expected ratio is about 2. The remaining factor of 3 must be attributed partly to difference in abundance, and partly to the corrective term under the radical in (7).

As Miss Fairfield[2] has shown, the width of the H lines is closely correlated with the Mount Wilson estimate of "n" or "s." As the

---

[1] The spectrum of η Orionis photographed with a dispersion of three prisms reveals the presence of very narrow lines of Si^{++} and of O^+, while the H lines are very broad and the He lines rather broad (A. Pogo, *Astrophysical Journal*, **68**, 312, 1928).

[2] *Harvard College Observatory Circular*, No. 264, 1924.

latter refers chiefly to the He lines, there can be no doubt that the broadening mechanism is the same in these two elements.

VI. LUMINOSITY EFFECT

We have mentioned in section IV that the observed dispersion in line-widths may be partly due to differences in the Stark effect. If we consider only one line, for example, $\lambda\,4472$, in a single spectral subdivision, the only quantity that can vary is F, which, as we have seen in (4), is proportional to W. For a single atom we have approximately

$$F = \frac{e}{x^2}\,,$$

where x is the distance to the nearest charge e. Assuming that there are n free electrons per cubic centimeter, we have

$$x = c \cdot n^{-1/3}\,,$$

and, consequently,

$$F = \text{Const. } e \cdot n^{2/3}\,.$$

Introducing the gas equation in the form

$$n \infty p'/T\,,$$

where p' is the pressure of the free electrons and T is the absolute temperature of the gas,

$$F = \text{Const. } e \cdot (p'/T)^{2/3}\,. \tag{12}$$

If we limit ourselves to a single spectral subdivision, we may put $T = \text{Const.}$, so that

$$F \infty (p')^{2/3}\,, \tag{13}$$

and, consequently,

$$W \infty (p')^{2/3}\,. \tag{14}$$

This relationship shows that a dispersion in W is equivalent to a dispersion in the partial electron pressures. As we shall see, this will have a marked influence upon the absolute magnitudes. The partial electron pressures may be assumed to be proportional to the total gas pressures, and the latter are, of course, related to the

[1] *Monthly Notices of the Royal Astronomical Society,* **85,** 782, 1925.

surface gravities. In fact, E. A. Milne[1] has shown that approximately

$$\frac{P_1}{P_2} = \left(\frac{g_1}{g_2}\right)^{1/2}\left(\frac{T_2}{T_1}\right)^2 .$$ (15)

There seems to be some slight uncertainty, however, as to the power $\frac{1}{2}$, and we shall for the present assume that, T_1 being equal to T_2,

$$p' \backsim g^m .$$ (16)

The surface gravity is given by

$$g = \frac{\gamma \cdot \mu}{R^2} ,$$ (17)

here γ is the constant of gravitation and μ and R are the mass and radius of the star. Consequently,

$$W = \frac{\text{Const. } \mu^{\frac{2m}{3}}}{R^{\frac{4m}{3}}} .$$ (18)

The radius of the star is readily expressed in terms of the absolute bolometric magnitude,[1] M:

$$log \ R = -0.2 \ M - 2 \ log \ T + 8.53 .$$ (19)

If we adopt Eddington's mass-luminosity relation, we can also express μ as a function of M and thus find the relationship between W and M. For simplicity we shall use Jeans's expression,[2]

$$M = 4.85 - 7.5 \ log \ \mu .$$ (20)

Substituting μ from (20) and R from (19) into (18), we get

$$log \ W = \frac{m}{5}M + \text{Const} .$$ (21)

Let us consider two groups of stars, the one with very narrow lines, W_1, and the second with very broad lines, W_2. Then

$$\Delta M = \frac{5}{m} \ log \ \frac{W_1}{W_2} .$$ (22)

[1] J. H. Jeans, *Astronomy and Cosmogony*, p. 47, 1928.

[2] *Ibid.*, p. 126.

Our actual numerical values in Table II suggest that for type B3:

$$W_1 = 1.4 \text{ A (narrow lines)} ,$$
$$W_2 = 2.8 \text{ A (wide lines)} .$$

The ratio W_1/W_2 is 0.5, and the corresponding value of ΔM is $+\dfrac{1.5}{m}$, in stellar magnitudes.

Assuming for m three different values, we find:

1. $m = \frac{1}{2}$; $\Delta M = +3.0$ mag.
2. $m = 1$; $\Delta M = +1.5$
3. $m = 2$; $\Delta M = +0.8$

(Stars with narrow lines should be more luminous than stars with broad lines.)

The first hypothesis agrees with Milne's theory (15). The second is that of Pannekoek. The third follows a recent suggestion of Miss Payne and F. S. Hogg,[1] who find empirically that m is greater than 1. The best representation of their observations, as a whole, is obtained with $m = 2$. It must be emphasized, however, that this representation is poorest for the earliest types, and that an extrapolation of their smooth curve would give approximately $m = \frac{1}{2}$ for stars of type B.

Observational evidence shows unmistakably that A stars and B stars with narrow lines are appreciably more luminous than stars with broad lines. This effect was discovered first by Adams and Joy[2] and independently by the writer.[3] A thorough discussion by Edwards[4] fully supports these results. In the case of the H lines, Miss Fairfield[5] found that their widths are correlated with reduced proper motion ($h = m + 5 + 5 \log$ [p.m.]), so that here, too, the requirements of the theory are fulfilled.[6]

[1] *Harvard College Observatory Circular*, No. 334, 5, 1928.

[2] *Astrophysical Journal*, **56**, 242, 1922; **57**, 294, 1923.

[3] Dissertation, University of Chicago, unpublished; *Abstracts of Theses, University of Chicago*, **2**, 1923.

[4] *Monthly Notices of the Royal Astronomical Society*, **87**, 365, 1927.

[5] *Harvard College Observatory Circular*, No. 264, 1924.

[6] This effect has also been verified by Miss A. V. Douglas in the spectra of A-stars (*Astrophysical Journal*, **64**, 262, 1926; *Journal of the Royal Astronomical Society of Canada*, **20**, 265, 1926). The relationship, discovered by Miss Payne and Miss Howe (*Harvard College Observatory Circular*, No. 287, 1925), between the number of Balmer lines visible in the spectrum and the absolute magnitude of the star, points in the same direction. Miss Payne has called my attention to this fact.

To obtain an entirely independent result I have computed the absolute magnitudes of the hotter B stars from the intensities of the detached (Hartmann) Ca^+ lines.[1] Using the relation $I_{ca} = f(\text{Dist.})$, the distance of the star is immediately found, and M is then computed by means of the apparent magnitude. It is not yet apparent whether these values are good approximations individually, but in the light of Eddington's hypothesis of interstellar matter, they should give useful results for the mean of many stars. The following values of ΔM were obtained from the Ca^+ intensities:

Spectrum	No. of Stars	ΔM	Remarks
O, Bo, B1, B2....	60	+1.2	Narrow—broad
B3.............	92	+1.9	Narrow—broad

The corresponding values of Edwards are:

Spectrum	ΔM	Remarks
Bo, B1, B2..............	+0.9	Narrow—broad
B3.....................	+1.5	Narrow—broad

For B3 the mean is $\Delta M = +1.7$ mag. Adams and Joy divided the stars into two groups only, viz., "n" and "s." Consequently, they obtained a smaller value for ΔM—viz., about one-half of that given above.

The observed value of ΔM apparently agrees best with the computed value if $m = 1$. However, I am inclined rather to accept Milne's result, $m = \frac{1}{2}$. It is only reasonable to assume that part of the dispersion in W is due to causes other than Stark effect. The observed difference $W_1 - W_2 = 1.4$ A is obviously a certain multiple of σ_{obs} in (11). If the contributory dispersions are all alike, i.e., if $\sigma_{\text{Stark}} = \sigma_{\text{rot}} = \sigma_{\text{abun}}$ we have

$$\sigma_{Stark} = \frac{\sigma_{\text{obs}}}{\sqrt{3}},$$

and we see, consequently, that we should have used a smaller difference between W_2 and W_1. The corrected values would have been

[1] *Astrophysical Journal*, **67**, 383, 1928.

$W_1 = 1.7$ A and $W_2 = 2.5$ A, so that the corrected ratio W_2/W_1 would have been 1.47. This would give $\Delta M = +1.7$ mag., provided $m = \frac{1}{2}$. This agrees exactly with the observations.

We see from the foregoing that the Stark effect gives a satisfactory explanation for the observed difference of luminosity between "s" and "n" stars.[1]

VII. FORBIDDEN LINES

In section II(c) it was stated that the presence of strong electric fields stimulates the appearance of spectral lines normally forbidden by the selection principle.[2] According to our estimate that the average intramolecular field in the reversing layers is of the order of 10^3–10^4 volts per centimeter and that fields up to 10^5 volts per centimeter are in some stars sufficiently numerous to produce appreciable absorption, it seems probable that some of the more prominent forbidden lines would be present. An inspection of the work of Takamine and Yoshida and of others shows that the near-diffuse lines of He make their first appearance at field intensities somewhat lower than 10^4 volts per centimeter.

Of the various elements only He seems to promise any results. Hydrogen shows no new lines outside the Balmer components, which are blended. All the other elements are either faint in the stars or not very susceptible to Stark effect. He, however, is both strong and very susceptible. A closer analysis is, therefore, justified.

In this connection the work of T. R. Merton[3] is particularly

---

[1] Some investigators have doubted the reality of this effect in the A stars and, in particular, Lindblad and Miss Fairfield have called attention to a systematic difference between the Mount Wilson and Harvard classifications for A stars. It is clear that the luminosity effect can either be blotted out or be spuriously introduced by an appropriate adjustment of the classification. The question reduces therefore to whether the Mount Wilson or the Harvard classification is preferable. I see no reason at present to discard the Mount Wilson system which is based upon slit-spectrograms and probably forms an excellent temperature sequence. Further investigation of this question is much needed.

[2] These forbidden lines violate the rule for the azimuthal quantum number: $\Delta k = \pm 1$. They do not involve transitions from metastable states and are therefore not restricted to low densities. While, according to Bowen, the forbidden nebular lines should be observable only in emission, there is no such restriction for lines stimulated in the electric field.

[3] *Proceedings of the Royal Society*, A, **95**, 30, 1919.

valuable. The Stark effect observed by him in a helium vacuum tube
excited by means of a fairly heavy discharge from an induction coil
was of the fluctuating type, and the general conditions were comparable to those in the stars. The accompanying sketch (Fig. 1),

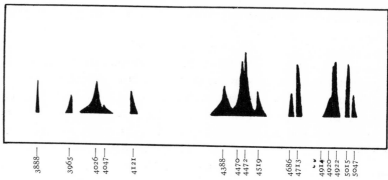

FIG. 1.—The helium spectrum in the electric field. (By R. T Merton)

TABLE III

HELIUM LINES IN ELECTRIC FIELD

Wave-Length	Series Relationship		Remarks
4472........	$2p^3 - 4d^3$	Diffuse triplet	Permitted
4470........	$2p - 4f$	Near-diffuse triplet	Forbidden
4922........	$2P - 4D$	Diffuse singlet	Permitted
4713........	$2p^3 - 4s^3$	Sharp triplet	Permitted
5015........	$2S - 3P$	Principal singlet	Permitted
4920........	$2P - 4F$	Near-diffuse singlet	Forbidden
4388........	$2P - 5D$	Diffuse singlet	Permitted
4026........	$2p^3 - 5d^3$	Diffuse triplet	Permitted
3888........	$2s^3 - 3p^3$	Principal triplet	Permitted
4519........	$2p - 4p$	Almost-sharp triplet	Forbidden
4908........	?	?	Forbidden
4121........	$2p^3 - 5s^3$	Sharp triplet	Permitted
3965........	$2S - 4P$	Principal singlet	Permitted
4047........	$2p - 5p$	Almost-sharp triplet	Forbidden

taken from Merton's paper, represents the *He* spectrum photographed through a neutral wedge, so that the heights of the lines
give an indication of their intensities. I have grouped them in
Table III, roughly in order of intensity.

The strongest forbidden line is λ 4470 (given as 4470 by Merton
and as 4469 by Stark), a component of the well-known line λ 4472,

which is particularly prominent in stellar spectra. The forbidden line, for which $\Delta k = 2$, was tentatively classified by Stark as belonging to the "near-diffuse" subordinate series of the helium triplet system. The next strongest forbidden line, λ 4920 ($\Delta k = 2$), lies in a region where the focus of our plates is not perfect and where we can hardly expect to see it separately from the diffuse singlet line λ 4922. There is a considerable decrease of intensity before we reach the next forbidden line, λ 4519, for which $\Delta k = 0$. We would hardly expect it to be represented in stellar spectra because it is so much fainter than λ 4470.

Hence, only one forbidden line of *He* would be expected to be present. It should lie nearly 2 A to the violet of λ 4472, and should, therefore, be separately visible in some stars. It is a fact that certain B-type stars show a faint but quite definite absorption line in that very position (λ 4470.046 A on Rowland's scale).[1] I have identified and measured this line in many stars. It appears to be strongest in or near subdivision B3, i.e., very near to the place where the ordinary *He* lines reach their maximum intensity. I have not been able to identify it with any other element, and Mr. F. E. Baxandall informs me that he too has looked in vain for a satisfactory identification.

While no final conclusions are as yet possible, it seems reasonable to identify, tentatively, the unknown line with the near-diffuse line of *He*.

Baxandall[2] lists two other unidentified lines in the spectrum of ϵ Orionis (Bo), λ 4070 and λ 4655. Neither of them agrees with any of the stronger forbidden lines of *He*. I have not tried to compare them with the forbidden lines of other elements. As was stated above, the fainter forbidden lines of *He* cannot be expected to be as easily visible as λ 4470, although there is perhaps a chance that λ 4920 might be found on particularly good plates.

VIII. PRESSURE-SHIFTS

In section II(b) we stated that almost all spectral lines, with the exception of H and He^+, show more or less asymmetry in their reso-

---

[1] *Astrophysical Journal*, **62**, 198, 1925.

[2] *Monthly Notices of the Royal Astronomical Society*, **83**, 166, 1923.

lution into components. The mol-electric fields in the reversing layers of the stars should therefore produce spurious shifts in the positions of certain lines. It is difficult to estimate the amount of such displacements without knowing exactly the average intensity of the field and the conditions under which the various components blend together. Take, for example, *He* 4472. Its components show, on the average, a strong shift toward the red. However, at a certain stage of field intensity (something less than 10,000 volts per centimeter) the near-diffuse line at λ 4470 makes its appearance. As the field increases, this new line also broadens, chiefly toward the violet. When the two lines merge together, we should observe a single broad line slightly shifted toward the violet.

There is little information available concerning wave-lengths in stellar spectra having broad lines. Attention has been called from time to time, by various investigators, to systematic differences in radial velocities from certain individual lines, but the data are too sporadic to be of much use in this connection.

A preliminary survey of the field may be made by using the wave-lengths derived by S. Albrecht[1] from the Yerkes three-prism measures of E. B. Frost and W. S. Adams. These measures are doubtless the most accurate ones available for B-type stars. They refer to spectra with good and relatively narrow lines for which the expected Stark effect is small. Nevertheless, the lines are appreciably broadened even in these stars, and it is probable that the average intensity of the mol-electric fields in their reversing layers is of the order of 10^3 volts per centimeter. An inspection of the data collected by Stark[2] suggests that the blended line *He* 4472 for a field of 30,000 volts per centimeter is displaced toward the red by about 0.5 A. This holds true only if it is not blended with the near-diffuse line at λ 4470. In the stars under consideration there is no danger of such a blending. For a field of 10^3 volts per centimeter the red shift would amount to approximately +0.02 A, a quantity that should be easily detectable in the measurements of Frost and Adams. The line *He* 4388 will show a similar red shift. *He* 4437 should also be shifted toward the red,

[1] *Astrophysical Journal*, **67**, 305, 1928.

[2] *Handbuch der Experimental-Physik*, **21**, 399, 1927.

but by a smaller amount, and this refers also to the line He 4713, which, as belonging to the sharp series, is little influenced by the Stark effect.

Of the other lines measured by Frost and Adams, those due to O^+ and N^+ are not appreciably affected by the Stark effect. Neither would we expect an appreciable shift in the case of Mg^+ 4481 or of the various lines of Fe^+, Ti^+, or Si^{++}.

Albrecht's results show conclusively that the two He lines, λ 4388 and λ 4472, are actually displaced toward the red. The four best-determined wave-lengths of his table give the following residuals from the normal laboratory wave-lengths:

$$
\begin{array}{ll}
He & 4388+0.015 \text{ A} \\
He & 4472+ .019 \\
Mg^+ & 4481- .012 \\
Si^{++} & 4552-0.005
\end{array}
$$

The displacement of the two He lines against the other two well-determined lines is of the order of $+0.03$ or $+0.02$ A, which agrees well with the predictions.

I have no doubt that these displacements are real, and Albrecht also considers them so. There remains, of course, a possibility that the adopted laboratory wave-lengths are in error or that there are downward convective currents of helium in the atmospheres of the stars. These possibilities should be further investigated. Unless one of them is confirmed, the observed residuals can be attributed to the Stark effect.[1]

IX. WIDTHS OF BALMER LINES

In section II(f) we stated that the Stark effect increases in size for the higher members of each series. In the case of hydrogen, equation (5) shows that

$$\Delta\lambda \infty \lambda_o^2 f \infty W . \tag{23}$$

[1] He λ 4437 also shows a shift toward the red as predicted by the theory, but λ 4713 is shifted to the violet, while the theory demands a small red shift. I do not believe that this disagreement is fatal to the Stark effect. The weight assigned to this line by Albrecht is small (only 58 as compared with 500, for λ 4472). But even more important is the fact that λ 4713 lies outside the region of good focus on most of our plates. I have frequently had occasion to convince myself that the precision of measurement in this region is greatly reduced by the instrumental errors.

Substituting for f the values of Vasnecov (4, 10, 18, 28 for $H\alpha$, $H\beta$, $H\gamma$, and Hδ, respectively) and for λ_0 the wave-lengths of the Balmer lines, we should expect to find the following ratios in the line-widths:

$$W_\alpha : W_\beta : W_\gamma : W_\delta = 0.6 : 1.0 : 1.4 : 2.0 .$$

C. T. Elvey's recent measurements[1] show that in Sirius the line-widths actually increase from $H\beta$ to $H\epsilon$.[2] However, the rate of this increase[3] is much smaller than that predicted:

$$W_\beta : W_\gamma : W_\delta : W_\epsilon = 1.00 : 1.10 : 1.04 : 1.16$$

This is a serious obstacle in the way of an explanation by Stark effect. However, it is perhaps not fatal since the laboratory experiments show a similar discrepancy. Indeed, Vasnecov has tabulated Hulburt's line-widths for the H lines:

p	25	50	75	100	125 mm
$H\beta$.........	8 A	24 A	46 A	76 A	101 A
$H\gamma$.........	6	28	61	92	118
$H\delta$.........	3.5	20	60	93	105

There is surprisingly little difference between $H\beta$ and Hδ, and yet there can be no doubt as to the character of the broadening. It is possible that line-intensity, and consequently abundance-broadening, should be taken into consideration. However, surprising as this may seem, Elvey's results indicate that the central intensity increases from $H\beta$ to Hϵ.

The idea suggests itself that the insufficiency in the increase of line-width may be due to the superposition of a mol-collisional broadening.[4] This tends to give a line-width proportional to λ_0^2, and

---

[1] *Astrophysical Journal*, 68, 145, 1928.

[2] Elvey expresses his line-contours by the equation $I = I_0 e^{-k(\lambda - \lambda_0)}$. Accordingly, $W = \frac{1}{k} \log_e I_0 / I'$. Thus roughly $W \propto 1/k$. The value of k tabulated by Elvey shows a distinct tendency to decrease from $H\beta$ to $H\epsilon$; consequently, W increases.

[3] There is also a tendency for W to increase in the other A stars.

[4] The discrepancy between observation and theory would be removed if the coefficient of general absorption of the stellar atmosphere has a minimum value near $H\beta$, and increases toward the ultra-violet. In such a case the $H\beta$ line would originate in deeper and denser layers than the rest of the Balmer lines, and consequently the increase in width from $H\beta$ to $H\epsilon$ would be less pronounced than is indicated by formula

the result will be a partial reversal of the increase due to f in formula (23).

X. CONCLUSIONS

The various tests that we were able to apply to the hypothesis of Stark broadening in stellar spectra of early type seem, in general, to give satisfactory results.

The broadening observed in the various elements is in agreement with the requirement of the Stark effect: that the neutral atoms of low atomic weight should show wider lines than the heavier atoms.

The Stark broadening requires that stars with narrow lines of any one spectral subdivision should be more luminous than stars with broad lines. This agrees well with the observations.

The existence of lines forbidden by the selection principle of the quantum theory is suggested by the close coincidence of an unidentified line in B stars with the strongest near-diffuse line of He, at $\lambda 4470$.

The possibility that certain line-shifts noted by Albrecht are due to an unsymmetrical Stark broadening is in agreement with our expectations.

The only apparent disagreement is found in the amounts by which the line-widths of the H lines increase. The observed increase from $H\beta$ to $H\epsilon$ is smaller than that required by the computations. This may be due to superimposed effects of abundance-broadening and of mol-collisional broadening, in the sense suggested by Stark.

The luminosity effect and the possibility that forbidden lines are present, as well as the pronounced relationship with atomic number, indicate that the mechanism of broadening is dominated by the pressure. This is characteristic of the mol-electric effect but also of the mol-collisional broadening. We conclude that the broadening mechanism probably depends upon some sort of interaction between the various atoms and free electrons in the reversing layers of these early type stars.

December 12, 1928

---

(23). An increase of the absorption coefficient from $H\beta$ toward the red would produce a narrower $H\alpha$ line. Unsöld's contours for the solar H lines seem to indicate such an effect. At a point on the contour where $I = 10$ per cent absorption, the widths are approximately as follows: $W\alpha = 4.7$A; $W\beta = 5.4$A; $W\gamma = 5.6$A; $W\delta = 3.4$A. The possible variation of the absorption coefficient with λ was investigated by R. Lundblad, and by A. S. Eddington (*Der innere Aufbau der Sterne*, p. 410, 1928).

STELLAR ANALYSIS BY MEANS OF THE HYDROGEN LINES

Erika Böhm-Vitense

Astronomy Department
University of Washington

While Struve, as he himself pointed out, was not the first to suggest that Stark-effect broadening might be important for the line profiles of the hydrogen and helium lines, he was able to show in 1929 that the dependence of their profiles on the temperatures and luminosities of the stars agreed qualitatively with this hypothesis: the half-widths of the H and He lines decrease with increasing luminosities and increasing temperatures, and a line at $\lambda 4470$ becomes more prominent when the electron pressure in the stellar atmospheres increases. The position of this line agrees with a forbidden He I transition which can be expected to occur only if electric fields are present. Furthermore the dependence of the line broadening upon the molecular weight and the degree of ionization supported the hypothesis of Stark broadening.

However there remained some problems with the H lines. The extremely small Balmer decrement seemed to be in conflict with the Stark-effect theory. Since the integral $\int \kappa_L \, d\nu$ taken over the line (κ_L is the line absorption coefficient) should be independent of the amount of broadening, it was expected that the equivalent width should not change when Stark broadening becomes large. However Elvey and Struve (1930) realized that the equivalent width of the line does indeed depend on the amount of broadening and on the depth dependence of the line absorption coefficient. They pointed out that the wings of the lines are mainly formed in deep layers with high electron pressure, while the top layers would only produce a narrow line. On the basis of very rough calculations, they were able to show that the equivalent width of the strong H lines would increase with increasing Stark broadening. Since the broadening is larger for the higher members of the Balmer series, the Balmer decrement would indeed be flattened and may even change sign. Of course we realize, as Elvey and Struve probably also did, that not only the depth dependence of the line absorption coefficient but also the depth dependence of the continuous absorption coefficient and of the source function are important for the accurate determination of the line profile. It is interesting to note that Struve in 1929 also pointed out that the possibility of deviations from thermodynamic equilibrium should be kept in mind.

These basic investigations of the Balmer lines then removed the last serious difficulty with the interpretation of the line widths of the H and He lines as being due to Stark broadening, and we can now use the H or He line profiles to determine important parameters of stellar atmospheres as, for instance, temperature and/or gravity. In fact the theory of Stark broadening has become one of the most effective tools for stellar analysis. The H lines can be measured in essentially all stars and there the entire series is always available for study. While in early-type stars they provide a simple method for determination of the electron pressure—in order to obtain a mean electron pressure we have only to count the number of visible Balmer lines and the Inglis-Teller formula tells us the mean electron density—they are also a sensitive measure of the temperature in stars of later spectral types.

Whether they are also indicators of other stellar parameters as, for instance, the chemical composition remains to be investigated. Knowing the broadening mechanism for the H and He lines, it seems possible that accurate line profiles alone might give us essential information on the structure of the atmosphere. But before we can derive this information, many problems have to be investigated. Particularly:

107

1. The theory of molecular Stark-effect broadening has to be evaluated to high accuracy in order to obtain results accurate enough for modern requirements.

2. Other broadening mechanisms as, for instance, turbulence and abundance broadening that were also discussed by Struve have to be studied in detail.

3. The theory of line formation, including the problem of radiative energy transfer in the lines has to be investigated.

4. The temperature and pressure stratification of an atmosphere having specified abundances, effective temperature T_e, and gravity acceleration g has to be investigated.

I. THE THEORY OF MOLECULAR STARK BROADENING

The theory of Stark broadening is naturally divided into three parts.

A. The determination of the Stark splitting of the atomic levels for a given electric field F, and the computation of the transition probabilities for the radiative transitions between these levels.

B. The probability distribution for the electric field strength.

C. The relationship between the statistical and the collisional broadening theory for ions and electrons.

Topic A is a quantum-mechanical problem. The theory and results are given by Bethe and Salpeter (1957). Topic B was investigated by Holtsmark (1919), while corrections for interactions between ions and electrons were given by Mozer and Baranger (1960). The first review of topics B and C was that of Unsöld (1943). The main conclusions of his discussion are still valid except that opinion about the contribution of the electrons to the broadening of H and He lines has changed. According to the theory as discussed by Unsöld, the collisional broadening theory which mainly describes the finite wave train during the time when the radiating atom is undisturbed, i.e., the time between collisions, controls the central parts of the line profile. The statistical theory describes the disturbance itself and should be used for the wings. The transition point occurs at $\Delta\omega = (v^n/2\pi Cc_n{}^n)^{1/n-1}$ in circular frequency units, where v is the relative velocity of radiating atom and perturbing particle, c_n is a numerical constant, C depends upon the nature of the colliding particles, and $n = 2$ for linear Stark-effect broadening. Because of this fact that $\Delta\omega \propto v^2$, the quantity $\Delta\omega$ is much larger for electrons, and in that case the transition from collisional to statistical broadening takes place far out in the wings. Because of the fact that collisional broadening theory yields much smaller half-widths than does the statistical theory, it seemed that the broadening contribution by electrons could be neglected except in the wings of the higher members of the Balmer series. When Griem (1954) found that the theory did not fit the line profiles measured in laboratory arcs, it appeared however that in fact the electrons cannot be neglected.

A series of both experimental as well as theoretical investigations then followed, to clarify the influence of the electrons. The results of this work are given in Griem's book *Plasma Spectroscopy* (1964). Discussions of the problem have also been given by K. H. Böhm (1960), Traving (1959), and more recently by van Regemorter (1965), who presents a clear account of theoretical developments and also the comparison with laboratory results. For the Balmer lines Hβ and Hγ, there appears to be good agreement between measured and theoretical line profiles. For the higher members of the Balmer series, there seems still to be some uncertainty in the theoretical predictions. This is also true for temperatures less than 5000° and for electron densities $n_e < 10^{15}$ cm^{-3}, as was pointed out by Pfennig in Heidelberg (1966).

It is of great advantage that not only is the broadening theory for the Balmer lines now rather well understood, at least for a certain range of temperatures and densities, but that also the line profiles for the He I lines $\lambda 4713$ and $\lambda 3965$ can be represented fairly well, including the contribution of their forbidden components. These may be helpful in studying the atmospheres of early-type stars.

It is still an open question whether the profiles given by Edmonds, Schlüter, and Wells (1967) should be preferred to those given by Griem. The theoretical profiles of $H\beta$ and $H\gamma$ may safely be used as tools to this end. These profiles are shown by Griem (1964, pp. 448–449). A profile for the He I line $\lambda 3965$ may also be found in Griem's book (p. 450).

It is hoped that in the near future we shall have more accurate theoretical line profiles to work with.

II. OTHER BROADENING MECHANISMS

Basically two broadening mechanisms have to be distinguished: First, broadening of the absorption coefficient, as for instance by radiation damping, thermal Doppler broadening and so-called microturbulence Doppler broadening.

Second, broadening of the line profiles themselves, for instance by rotation and macroturbulence (subsection 2 below); a broadening of the line profile also occurs due to abundance broadening (subsection 3, below).

As is well known, the effect of radiation damping is negligible in comparison with Stark broadening for H lines (see for instance Mihalas 1965). Let us consider these points in detail.

1. Doppler Broadening of the Absorption Coefficient

Doppler broadening occurs on account of the thermal velocity field which is always present. There may also be an additional velocity field with a scale much smaller than the equivalent height, which is called microturbulence.

Thermal broadening is determined if the temperature can be estimated. For a temperature of $10,000°$, as in an A0 star, we find the Doppler width of a H line to be $\Delta\lambda_D = 4.3 \times 10^{-5} \lambda = 0.19$ Å for $H\gamma$. For an O star, we find about twice this value. Since the half-width due to Stark broadening is five to ten times as large, we need not be concerned about thermal Doppler broadening except for the central parts of the line where the final shape of the absorption coefficient can be obtained by folding the Stark profile into a Doppler profile.

Turbulence broadening is much more difficult to handle since we hardly know anything about the probability distribution of the velocity field, especially since for early-type stars the reasons for the occurrence of turbulence are not understood. Fortunately we may assume that microturbulence will most probably not occur with large Mach numbers, that is, with velocities much larger than thermal, and will therefore not influence the H line profiles seriously except perhaps in the central parts. We can estimate the influence of microturbulence from a curve of growth analysis of the metal lines. However the depth dependence will remain very uncertain. For large microturbulence, the central parts of the H lines can probably not be used for the analysis.

2. Broadening of the Line Profiles by Velocity Fields

A broadening of the line profile without a corresponding broadening of the absorption coefficient may occur on account of the presence of large-scale velocity fields, due to macroturbulence or rotational broadening. At each point of the star's surface the profile of the absorption coefficient is independent of these velocity fields and can be computed *a priori*. The final line profile is then obtained by superimposing the line profiles of the different points which are shifted with respect to each other in wavelength by amounts determined by the line of sight component of the velocity at the particular part of the surface.

For a rotating star with a given $v \sin i$, which can be determined from the profiles of weak metal lines, the velocity field and the line of sight components at different points are well known and the final line profile can easily be obtained.

Macroturbulence is more difficult to handle since again we do not know the velocity field. Observations of supergiants seem to show that macroturbulence may occur with rather large Mach numbers. In these cases we have to estimate the velocity field from the profiles of the metal lines. The uncertainties in the analysis of the H lines will be fairly large for large macroturbulence, as in early-type supergiants, if we cannot perform the analysis from the wings alone.

3. Saturation Broadening

It is now well-known from curve-of-growth studies that lines strong enough to appear on the flat part of the curve of growth are broadened because they become optically thick in the line wings. If $\kappa_L/\kappa_c \gg 1$ at a certain wavelength $\Delta\lambda$ from the line center, the intensity at that wavelength will approach the central intensity of the line no matter how far out in the wing of the absorption coefficient $\Delta\lambda$ is. Since H and He are the most abundant elements, the number of atoms absorbing in those lines is quite large and therefore saturation broadening is quite pronounced.

The radiative flux observed in the line $F_\lambda(\Delta\lambda)$ can be computed according to

$$F_\lambda(\Delta\lambda) = 2 \int_0^\infty S_\lambda(\tau_\lambda) K_2(\tau_\lambda) \, d\tau_\lambda \tag{1}$$

where

$$K_2(x) = \int_1^\infty \frac{e^{-xw}}{w^2} \, dw = x \int_x^\infty \frac{e^{-u}}{u^2} \, du$$

and where τ_λ is the optical depth in the line. If τ_L is the optical depth for the line absorption, τ_c is the optical depth for the continuous absorption, $\tau_\lambda = \tau_L + \tau_c$,

$$\tau_\lambda = \int_0^t (\kappa_L(\Delta\lambda) + \kappa_c) \, dt = \int_0^{\tau_c} \left(1 + \frac{\kappa_L(\Delta\lambda)}{\kappa_c} \right) d\tau_c, \tag{2}$$

where t is geometrical depth, and S_λ is the source function. It is a weighted mean of the source function for the continuum S_c and that for the line emission S_L, namely

$$S_\lambda = \frac{\kappa_L S_L + \kappa_c S_c}{\kappa_L + \kappa_c} = \frac{S_c + (\kappa_L/\kappa_c) S_L}{1 + \kappa_L/\kappa_c}. \tag{3}$$

In order to study the abundance broadening, let us assume for the moment that

$$S_L = S_c = S = a + b\tau. \tag{4}$$

Let us further assume that

$$\frac{d}{d\tau_c} \left(\frac{\kappa_L}{\kappa_c} \right) = 0. \tag{5}$$

We then find (see Unsöld [1955], p. 154)

$$\frac{F_c - F_\lambda}{F_c} = \frac{S(\tau_c = \tfrac{2}{3}) - S(\tau_\lambda = \tfrac{2}{3})}{S(\tau_c = \tfrac{2}{3})}, \tag{6}$$

where

$$\tau_\lambda = \frac{\kappa_c + \kappa_L}{\kappa_c} \tau_c, \tag{7}$$

or

$$\tau_c = \tau_\lambda - \Delta\tau \quad \text{with} \quad \Delta\tau = \tau_\lambda \cdot \frac{\kappa_L}{\kappa_L + \kappa_c}. \tag{8}$$

Since

$$S(\tau_\lambda = \tfrac{2}{3}) = S(\tau_c = \tfrac{2}{3} - \Delta\tau) = S(\tau_c = \tfrac{2}{3}) - \left[\Delta\tau \cdot \frac{dS}{d\tau_c}\right]_{\tau_c=2/3} \tag{9}$$

it follows that

$$\frac{F_c - F_\lambda}{F_c} = \left[\Delta\tau \frac{d \ln S}{d\tau_c}\right]_{\tau_c=2/3} = \left[\frac{2}{3}\frac{\kappa_L}{\kappa_c + \kappa_L}\frac{d \ln S}{d\tau_c}\right]_{\tau_c=2/3} \tag{10}$$

For $\kappa_L \ll \kappa_c$ everywhere in the line,

$$R = \frac{F_c - F_\lambda}{F_c} = \left[\frac{2}{3}\frac{\kappa_L}{\kappa_c}\frac{d \ln S}{d\tau_c}\right]_{\tau_c=2/3}, \tag{11}$$

and the line profile will resemble the profile of κ_L.

For $\kappa_L \geq \kappa_c$ we write

$$\kappa_L = \kappa_{L_0} f(\Delta\lambda), \tag{12}$$

where $f(0) = 1$ and from equation (10) we find

$$R = \frac{F_c - F_\lambda}{F_c} = \frac{2}{3}\frac{\kappa_{L_0} f}{\kappa_c + \kappa_r}\frac{d \ln S}{f \, d\tau_c}, \tag{13}$$

and

$$R_0 = R(\Delta\lambda = 0) = \frac{2}{3}\frac{\cdot}{\cdots_c + \kappa_{L_0}}\frac{d \ln S}{d\tau_c}. \tag{14}$$

The half width $\Delta\lambda(\tfrac{1}{2})$ can then be determined from

$$R(\Delta\lambda(\tfrac{1}{2})) = \tfrac{1}{2}R(0), \tag{15}$$

or

$$\frac{2}{3}\frac{\kappa_{L_0}f(\tfrac{1}{2})}{\kappa_c + \kappa_{L_0}f(\tfrac{1}{2})}\frac{d \ln S}{d\tau_c} = \frac{1}{3}\frac{\kappa_{L_0}}{\kappa_c + \kappa_{L_0}}\frac{d \ln S}{d\tau_c}. \tag{16}$$

From this we obtain

$$f(\tfrac{1}{2}) = \left(2 + \frac{\kappa_{L_0}}{\kappa_c}\right)^{-1}. \tag{17}$$

For a Doppler profile

$$f = \exp\left[-(\Delta\lambda/\Delta\lambda_D)^2\right]. \tag{18}$$

For a dispersion profile

$$f = \frac{1}{1 + (\Delta\lambda/A)^2}, \tag{19}$$

where

$$A = \lambda^2\gamma/4\pi c.$$

For a Stark broadened profile in the wings we obtain roughly

$$f = \frac{C F_0^{2.5}}{\Delta\lambda^m} \tag{20}$$

with $F_0 = 1.25 \times 10^{-9} n_e^{2/3}$ [c.g.s.] and $2 < m < 2.5$.

$$\text{For } H\beta: \ C = 35.7 \ R(n_e, T)$$
$$\text{For } H\gamma: \ C = 60.0 \ R(n_e, T)$$

where $R(n_e, T) \approx 0.5$ (Griem 1964, pp. 528, 92). If we assume $m = 2$ for a order-of-magnitude estimate and if we replace A^2 by $CF_0^{2.5}$, then the line profile for large $\Delta\lambda$ is analogous to (19). In Table 1 the half-widths of lines as a function of κ_{L_0}/κ_c as derived from equation (17) are given, as well as the ratios of the dispersion- to the Doppler half-widths obtained from

$$RW = \frac{\Delta\lambda(\tfrac{1}{2})_{\text{Dis.}}}{\Delta\lambda(\tfrac{1}{2})_{\text{Dop.}}} = \left(\frac{\Delta\lambda(\tfrac{1}{2})_{\text{Dis.}}}{A}\right) \Bigg/ \left(\frac{\Delta\lambda(\tfrac{1}{2})_{\text{Dop.}}}{\Delta\lambda_D}\right) \cdot \frac{A}{\Delta\lambda_D} \text{ for a line at 4300 Å}, \quad (21)$$

where

$$\frac{\Delta\lambda_D}{A} = \frac{4\pi\lambda v}{0.22} \tag{22}$$

if we take

$$\gamma = \gamma_{\text{cl.}} = 8\pi e^2/3mc\lambda^2,$$

and $v = 10^6$ cm/sec, corresponding to the thermal velocity of H at $T \approx 10^4\,^\circ$K (Unsöld 1955, p. 275).

It is obvious from Table 1 that even though the half-widths of a dispersion profile increase much more rapidly than for a Doppler profile, the half-widths of lines

TABLE 1

HALF-WIDTHS OF LINES AS A FUNCTION OF THE ABSORPTION
COEFFICIENT IN THE LINE CENTER

	κ_{L_0}/κ_c	$\ll 1$	10	10^2	10^4	10^8
Doppler profile	$\Delta\lambda(\tfrac{1}{2})/\Delta\lambda_D$	0.83	1.58	2.15	3.04	4.29
Dispersion profile	$\Delta\lambda(\tfrac{1}{2})/A$	1	3.32	10.05	10^2	10^4
RSt		1.6×10^2	2.78×10^2	6.19×10^2	4.11×10^3	3.09×10^5
RW		4.9×10^{-4}	8.6×10^{-4}	1.9×10^{-3}	1.35×10^{-2}	9.5×10^{-1}

with a Voigt profile (Doppler core and dispersion wings) will be determined by the Doppler core unless $\kappa_{L_0}/\kappa_c > 10^8$ and $\Delta\lambda(\tfrac{1}{2}) > 4.3\,\Delta\lambda_D$. For a Stark-broadened line we have to replace A by

$$B = (CF_0^{2.5})^{1/2}. \tag{23}$$

With $n_e = 10^{15}$, we obtain $F_0 = 1.25 \times 10^{-9}$ and for Hγ we find for

$$v = 10^6 \text{ cm/sec}, \frac{B}{\Delta\lambda_D} = 1.3 \times 10^2. \tag{24}$$

In Table 1 we also give

$$RSt = \frac{\Delta\lambda(\tfrac{1}{2})_{\text{St}}}{\Delta\lambda(\tfrac{1}{2})_{\text{Dop.}}} = \frac{\Delta\lambda(\tfrac{1}{2})_{\text{Dis.}}}{B} \frac{\Delta\lambda_D}{\Delta\lambda(\tfrac{1}{2})_{\text{Dop.}}} \frac{B}{\Delta\lambda_D} \tag{25}$$

For $n_e \geq 10^{14}$, the half-width for a Stark-broadened line is always determined by the Stark contour. Since

$$f(\tfrac{1}{2}) = \frac{CF_0^{2.5}}{\Delta\lambda(\tfrac{1}{2})^2}, \tag{26}$$

we find

$$\Delta\lambda(\tfrac{1}{2})^2 = CF_0^{2.5}f(\tfrac{1}{2})^{-1} = (2 + \kappa_{L_0}/\kappa_c)CF_0^{2.5} \tag{27}$$

while for $\kappa_{L_0}/\kappa_c \gg 2$,

$$\Delta\lambda(\tfrac{1}{2})^2 = \frac{\kappa_{L_0}}{\kappa_c} C F_0^{2.5}. \tag{28}$$

Since $\kappa_{L_0} \propto \dfrac{n}{F_0}$, where n is the number of atoms absorbing in the line,

$$\Delta\lambda(\tfrac{1}{2})^2 \propto \frac{n}{\kappa_c} C F_0^{1.5} \propto \frac{n}{\kappa_c} n_e. \tag{29}$$

The half widths for the H lines are determined by the product $\dfrac{n \cdot n_e}{\kappa_c}$. An increase of n/κ_c will have the same effect as an increase of n_e, so that the line profile can give us information only about the product $\dfrac{n \cdot n_e}{\kappa_c}$.

If κ_L/κ_c does depend on the optical depth, the line profile will give us information about a mean value of $\dfrac{n \cdot n_e}{\kappa_c}$ and possibly also about its depth dependence.

If the source function S is not a linear function of τ, we still can compute $F_\lambda(\Delta\lambda)$ from equation (1). There will be small correction terms required in equation (6) due to the higher-order derivatives of S with respect to τ but the line profile will still depend on $\dfrac{\kappa_{L_0}}{\kappa_c} \propto \dfrac{n \cdot n_e}{\kappa_c}$, and no separation of n and n_e will be possible.

III. THEORY OF LINE FORMATION

So far we have always made the assumption (eq. 4) that $S_L = S_c$. Let us now consider the situation if $S_L \neq S_c$. We shall however assume S_L to be independent of $\Delta\lambda$ within a line; i.e., we assume the line emission to be proportional to the line absorption coefficient κ_L. This is a good approximation for pressure-broadened lines. For simplicity, we still assume a linear depth-dependence of S_c and S_L, and that κ_L/κ_c is independent of depth.

We then have to replace equation (6) by

$$\frac{F_c - F_\lambda}{F_c} = \frac{S_c(\tau_c = \tfrac{2}{3}) - S(\tau_\lambda = \tfrac{2}{3})}{S_c(\tau_c = \tfrac{2}{3})}, \tag{30}$$

and while equations (7) and (8) remain the same, (9) can be rewritten for S:

$$S(\tau_\lambda = \tfrac{2}{3}) = S(\tau_c = \tfrac{2}{3} - \Delta\tau) = S(\tau_c = \tfrac{2}{3}) - \left[\Delta\tau \cdot \frac{dS}{d\tau_c}\right]_{\tau_c=2/3} \tag{31}$$

and

$$\begin{aligned}
\frac{F_c - F\lambda}{F_c} &= \frac{S_c(\tau_c = \tfrac{2}{3}) - S(\tau_c = \tfrac{2}{3}) + \left[\Delta\tau \dfrac{dS}{d\tau_c}\right]_{\tau_c=2/3}}{S_c(\tau_c = \tfrac{2}{3})} \\
&= \frac{\left[S_c - S\right]_{\tau_c=2/3} + \Delta\tau \dfrac{dS}{d\tau_c}\Big]_{\tau_c=2/3}}{S_c(\tau_c = \tfrac{2}{3})}
\end{aligned} \tag{32}$$

The influence of $S_L \neq S_c$ is easiest to examine in the wings for $\kappa_L \ll \kappa_c$. From equation (3) we get

$$S = S_c + \frac{\kappa_L}{\kappa_c}(S_L - S_c)$$

if we carry only first order terms in $\frac{\kappa_L}{\kappa_c}$. Equation (31) then becomes

$$S(\tau_\lambda = \tfrac{2}{3}) = S_c + \frac{\kappa_L}{\kappa_c}(S_L - S_c)_{\tau_c=2/3} - \Delta\tau\frac{dS_c}{d\tau_c} \tag{33}$$

and

$$R = \frac{F_c - F_\lambda}{F_c} = \Delta\tau\frac{d\ln S_c}{d\tau_c} - \frac{\kappa_L}{\kappa_c}\left(\frac{S_L - S_c}{S_c}\right) = \frac{\kappa_L}{\kappa_c}\left(\frac{d\ln S_c}{d\ln \tau_c} - \frac{S_L - S_c}{S_c}\right)_{\tau_c=2/3}. \tag{34}$$

The first term on the right-hand side is the same as in the case $S_L = S_c$, while the second term gives the correction for $S_L \neq S_c$. The greatest depth of the line is obtained for $S_L \leq 0$, namely

$$R = \frac{\kappa_L}{\kappa_c}\left(\frac{2}{3}\frac{d\ln S_c}{d\tau_c} + 1\right)_{\tau_c=2/3}. \tag{35}$$

Since $\frac{d\ln S_c}{d\ln \tau_c} \approx 1$, this corresponds to an increase in line depth by about a factor 2 at the most.

If on the other hand $S_L > S_c$, we obtain $R \approx 0$ for $S_L \approx 2S_c$ and get emission lines for $S_L > 2S_c$.

If $\kappa_L \gtrsim \kappa_c$, we can still study the special cases $S_L \to 0$ and $S_L \sim 2S_c$. For $S_L \to 0$ we find

$$S = S_c \cdot \frac{\kappa_c}{\kappa_L + \kappa_c}. \tag{36}$$

Inserting this in equation (34), we get

$$R = \frac{\kappa_L}{\kappa_L + \kappa_c}\left[1 + \frac{\kappa_c}{\kappa_L}\cdot\Delta\tau\cdot\frac{d\ln S_c}{d\tau_c}\right] = \frac{\kappa_L}{\kappa_L + \kappa_c}\left[1 + \frac{2}{3}\frac{\kappa_c}{\kappa_c + \kappa_L}\cdot\frac{d\ln S_c}{d\tau_c}\right]. \tag{37}$$

If $\kappa_L \ll \kappa_c$, we recover equation (35), while for $\kappa_L \gg \kappa_c$, we derive

$$R = 1 + \frac{2}{3}\frac{\kappa_c}{\kappa_L}\cdot\frac{d\ln S_c}{d\tau_c} \to 1, \tag{38}$$

that is, $F_\lambda = 0$ as would be expected. For $S_L = 2S_c$, we find

$$S = S_c\left[\frac{2\kappa_L + \kappa_c}{\kappa_L + \kappa_c}\right], \tag{39}$$

and again from equation (34)

$$R = \frac{2\kappa_L + \kappa_c}{\kappa_L + \kappa_c}\Delta\tau\frac{d\ln S_c}{d\tau_c} - \frac{\kappa_L}{\kappa_L + \kappa_c} = \Delta\tau\frac{d\ln S_c}{d\tau_c} + \frac{\kappa_L}{\kappa_L + \kappa_c}\left(\Delta\tau\frac{d\ln S_c}{d\tau_c} - 1\right). \tag{40}$$

for $\kappa_L \ll \kappa_c$, we of course obtain again

$$R = \frac{\kappa_L}{\kappa_c}\left(\Delta\tau\frac{d\ln S_c}{d\tau_c} - 1\right). \tag{41}$$

On the other hand for $\kappa_L \gg \kappa_c$ we obtain

$$R = 2\Delta\tau\frac{d\ln S_c}{d\tau_c} - 1, \tag{42}$$

while

$$F_\lambda = 2S_c(\tau_c = 0) \tag{43}$$

as is obvious.

From the central intensities of the Balmer lines in the solar spectrum we find that $S_L(\tau_\lambda = \tfrac{2}{3}) \approx B_v$ (3800°), which (Vitense 1953) corresponds to the low chromosphere, while from the center-to-limb variation in the continuum one finds that $S_C \approx B_v$ (4500°) for the temperature minimum (Böhm and Böhm-Vitense 1959). Here, B_v represents the Planck function. The difference between S_C and S_L will surely be much smaller for deeper layers, and will be zero for layers where the line radiation cannot escape. Since for a wavelength of 4500 Å, B_v (3800°) = 0.27 B_v (4500°), we find for the Balmer lines in the sun that

$$S_C \geq S_L \geq S_C/4.^1 \tag{44}$$

Thus the corrections in the line wings due to the fact that $S_L \neq S_C$ will be smaller, and probably much smaller, than a factor 2.

IV. THE TEMPERATURE AND PRESSURE STRATIFICATION OF THE ATMOSPHERE

According to present-day knowledge, the state of a stellar atmosphere is completely determined by a relatively few parameters: the radiative flux $\pi F = \sigma T_e^4$ wherein T_e is the effective temperature, the gravity g, and the chemical composition. For normal stars, other parameters such as the rotation or the magnetic field are regarded as of minor influence.

To determine these key parameters from the H lines, we have to calculate the line profiles according to equation (1). This means that we must know S as a function of the geometrical depth t or as a function of a mean optical depth $\bar{\tau}$. We must also know the wavelength-dependent optical depth in the line, τ_λ, as a function of optical depth in the continuum. In other words, we must know the line and continuous absorption coefficient as a function of the mean absorption coefficient. The source function S_λ and the absorption coefficient κ_λ can in principle be computed if the electron temperature T and the particle density n are known as a function of depth. The run of $n(\bar{\tau})$ and $T(\bar{\tau})$ are given by the hydrostatic (or possibly hydrodynamic) equilibrium and by the radiative (or radiative and convective) equilibrium, which combine the transfer equations for all wavelengths. Unfortunately we do not yet possess the mathematical tools to solve simultaneously an infinite number of transfer equations and an infinite number of statistical equilibrium equations, so certain approximations have to be introduced. One basic approximation that has already been used, although this may not always have been justified, is that $S_C = B_\lambda.^2$ For example, if the degree of ionization of the element responsible for the continuous absorption does not agree with the thermal equilibrium value from the Saha equation but is smaller by a factor C, then the source function also is smaller by this same factor (see for example, Thomas and Athay 1961). The particles responsible for the continuous absorption in most stellar atmospheres are H^- and H. It is still an open question how far out in the atmospheres these particles are ionized thermally.

[1] In the general case of course, S_L has to be determined from the equations of statistical equilibrium if the radiation field and the electron temperature are known. This however requires the simultaneous solution of the radiative transfer problem in the lines and in the continuum, which has not yet been accomplished.

[2] Cases where

$$S_c = \frac{\kappa}{\kappa + \sigma} B_\lambda + \frac{\sigma}{\kappa + \sigma} \int I_\lambda \frac{d\omega}{4\pi},$$

wherein κ is the absorption and σ the scattering coefficient, both being of significance, have also been treated (Traving 1955, Böhm and Deinzer 1965).

The second assumption that has been extensively used to compute $T(\bar{\tau})$ is that of a grey atmosphere: the assumption that the continuous absorption coefficient is independent of wavelength. Although this can be avoided, let us consider this case first in order to study the influence of $S_C \neq B_\lambda$. For a grey atmosphere, the condition of radiative equilibrium yields (Unsöld 1955, p. 129)

$$\mathfrak{J} = \int \mathfrak{J}_\nu \, d\nu = \int S_\nu \, d\nu = S, \tag{45}$$

and the solution of the transfer equation yields

$$\mathfrak{J}(\tau) = \tfrac{3}{4} F(\tau + q(\tau)) = S(\tau). \tag{46}$$

If

$$S(\tau) = B = \frac{\sigma}{\pi} T_{\mathrm{th}}{}^4, \tag{47}$$

then

$$T_{\mathrm{th}}{}^4 = \tfrac{3}{4} T_e{}^4(\tau + q(\tau)), \tag{48}$$

as is well known. If on the other hand

$$S(\tau) = \frac{1}{C} \cdot B(T_{N\mathrm{th}}) \tag{49}$$

then

$$T_{N\mathrm{th}}^4 = C \cdot \tfrac{3}{4} T_e{}^4(\tau + q(\tau)) = C \cdot T_{\mathrm{th}}{}^4 \tag{50}$$

If $C \gg 1$, then $T_{N\mathrm{th}} \gg T_{\mathrm{th}}$. This effect has been studied by Cayrel (1964) for the case of the sun and the ionization of H⁻.

The condition of radiative equilibrium determines the relation between $\mathfrak{J}$, s and τ. The relation between S and T is a matter of statistical equilibrium. However in a non-grey atmosphere the wavelength dependence of S_ν enters into the determination of S by means of the condition

$$\int \kappa_\nu \mathfrak{J}_\nu \, d\nu = \int \kappa_\nu S_\nu \, d\nu = \bar{\kappa}_s \cdot S \qquad \text{with} \qquad \bar{\kappa}_s = \frac{\int \kappa_\nu S_\nu \, d\nu}{\int S_\nu \, d\nu} . \tag{51}$$

If $\bar{\kappa}_s$ does not depend upon the wavelength dependence of S_ν (for instance, in a picket-fence model) the assumption $S_\nu = B_\nu$ will be of no influence upon the determination of $S(\tau)$ and we can consider $S(\tau)$ as being well known, at least as long as the influence of the lines upon the continuum can be neglected.

For our problem however we need to know S_ν itself, and the relation between S and S_ν depends on T. From equations (48) and (50) we obtain

$$T_{N\mathrm{th}} = C^{1/4} T_{\mathrm{th}}, \tag{52}$$

and the corresponding ratios of S_ν are

$$RS = \frac{S_{\nu\mathrm{th}}}{S_{\nu N\mathrm{th}}} = C \exp\left[-\frac{h\nu}{k}\left(\frac{1}{T_{\mathrm{th}}} - \frac{1}{T_{N\mathrm{th}}}\right)\right] = C \exp\left[-\frac{h\nu}{kT_{\mathrm{th}}}(1 - C^{-1/4})\right] \tag{53}$$

For $C = 2$ we find

$$RS = 2 \exp\left[-0.16 \, h\nu/kT_{\mathrm{th}}\right], \tag{54}$$

which for $T_{\text{th}} = 5040°$ yields $RS(\text{H}\beta) = 0.782$

and $\qquad\qquad\qquad\qquad\quad RS(\text{H}\alpha) = 1.00,$

while for $T_{\text{th}} = 10800°$ $\qquad RS(\text{H}\gamma) = 1.18,$

$\qquad\qquad\qquad\qquad\qquad\quad RS(\text{H}\beta) = 1.25,$

$\qquad\qquad\qquad\qquad\qquad\quad RS(\text{H}\alpha) = 1.41.$

For stars of solar type the uncertainties in S_{vc} will be smallest for Hα, while for A-type stars the continuum at Hγ or even higher members of the Balmer series will be less sensitive to deviations from thermodynamic equilibrium. Of course only $d \ln S_v/d\tau$ enters into the line profile. If C would remain constant over an interval $\Delta\tau = 1$ it would not do any harm, but if C should be $\neq 1$ in high layers it will probably change to $C = 1$ at $\tau \geq 1$.

So far we have not considered the influence of the lines upon S_c. The transfer- and radiative-equilibrium equations primarily force S_L down in those layers where the line becomes transparent. The effect is largest for the strongest lines which means it will mainly be effective in very high layers. If thermodynamic equilibrium is required, we make $S_c = S_L$ and therefore force S_c down by the same amount, yet the equation of radiative equilibrium $\int \kappa_v \mathfrak{J}_v \, dv = \int \kappa_v S_v \, dv$ does not tell us much about S_c since the continuum hardly contributes to these integrals if strong lines are present. Generally $S_c \neq S_L$ and the changes in S_c are completely determined by the interaction of S_L with S_c which may be very weak. If we consider for instance the Ca II K line in the sun, then a low S_L gives us low excitation of the upper level. This will have hardly any influence on the continuum, because neither the electron temperature nor the degree of ionization of H^- will be influenced. On the other hand if we consider Lα in an A star, then the low excitation of the second quantum level will influence the degree of ionization of H and therefore change S_c, especially in the Balmer continuum. There may also be other ways of interaction with the continuum, for example, through collisions, but generally the influence will not be so strong as to make $S_c = S_L$, so that usually we will find $S_L < S_c$.

The conclusion then is that the lines will influence the source function of the continuum in high layers, but the effect will be less than that computed on the assumption of thermodynamic equilibrium. This effect may quite often be completely masked by a small mechanical energy input into the high layers, for instance from a convection zone if one is present. If the atmosphere has to dispose of a small amount of excess energy, the increase of the source function has to be approximately (Unsöld 1955, p. 162)

$$\Delta S = -\frac{1}{4} \frac{dF(\bar{\tau})}{d\bar{\tau}}. \tag{55}$$

If there is for instance an energy input of only $0.001 \cdot F$ in a range $0.001 < \bar{\tau} < 0.002$, then

$$\frac{dF}{d\tau} \approx -F \qquad \text{and} \qquad \Delta S \approx \tfrac{1}{4}F$$

Since $S(0.001) \approx \tfrac{1}{2} \cdot F$, this means that $\Delta S \approx \tfrac{1}{2}S$.

This increase of S due to a small input of energy is well known in the solar chromosphere.

Because of this uncertainty in the amount of possible non-radiative energy sources and also because of the uncertainty of the influence of the lines, we cannot yet say much about S_c in the high layers. It may therefore be safer to consider only those parts of the lines for our analysis which are formed below $\tau_c = 0.1$, where these uncertainties are probably not serious, and where we can trust the computations made with the assumption $S_{vc} = B_\nu$ for the determination of S_c as long as there are no large inhomogeneities due to convective currents.

What alterations does convection introduce? We know that in deep layers it will cause the temperature gradient to flatten, but this will usually occur below $\tau = 2$ and therefore will not have a large influence on the H lines.

There is a small amount of convective energy transport πF_c also in the visible atmosphere which for the sun can be easily computed from

$$\pi F_c = \rho c_p T \bar{v} \frac{\Delta T}{T} . \tag{56}$$

With $v = 2$ km/sec, $\Delta T/T \approx 0.05$, $P_g \approx 10^5$ dynes cm^{-2}, we obtain $\pi F_c = 2.5 \times 10^9$ [cgs], so that F_c is about 4 percent of the radiative flux. This is a rather small amount and its influence on $S_c(\tau)$ will depend upon where it is finally dumped. If it is released, say between $0.1 < \tau < 1$, then $dF/d\tau \approx 0.04 \ F$ and $\Delta S \approx 0.01 \ F \approx 0.01 \ S$. Uncertainties of 1% we expect to encounter in any case. If the F_c is transferred into heat in higher layers, its influence might be serious as already discussed. In the region around $\tau = 1$ the most serious effect of convection are the inhomogeneities in the horizontal layers. Small scale circulations may also distort the radiative equilibrium stratification in the vertical direction. Center-to-limb variation of the solar continuum does not show a distortion of the radiative equilibrium stratification in the mean but the continuum is not very sensitive to it. The solar granulation clearly shows the inhomogeneities in the horizontal planes; we do not yet know the distribution of $\Delta T/T$. Since for convective stars the excitation of the second quantum level of H is very sensitive to T, this really introduces a great uncertainty in the interpretation of the Balmer lines. All we can do now is to use a two- or three-stream model which is rather unsatisfactory, but will give us some insight into the uncertainties encountered.

In summary, four major uncertainties remain to influence the source function of the continuum: the relation of S_ν/S for non-local thermodynamic equilibrium, the influence of S_L on S_c, the influence of the inhomogeneities due to convection as well as the influence of mechanical energy input into high layers.

V. THE LINE PROFILES

Extensive computations of H line profiles have been published by Mihalas (1965). (Helium line profiles have also been computed, but are not published.) They cover the range $0.101 < \theta_e < 0.7$, i.e., $7200° < T_e < 49900°$K, and $1 < \log g < 4.5$ for the lower temperatures while smaller ranges in $\log g$ are considered for the high temperatures. The computations were made with the assumption $S_\nu = B_\nu$ except that coherent electron scattering in the continuum has been allowed for (see footnote 2). The possible influence of the lines upon S_c has not been taken into account. The absorption contributions from the elements heavier than He have also been neglected. Within these limitations the models have been constructed with the best methods now available and certainly do provide a very helpful tool for the analysis of early-type spectra. According to equation (11) the line depth R in the wings is proportional to $(\kappa_L/\kappa_c)(d \ln S/d\tau)$. In the zero-order approximation $d \ln S/d\tau$ is independent of T_e and $\log g$, so $R \propto (\kappa_L/\kappa_c)$. Now $\kappa_L \propto n_e \cdot n(H) \exp (-\chi_2/kT)$ (χ_2 is the excitation energy for the second quantum level of hydrogen), while the continuous absorption is due to the Paschen continuum, which is

$$\kappa_c \propto n(\text{H}) \exp (-\chi_3/kT)$$

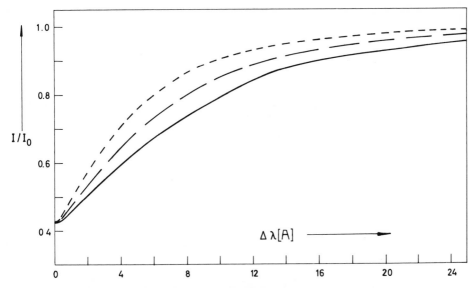

FIG. 1.—The dependence of the line profiles of Hγ on the gravity g for early-type stars with
$\theta = 0.32$
———— $\log g = 4.5$; —— —— $\log g = 4.0$; – – – – $\log g = 3.5$.

for hot stars as long as electron scattering is negligible. Therefore

$$R \propto \frac{\kappa_L}{\kappa_c} \propto \exp\left[(\chi_3 - \chi_2)/kT\right] \cdot n_e. \tag{57}$$

Under these circumstances the depth in the wings is a very weak indicator of the temperature but a strong indicator of the electron density n_e, which is determined by the pressure and is therefore roughly proportional to $g^{1/2}$ (see for instance Unsöld 1955 p. 206). This is demonstrated by Figure 1, for which the data have been taken from Mihalas (1965). It is the well-known luminosity effect discussed by Struve in Figure 2 of the accompanying paper, and shows that the dependence on T_e is very weak. If one would like to determine T_e by means of the H lines, an error in $\log g$ by a factor 2 would cause an error in θ by 0.04. Oke (1967) therefore uses the energy distribution in the continuum for the determination of T_e. However for very hot stars this method also becomes insensitive, and the intensities of the He II lines are probably a better indicator of T_e, as shown by Mihalas in his Figures 13 and 14. Mihalas' Figure 14 is reproduced here in our Figure 3.

If electron scattering becomes important, then

$$\frac{\kappa_L}{\kappa_c} \propto \frac{n_e \cdot n(H) \exp\left(-\chi_2/kT\right)}{n_e} = n(H) \exp\left(-\chi_2/kT\right), \tag{58}$$

and under most circumstances H is then mainly ionized, so the number density of hydrogen $n(H) \propto n_e^2 T^{-3/2} \exp\left(\chi_i/kT\right)$, where χ_i is the ionization energy of H, and

$$\frac{\kappa_L}{\kappa_c} \propto n_e^2 \, T^{-3/2} \exp\left[(\chi_i - \chi_2)/kT\right]. \tag{59}$$

Again, the line depths in the wings should be a sensitive $\log g$ indicator.

If however the continuous absorption coefficient is due to H$^-$, as in late-type stars,

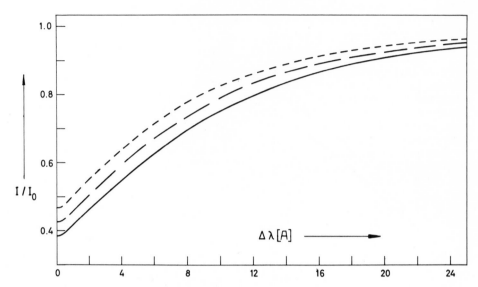

FIG. 2.—The weak dependence of the profile of Hγ on the effective temperature T_e for stars with high temperatures is shown; log $g = 4.5$.
———— $\theta = 0.36$; ———— $\theta = 0.32$; ———— $\theta = 0.28$.

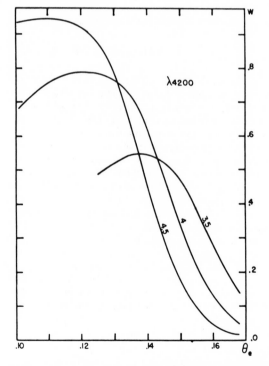

FIG. 3.—The dependence of the intensities of the HeII lines on the effective temperature T_e of the stars for a helium abundance of $n(\text{He})/n(\text{H}) = 0.15$.

then

$$\frac{\kappa_L}{\kappa_c} \propto \frac{n(\mathrm{H})n_e \exp{(-\chi_2/kT)}}{n(\mathrm{H})n_e} = \exp{(-\chi_2/kT)}. \tag{60}$$

Under these conditions the H lines are a sensitive indicator of the temperature, because for low T the exponential function depends strongly on T, but the H lines don't tell us anything about the pressure and therefore about log g. There is no luminosity effect except through the fact that for the same spectral type the temperatures are lower for a higher-luminosity star (Vitense 1951).

The temperature dependence of the hydrogen line profiles for the lowest temperature investigated by Mihalas is shown in Figure 4 while the log g independence is demonstrated in Figure 5.

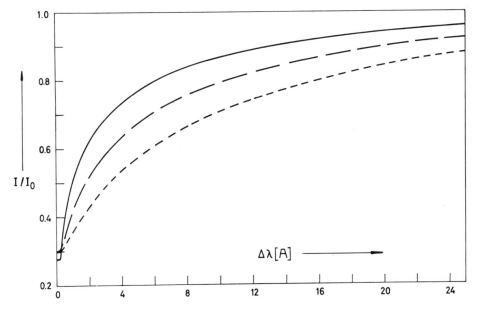

Fig. 4.—The dependence of the profile of Hγ on T_e for temperatures around 8000°K; log $g = 4.44$.
——— $\theta = 0.7$; ——— $\theta = 0.65$; ---- $\theta = 0.6$.

For temperatures around 10,000°, the H lines have their maximum intensity and are not very sensitive to anything. For slightly higher temperatures the hydrogen lines look the same on a certain curve $\theta(g)$ except for the central intensities, which are uncertain in any case, as is shown in Figure 6. So it is not possible to determine g or T without knowing one of them independently.

The influence of the He abundance on the H lines of hot stars has also been studied by Mihalas (1965). It is obvious that there is hardly any difference between the different profiles as long as the He abundance is not changed by a large factor. For lower temperatures the influence of the He abundance on the H lines has been studied by the author (Böhm-Vitense 1967).

In the determination of the H line profiles the He abundance enters in two ways: first it contributes to the continuous absorption coefficient, and second it influences the molecular weight μ and therefore the gas pressure P. When discussing the influence of the He abundance both effects have to be taken into account simultaneously. The

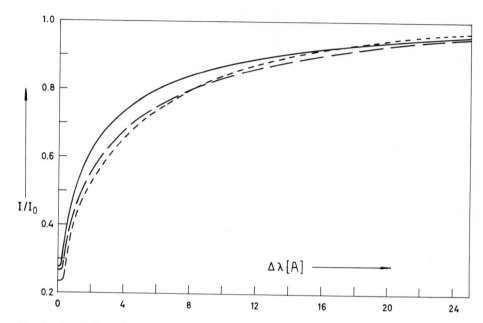

FIG. 5.—The independence of the profile of Hγ on the gravity g is demonstrated for $\theta = 0.7$.
————— $\log g = 4.44$; ————— $\log g = 3.0$; — — — — $\log g = 2.0$.

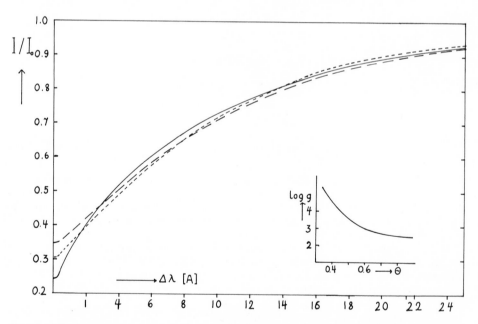

FIG. 6.—The approximate equality of the profiles of Hγ for values of θ and g lying on the curve
given in the lower right corner of the figure is shown.
————— $\theta = 0.6$, $\log g = 3.0$; ————— $\theta = 0.4$, $\log g = 4.5$; — — — — $\theta = 0.45$, $\log g = 4.0$.

pressure stratification is determined by

$$\frac{dP}{d\tau} = \frac{g \cdot \rho}{\kappa_c} = g \cdot \frac{P \cdot \mu}{\kappa_c \cdot R_g \cdot T} \tag{61}$$

where R_g is the gas constant. An increase in μ/κ_c acts in the same way as does an increase of g so long as the change of μ/κ_c is depth independent.

The influence of the He abundance on μ and κ_c will be different under different conditions. In the following we shall discuss different cases. We denote the ratio of the He abundance to H by Y:

$$Y = \frac{n(\mathrm{He}) + n(\mathrm{He^+}) + n(\mathrm{He^{++}})}{n(\mathrm{H}) + n(\mathrm{H^+})}. \tag{62}$$

We then have to distinguish between the cases when H is still quite abundant and determines κ_c, that is, when $Y \leq 1$, and the cases where κ_c is determined by He, e.g., $Y \gg 1$.

On the other hand, the situation will also be different for different degrees of ionization, that is, for different temperatures. The following Table demonstrates the situation. We compare the physical parameters at equal optical depth τ (equal temperatures) for atmospheres with different Y:

A. LOW TEMPERATURE

H is mainly neutral, $Y = \dfrac{n(\mathrm{He})}{n(\mathrm{H})}$, $n_e = n(\mathrm{H^+})$, $\mu = \dfrac{1 + 4Y}{1 + Y}$

Case 1	*Case 2*
$Y \leq 1$	$Y \gg 1$
$\kappa_c = \kappa_c(\mathrm{H^-})$	$\kappa_c = \kappa_c(\mathrm{He^-})$
$R \propto \dfrac{\kappa}{\kappa_c} \propto \dfrac{n(\mathrm{H})n_e}{n(\mathrm{H})n_e}$	$R \propto \dfrac{\kappa}{\kappa_c} \propto \dfrac{n(\mathrm{H})n_e}{n(\mathrm{He})n_e} \propto \dfrac{n(\mathrm{H})}{n(\mathrm{He})}$
i.e., independent of Y	$R \propto Y^{-1}$

B. HIGH TEMPERATURE

H is ionized, He is neutral.

$$n_e = n(\mathrm{H^+}), \ Y = \frac{n(\mathrm{He})}{n(\mathrm{H^+})}, \ P = n(\mathrm{H^+})(2 + Y), \ \mu = \frac{1 + 4Y}{2 + Y}$$

Case 3	*Case 4*
$Y \leq 1$	$Y \gg 1$
$\kappa_c = \kappa_c(\mathrm{H})$	$\kappa_c = \kappa_c(\mathrm{He})$
$\kappa_c \propto n(\mathrm{H}) \propto n_e^2 \propto \dfrac{P^2}{(2 + Y)^2}$	$\kappa_c \propto \dfrac{YP}{2 + Y}$
$\dfrac{dP}{d\tau} \propto \dfrac{P\mu}{\kappa_c} \propto \dfrac{P(1 + 4Y)}{(2 + Y)}\dfrac{(2 + Y)^2}{P^2}$	$\dfrac{dP}{d\tau} \propto \dfrac{P(1 + 4Y)(2 + Y)}{(2 + Y)YP} = \dfrac{(1 + 4Y)}{Y}$
$\quad = \dfrac{(1 + 4Y)(2 + Y)}{P}$	
	$P \propto \dfrac{1 + 4Y}{Y}$
$P^2 \propto (1 + 4Y)(2 + Y)$	
$\kappa_c \propto n_e^2 \propto \dfrac{1 + 4Y}{2 + Y}$	$\kappa_c \propto \dfrac{1 + 4Y}{2 + Y}$
$R \propto \dfrac{\kappa_L}{\kappa_c} \propto \dfrac{n(\mathrm{H})n_e}{n(\mathrm{H})} = n_e$	$R \propto \dfrac{\kappa_L}{\kappa_c} \propto \dfrac{n(\mathrm{H})n_e}{n(\mathrm{He})} = \dfrac{n(\mathrm{H})}{Y} \propto \dfrac{n_e^2}{Y}$
$R \propto \left(\dfrac{1 + 4Y}{(2 + Y)}\right)^{1/2}$	$R \propto \dfrac{(1 + 4Y)^2}{Y^3(2 + Y)^2} \approx \dfrac{1}{Y^3}$

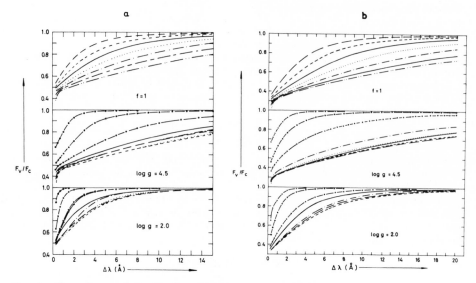

FIG. 7.—Dependence of the $H\gamma$ line profile on the hydrogen abundance $NH = 0.85/f$ for $\log g = 2.00$ and $\log g = 4.5$.
Top: Dependence of $H\gamma$ on $\log g$ for normal hydrogen abundance $(f = 1)$.
Left: $T_e = 12,900°$, right $T_e = 9,900°$.
Curves are as follows: Top: $\log g = 2.00$, –––––; $\log g = 2.5$, – – – –; $\log g = 3.0$, ———; $\log g = 3.5$, · · · · ·; $\log g = 4.0$, · – · – ·; $\log g = 4.5$, – – –; $\log g = 5.0$, – ·· – ·· –;
Center and bottom: $f = 1$, ———; $f = \sqrt{10}$, – – – –; $f = 10$, – – – –; $f = \sqrt{10^3}$, · · · · ·; $f = 10^2$, – · – · –; $f = 10^3$, – ◯ – ◯ – (for $\log g = 2.0$, $T_e = 12,900°$ for two different temperature stratifications in the top layers); $f = 10^4$, — ☐ — ☐ —; $f = 10^5$, –●●●.

From this we may conclude that:
For $Y \leq 1$, the H line intensities will increase with decreasing H abundance (if they change at all), due to the increase of the gas pressure P.
For $Y \gg 1$, the H line intensities will decrease rather steeply with decreasing H abundance.

This is illustrated in Figure 7. In the same Figure we show at the top also the dependence of the H line profile on $\log g$ for $Y = 0.176$. For case 1, we find R independent of g but $P \propto g^{2/3}$ and $n_e \propto g^{1/3}$. For case 3, we derive $P^2 \propto g$ and $R \propto n_e \propto P \propto g^{1/2}$.

In Table 2 the ratios of the gas pressures and electron densities for $Y = 1$ and 9 to those for $Y = 0.11$ are given. Also given is the ratio of gravities which would produce the same increase in P, n_e, and R.

It turns out that in the transition region between case 1 and case 3, $e.g.$, in the temperature range where H is partly ionized, the changes in the H line profiles are

TABLE 2

GAS PRESSURES AND ELECTRON DENSITIES FOR VARIOUS He ABUNDANCES

		1	9		1	9		1	9
	Y			Y			Y		
Case 1	P/P_0	2.8	18.1	n_e/n_{e0}	1.25	1.42	g/g_0	1.9	2.9
Case 3	P/P_0	2.2	6.0	n_e/n_{e0}	1.55	1.16	g/g_0	2.4	1.34

also the same for an increase in Y and an increase in g. (Böhm–Vitense 1967). There-fore, unfortunately there is no possibility to determine the He abundance by means of the H lines unless one knows g very accurately.

Different abundances of the metals will hardly affect the H lines. For low tempera-tures, when the metal content will influence the electron densities, the hydrogen lines do not depend upon n_e. For higher temperatures the electrons come from the H, so the metal content will not influence the electron density.

VI. CONCLUSIONS

From these discussions, it is clear that the Stark broadened wings of the hydrogen lines are a helpful tool to determine the electron densities for hot stars, where the opacity is due to H, while in cooler stars where the opacity is due to H⁻, they are good indicato: s of the temperature.

However, at the moment we have to keep in mind various uncertainties in the expected H line profiles. They originate mainly from lack of knowledge about the line source function S_L and the continuous source function S_c in high layers of the stellar atmosphere. In cooler stars, the H line wings are formed in rather deep layers, so these uncertainties are unimportant. Therefore the temperature determination for cool stars should be safe except for the influence of the inhomogeneities caused by convection.

For stars of spectral types A and earlier however, in which the hydrogen lines are formed in high layers, the total uncertainty of S_L and S_c will enter. An error of a factor 2 in the expected line depth yields an error in n_e by the same factor, and since $n_e \propto g^{1/2}$, this causes g to be in error by a factor 4. Even with this uncertainty we can still distinguish between main sequence stars, giants, and supergiants, but it prevents any further conclusions from being drawn from g, as for instance with respect to the mass or radius of the star. We hope that our knowledge of S_L and S_c can be improved in the near future.

The uncertainties of S_L and S_c will not be serious for the He lines, which are formed deeper. However the wings of the He lines are determined by both the electron density and the He abundance, which is still subject to some uncertainty. In addition the He lines are quite temperature sensitive. Also the theory of the Stark broadening of the He lines has not been developed as well as that for the H lines, but hopefully will be soon.

With respect to abundance determination from the H lines, we cannot expect to see differences in the metal abundance. However large He abundances may be detected in cooler stars with no visible He lines, from the H line profiles. Chances of this appear to be best for A stars, or in stars whose g-values are well known from other sources.

REFERENCES

Bethe, H., and Salpeter, E. 1957, *Quantum Mechanics of One- and Two-electron Atoms* (Berlin: Springer Verlag).
Böhm, K. H. 1960, *Stellar Atmospheres*, ed. J. L. Greenstein (Chicago: Univ. Chicago Press), p. 88.
Böhm, K. H. and Böhm-Vitense, E. 1959, *Zs. f. Ap.*, **50**, 69.
Böhm, K. H. and Deinzer, W. 1965, *Zs. f. Ap.*, **61**, 1.
Cayrel, R. 1964, *Smithson. Inst. Ap. Obs. Res. in Space Sci., Spec. Rep.*, No. 167, p. 169.
Edmonds, F. N., Jr., Schlüter, H., and Wells, D. C. 1967, *Mem. R.A.S.*, **71**, 271.
Griem, H. 1954, *Zs. Phys.*, **137**, 280.
———. 1964, *Plasma Spectroscopy* (New York: McGraw-Hill).
Holtsmark, J. 1919, *Phys. Zs.*, **20**, 162.
Inglis, D. R. and Teller, E. 1939, *Ap. J.*, **90**, 439.
Mihalas, D. 1965, *Ap. J. Suppl.*, **9**, 321.
Mozer, B. and Baranger, M. 1960, *Phys. Rev.*, **118**, 626.

Oke, J. B. 1967, *Aerodynamic Phenomena in Stellar Atmospheres* (IAU Symposium No. 28), ed.
 R. N. Thomas (New York: Academic Press), p. 179.
Pfennig, H. 1966, *J. Quant. Spectr. Rad. Transfer*, **6**, 549.
Regemorter, H. van. 1965, *Ann. Rev. Astr. Ap.*, **3**, 71.
Thomas, R. W. and Athay, R. G. 1961, *Physics of the Solar Chromosphere* (New York: Interscience).
Traving, G. 1955, *Zs. f. Ap.*, **36**, 1.
————. 1959, *Mitteil. d. Astr. Gesell.*, Sond. No. 1 (Karlsruhe: Braun).
Unsöld, A. 1943, *Viertel. d. Astr. Gesell.*, **78**, 213.
————. 1955, *Physik der Sternatmosphären* (Berlin: Springer Verlag).
Vitense, E. 1951, *Zs. f. Ap.*, **29**, 73.
————. 1954, *Zs. f. Ap.*, **34**, 209.

5. SHELL SPECTRA

O. STRUVE and K. WURM: The Excitation of Absorption Lines in Outer Atmospheric Shells of Stars. *Astrophysical Journal*, **88**, 84, 1938.

A. B. UNDERHILL: Dilution Effects in Extended Atmospheres.

THE EXCITATION OF ABSORPTION LINES IN OUTER ATMOSPHERIC SHELLS OF STARS

O. STRUVE AND K. WURM

ABSTRACT

I. The *He* I lines show abnormal intensities in the outer shells of peculiar B stars. In ζ Tauri and φ Persei *He* I λ 3965 (2^1S-4^1P) is narrow and sharp, while all lines arising from the levels 2^1P and 2^3P are broad and diffuse. This is attributed to the dilution of the radiation from the B star in the shell. The anomalous intensities of *He* I are investigated in the entire sequence of B stars having outer shells.

II. The theory of the excitation of absorption lines in a field of diluted radiation is developed and applied to six levels of *He* I: 1^1S, 2^1S, 2^1P, 2^3S, 2^3P, and the state of ionization. The populations in the various states have been computed for dilution factors 0.1, 0.02, and 0.01, and for temperatures of 10^4 and 2.5×10^4 degrees.

III. The theoretical populations for a dilution factor of the order of 0.01 agree with the observations. Accordingly, the radius of the shell is approximately five times the radius of the star. The entire *He* I anomaly is accounted for by the theory.

IV. The absence of rotational broadening in lines produced in the shells shows that these shells do not rotate as solid bodies with the photospheres. The latter are rotating rapidly in most of the stars investigated. Some lines, such as *Si* II in ζ Tauri, show broadening intermediate between the normal *He* I lines in the deepest layers and the λ 3965 line of *He* I in the shell. Hence, the elements are stratified. The ionization decreases outward in all stars investigated except β Lyrae. The strength of the *H* cores in the shells is explained by the metastability of the 2S-level. The weakness of *Mg* II 4481 is explained by the fact that the lower level has strong transitions downward.

V. The *He* I anomaly in ordinary giants and dwarfs of class B bears a strong resemblance to the anomaly in the shells. The explanation is doubtless similar, but there are complications due to Stark effect, collisional damping, etc.

I. THE HELIUM ANOMALY IN THE OUTER SHELLS OF PECULIAR B STARS

ζ Tauri.—This star is representative of a small group of B stars which have extremely sharp and deep absorption lines of hydrogen, between symmetrically spaced emission components of relatively low intensity. These emission components are conspicuous in *Hα* and *Hβ*; they are weak in *Hγ* and are absent in *Hδ* as well as in all higher members of the Balmer series. The hydrogen lines show also fairly conspicuous broad absorption wings, suggestive of the normal Stark effect in early-type stars. The *He* I lines λλ 4472, 4388, 4144, 4121, 4026, and 4009 are broad and very diffuse. They suggest rapid axial rotation—of the order of 200 km/sec at the equator, for $v_{rot} \sin i$.

The star is a well-known spectroscopic binary, having been discovered to be such by Frost and Adams[1] in 1903. The period is 138 days according to Adams,[2] or 133 days according to Miss Losh.[3] The total range in velocity is 22 km/sec, but the results from different lines are not entirely consistent.[3] Miss Losh concludes: ". . . . the velocity or wave length variations of ζ Tauri can not be explained fully by orbital motion alone. While these curves undoubtedly result largely from orbital motion, clearly there are other factors involved which can not now be identified. . . . ."

The rotational explanation of the great widths of the He I lines listed above is so well supported by their contours that it can be safely trusted in our discussion, despite the fact that the absorption lines of Fe II, Ni II, Si II, and the cores of the Balmer lines are very much narrower. In order to strengthen this conclusion we have examined the spectra of all B-type stars for which spectrograms are available at the Yerkes Observatory. In all normal stars—several hundred in number—the dish-shaped rotational contours are observed in all lines of all elements. This rule is violated only in a small group of stars—all of peculiar spectrum—which have emission lines of hydrogen, narrow absorption cores between the double emission components of hydrogen, narrow lines of Fe II, which sometimes lie between double emission components, strong narrow lines of Ni II and weak ones of Si II, occasionally very faint narrow Mg II 4481, but always broad and diffuse lines of He I, λλ 4472, 4388, 4026, 4009, etc. The best-known representatives of this group of stars are ζ Tauri, φ Persei, 48 Librae, ε Capricorni, o Aquarii, +47°3985, and β Monoceratis A.

Since the presence of emission lines demonstrates the existence of a large outer shell, the simultaneous appearance of narrow absorption lines of Fe II, Ni II, H, etc., with broad He I lines raises no serious difficulty in regard to the rotational interpretation of the broad He I lines. The broad lines are believed to come from the true reversing layer of the star, while the narrow lines originate in the outer shells.

However, the spectrum of ζ Tauri reveals a phenomenon of very

[1] *Ap. J.*, **17**, 151, 1903.

[2] *Ibid.*, **22**, 115, 1905.   [3] *Pub. Obs. U. Michigan*, **4**, 21, 1932.

great interest. While all other He I lines which we are able to observe on our plates between λ 3900 and λ 5000 are rotationally broadened, the line He I 3965 is exceedingly narrow and sharp (Pl. II). Since we cannot reasonably question the rotational theory of the broad He I lines, we are compelled to attribute the origin of the sharp λ 3965 to the same outer shell which gives rise to the narrow cores of the hydrogen lines.

A clue to the explanation of this phenomenon is found in the metastability of the 2^1S-level of the He I atom from which the line λ 3965 arises. It is due to the transition 2^1S $-$ 4^1P. An inspection of the Grotrian diagram for He I shows that the only other member of the series 2^1S $-$ n^1P which can be observed with our spectrograph, λ 5016, is in the green part of the spectrum, where our photographic emulsion was not sensitive. It is, however, quite certain from evidence supplied by other stars that all members of the series 2^1S $-$ n^1P behave in substantially the same manner as λ 3965.

It is evident that the relatively greater strength of λ 3965 in the shell than in the normal reversing layer is caused by the effect of dilution of radiant energy. The interpretation is thus substantially similar to that which Bowen has advanced for the explanation of large populations in metastable levels in gaseous nebulae. In the shell the dilution of the radiation is quite pronounced and the Rosseland cycle goes into operation. If the radiation is sufficiently diluted, the atoms accumulate in the metastable level 2^1S. Hence, in the shell λ 3965 is strong. In the star proper, where rotation is large, λ 3965 is weak because we have essentially conditions of thermodynamic equilibrium.

The spectrum of He I possesses another metastable level, namely, 2^3S. Of the series 2^3S $-$ n^3P, only λ 3889 falls in the region which can be photographed with our spectrograph, and this line is hopelessly blended with $H\zeta$. It is, however, probable from evidence supplied by other stars (β Lyrae) that this line behaves substantially as does λ 3965. A detailed study is impossible.[4]

The He I series (2^1P $-$ n^1S) and (2^1P $-$ n^1D) contain many observable lines. In ζ Tauri all of them belong to the reversing layer

---

[4] It is suggestive that O. C. Wilson has found He I 3889 as a moderately strong absorption line in the Orion nebula (*Pub. A.S.P.*, **49**, 338, 1937).

132

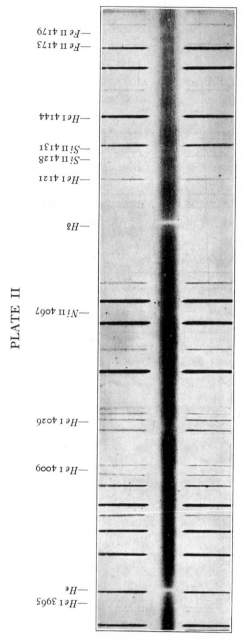

PLATE II

The Spectrum of ζ Tauri

(1937 Sept. 21; 11ʰ03ᵐ G.C.T.)

The *He* I line λ 3965 is sharp and narrow, while the other lines of *He* I are very broad and diffuse

of the star, not to the shell. The triplet series $(2^3P - n^3S)$ and $(2^3P - n^3D)$ in ζ Tauri also belong to the reversing layer.

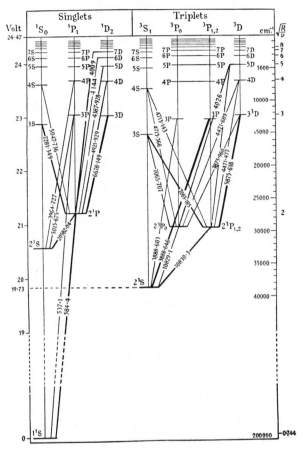

FIG. 1.—Grotrian diagram of *He* I

It is obvious that the population of the level 2^1S is greatly increased in the shell with respect to the levels 2^1P and 2^3P. How much this increase amounts to cannot be ascertained from ζ Tauri alone, since the shell shows no lines originating from levels 2^1P and 2^3P, while the reversing layer shows no line originating from 2^1S. In the following discussion we shall supply additional information from other stars.

The unusual interest of this problem is enhanced by the fact that the metastable $He\,\textsc{i}$ levels lie very high: 2^1S at 20.5 volts and 2^3S at 19.7 volts. The ionization potential of $He\,\textsc{i}$ is 24.5 volts. In this respect the case under consideration differs materially from Bowen's case of the nebular lines of $[O\,\textsc{iii}]$: there the metastable levels are close to the ground-level, and their populations are largely produced by collisions.

If we consider in our He problem only three states, namely, the ground state, the metastable state, and a third state which combines directly with the ground state (this may be the 2^1P-level or the state of ionization), then Rosseland's formula may be used to compute the relative populations. Disregarding for the moment the statistical weights, we have

$$\frac{n_2}{n_1} = e^{-h\nu_{12}/kT}\left[\frac{1 - (1 - W)e^{-h\nu_{23}/kT}}{1 - (1 - W)e^{-h\nu_{13}/kT}}\right],$$

where W is the dilution factor $R^2/4r^2$.

This formula is exact. If the metastable-level is close to the ground-level, $\nu_{23} \approx \nu_{13}$ and the expression in square brackets drops out. We then have pure Boltzmann distribution for n_2. In our case, $h\nu_{23} \neq h\nu_{13}$, and the expression in brackets must be retained. However, for moderately large dilutions, W is so close to zero that it affects the expression very little. In our case, for $T = 20,000°$ we have approximately

$$\left[\frac{1 - (1 - W)e^{-h\nu_{23}/kT}}{1 - (1 - W)e^{-h\nu_{13}/kT}}\right] = 0.68 .$$

On the other hand, the ratio

$$\frac{n_3}{n_1} = We^{-h\nu_{13}/kT}\left[\frac{1}{1 - (1 - W)e^{-h\nu_{13}/kT}}\right] \approx \frac{1}{0.999}We^{-h\nu_{13}/kT} .$$

It is obvious that if we can obtain from the observations satisfactory estimates of n_3/n_1 and n_2/n_1, we should, in principle, be able to determine W. In practice we shall use the following observed quantities

$$\frac{\text{Int. 3965 (shell)}}{\text{Int. 3965 (star)}} \cdot \frac{\text{Int. 4009 (star)}}{\text{Int. 4009 (shell)}} .$$

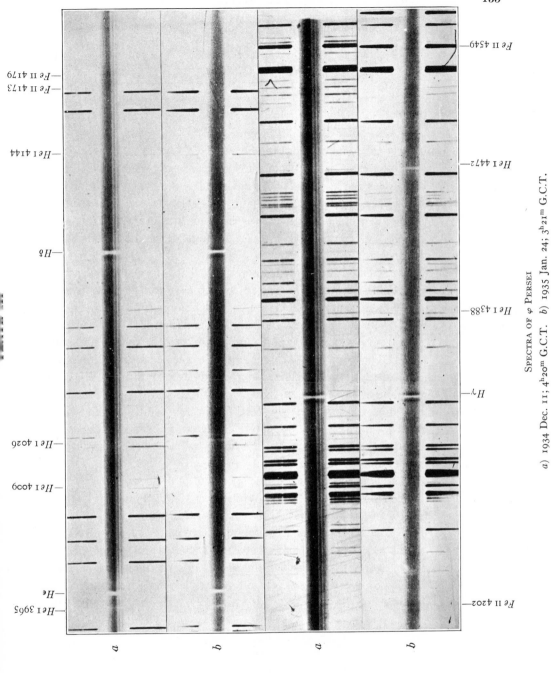

SPECTRA OF φ PERSEI

a) 1934 Dec. 11; 4ʰ20ᵐ G.C.T. *b)* 1935 Jan. 24; 3ʰ21ᵐ G.C.T.

Within the precision of the three-cycle theory this quantity should be equal to (disregarding the statistical weights)

$$\frac{n_0 \text{ (shell)} \, 0.68 \, e^{-h\nu_{12}/kT}}{n_0 \text{ (star)} \, e^{-h\nu_{12}/kT}} \times \frac{n_0 \text{ (star)} \, e^{-h\nu_{13}/kT}}{n_0 \text{ (shell)} \, \dfrac{1}{0.999} \, W e^{-\frac{h\nu_{13}}{kT}}}.$$

It is clear that all values of n_0 and $e^{-h\nu/kT}$ drop out, leaving only a numerical constant close to one and $1/W$.

However, in view of the fact that the three-cycle theory of Rosseland does not allow us to investigate the simultaneous behavior of the singlets and triplets, we have, in Section II, analyzed the corresponding problem of six states: the ground-level 1^1S, the singlet levels 2^1S and 2^1P, the triplet levels 2^3S and 2^3P, and the state of ionization.

φ Persei.—A star which resembles ζ Tauri to a surprising degree is φ Persei. It, too, is a spectroscopic binary of fairly long period—126.6 days. The spectral features, which are variable, have been described by several astronomers.[5] The emission lines of H and Fe II are stronger than in ζ Tauri, and so are also the central absorption cores. The He I lines λλ 4472, 4388, 4144, 4026, and 4009 are almost always exceedingly broad and diffuse, suggesting an equatorial component in the line of sight of the rotational velocity amounting to about 250 km/sec. The He I line λ 3965 is strong and sharp—as in ζ Tauri. It may possibly be superposed over a background of a broad and diffuse line, but this line is too faint to be identified with certainty.

A striking peculiarity of the spectrum of φ Persei is the variation of the He I absorption lines. While λ 3965 is always sharp, the other He I lines are broad and diffuse, except at certain phases—especially from 20 to 40 days after maximum positive velocity.[6] Plate III shows at the top (*a*) the normal spectrum, at the bottom (*b*) the spectrum taken during the interval of phase stated in the foregoing. The He I lines in (*b*) show sharp cores for the triplets λλ 4472 and 4026 but no cores for the singlets λλ 4388 and 4009. The singlets remain per-

---

[5] See H. F. Schiefer, *Ap. J.*, **84**, 568, 1936, where complete references to other papers are given. See also Struve and Swings, *ibid.*, **75**, 161, 1932.

[6] This feature was detected by F. C. Jordan, *Pub. Alleghany Obs.*, **3**, 34, 1913.

fectly normal while the triplets develop cores which are superposed over the broad rotational lines. At the same time the singlet λ 3965 is strengthened.

It is not yet possible to advance a complete physical interpretation of the spectrum of φ Persei. But the phenomenon resembles in a striking manner the variation of the He I lines in μ Sagittarii discovered by Morgan.[7] There, too, the strengthening of the lines is limited to the He I triplets, and the entire phenomenon takes place within a short interval of time, on the descending branch of the velocity curve, not far from the place where an eclipse would occur if the inclination were suitable.

There is also a distinct resemblance between the phenomena in these two stars and the spectroscopic features of ζ Aurigae, ϵ Aurigae, and VV Cephei immediately preceding and following eclipse. It would be profitable to pursue this matter and to investigate whether eclipses by the atmospheres of spectroscopic components are really responsible for the phenomenon.[8]

It is sufficient for our present purpose to assume, as seems reasonable, that in the phase interval of 20–40 days after maximum the amount of He I gas in a field of diluted radiation is greater than normal. We conclude that although λ 3965 ($2^1S - 4^1P$) is, among the observable lines, the one that is most enhanced in the presence of diluted radiation, the triplet lines λ 4472 ($2^3P - 4^3D$) and λ 4026 ($2^3P - 5^3D$) are also enhanced relative to the singlet lines λ 4388 ($2^1P - 5^1D$), λ 4144 ($2^1P - 6^1D$), and λ 4009 ($2^1P - 7^1D$).

The other triplet lines are not suitable for a comparison; λ 4121 ($2^3P - 5^3S$) may be present, but it is very weak. The normal spectrum of φ Persei reveals, upon close inspection, an exceedingly faint core in λ 4026 and perhaps also in λ 4472. This would, if real, show that the phenomenon is naturally present in the outer shell.

Other stars having outer shells.—It is evident that the spectral type of ζ Tauri and φ Persei is composite. Normal B stars do not show the absorption lines of Ni II and Fe II, except in the latest

[7] $Ap. J.$, **75**, 407, 1932.

[8] We are indebted to Dr. W. W. Morgan for having called our attention to this possibility in the case of μ Sagittarii. At his suggestion Dr. C. T. Elvey is now observing μ Sagittarii with the photoelectric photometer at the McDonald Observatory.

138

PLATE IV

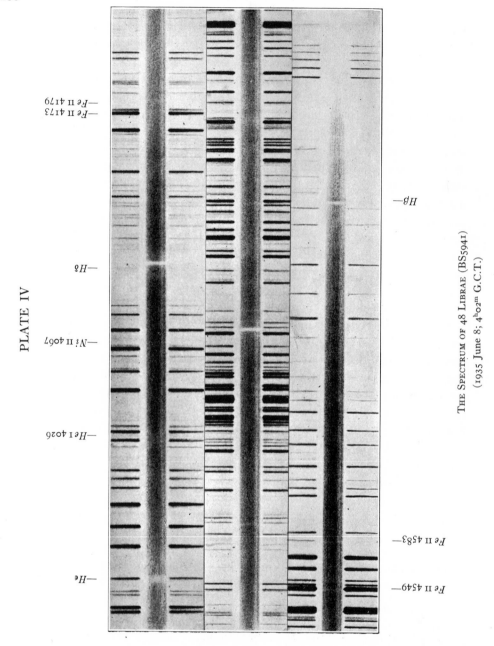

—Fe II 4179
—Fe II 4173

—Hβ

—Hδ

—Ni II 4067

—He I 4026

Fe II 4583—

—He

Fe II 4549—

THE SPECTRUM OF 48 LIBRAE (BS5941)
(1935 June 8; 4ʰ02ᵐ G.C.T.)

subdivisions where the *He* I lines are weak and the Stark wings of the *H* lines are strong. For purposes of classification it will be best to treat separately the normal rotational reversing layer and the outer shell. Although such a division is somewhat arbitrary—the two types of structure doubtless merge into one another—the classification becomes more consistent. The spectral type of φ Persei is given as B9pe by Schiefer,[5] and that of ζ Tauri is given as B3p in the *Henry Draper Catalogue*. From the fairly large intensities of the rotational *He* I lines in ζ Tauri it would appear that the spectral class is about B3. In φ Persei the rotational *He* I lines are somewhat weaker than in ζ Tauri, but the characteristic stellar lines of the early subdivisions of class B, such as *Si* III, *Si* IV, *O* II, etc., are not visible, perhaps because of the large rotational broadening. Neither is there any indication of a broad *Mg* II 4481, which should be present if the spectrum were B5 or later. The Stark wings of *H* are fairly well pronounced, however, and it seems reasonable to attribute class B3 to the reversing layer of φ Persei.

It will be of interest to consider several other stars of similar character. ϵ Capricorni[9] (Pl. IV) has strong rotational lines of *He* I, variable sharp lines of *Fe* II in absorption, and strong cores of *H* superposed over fairly strong Stark wings. *Mg* II 4481 is present as a faint absorption line of intermediate width, probably variable in intensity. The spectral class from the broad *He* I lines may be B5. The line *He* I 3965 is not visible, but our plates are not quite conclusive in this respect. It is fairly certain, however, that λ 3965 is not as strong as in φ Persei. The spectral type of the shell is definitely later than that of φ Persei.

The spectrum of 48 Librae resembles that of ϵ Capricorni. The broad *He* I lines are very weak, while the sharp *Fe* II lines are quite strong. There is probably a very weak broad line of *Mg* II 4481 (but its width seems to be less than that of *He* I 4472). The *Ni* II line λ 4067 is sharp. The *Si* II lines 4128.0 and 4131.9 are moderately sharp and very faint. The wings of the *H* lines are faint. There is a slight suggestion of a faint, narrow line of *He* I 3965, but

[9] *Ap. J.*, **75**, 176, 1932. Numerous other stars of similar character are known from the work of Merrill, Harper, and others. We shall confine ourselves here to a few typical stars for which spectrograms have been secured at the Yerkes Observatory.

the evidence is not conclusive. The spectral class of the reversing layer is about B5.

The sequence of peculiar stars with large outer shells continues to cooler objects such as 17 Leporis[10] and the I component of ε Aurigae.[11] The former possesses a rotationally broadened line of *Mg* II 4481 but fails to show the lines of *He* I. The wings of the *H* lines are strong. Accordingly, the spectral type of the reversing layer is A. The I component of ε Aurigae shows the spectrum of its outer shell several months prior to the first contact of the eclipse and a similar length of time following the last contact of the eclipse. Its spectral type is F.

On the hot side of ζ Tauri our information concerning the absorption spectra of shells is limited to the B5 component of β Lyrae[12] and the expanding shell of P Cygni.[13] The latter shows only the spectrum of the shell. There is no indication of Stark wings of *H*, or of normal, broadened *He* I lines. The so-called B5 component of β Lyrae is produced by absorption in the expanding shell which surrounds the binary system composed of the strong B8 component and the hypothetical A component. The expanding shells of both stars show very strong *He* I 3965 and strong triplet *He* I lines; the singlet lines λλ 4009, 4144, and 4388 are also present, and are moderately strong in absorption but are weak in emission. Both shells correspond to much earlier types than the shell of ζ Tauri. Hence, it would appear that at least on the hot side of the sequence the normal *He* I singlet level 2^1P is fairly well populated, although there is a deficiency as compared with the populations of the corresponding triplet levels.

Summarizing our observations concerning the *He* lines, we establish the following properties:

1. In the cooler stars, λ 3965 ($2^1S - 4^1P$) is the first line to appear in the shell.

2. The triplet lines ($2^3P - n^3D$) appear in the shell when (ζ Tauri

---

[10] *Ap. J.*, **76**, 85, 1932.   [11] *Ibid.*, **86**, 570, 1937.

[12] Struve, *Observatory*, **57**, 265, 1934; Baxandall-Stratton, *Annals Solar Phys. Obs.* (Cambridge), **2**, Part 1, 1930.

[13] *Ap. J.*, **81**, 66, 1935.

and abnormal stage of φ Persei) the number of atoms is large, but for the cooler stars λ 3965 is stronger than λ 4026.

3. When the excitation of the shell is larger, so that Fe II no longer appears in absorption, the singlet lines ($2^1P - n^1D$) begin to appear faintly in the spectrum of the shell. In β Lyrae[14] the estimated intensities of three lines are: λ 3965, 10; λ 4009, 2; λ 4026, 8. In a normal B5 giant, 67 Ophiuchi, the corresponding intensities are: λ 3965, 8; λ 4009, 9; λ 4026, 12. In the shell of β Lyrae, λ 3965 remains the strongest of the three lines.

4. When the excitation in the shell is very high, the triplet lines become relatively stronger and λ 4026 surpasses λ 3965. The intensities in P Cygni[15] are: λ 3965, 10; λ 4009, 8; λ 4026, 15. Although the singlet lines ($2^1P - n^1D$) remain considerably weaker than the triplets ($2^3P - n^3D$), the difference is less striking than in shells of lower excitation.

II. POPULATIONS OF THE SINGLET AND TRIPLET LEVELS

In regard to the variation of the absorption lines from star to star, as described in the preceding part of this paper, it is quite reasonable to seek the causes in a variation of the relative populations of the states involved. We may expect that in passing within the absorbing layer to greater distances from the photospheric region there will be more or less marked deviations from thermodynamical equilibrium, which may give a distribution of the atoms over the various states that is very different from a Boltzmann distribution for any temperature. As is well known, there exists no adequate theoretical formula for such conditions by means of which we can compute the ionization and excitation. It is not to be expected that such a general formula can be derived. It is not difficult to see that the excitation properties, especially the relative populations of the energy levels, will depend strongly on the arrangement of these levels within the individual atom. Therefore, every atom must be treated more or less separately, even though we have some general rules which have been derived to explain the appearance of bright lines, especially forbidden lines in stars with extended atmospheres of low pressure. But in our case such a general rule cannot

[14] Pillans, *ibid.*, **80**, 57, 1934.   [15] *Ibid.*, **81**, 69, 1935.

give a satisfactory explanation of the different features of the observations, and we shall therefore try to obtain more accurate information.

Since approximate values of the transition probabilities for the strongest He I transitions are known from a quantum-mechanical computation of Hylleraas,[16] it is possible to carry out at least an approximate computation of the relative populations of the most important He I levels, if we can neglect the influence of collisions. To carry through this computation we shall use a method suggested by Rosseland (theory of cycles) and worked out in detail by Woolley[17] in connection with the fluorescence in $H\alpha$ and $H\beta$ in an atmosphere of diluted radiation; for this problem Woolley took into account the simplest possible transition schemes involving only four levels in each case. Our problem is somewhat more complicated because we are interested in the two He I systems (singlets and triplets). We shall, therefore, take into consideration six states, namely, 1^1S, 2^1S, 2^1P, 2^3S, 2^3P, and the ground state of He II. The limitation to these six states naturally disregards higher quantum states in both systems from which strong transitions start toward the lower levels. But we may expect that these higher transitions do not appreciably change the relative properties for singlets and triplets, especially in view of the fact that the two systems are arranged in an entirely similar manner. We shall discuss the influence of collisions later. It is possible to predict approximately how these, if they are present at all, will change the distribution created by radiation alone.

The population of a particular state will be determined by certain coefficients A_{ik} and A_{ki}, which are the expressions for the probability that an atom makes a transition per second, as indicated by the subscripts. Using the notation of Woolley's paper, which is identical with that of Eddington in his *Internal Constitution of the Stars*, we have ($i < k$):

$$A_{ik} = a_{ik}\rho(\nu) ; \qquad A_{ki} = b_{ki} + a_{ki}\rho(\nu) , \qquad (1)$$

where $\rho(\nu)$ is the energy density of radiation, b_{ki} is the probability coefficient for spontaneous transitions, and a_{ik} and a_{ki} are the coef-

[16] *Zs. f. Phys.*, **106**, 395, 1937.   [17] *M.N.*, **94**, 631, 1934.

ficients for absorption and for induced emission. The a's and the b's are connected by the well-known relations

$$a_{ik}\rho(\nu) = \frac{q_k}{q_i} \times \frac{b_{ki}}{e^{h\nu/kT} - 1} \quad \text{and} \quad a_{ki}\rho(\nu) = \frac{b_{ki}}{e^{h\nu/kT} - 1}, \qquad (2)$$

where q designates the statistical weight of the term and $\rho(\nu)$ is now the equilibrium value at the temperature T. If we deal with diluted radiation, the equilibrium value of $\rho(\nu)$ has to be replaced by $\rho'(\nu) = W\rho(\nu)$, where W is the dilution factor.

TABLE 1

	1^1S	2^1S	2^1P			2^3S	2^3P
2^1P............	0.349	0.39	....	2^3P............		0.55	....
Cont............	1.55	1.38	0.19	Cont............		1.36	0.19

With the exception of A_{6i}, the index 6 referring to the ionized state, all A_{ik} and A_{ki} can be computed with the use of the values of the oscillator strength, f, which were computed by Hylleraas and are listed in Table 1.

The values of b_{ik} can be computed with the aid of the relation

$$f_{ik} = \frac{1}{3} \times \frac{q_k}{q_i} \times \frac{b_{ki}}{\omega}, \qquad (3)$$

where ω is the classical damping constant equal to $8\pi^2 e^2 \nu^2 / 3mc^3$. The probability coefficients for transitions into the ionized state can be evaluated from

$$A_{i6} = \frac{3 f_{i6}\omega}{e^{h\nu/kT} - 1} \left[= \frac{\pi e^2}{mh\nu} \times f\rho(\nu) \right]. \qquad (4)$$

The probability coefficients for the opposite jumps, A_{6k}, cannot be determined directly. But we require only their relative values, which can be found by applying the principle of detailed balancing in the thermodynamical equilibrium, where the following relation must hold:

$$A_{61} : A_{62} : A_{63} \ldots = n_1 A_{16} : n_2 A_{26} : n_3 A_{36} \ldots . \qquad (5)$$

The values of n are here given by the Boltzmann formula.

With respect to the recombination, we must take into account the fact that the temperature of the recombining electrons may be different from the temperature T which we assume for the radiation. But the values of A_{6k} in Table 3 show that over a large range of temperature there is no strong dependence of the values of A_{6k} upon temperature.

In a steady state we have

$$\sum_k n_i A_{ik} = \sum_k n_k A_{ki} , \tag{6}$$

which means that the number of transitions ending in a state is equal to the number which start from it. As was pointed out by Woolley, we can include in the states considered in (6) the ionized state, without specifying a definite electron pressure, if we do not wish to determine n_6/n_k.

If we have to deal with r different states, including the ionized state, then, according to (6), we have r different equations with $r - 1$ unknowns: $n_2/n_1, n_3/n_1, \ldots, n_r/n_1$. In our case r is equal to six. Using the abbreviations

$$\left.\begin{aligned}
A_{11} &= -(A_{12} + A_{13} + \ldots A_{1r}) , \\
A_{22} &= -(A_{21} + A_{23} + \ldots A_{2r}) , \\
\end{aligned}\right\} \tag{7}$$
$$\cdot \ \cdot \ \cdot \ \cdot \ \cdot \ \cdot \ \cdot \ \cdot \ \cdot \ \cdot \ \cdot \ \cdot \ \cdot$$

the determinant

$$D = \begin{vmatrix}
A_{11} & \ldots & A_{r1} \\
A_{12} & \ldots & A_{r2} \\
\ldots & \ldots & \ldots \\
\ldots & \ldots & \ldots \\
A_{1r} & \ldots & A_{rr}
\end{vmatrix} \tag{8}$$

vanishes and the values of $n_1, n_2, n_3, \ldots$, must be proportional to the minors $A_{1i}, A_{2i}, A_{3i}, \ldots$.

We have evaluated the minors A_{21}, A_{31}, A_{41}, and A_{51} for the levels 2^1S, 2^1P, 2^3S, and 2^3P, in which we are interested, for two different temperature values, $T = 10^4$ and $T = 2.5 \times 10^4$, and for the dilution factors $W = 1$, $1/10$, $1/50$, and $1/100$. The constants used are listed in Tables 2 and 3. The values of A_{ik} and A_{ki} between singlets and triplets are all set equal to zero because these intercombinations are forbidden or are extremely weak. Table 4 gives the results of

TABLE 2
LINE TRANSITIONS

TRANSITION	VOLTS	$e^{h\nu/kT}$ $T=10^4$	$e^{h\nu/kT}$ $T=2.5\times10^4$	f_{ik}	q_i/q_k	ω	b_{ki}
1¹S→2¹P	21.13	4.37×10^{10}	1.80×10^4	0.349	1/3	6.44×10^9	2.25×10^9
2¹S→2¹P	0.60	2.01	1.32	0.390	1/3	5.19×10^6	2.02×10^6
2³S→2³P	1.14	3.74	1.70	0.550	3/9	1.88×10^7	1.03×10^7

TRANSITION	$A_{ik};\ W=1$ $T=10^4$	$T=2.5\times10^4$	$A_{ik};\ W=1/10$ $T=10^4$	$T=2.5\times10^4$	$A_{ik};\ W=1/50$ $T=10^4$	$T=2.5\times10^4$	$A_{ik};\ W=1/100$ $T=10^4$	$T=2.5\times10^4$
1¹S→2¹P	1.54×10^{-1}	3.75×10^5	1.54×10^{-2}	3.75×10^4	3.08×10^{-3}	7.50×10^3	1.54×10^{-3}	3.75×10^3
2¹S→2¹P	6.00×10^6	1.89×10^7	6.00×10^5	1.89×10^6	1.20×10^5	3.78×10^5	6.00×10^4	1.89×10^5
2³S→2³P	1.13×10^7	4.41×10^7	1.13×10^6	4.41×10^6	2.26×10^5	8.82×10^5	1.13×10^5	4.41×10^5

TRANSITION	$A_{ki};\ W=1$ $T=10^4$	$T=2.5\times10^4$	$A_{ki};\ W=1/10$ $T=10^4$	$T=2.5\times10^4$	$a_{ki};\ W=1/50$ $T=10^4$	$T=2.5\times10^4$	$A_{ki};\ W=1/100$ $T=10^4$	$T=2.5\times10^4$
1¹S→2¹P	2.25×10^9	2.25×10^9	2.25×10^9	2.25×10^9	2.25×10^9	2.25×10^9	2.25×10^9	2.25×10^9
2¹S→2¹P	4.02×10^6	8.33×10^6	2.22×10^6	2.64×10^6	2.06×10^6	2.14×10^6	2.03×10^6	2.08×10^6
2³S→2³P	1.41×10^7	2.60×10^7	1.07×10^7	1.17×10^7	1.03×10^7	1.06×10^7	1.03×10^7	1.05×10^7

TABLE 3
TRANSITIONS INTO CONTINUUM

TRANSITION	VOLTS	$e^{h\nu/kT}$ $T=10^4$	$T=2.5\times10^4$	f_{ik}	q	ω	$A_{i6};\ W=1$ $T=10^4$	$T=2.5\times10^4$
1¹S→Cont.	24.47	2.11×10^{12}	8.50×10^4	1.55	1	8.66×10^9	1.91×10^{-2}	4.74×10^5
2¹S→Cont.	3.94	9.56×10^1	6.17	1.38	1	2.26×10^8	9.90×10^6	1.81×10^8
2¹P→Cont.	3.36	4.94×10^1	4.76	0.9	3	1.63×10^8	1.92×10^6	2.47×10^7
2³S→Cont.	4.73	2.42×10^2	9.02	1.36	3	3.26×10^8	4.79×10^6	1.66×10^8
2³P→Cont.	3.60	6.46×10^1	5.31	0.19	9	1.88×10^8	1.70×10^6	2.46×10^7

TRANSITION	$A_{i6};\ W=1/10$ $T=10^4$	$T=2.5\times10^4$	$A_{i6};\ W=1/50$ $T=10^4$	$T=2.5\times10^4$	$A_{i6};\ W=1/100$ $T=10^4$	$T=2.5\times10^4$	A_{6k} $T=10^4$	$T=2.5\times10^4$
1¹S→Cont.	1.91×10^{-3}	4.74×10^4	3.82×10^{-4}	9.48×10^3	1.91×10^{-4}	4.74×10^3	1000.00	1000.00
2¹S→Cont.	9.90×10^5	1.81×10^7	1.98×10^5	3.63×10^6	9.90×10^4	1.81×10^6	23.30	28.10
2¹P→Cont.	1.92×10^5	2.47×10^6	3.84×10^4	5.94×10^5	1.92×10^4	2.47×10^5	6.89	8.68
2³S→Cont.	4.79×10^5	1.66×10^7	9.58×10^4	3.32×10^6	4.79×10^4	1.66×10^6	86.20	111.11
2³P→Cont.	1.70×10^5	2.46×10^6	3.40×10^4	4.92×10^5	1.70×10^4	2.46×10^5	24.80	29.40

our computation. Judging from the numerical values of the differences appearing in the members of the minors, the accuracy of the computation is of the order of 5 per cent. To obtain a higher accuracy, three decimals in the A_{ik} and A_{ki} should be used instead of only the two used here.

As is to be expected, for $W = 1$ we obtain a Boltzmann distribution. This computation was carried out for the purpose of checking our results. What is demonstrated most strikingly in Table 4 is the

TABLE 4

Relative Populations in Four Levels of *He* I

$T = 10^4$

	Boltzmann Distribution at $T = 10^4$	$W = 1$	$W = 1/10$	$W = 1/50$	$W = 1/100$	Term
n_2.........	100	100	100	100	100	2^1S
n_3.........	150	152	11	5	2	2^1P
n_4.........	753	756	814	1140	1150	2^3S
n_5.........	610	602	75	30	15	2^3P

$T = 2.5 \times 10^4$

	Boltzmann Distribution at $T = 2.5 \times 10^4$	$W = 1$	$W = 1/10$	$W = 1/50$	$W = 1/100$	Term
n_2.........	100	100	100	100	100	2^1S
n_3.........	226	227	45	5	3	2^1P
n_4.........	428	420	501	545	538	2^3S
n_5.........	703	710	321	70	40	2^3P

preference of the metastable states for higher dilutions. But the two metastable levels 2^1S and 2^3S behave in a slightly different manner: the relative population of the 2^3S state increases somewhat more strongly than that of the 2^1S state. This may be expressed by saying that the metastability of the 2^3S-level is somewhat higher than that of the 2^1S-level. This is explained by the fact that the depopulation of the former can go in only one direction, namely, over the ionized state, while for the 2^1S-level the atoms can also escape over the very short-lived 2^1P-level to the ground state. That the 2^1P term shows the strongest decrease is quite consistent with the general rule hereinbefore mentioned. The table shows a number of other interesting features which we shall discuss in connection with the observational data in Section III.

We shall now discuss the influence of collisions. Since the atmosphere which we are considering is in a high state of ionization, we can neglect the atomic and ionic collisions and direct our attention only to electron collisions. Analogous to our A_{ik} and A_{ki} for the radiation jumps, we define certain coefficients a_{ik} and a_{ki} for collision transitions, which designate the probability that an atom will make, per second, a transition, as indicated by the subscripts. Naturally, our a_{ik} and a_{ki} depend on the electron density n_e, as well as on the velocity distribution of the electrons (temperature) and the excitation and de-excitation cross-sections σ_{ik}^2 and σ_{ki}^2. Strictly speaking, our a_{ik} and a_{ki} would be determined by [18]

$$a_{ik} = n_e \int_{\eta = \eta_0}^{\infty} S_i^k(\eta) \left(\frac{2\eta}{m}\right)^{1/2} \mu(\eta) d\eta \; ;$$

$$\left[\mu(\eta) d\eta = \frac{2\pi}{(\pi kT)^{3/2}} e^{-\eta/kT} \eta^{1/2} d\eta\right] , \tag{9a}$$

$$a_{ki} = n_e \int_0^{\infty} S_k^i(\eta) \left(\frac{2\eta}{m}\right)^{1/2} \mu(\eta) d\eta , \tag{9b}$$

where η is the energy of the relative motion of atom and electron, η_0 is equal to the excitation energy of the level, S_i^k and S_k^i are the excitation and de-excitation cross-sections, n_e is the electron density, and m is the mass of the electron.

Since we do not have any information on the functions $S(\eta)$, the best we can do is to replace them by an average value of the excitation cross-section σ_{ik}^2. Even concerning σ_{ik}^2 we have very little information. From the experimental results on the determination of σ^2 for different elements, we can only conclude that it varies from 10^{-16} to 10^{-18} cm² and to even smaller values. Assuming a constant σ^2, equation (9) gives, after integration,[19]

$$a_{ik} = n_e \sigma_{ik}^2 2 \left\{\frac{2\pi kT}{m}\right\}^{1/2} e^{-\eta_0/kT} , \tag{10a}$$

$$a_{ki} = n_e \sigma_{ki}^2 2 \left\{\frac{2\pi kT}{m}\right\}^{1/2} . \tag{10b}$$

[18] See R. H. Fowler, *Statistical Mechanics*, 2d ed., p. 680, Cambridge, 1936.

[19] *Ibid.*, pp. 703–706.

There are some indications that for higher values of the excitation energy, σ^2 is nearer to 10^{-18} than to 10^{-16} cm². Assuming $\sigma^2 = 10^{-18}$, we find for the deepest upper level, 2^3S,

$$a_{14} = \begin{cases} n_e \times 10^{-18} & \text{for} \quad T = 10^4, \\ n_e \times 10^{-12} & \text{for} \quad T = 2.5 \times 10^4. \end{cases} \tag{11}$$

These values of a_{14} we compare with our A_{ik} in Tables 2 and 3 for transitions to 2^1P and to the continuum. Regarding the case $W = 1$, we have $A_{13} = 1.5 \times 10^{-1}$ and $A_{16} = 2 \times 10^{-2}$ for $T = 10^4$ and $A_{13} = 4 \times 10^5$ and $A_{16} = 5 \times 10^5$ for $T = 2.5 \times 10^4$. There will be an appreciable additional population by collisions from the ground state only if a_{14} becomes of the same order as the values of A_{ik}. This means that the electron density must be

$$n_e \gtrsim 10^{18} \text{ cm}^{-3} \quad \text{for} \quad T = 10^4 \text{ and } T = 2.5 \times 10^4. \tag{12}$$

Even if we should assume the higher value of σ^2 ($= 10^{-16}$), which gives $n_e \gtrsim 10^{16}$, it is evident that collisions can play no role because we cannot expect such electron densities to exist in our shells. The analysis of stellar atmospheres points to an electron pressure of between 1 to 100 bar for normal layers, which corresponds to values of n_e between 10^{12} and 10^{14}.

For higher dilutions there will be no difference if the density decreases in the same proportion as the dilution increases ($W\rho = $ const.); this condition may be approximately fulfilled in most cases. If the density decreases much more slowly than the radiation density, we may arrive at a stage where collisions begin to have a slight influence. But there is another point to take into consideration which has already been mentioned in connection with the recombination processes. In carrying out our estimates we have assumed that the temperature of the electrons is identical with the radiation temperature T. This will probably not always be the case, as is evident from the following consideration: the average velocity of the photoelectrons after leaving the atoms will have a temperature distribution which is identical with the radiation temperature (color temperature), as has been pointed out by Eddington. The distribution will be preserved if the electrons, before recombination, have no opportunity to give off energy in collisions. But this will strongly depend on the atomic composition of the atmosphere. Independently of the pressure the electrons will give off energy in collisions before recombining, if the atmosphere contains atoms with low-lying energy levels, because, generally, the excitation cross-sections are much larger than the recombination cross-sections. The former range from 10^{-16} to 10^{-18} cm², the latter[20] from 10^{-18} to 10^{-21} cm². The appearance of the well-known forbidden lines in the nebular spectra, the excitation of which is, according to Bowen, created by electron impact, shows that such loss of energy by the photoelectrons takes place even in stellar atmospheres of high ionization and of extremely low pressure.[21] Such a drop of the electron temperature will, even if slight, further diminish our values of a_{14} in equation (10).

Besides the collisions from the ground state, we have still to consider those collisions which can produce transitions of the atoms between the different higher levels. For these small differences in energy we may assume a value of σ^2 of the order of 10^{-16} cm². Since the factor $e^{-h\nu/kT}$ does not change appreciably for the four different levels, we may neglect it. Then we have

$$a_{ik} \approx n_e \times 10^{-8} \quad \text{for} \quad T = 10^{-4} \quad \text{and} \quad T = 2.5 \times 10^{-4} . \quad (13)$$

The average values of A_{ik} (see Tables 2 and 3) for the transitions involved are of the order of 10^7, for $W = 1$. Hence, we find for the condition that the collisions will have an influence:

$$n_e \gtrsim 10^{15} \text{ cm}^{-3} . \qquad (W = 1) \quad (14)$$

The lower limit of n_e in (14), as compared with (12), originates chiefly from the higher assumed value of σ^2, whereby we come much nearer to the critical value of n_e (10^{12} to 10^{14}). Going to higher dilutions, the influence of collisions does not increase if we have $W\rho = $ constant. But there is a difference between the low and the high excitations if we consider the case in which ρ decreases more slowly than the dilution increases and in which we have at the same

[20] F. L. Mohler, Rev. Mod. Phys., 1, 216, 1929.

[21] Naturally, the process described is independent of pressure; but, as has already been explained, it depends strongly upon the atomic composition.

time a drop of the electron temperature. With decreasing T the factor $e^{-\eta_0/kT}$ in formula (10a) decreases only slightly for small values of η_0, that is, for excitation energies of 1–5 volts. But for the 20 volts of excitation from the ground state the factor $e^{-\eta_0/kT}$ decreases very much. This effect again suggests that we must at first expect an influence of collisions between closely spaced levels. Since our estimates show that we are near the critical values of n_e when an influence of collisions for the higher He I levels begins to appear, we may further examine how the collisions change the populations created by absorption. As was pointed out by one of the present writers,[22] the collisions in an atmosphere of diluted radiation create a very similar distribution of the atoms over the various energy levels of different lifetimes, as does the diluted radiation. In both cases there results an accumulation in metastable states. In general, the shorter the lifetime of a level, the greater must be the electron density (as in the case of absorption excitation, the greater must be the radiation density) in order to produce a similar approximation toward a Boltzmann distribution. Therefore, the first influence of collisions will express itself in a tendency to give an approach to a Boltzmann distribution (corresponding to the electron temperature T_e) for the two metastable levels 2^3S and 2^1S (but see below). The second stage will show such an approach for the 2^3P-level ($\tau = 10^{-7}$ sec) which may be reached at electron densities of $n_e = 10^{15}$, provided that $a_{ik} \gtrsim A_{ik}$ for this transition.[23] To establish a Boltzmann distribution for the very short-lived 1^1P-level ($\tau = 10^{-9}$ sec), we must have at least an electron density of $n_e = 10^{17}$ cm^{-3}. To obtain more accurate information we need more knowledge concerning the differences in the cross-sections σ_{ik}^2 and σ_{ki}^2 of excitation and de-excitation for the different transitions 2^3S $\leftrightarrow 2^1S$, 2^3P, 2^1P; 2^3S $\leftrightarrow 2^1$P, 2^3P; 2^3P $\leftrightarrow 2^1$P. We can expect that within a system (singlets or triplets) σ_{ik}^2 is somewhat larger than for intercombination transitions. This will somewhat diminish the approach toward temperature distribution between the two systems. Since collisions from transitions 2^1S $\rightarrow 2^1$P occur more frequently

[22] K. Wurm, *Zs. f. Ap.*, **14**, 321, 1937.

[23] If $a_{ik} < A_{ik}$, then, naturally, the collisions cannot appreciably alter the populations created by absorption.

than transitions from 2^3S to 2^1S, there would be an additional de-population of the singlet system as compared with the 2^1P state, and hence a stronger preference of the triplet levels. This condition we may expect when the influence of collisions first appears.

In concluding this section it may be pointed out that for levels which lie near the ground state, as we find them in Fe II, Ti II, Cr II, and other atoms, the collisions will generally act in quite the same manner as has been shown for the higher He I states. This problem has been already treated extensively by one of us in an earlier paper,[22] in connection with the problem of Fe II emission lines in stars with extended atmospheres. It was shown there that even relatively low electron densities are able to create an approximate Boltzmann distribution if the levels are metastable.

III. COMPARISON OF THE THEORY WITH OBSERVATIONS

Table 4 refers the populations of the various levels to that of state 2^1S as standard. In comparing the observations of Section I with the tabular values, the following considerations are important:

1. The theory of the transfer of radiation has been worked out so far for only three states.[24] We shall therefore follow the usual practice and assume that the intensity at any point within the line is given by $I = I_0 e^{-\sigma H}$, where σ is the absorption coefficient and H is the thickness of the shell.

2. The absorption coefficient in the shells is determined largely by turbulence. Stark effect is absent, and the characteristic bell-shaped contours of the H cores show conclusively that turbulence is effective.

3. The total absorption[25] within a line is equal to

$$A = 2 \int_0^\infty (1 - e^{-a e^{-x^2/b^2}}) dx = \sqrt{\pi} ab \left\{ 1 - \frac{a}{2! \sqrt{2}} + \frac{a^2}{3! \sqrt{3}} \cdots \right\},$$

$$a = 10^{-13} Nf \frac{H}{b}, \qquad b = \lambda_0 \frac{v_0}{c},$$

[24] Chandrasekhar, $Zs.\ f.\ Ap.$, 9, 266, 1935; Ambarzumian, $Bull.\ Obs.\ Poulkovo$, 13, No. 3, 1933; $Poulkovo\ Circ.$, No. 6, 10, 1933; $M.N.$, 93, 50, 1931.

[25] $Ap.\ J.$, 79, 430, 1934.

where b is the Doppler shift corresponding to the most probable turbulent velocity v_0. Thus, A is proportional to f if the former is small. We shall assume that the values of f are independent of temperature and of the degree of dilution.

4. For faint lines the total absorptions of the lines should then be directly proportional to the values of n in Table 4. For strong lines the total absorptions will vary little with n.

5. From Table 4 we predict that in all shells the triplet series $(2^3S - n^3P)$ must experience the greatest enhancement. Since the only suitable line, λ 3889, is blended with $H\zeta$, we are unable to test this prediction.

6. For all shells the lines $(2^3P - n^3D)$, $(2^3P - n^3S)$, $(2^1P - n^1D)$, and $(2^1P - n^1S)$ are greatly reduced in strength.

7. This reduction is greatest for the singlet lines $(2^1P - n^1D)$ and $(2^1P - n^1S)$, and the degree of this reduction does not depend appreciably upon the temperature.

8. The triplet lines are reduced in intensity about half as much as the singlets in the cool shells. Thus, even in the cool shells the triplet lines $\lambda\lambda$ 4026, 4472, etc., will appear relatively stronger than the singlet lines. This is in agreement with conclusion (2) of Section I.

9. In the hotter stars the triplet lines are reduced by only a factor of 18, when W changes from 1 to 0.01, while the singlets are reduced by a factor of 76. Thus, we should expect that $\lambda\lambda$ 4026, 4472, etc., will gain in intensity relative to λ 3965, as well as relative to $\lambda\lambda$ 4009, 4144, 4388, etc. This is essentially in agreement with conclusions (3) and (4) of Section I.

10. There remains the question why in P Cygni the singlet lines $(2^1P - n^1D)$ are not as faint, relatively to the triplet lines $(2^3P - n^3D)$, as in shells of lower excitation (conclusion 4 of Sec. I). The answer is almost certainly found in the effect of turbulence which makes A insensitive to n when α is large. That this interpretation is correct follows from the intensities of the emission lines in P Cygni, which are very weak in series $(2^1P - n^1D)$ but are strong in λ 3965 as well as in series $(2^3P - n^3D)$.

PLATE V

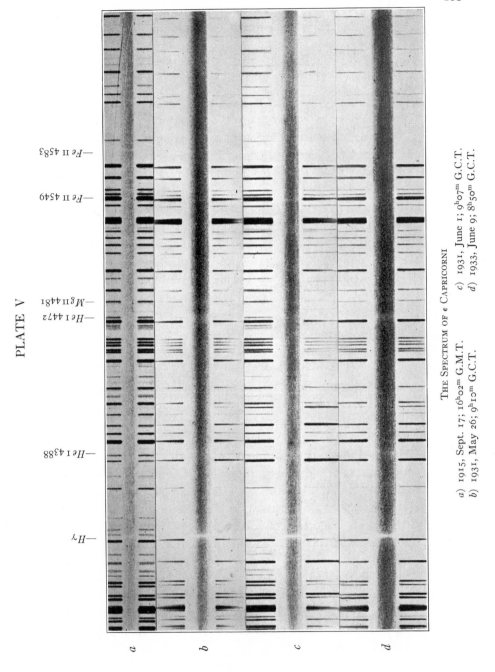

—Fe II 4583

—Fe II 4549

—Mg II 4481
—He I 4472

—He I 4388

—Hγ

a

b

c

d

THE SPECTRUM OF ε CAPRICORNI

a) 1915, Sept. 17; 16ʰ02ᵐ G.M.T.   *c)* 1931, June 1; 9ʰ07ᵐ G.C.T.
b) 1931, May 26; 9ʰ10ᵐ G.C.T.   *d)* 1933, June 9; 8ʰ50ᵐ G.C.T.

It is of some interest to evaluate the quantity

$$\frac{\text{Int. 3965 (shell)}}{\text{Int. 3965 (normal)}} \times \frac{\text{Int. 4009 (shell)}}{\text{Int. 4009 (normal)}} .$$

Unfortunately, we cannot make an accurate evaluation because in normal B stars λ 4009 is greatly broadened by Stark effect, while λ 3965 is not sensitive to Stark broadening. Nevertheless, from the combined results of our observations we estimate that this ratio is between 10 and 100. This fixes the dilution factor at approximately $W = 0.01$. But

$$W = \frac{1}{4}\frac{R^2}{r^2} + \frac{1}{16}\frac{R^4}{r^4} + \ldots ,$$

where R is the radius of the star and r is the effective radius of the shell. Accordingly,

$$r \approx 5R .$$

This gives a good estimate of the order of size of the average shell.

IV. PHYSICAL CONDITIONS IN OUTER SHELLS

The spectral lines which originate in the outermost shells of ζ Tauri, φ Persei, 48 Librae, ε Capricorni, etc., are relatively sharp. They suggest the presence of a moderate amount of turbulence (17 Leporis) but give no indication of rotation. This suggests that the shell does not rotate as a solid body with the star. Otherwise the rotational component in the line of sight would be constant and all lines would show the same broadening. We conclude that the angular velocity of rotation decreases with r.

A close inspection of Plate V shows that in ζ Tauri the Si II lines are appreciably broadened, and that even the Fe II lines are not as sharp as He I 3965 or Ni II 4067. We conclude that the atoms in the shell are stratified. The normal He I lines and the H wings are produced in the lowest levels, which we have referred to as the "reversing layer." Si II occurs higher in the shell, Fe II lies still higher, and the cores of H, Ni II, and He I 3965 occupy the highest levels.

A similar picture of stratification can be outlined for 48 Librae and ϵ Capricorni. The result is in harmony with the stratification formerly found in P Cygni.

The decrease in angular velocity of rotation with r is puzzling. It may be due, as Chandrasekhar suggests,[26] to the conservation of angular momentum. If $rv = $ constant, we should find that for $r = 5R$, $v_{\text{shell}} = 0.2v_{\text{rot}}$. If the star and its shell rotated as a solid body, we should have $v_{\text{shell}} = 5v_{\text{rot}}$. A decrease in the rotational broadening by a factor of 25 would suffice to make the lines appear entirely narrow. Incidentally, this line of reasoning serves to confirm the general order of magnitude found for W.

The stage of ionization and excitation in the shells considered here (with the exception of β Lyrae) is considerably lower than in the corresponding reversing layers. Thus, the shells show lines of Fe II requiring an ionization energy of 7.8 volts for the ionization of Fe I and an additional energy of excitation amounting to 3 volts for the lower levels and 6 volts for the upper levels. For Ni II an ionization energy of 7.6 volts plus excitation energies of 4 and 7 volts are required. In the reversing layer we have predominantly He I, requiring excitation energies of over 20 volts. In terms of ordinary spectral classes we should attribute class A to the shells of ζ Tauri, φ Persei, 48 Librae, and ϵ Capricorni, and one of the intermediate subdivisions of class B to the reversing layers.

A similar result has already been obtained in the case of P Cygni,[13] where the stage of ionization decreases with the height in the reversing layer. This tendency for the ionization to decrease outward implies[27] that the pressure decreases outward less rapidly than $1/r^2$.

The spectra of the shells present one conspicuous anomaly: the line Mg II 4481 is either absent or very weak. Incidentally, this anomaly has already been noticed in the emission spectra of Be stars.[28] In the light of the preceding discussion the unusual weakness of λ 4481 is easily explained. This line originates between the lowest ^2D term and the lowest ^2F term of the Mg II atom. The ^2D

[26] Private communication. We are also indebted to Dr. L. H. Thomas for a discussion of this question.

[27] *Ap. J.*, **81**, 87, 1935.   [28] Struve and Swings, *ibid.*, **75**, 165, 1932.

term is not metastable, but connects with the lowest ^{2}P term, which in turn connects with the ground-level ^{2}S. By line absorption two processes are required to lift the atom from the ground-level into level ^{2}D. The population will, therefore, be proportional to W^2. This will be modified by recombinations, but even in the most favorable case the level ^{2}D will be strongly depopulated by any appreciable dilution of the radiation.

The Balmer lines are strong in all shells, their intensities bearing about the normal ratio to the intensities of the He I triplets. This may be due to the metastability of the 2S-level of H. The shell should, strictly speaking, show only the series (2S − nP) of H, because the 2P-level is not metastable. In principle, this should lead to an abnormal Balmer decrement, since the f-values of the (2S − nP) series differ from those of the combined Balmer lines.[29] But for practical purposes only the ratio $H\alpha/H\beta$ is suitable. In the shell $f\alpha/f\beta = 0.425/0.102$, while in a normal star $f\alpha/f\beta = 0.637/0.119$. No observations of $H\alpha$ are available.

Another fairly conspicuous feature is the weakness of Si II in most shells. This is also due to the fact that the lower level of the lines λ 4128 and λ 4131 has a strong transition toward the ground-level. We expect a similar behavior of Si II λ 4128 and λ 4131 and Mg II 4481, since these lines have not only similar excitation potentials but also nearly the same first and second ionization potentials:

	First Ionization Potential	Second Ionization Potential	Excitation Potential
Mg..........	7.61	14.97	8.83 Mg II λ 4481
Si..........	8.12	16.27	9.79 Si II $\begin{cases}\lambda\ 4128 \\ \lambda\ 4131\end{cases}$

On the other hand, the great relative strengths of the Fe II lines are caused by the metastability of all lower levels of the observable lines of ionized iron. In the case of lines like Fe II and Ti II, which arise from low levels—up to about 5 volts above the ground state—

[29] See A. Unsöld, *Physik d. Sternatmosphären*, pp. 187, 188, Berlin, 1938.

we have to take into consideration the additional population from collisions, which is then very effective (see Sec. II).

The astronomical evidence strongly suggests that the lower level of the Ni II line λ 4067 must also be metastable. This line is due to the transition $a^2G - z^2D°$. The lower level, a^2G, is even, as is the ground level, ^2D. Since it lies lower than the lowest odd level of Ni II, it is metastable.

In φ Persei and in most other stars surrounded by shells, including ordinary Be stars, the broadened absorption lines of He I are weaker than can be accounted for by rotation alone. Although this could in part be caused by incipient emission in the lines, it does not seem probable that the shells of φ Persei, ζ Tauri, etc., give rise to strong emission in helium. In these stars even the H emissions are weak. It is difficult to avoid the conclusion that the shell produces a screen of continuous emission which weakens the underlying absorption lines.

V. THE He I ANOMALY IN NORMAL B STARS

The He I anomaly in the shells of peculiar B stars strongly resembles the He I anomaly in ordinary B stars—especially in the supergiants.[30] In stars of low luminosity the intensities are complicated by the normal Stark effect and by collisional damping.[31] Nevertheless, the decrease of the singlets ($2^1P - n^1D$) relative to the triplets ($2^3P - n^3D$) in such stars as 10 Lacertae and τ Scorpii suggests that the dilution of the radiation is already appreciable in the reversing layers of these objects.

The explanation of the He I anomaly in normal B stars and in supergiants proposed some time ago[30] agrees in all essential features with that presented in this paper. It is clear from Table 4 that even for relatively inappreciable dilutions the 2^1P level is greatly depopulated. Under favorable circumstances the equivalent breadths of absorption lines should easily reveal differences corresponding to a factor of 2 in the numbers of atoms. It is, therefore, entirely reasonable that departures from thermodynamic equilibrium will be observable in the supergiants. This is especially true because Stark

[30] Struve, $Ap.\ J.$, **82**, 256, 1935, and earlier papers.

[31] $Observatory$, **61**, 53, 1938.

effect and collisional damping are negligible in these stars. But it would be incorrect to conclude that the shells of peculiar B stars and the atmospheres of recognized supergiants of class B are identical. Not only are the departures from a Boltzmann distribution in He I less pronounced, but there are other important spectroscopic differences: the lines of Mg II 4481 and of Si II 4128 and 4131 are very strong in the supergiants. Evidently, the dilution is not sufficient to weaken appreciably these lines in comparison with the same lines in dwarfs. But the effect of turbulence in the supergiants makes a comparison with the dwarfs somewhat unreliable.

This difficulty is also present in the case of He I. It will be necessary to investigate the curves of growth for the individual lines in giants and in dwarfs before we shall be able to determine the numerical value of W.

YERKES OBSERVATORY
May 1938

DILUTION EFFECTS IN EXTENDED ATMOSPHERES

Anne B. Underhill

Sonnenborgh Observatory
Utrecht

I. INTRODUCTION

The interpretation of stellar spectra in terms of the relative dimensions of a star and of the physical conditions in the atmosphere of the star is based upon the fact that details of the line spectrum may be used as diagnostic tools to determine the state of the atmosphere. At the time Struve and Wurm wrote their paper about dilution effects in stellar spectra (Struve and Wurm 1938), the methods of analysis based upon the theory of the curve of growth in a single layer at one temperature and one pressure were proving very effective in giving an understanding of the empirically ordered sequence of spectral types. It was evident that if one assumed that the radiation and matter in stellar atmospheres interacted as though there was a state of local thermodynamic equilibrium (LTE), not only the broad trends seen in the spectral sequence could be understood but also a rather consistent interpretation could be found for the strengths of most of the absorption lines in normal stellar spectra.

However, the spectra of a few stars do not fit into the normal pattern. Some of these peculiar stars (shell stars) have basically a B-type spectrum, but in addition there is a set of rather sharp absorption lines which simulates, more or less, the spectrum of a supergiant of later spectral type than is suggested by the basic B-type lines. Struve and Wurm directed attention to the unusual pattern of intensities found in the additional set of lines and noted that in an atmosphere which was not in LTE, the distribution of the atoms over the various states might be very different from a Boltzmann distribution. They called these anomalies dilution effects.

In their theoretical treatment of the problem they represented the fact in a shell the radiation field and the atoms and ions were in LTE by writing the radiation field as $WB_\nu(T)$. Here W is the dilution factor, W being less than unity, and $B_\nu(T)$ is the Planck function at the local temperature T. Because the extended atmosphere is transparent, some radiation escapes and the radiation density is less than expected in TE. The geometric dilution factor at any point in an extended atmosphere is equal to the ratio of the solid angle subtended by the photosphere at that point to 4π.

The achievements of Struve and Wurm were the following:

(a). They recognized specific spectroscopic anomalies (called dilution effects) which indicate the presence of an extended atmosphere around a B star. The primary anomalies are in the spectrum of He I. In an extended atmosphere where the radiation field is dilute, lines from the metastable levels 2^3S and 2^1S are strengthened with respect to other lines in the He I spectrum. At rather high temperatures, lines from the 2^3P levels are also strengthened while lines from the 2^1P and 2^1S levels decrease in strength. These anomalies are seen in the spectra of shell stars, of supergiants, and of stars of types B0 and earlier. Secondly, in shell spectra which simulate the spectra of A-type supergiants, the lines Mg II $\lambda4481$ and Si II $\lambda\lambda4128$, 4130 often are rather weak in comparison to what would be expected according to the strength of lines of Fe II, Cr II, and Ni II. The latter effects may be termed "relative spectroscopic anomalies" because they concern the relative intensities of lines from two or more ions. Simple dilution effects refer to spectroscopic anomalies in the spectrum from one specific atom in one stage of ionization.

(b). They noted that the anomalous pattern of line strengths implies that the relative populations of the levels from which the observed lines arise are not those corresponding to thermal equilibrium at a temperature T. They wrote down the equations of statistical equilibrium for six states of He I and obtained solutions for the case of a diluted radiation field and no transitions caused by collisions. They found varying degrees of over-population of the metastable levels of He I and of the levels lying immediately above the metastable levels. The populations depend on the values adopted for the temperature and the dilution factor. Struve and Wurm also discussed qualitatively the changes which would result if transitions resulting from collisions with electrons were included in the equations of statistical equilibrium.

(c). They applied their theory of the dilution effects in He I to interpretation of the observed absorption line intensities in shell stars and concluded that the average value of the dilution factor in shells is about 0.01, i.e., that the radius of a shell is about five times that of the star. Following their discussion of the He I spectrum, they discussed relative spectroscopic anomalies in Mg II and Si II and concluded that the line formation process in an extended atmosphere could not be well represented by theories which assume LTE everywhere in the atmosphere.

(d). They hinted in their paper that dilution effects would be significant for the interpretation of the spectra of B-type supergiants and stars of spectral types B0 and O, but gave no specific conclusions on this matter.

If one assumes that line formation in stellar atmospheres may be treated as though LTE exists, the monochromatic equation of transfer for a beam inclined at an angle ϑ to the normal in a plane parallel atmosphere is

$$\cos \vartheta \, \frac{dI_\nu(z, \vartheta)}{dz} = -(\kappa_\nu(z) + l_\nu(z))I_\nu(z, \vartheta) + (\kappa_\nu(z) + l_\nu(z))B_\nu(z). \tag{1}$$

Here $B_\nu(z)$ is the Planck function at depth z in the atmosphere, $\kappa_\nu(z)$ is the continuous absorption coefficient and $l_\nu(z)$ is the line absorption coefficient. The line absorption coefficient and the continuous absorption coefficient can be evaluated given the temperature and pressure at depth z in the atmosphere. In particular

$$l_\nu(z) = N(z) \, a_\nu(z), \tag{2}$$

where $a_\nu(z)$ is the atomic absorption coefficient at depth z (a function depending on the temperature and on the density if Stark broadening or collisional broadening is important), and $N(z)$ is the number of atoms excited to the level from which the spectral line arises. If there is TE, N is a function of the electron temperature only and it may be calculated using Boltzmann's law.

Struve and Wurm did not consider the possibility that the form of the emissivity in an extended atmosphere (the last term in equation (1) might be different from that valid in TE. Rather they focussed attention on the point that for some lines $N(z)$ and thus l_ν, would be changed in an extended atmosphere from its value in a normal stellar atmosphere. They showed, using somewhat intuitive reasoning, how the observations of the anomalous intensities of spectral lines in shell spectra could be interpreted in terms of changed level populations.

The populations, N_i, of a series of levels i of an atom or ion may be found from the equations of statistical equilibrium. For every level it is stated that the number of atoms entering that level by all possible processes per unit time is equal to the number of atoms leaving that level per unit time. In general atoms can enter a level from above as a result of spontaneous emission of radiation, as a result of induced emission caused by the radiation field, and as a result of collisions. The atoms may enter the level from below as a result of the absorption of radiation, and as a result of collisions. An atom may leave a level by the same processes.

Consider level j of an atom. Let all states (including the ionized state) lying above j be indicated by the subscript u and all states lying lower than j by the subscript l. Then for every level the following equation is valid if statistical equilibrium exists:

$$N_j\left[\sum_l A_{jl} + \sum_u B_{jl}I_{jl} + \sum_u B_{ju}I_{ju} + \sum_l \sum_p C_{jl}(p)N_p + \sum_u \sum_p C_{ju}(p)N_p\right]$$
$$= \sum_u N_u(A_{uj} + B_{uj}I_{uj}) + \sum_l N_l B_{lj}I_{lj} + \sum_l \sum_p C_{lj}(p)N_p + \sum_u \sum_p C_{uj}(p)N_p. \quad (3)$$

Here the order of the subscripts indicates the direction of the transition. The terms I_{jl} and I_{ju} represent the radiation field in the frequency corresponding to the transition from state j to l (likewise from state j to state u). The quantity I_{jl} is the same as I_{lj}. The A_{jl} and the A_{uj} are the coefficients for spontaneous emission while the B_{jl} and B_{ju} represent respectively the coefficients for stimulated emission and for absorption. The collision cross-sections with perturbers of type p are represented by $C_{jl}(p)$ and $C_{ju}(p)$. The density of perturbers of type p is given by $N(p)$.

An equation similar to equation (3) exists for each state of the atom. The level populations may be found by solving the complete set of equations, given the radiation field. However, in a stellar atmosphere or in a shell, one does not know the radiation field *a priori* and one must solve the equations of statistical equilibrium together with the equation of transfer in order to describe the situation. This is the full problem of the formation of lines in an extended atmosphere under conditions when LTE does not exist.

In their first survey of the problem, Struve and Wurm made the approximation that for each transition $j \to l$ or $j \to u$

$$I_{jl} = WB_\nu(T), \quad (4)$$

where ν is the frequency corresponding to the transition $j \to l$ or $j \to u$. They neglected all transitions caused by collisions, and they restricted the helium atom to six states, namely the ground state 1^1S, the four excited levels with $n = 2$ (2^3S, 2^1S, 2^3P, and 2^1P) and the once-ionized continuum. They assumed that the relative number of recombinations to each of the selected states was as in TE and they showed that these numbers do not depend strongly on temperature.

It is probably acceptable to neglect transitions which are caused by collisions when one is studying an extended atmosphere where the density is low. However, the effects of collisions should be considered if the theory is to be applied to stars with atmospheres of moderate density. Struve and Wurm made use of the best available estimates of the radiative transition cross-sections and reached the conclusions already listed. They demonstrated clearly that the metastable levels of He I would become overpopulated relative to the other levels in the He I atom in comparison to what would be expected for TE.

II. FURTHER DEVELOPMENTS OF THE SIMPLE THEORY OF DILUTION EFFECTS IN He I

Wellmann (1952) carried the discussion initiated by Struve and Wurm further and calculated the relative level populations in He I taking account of levels with $n = 1$ to 4 as well as the singly ionized state. He did not allow for collisional transitions, but he did consider the possibility that the electron temperature which is important for calculating the rates of recombination might be different from the radiation temperature which is used in calculating the Planck function. Ghobros (1962) extended Wellmann's calculations by taking into account separately all levels having $n = 1$ to 5. Both Wellmann and Ghobros express the changes in the level populations from those to be expected in LTE in terms of numbers b_j which give the ratio of the computed population to that to be expected in TE at the electron temperature.

Both Wellmann and Ghobros give tables of b_j for a selection of radiation and electron temperatures and values of the dilution factor, W. Their results may be summarized as follows for values of temperature and W of interest for the interpretation of B-type spectra.

1. The relative populations of the levels 1^1S and 2^1S vary about as W^{-1} for all temperatures investigated. The population of 2^3S increases even more rapidly as W becomes small. Thus in a diluted radiation field the atoms tend to concentrate in these three levels.

2. The populations of the ^1P, ^3S (case $n > 2$), and ^3D levels decrease about as W; thus the populations of these levels become small in a diluted radiation field.

3. The populations of the 2^3P and 3^3P levels increase strongly as W becomes small.

Wellmann finds that taking into account the possible radiative transitions among upper levels reduces the overpopulation of the 2^3S and 2^1S levels which was predicted by Struve and Wurm. However, it should be recalled that when W is of the order of 0.1, collisions surely will be important. Struve and Wurm concluded that the first influence of collisions will express itself in a tendency for the populations of the two metastable levels 2^3S and 2^1S to approach the Boltzmann distribution corresponding to the electron temperature. The second stage will show such an approach for the 2^3P level. This may be reached at electron densities of 10^{15} cm^{-3} provided that the collisional transition probabilities are about equal to the radiative transition probabilities for the relevant transitions. Furthermore they state that when the influence of collisions first appears, one may expect a strong preference for the triplet levels. For levels lying relatively near the ground state (such as the Fe II metastable levels) relatively low electron densities are able to create an approximate Boltzmann distribution (Wurm 1937). Clearly it is important to look at the problem of dilution effects in the He I spectrum once more, introducing modern estimates of the collision cross-sections and, if possible, making a more internally consistent estimate of the radiation field than has been done so far.

Mathis (1957) has discussed the He I spectrum in connection with departures from LTE in a gaseous nebula. Here the dilution factor is very small and collisions can be neglected. In this case two-photon transitions can be important for depopulating 2^3S. Robbins (1968) has indicated that radiative transitions from 2^3S to doubly excited states followed by autoionization will be an important process in planetary nebulae.

III. DILUTION EFFECTS IN SPECTRA OF SIMILAR STRUCTURE TO He I

In an atom or ion where two electrons lie outside a tightly bound nucleus one will obtain singlet and triplet terms. The probability of intercombinations between the two sets of terms will be small so long as the inner electrons and the nucleus form a tightly bound unit. Consequently the lowest triplet levels will be metastable and dilution effects similar to those observed in the He I spectrum may be expected. The ions of the He I isoelectronic sequence are not interesting so far as interpreting stellar spectra is concerned because they produce lines in the far ultraviolet. The C III, N IV, and O V spectra have singlet and triplet terms like those of He I, the ground state being $1s^2\,2s^2$ instead of $1s^2$. Since for C III, N IV and O V the $1s^2$ electrons are tightly bound, they may be considered as forming a unit with the nucleus. No strong intersystem combination lines have been discovered for C III, N IV and O V, thus the lowest triplet level in each case is metastable just as in the case of He I and one may expect that the next highest triplet level will also be rather strongly populated in an extended atmosphere where the temperature is high, while the singlet levels are underpopulated, because they can combine with the ground state.

A similar situation exists for Si III. In this case the configuration of the ground state is $1s^2 2s^2 2p^6 3s^2$. The first ten electrons together with the atomic nucleus form a tightly bound unit. The spectroscopic terms are singlets and triplets. It is doubtful if any strong intersystem combinations exist, thus the lowest triplet term is metastable and one may expect this term and the terms just above it to be overpopulated in an extended atmosphere where the temperature is around 20,000° and the density is low.

To appreciate the similarities between the spectra of He I, C III, N IV, O V and Si III one should look at the lowest triplet terms which are built on the ^2S term of the next higher ion. They are listed below. The configuration of the two outer electrons is indicated in each case.

He I	C III, N, IV, O V	Si III
$1s2s\ ^3S$	$2s2p\ ^3P^0$	$3s3p\ ^3P^0$
$1s2p\ ^3P^0$	$2s3s\ ^3S$	$3s3d\ ^3D$
$1s3s\ ^3S$	$2s3p\ ^3P^0$	$3s4s\ ^3S$
$1s3p\ ^3P^0$	$2s3d\ ^3D$	$3s4p\ ^3P^0$

The dilution effects noted by Struve and Wurm for the case of He I are that the multiplets arising from the lowest levels (2^3S) are particularly strong in an extended atmosphere of high temperature, that is for stars of type B2 and earlier. In stars of types B0 and O, the lines arising from the second lowest triplet term (2^3P) also become strong.

In the spectra listed above the multiplet connecting the two lowest triplet terms has the following wavelengths:

He I	2^3S–2^3P^0	10830 Å
C III	2^3P^0–3^3S	538
N IV	2^3P^0–3^3S	322
O V	2^3P^0–3^3S	215
Si III	3^3P^0–4^3S	997

Only the He I multiplet can be observed from the surface of the earth. It is very strong in absorption in B stars with extended atmospheres. The multiplets from the next lowest triplet terms may also be expected to show dilution effects. The most readily observed multiplets are the following:

He I	2^3P^0–4^3D	4471 Å
C III	3^3S–3^3P^0	4650
N IV	3^3S–3^3P^0	3478
O V	3^3S–3^3P^0	2790
Si III	4^3S–4^3P^0	4552

The strengthening of the He I triplets from 2^3S and from 2^3P^0 with respect to the singlets in stars of spectral type B2 and earlier is a well-known phenomenon. In fact it is used as a secondary criterion in classifying stars of type B2 and earlier. Struve and Wurm have pointed out its significance. The C III multiplet $\lambda4650$ is particularly strong in absorption in supergiants of types B0 and O9. In WC stars the emission feature at 4650 Å has a steep shortward side which is very suggestive of the presence of a strong, shortward displaced absorption component formed in the expanding shell around the emitting envelope. In WN stars a similar steep shortward side is seen on the emission feature due to the N IV multiplet at 3478 Å. It is presumably due to absorption in the expanding, extended atmosphere around the emitting envelope. The Si III multiplet is particularly strong in absorption in stars of type B1 (Underhill 1963). It seems most likely that this also is a dilution effect. This interpretation then gives a straight-forward explanation of the observation of Struve and Roach (1939) that the emission lines at Si III λ4552, λ4568, and λ4574 are unexpectedly weak in the spectrum of P Cyg. The expected emission lines are largely swallowed up by the extra-strong absorption feature formed in the extended atmosphere of P Cyg.

This discussion of dilution effects in He I-like spectra is qualitative. It indicates that the level populations of the ions C III, N IV, O V and Si III in the extended atmospheres around stars of types B2 and earlier as well as those of He I probably are not what would arise in TE. Until an adequate study of the coupled problems of statistical equilibrium and radiative transfer in extended atmospheres of high temperature has been made, lines from these spectra should not be used to estimate abundances, at least in O- and early B-type stars.

IV. THE MEANING OF RELATIVE SPECTRAL ANOMALIES

Struve and Wurm also considered qualitatively the apparent strengths of the absorption lines Mg II λ4481 and Si II λλ4128, 4130 in shell spectra. They remarked: "The spectra of the shells present one conspicuous anomaly: the line Mg II λ4481 is either absent or very weak," and "another fairly conspicuous feature is the weakness of Si II in most shells." One must presume that Struve and Wurm meant that these lines were weak relative to the Fe II absorption lines in comparison to what is seen in the spectra of A-type supergiants. Struve and Wurm indicated that these anomalies could be understood because the Fe II lines all arise from relatively low-lying meta-stable levels which are well populated in supergiant atmospheres and in extended atmospheres, whereas the Mg II and Si II lines arise from levels with high excitation potentials. In an extended atmosphere where collisions are not important, it may be expected that the population of the lower levels of Mg II λ4481 and of Si II λλ4128, 4130 are determined by radiative processes. In a diluted radiation field the necessary radiative processes will not occur frequently.

Before discussing this subject further, it is useful to look at some data about the equivalent widths of lines of Fe II, Si II, and Mg II in the spectra of supergiants and shell stars. Little quantitative information is available, however. The measurements by Groth (1960) in the spectrum of α Cyg may be taken as representative of line strengths in an A-type supergiant. The line strengths vary in shell spectra. The average values for 48 Lib in the years 1950 to 1955 inclusive (Underhill 1954, and unpublished) are given in Table 1 together with some measurements for ζ Tau in different years

TABLE 1

THE EQUIVALENT WIDTHS (IN Å) OF SOME LINES OF FE II, MG II AND SI II IN α CYG AND IN SHELL STARS.

Line	α Cyg A2 Ia	48 Lib 1950–1955	ζ Tau 1950	ζ Tau 1956	ζ Tau 1958	ζ Tau 1960	ζ Tau 1964
Fe II λ4351	0.48	0.52	0.32	0.52	1.0	0.19	0.34
4508	0.36	0.46	0.31		0.50	0.23	
4515	0.33	0.41	0.26		0.42	0.19	
Mg II 4481	0.68	0.20	0.2:		0.90		
Si II 6371	0.49		0.36				0.55
6347	0.60		0.7:				0.69
4130	0.25	0.21	0.22	0.25	0.46		0.27
4128	0.25	0.20	0.19	0.23			0.21
3862	0.36			0.18	0.42	0.22	0.24
3856	0.44			0.25	0.49	0.24	0.32
3853	0.23			0.10			

(Underhill 1952, Hack 1957, Gökgöz, Hack and Kendir 1962, 1963 and Underhill and Van der Wel, unpublished). Since the strengths of the shell lines of 48 Lib did not vary greatly over the years 1950 to 1955, an average value gives a fair representation of the observations. The strengths of the shell lines of ζ Tau have varied quite considerably between 1950 and 1964 as the work of Mrs. Hack and her colleagues has shown. Most of Mrs. Hack's results are published only graphically. I am indebted to her for sending me the data in numerical form.

There is a definite indication that often the Mg II line λ4481 is weak in shell stars in comparison to the Fe II lines. Gökgöz, Hack and Kendir (1962) have shown that the Mg II line in ζ Tau has increased in strength more rapidly than have most of the lines in the shell spectrum. They show that the He I lines, particularly those from 2^3P, also have increased greatly in strength between 1952 and 1960. Strong shell cores are seen to be present for these He I lines on the 1964 spectra of ζ Tau. The relative weakening of Mg II in shell spectra pointed out by Struve and Wurm appears to be conspicuous only when the excitation temperature in the shell corresponds to that in a late B- or early A-type atmosphere. The lower levels of the lines at 4481 Å have an excitation potential of 8.63 volts and these levels can be reached from the ground state only by two successive radiative transitions.

The line Mg II λ4481 is remarkable in that it persists at moderate strength, $W_\lambda \approx 0.1$ Å, to at least spectral type O9 although the Si II lines at λ4128 and λ4130, which come from energy levels of slightly higher excitation potential, have disappeared. This persistence of λ4481 occurs because at temperatures in the neighborhood of 20,000° to 25,000° and moderate electron pressures the ionization of magnesium does not proceed far past Mg^+, whereas the ionization of silicon can proceed readily through Si^{++} to Si^{+++} (Underhill 1965a).

The data in Table 1 do not support the remark of Struve and Wurm that the lines of Si II are weak in most shells. The Si II lines, particularly those at 6371 Å and 6347 Å are strong in the shell spectrum of ζ Tau and they appear in spectra of other shell stars. The lower levels of λλ4128, 4130 of Si II, and of all the Si II lines listed in Table 1, can be reached by a single radiative transition from the ground state of Si II. Thus the situation for Si II is different from the case for Mg II.

There is indeed something unusual about the lines Si II λλ4128, 4130. Beals (1950) has noted that in the extended atmosphere of P Cyg, absorption and emission components are seen at all the normally observed Si II lines except λλ4128, 4130. These lines are present neither in absorption nor in emission according to Beals. This observation has been confirmed from a survey of a sequence of high-dispersion spectra of P Cyg covering the spectral region Hα to 3600 Å (de Groot 1965). Usually emission and absorption components are seen at the positions of all the Si II lines except λλ4128, 4130. These lines appear weakly in absorption or are absent; on occasional spectrograms a faint emission feature is seen beside the weak absorption line. Beals (1950) quotes Merrill as having observed the same phenomenon in the spectrum of HD 207757. In normal absorption line spectra of type B8 to B5, the lines λλ4128, 4130 are strong in absorption and there are no remarks in the literature about their strengths relative to those of other Si II lines being unusually low.

There are some difficulties with the gf values of Si II λλ4128, λ4130. Configuration interaction is an important factor in the theoretical representation and it has been explored in detail by Charlotte Froese Fischer (1968) who has predicted a set of gf values for Si II lines. However, use of these gf-values with normal model atmosphere predictions of the strengths of the Si II lines in B stars does not yield results in close accord with observation (Underhill 1968a). Generally λ4128 and λ4130 are observed to be weaker than predicted.

The discussion to this point has turned around spectroscopic anomalies which occur when the radiation field in selected frequencies is less than the radiation field expected in TE and the excited levels are populated chiefly as a result of radiative transitions.

However, in an extended atmosphere the radiation field may also be strongly augmented in selected frequencies by the presence of strong emission lines. Two lines expected to be strong in early B-type stars and in O stars are Lyman α of hydrogen (1215 Å) and Lyman α of He II (303.8 Å). These monochromatic radiations can excite atoms and ions to particular states for which the energy difference of the transition from an abundantly populated state of the ion or atom happens to coincide closely with the energy of the photons. The stimulation of particular lines in emission in extended atmospheres by such fluorescent processes is well known. One example is the occurrence of the multiplet of N III at $\lambda\lambda4634, 4640, 4641$ in emission in Of stars and in planetary nebulae. Bowen (1935) has pointed out that the observed N III emission lines occur as a result of the coincidence in energy between He II $\lambda3303.8$ and an O III line and a following coincidence in energy between the O III line $\lambda374$ stimulated in this way with a line of N III. The occurrence of He II $\lambda4686$ in emission in Of atmospheres is believed to be due to a coincidence in energy between Lα of hydrogen and the 2–4 transition of He II. This coincidence leads to an overpopulation of $n = 4$ level of He II and the appearance of He II $\lambda4686$ in emission as the ions cascade to states of lower energy.

Ionization from excited levels as well as from the ground level of an atom or ion may occur as a result of a strong flux of monochromatic radiation. Such ionizations should, in principle, be taken into account in solving the equations of statistical equilibrium for level populations in an extended atmosphere. It is probable that the appearance of C III $\lambda5696$ in emission in Of stars (and of no other C III lines) is due to an ionization caused by He II $\lambda3303.8$ which leaves the C^{++} ion in the upper state of $\lambda5696$ (Underhill 1957). Collision with perturbers in an excited state and the consequent transfer of energy can also provide particular ionization preferences (Underhill 1968b).

Emission peaks are observed in the cores of the H and K lines of Ca II in many stars of types G and K. Some theories attempt to account for the strength of these emission peaks by taking account of the collisional transitions which may occur from the metastable 3^2D levels of Ca II to 4^2P^0, which are the upper states of the transitions giving the H and K lines. When the populations of the lowest levels of the Ca$^+$ ion are found by solution of the equations of statistical equilibrium, one should also take account of the fact that the energy of Lα of hydrogen is just sufficient to cause ionization from the metastable 3^2D levels. A strong flux of Lα may be expected in the chromosphere of the sun when conditions are such that the level populations are not those found in thermal equilibrium. The extra ionization caused by the Lα radiation would tend to increase the strength of the emission peaks at H and K because ions which would otherwise be "out of play" because they were in the metastable 3^2D levels would be distributed into states which would lead to the emission of H and K. Whether collision processes or ionization by Lα are most important for depopulating the metastable 3^2D levels and overpopulating 4^2P^0 would depend upon the local electron density and radiation field.

V. CLASSIFICATION PROBLEMS RAISED BY THE OCCURRENCE OF DILUTION EFFECTS

The spectroscopic criteria which are used for determining spectral types have been selected from the strong lines in stellar spectra by an empirical process. The relative strengths of certain lines are used to define the spectral type while the strength of other lines are used as luminosity criteria. By means of the latter one may find the visual absolute magnitude of the star and finally estimate its bolometric luminosity. Thus from studying the spectrum of a star one may place the star rather accurately in the Hertzsprung-Russell diagram. In order to map an observed H-R diagram uniquely into a theoretical luminosity-log T_e diagram in which the evolutionary tracks of stars are displayed, it is necessary to have single-valued relations between spectral type and T_e on the one hand and between M_v and M_{bol} or radius on the other hand.

It has been possible to show that the dominant factor which governs the varying strengths of the spectral-type classification lines over most of the range of spectral types is the total radiation field of the star. Thus it has been possible to establish a monotonic relationship between spectral type and effective temperature valid for most spectral types. This empirical relationship represents the fact that the populations of the levels from which the selected classification lines arise vary in a unique and single-valued way with increasing temperature.

The lines of He I are important spectral-type classification criteria for the stars of types B and O. In particular at types B2 and earlier it is noted that the lines from the 2^3P^0 levels are stronger with respect to the lines from the 2^1P^0 levels than is the case in late B-type spectra. The lines from the 2^3S and 2^1S levels are not important as spectral-type criteria for normal stars. They serve, however, to distinguish some peculiar stars. The strengthening of the lines of the triplet series with respect to the lines of the singlet series is a dilution effect. It indicates that the populations of the various levels of the He I atom depend not only on the T_e of the star but also on the extent of the atmosphere, that is on the size of the departures from LTE which occur in the line-forming regions. Thus at types B2 and earlier, one begins to find that the strengths of the empirically-selected spectral-type classification criteria do not depend dominantly on T_e of the star but are controlled by other factors.

In the O stars an important spectral-type criterion is the relative intensity of the He II lines at 4541 Å and 4200 Å to the He I line at 4471 Å. Now, the apparent strength of the He II lines is largely determined by the strength of the absorption in the cores of these lines. It turns out that the strength of the core is strongly dependent on the state of motion in the outer parts of the atmosphere and not on the temperature because even at type O9 the He$^+$ ions are so abundant, owing to the large abundance of helium, that the centers of the lines of the 4-n series are saturated. Relative motion of the ions due to the field of velocities in the atmosphere tends to reduce the saturation, and thus in a turbulent atmosphere one obtains stronger lines for a given abundance of He$^+$ ions than one would obtain in a quiescent atmosphere. The Stark-broadened wings of He II $\lambda4541$ and He II $\lambda4200$ are formed deep in the atmosphere and their strength is closely related to T_e. However, the wings alone do not determine the equivalent widths of the observed lines. The equivalent widths of the He II lines in O stars are, thus, not uniquely determined by T_e. They depend in some degree on the extent of the atmosphere and on its state of motion. These remarks can readily be verified from the calculations by Underhill and de Groot (1964) of lines formed in a model atmosphere which, so far as the continuous spectrum is concerned, gives a good representation of an O9 star. Since dilution effects are present in the spectrum of He I in the O types, the classification ratio He II/He I thus turns out to be a function of the extent of the atmosphere and of the state of motion in the outermost layers, as well as of the effective temperature of the star.

The strength of the Si IV lines $\lambda4088$ and $\lambda4116$ are also used in classifying O-type spectra. The equivalent widths of these lines are largely determined by the broadening of the cores resulting from motions of the Si^{+++} ions. In fact, calculation of theoretical spectra using model atmospheres demonstrates that the strength of all the empirically selected spectral-type classification lines for O stars depend not only on T_e but also on the extent of the atmosphere and on its field of motion. The result is that one cannot say with confidence that the O stars are in every case arranged according to increasing T_e when one arranges them in order of the assigned types from O9 to O5. Particular attention must be paid to fine details in the spectrum when interpreting the general appearance of the spectrum in order to separate dilution effects and the effects of motion in an extended atmosphere from the effects due to increasing effective temperature. The relationship between spectral type and effective temperature must be single-valued and clear if one is to map the H-R diagram accurately on to the theoretical luminosity vs. $\log T_e$ diagram. In the O spectral types this condition is not met.

Spectroscopic luminosity criteria have been selected empirically, and are calibrated in terms of visual absolute magnitude by means of standard objects whose luminosities are known from geometrical or other considerations. Two stars which have the same spectral type, thus the same T_e, but which differ in luminosity must differ in radius. The more luminous star has the greater radius. Spectroscopic luminosity criteria, therefore, should be sensitive chiefly to the radius of the star. Now, the luminosity of a star is determined by the radius of the photosphere, that is, of the region where an optical depth of about 0.5 is reached in the continuous spectrum. It is not necessarily determined by the radius of the line-forming region. On the other hand, spectroscopic criteria having to do with the intensities of lines determine the radius of the line-forming region. Nevertheless, spectroscopic luminosity based on the characteristics of the line-forming region can be used to infer M_v so long as the relationship between the radius of the line-forming region and the radius of the photosphere remains constant in the group of objects being classified.

An example will make these statements clear. In B-type stars the chief spectroscopic criterion for luminosity is the Stark broadening of the hydrogen lines. In the line-forming region of the atmosphere of a luminous star, the Stark broadening is less than it is in the line-forming region of a main-sequence star. It follows that the electron pressure is less there and that the radius of the supergiant is large. An important point is that the continuous opacity in a supergiant atmosphere is sufficiently great, at least in the normally observed wavelength range, that one observes only a more or less homogeneous, low-pressure section of the atmosphere. One does not see through to depths where the pressure is large. In a shell star, on the other hand, although the absolute strengths of some of the lines are similar to those in supergiants, the continuous opacity is much reduced and one can see through to the deeper layers of the star where the Stark broadening is large. It is found that the luminosity of a shell star is that of a main-sequence star even though some of the lines formed in the extended atmosphere are very like those formed in supergiant atmospheres. One must exclude the lines formed in a shell from the criteria before using a spectroscopic method based on line strengths to estimate the M_v's of O and B stars. This is necessary because in shells the ratio of the line opacity to the continuous opacity does not fall within the limits set by normal stars. Each luminosity criterion is calibrated in terms of luminosity by means of normal stars and thus is valid only for them.

The problem of separating supergiants from shell stars can be solved easily at type B when the spectra are studied, for one may readily see that the continuous absorption in the shell is much less than it is in a supergiant atmosphere which produces lines of Fe II, Ti II, Cr II, etc, and sharp hydrogen lines of the same strength as the shell lines. This state of affairs, $i.e.$, that objects of different characteristics may have lines of the same strength in their spectra, arises because the continuous opacity between 3650 Å and 8206 Å in B-type supergiant atmospheres and in shell stars comes from two sources: from absorption in the Paschen continuum of H I and from electron scattering. Depending on the extent of the atmosphere one or the other source may dominate. If at a certain range of spectral type the opacity comes dominantly from one source, be it electron scattering or be it absorption in continua or in lines, then all the stars with extended atmospheres giving absorption lines of a certain strength will correspond to about the same ratio of line to continuous absorption coefficient and they will all either correspond to shell stars, l_v/κ_v having a certain value but κ_v small so that the photosphere is small, or to supergiants where l_v/κ_v will have about the same value but κ_v is larger and thus the photosphere is large.

The above discussion can be summarized by saying that dilution effects in line spectra and the Stark broadening of the spectra of H, He I and He II can be used to distinguish objects in which the density is low in the line-forming region of the stellar atmosphere. This information must be coupled with information about the opacity in the continuous spectrum from the same layers in order to decide whether the star with an extended atmosphere is a supergiant or a shell star. Both shell stars and supergiants occur in B types, whereas among the stars of type A and later the stars with extended atmospheres all appear to be supergiants; among the O types the stars with extended atmospheres can best be classified as shell stars. The evidence for the last statement is outlined in the next section.

The importance of the continuous opacity in extended atmospheres for determining whether a star falls into the class of supergiants or into the class of shell stars was first clearly expressed by Struve (1942) in a review paper. He found that the observational data on supergiants and shell stars of types O and B could be understood by postulating that the atmosphere of each star consisted of three layers: (1) A main-sequence reversing layer which does not expand appreciably, (2) An inner stationary shell, which Struve designated as the chromosphere and which has the same velocity as the reversing layer, and (3) An outer expanding shell which Struve called a corona and which produces the P Cygni-type lines seen in many supergiants, and occasionally in shell stars.

What one sees of each of these parts of the atmosphere determines the spectral classification of the star. The theoretical interpretation of the observed line profiles and intensities must be based on a detailed study of the particular radiative and collisional transitions which may occur in the extended atmosphere. This art has progressed little since Struve first enunciated his ideas.

The usefulness of the proposed classification scheme depends upon the tacit assumption that the three layers are continuous and overlie one another so that we are not seriously concerned with a field of radially moving prominences which are essentially opaque to the underlying gases while the stationary reversing layer is seen between the prominences. In the case of shell stars where the underlying star is rapidly rotating, one tends to think that the shell will be confined closely to the equatorial plane and that the unshadowed B star is seen on either side of the shell. However, it cannot be entirely so, for the centers of strong lines like the shell-formed cores of the Balmer lines and the extraordinary strong NaI D lines seen in the spectrum of ζ Tau in October 1964 (Underhill 1965b) are very deep. This means that the absorbing atoms must be in front of most of the stellar disc.

The weakness of the continuous absorption from the $n = 3$ level of H I in the shells around B stars, in comparison to its strength in supergiants of types B8 to A2, may be due to departures of the hydrogen level populations from TE values. In supergiants there appear to be enough collisions to maintain an adequate population of the level $n = 3$ to provide considerable opacity, but in shells the third level is depopulated by radiative transitions $n = 3 \to 2$, which give Hα in emission, and it seems that the density is too low to permit sufficient collisions to repopulate the level to produce a large opacity in the Paschen continuum.

VI. THE CLASSIFICATION OF THE O-TYPE STARS

It has already been indicated in the preceding section that although it is possible to arrange the O stars in a sequence O5 to O9, one may not conclude that the effective temperature of these stars decreases monotonically along the sequence, for the chief spectral-type classification criteria are affected by dilution effects and by the state of motion in the extended atmospheres that appear to be present. Morgan and Keenan in setting up the MK system of spectral classification very wisely gave no luminosity classes for stars of spectral type O8 and earlier because they felt they could not define spectroscopic criteria that isolated groups of O stars of differing luminosities by the same T_c's. This point of view is supported by the discussion of the previous section. Although one may find criteria, based chiefly on dilution effects and on the state of motion of the extended atmosphere, which permit one to infer the presence of such an envelope, one finds no grounds for believing that the stars with extended atmospheres which are opaque in line frequencies have significantly larger photospheres than do those with less extended atmospheres.

The O stars are rare, and not many are bright enough nor close enough for one to find many standard stars of differing, accurately known absolute magnitudes for use in calibrating a selected spectroscopic criterion in terms of M_v. The survey by Underhill (1955) showed that there appeared to be little range in M_v among the northern O stars known to be members of clusters (and thus at known distances so that their M_v's could be found), and that the strengths of the hydrogen lines were not a useful criterion of M_v among the early O-type stars. Her conclusion was that the stars of types O8 and earlier all had about the same M_v. Further study has led to the conclusion that the Of stars are the equivalent of shell stars. The line spectra of these stars show dilution effects, and particular fluorescent emission lines indicate the presence of an extended low-density atmosphere opaque in line frequencies, but there is no evidence of the presence of a large photosphere as in a supergiant.

The problem of the luminosities of the O-stars has continued to receive attention. Botto and Hack (1962) and Hack (1963) have selected the following criteria as a mean by which to assign "luminosity classes" I to V as in the work of Morgan and Keenan:

1. the equivalent width of Hγ,

2. the ratio

$$r = \frac{W(\text{CIII } \lambda 4650 + \text{NIII } \lambda 4097 + \text{Si IV } \lambda 4088 + \text{Si IV } \lambda 4116)}{W(\text{H}\gamma + \text{H}\delta)},$$

3. the ratio $R_c/W(\text{H}\gamma)$, where R_c is the central absorption of Hγ, and

4. the quantum number n_m of the last visible hydrogen line.

The discussion of Sections 3 and 5 leads us to expect that the numbers found from criteria 2, 3, and 4 will increase as the atmosphere becomes more extended and less dense. These criteria indicate the presence of an extended line-forming region at low density. None of the above criteria indicate that the photospheric radius of the star is increased significantly over that of an O star having small values for these criteria. It is the author's experience with the line Hγ, that in the early O-type stars the equivalent width is more strongly affected by the state of motion in the outer atmosphere than by Stark effect. There is also the problem of separating the blend of Hγ with He II $\lambda 4338$. The writer considers the strength of Hγ to be of little value for indicating the electron pressure in an early O-type atmosphere.

Botto and Hack and Hack assign "luminosity classes" according to the values of the above criteria. It is regrettable that they have used this terminology because the words luminosity class when used with stars of types B0 and later are clearly related to the size of the photosphere and thus to the M_v of each star. Among the early O-type stars a similar correlation cannot be made. Botto and Hack and Hack do not attempt to find the distances of individual O stars by secure, well-established methods and then from such data to find the absolute magnitudes of enough standard O stars to give a calibration of their "luminosity classes" in terms of M_v. Rather they adopt the M_v's given by Underhill (1955) and by Kopilov (1958), and make their calibration. It is the writer's opinion that her own M_v's give no grounds for such a calibration, and that Of stars, that is, shell stars of type O, should not be confused with true supergiants which are stars of great luminosity. Supergiants have large photospheres in comparison to the photospheres of main-sequence stars of about the same T_e.

VII. CONCLUDING REMARKS

In the previous sections, attention has been focused on particular spectroscopic phenomena which indicate that a treatment of the problem of line formation as though the radiation field and atoms and ions were in thermodynamic equilibrium is not appropriate to the study of extended stellar atmospheres. It appears that all supergiants may be classified as stars with extended atmospheres, and that the main-sequence stars of types B2 and earlier possess sufficiently extended atmospheres that the results of departures from TE are seen in their spectra. Interpretation of the chief lines in the spectra of Wolf Rayet stars, of O stars, both normal and Of objects, and of Be and shell stars cannot proceed without a better theoretical treatment of line formation than that involving the assumption of LTE. Thus there is a pressing need for the development of a more realistic theory of line formation. This need is pressing also for the cooler stars because spectral classification systems are usually based on the intensities of the strong lines and it is just these lines which are formed sufficiently high in the atmosphere that level populations and radiative transfer processes are not those which occur in TE. Deductions made from the strong lines by standard LTE theories may be in error; yet it is on just such deductions that ideas concerning the evolution and development of stars are based.

A qualitative discussion of dilution effects in extended atmospheres and a glance at the major characteristics of O and B-type spectra of supergiants and shell stars indicates that the following two points should be taken into account in an improved theory of line formation.

(a). The detailed balance between collisional processes and specific radiative processes must be considered when computing the populations of the excited states of atoms and ions. The density in the line-forming region is low in the objects under discussion. The observation of Ca^+ "clouds" in the atmospheres of K giants such as 31 Cyg (see, for instance Wright and Odgers 1962) raises the suspicion that in some stars with extended atmospheres an irregular density structure might occur. However, taking account of this possibility can probably be left to a later improvement of the theory.

(b). The gases in the outer atmospheres of stars seem always to be in motion. In some cases there appears to be a general inflow or outflow. Always there is a randomly directed field of motion which is usually called turbulence. The field of motion, whether directed or chaotic, usually is not constant in its characteristics. Irregular fluctuations occur with time. These fluctuations may be connected with the small, irregular fluctuations in light which are shown by many early-type stars with extended atmospheres.

It thus appears that a rather complicated and sophisticated theory of line formation is required if quantitative information is to be deduced from the strengths and shapes of the lines formed in extended stellar atmospheres. Even rather general methods of taking into account turbulence (chaotic motions) and dilution effects would lead to significant improvements in estimates of the abundances of the light elements from analyses of line strengths in stellar spectra. It is a fact that the helium abundance has been estimated from the spectra of early B-type or late O-type stars ignoring dilution effects and the lifting of the curve of growth which results from the presence of a chaotic field of motion. Closer inspection of the spectra of the stars which have been studied shows that both chaotic motion and dilution effects occur. Inclusion of either or both of these effects in the theory would lead to a reduction in the estimated abundance of helium. Changes would also result in estimates of the abundances of other elements.

The problem of the formation of the strong lines in stellar atmospheres can perhaps be aided by the study of radiative transfer problems in plasmas in which shocks occur. Many of the spectroscopic phenomena which occur are the same as in stars, and comparable densities and temperatures are encountered in these laboratory situations to those that appear to be present in stars.

REFERENCES

Beals, C. S. 1950, *Pub. Dom. Ap. Obs. Victoria*, **9**, 1.
Botto, P. and Hack, M. 1962, *Mem. Soc. Astr. Ital.*, **32**, 159; *Contr. Milano-Merate Obs.* No. 194.
Bowen, I. S. 1935, *Ap. J.*, **81**, 1.
Froese Fischer, C. 1968, *Ap. J.*, **151**, 759.
Ghobros, R. A. 1962, *Zs. f. Ap.*, **56**, 113.
Gökgöz, A., Hack, M., and Kendir, I. 1962, *Mem. Soc. Astr. Ital.*, **33**, 239; *Contr. Milano-Merate Obs.* No. 196.
———. 1963, *Mem. Soc. Astr. Ital.*, **34**, 87; *Contr. Milano-Merate Obs.* No. 206.
de Groot, M. 1965, private communication.
Groth, H. G. 1960, *Zs. f. Ap.*, **51**, 206.
Hack, M. 1957, *Mem. Soc. Astr. Ital.*, **28**, 71; *Contr. Milano-Merate Obs.* No. 104.
———. 1963, *Mem. Soc. Astr. Ital.*, **34**, 1; *Contr. Milano-Merate Obs.* No. 202.
Kopilov, I. 1958, *Pub. Crimea Ap. Obs.*, **20**, 156.
Mathis, J. S. 1957, *Ap. J.*, **125**, 318.
Robbins, R. R. 1968, *Ap. J.*, **151**, L 35.
Struve, O. 1942, *Ap. J.*, **95**, 134.
Struve, O. and Roach, F. E. 1939, *Ap. J.*, **90**, 727.
Struve, O. and Wurm, K. 1938, *Ap. J.*, **88**, 84.
Underhill, A. B. 1952, *Pub. Dom. Ap. Obs. Victoria*, **9**, 139.
———. 1954, *Pub. Dom. Ap. Obs. Victoria*, **9**, 363.
———. 1955, *Pub. Dom. Ap. Obs. Victoria*, **10**, 169.
———. 1957, *Ap. J.*, **126**, 28.
———. 1963, *Bull. Astr. Inst. Neth.*, **17**, 161.
———. 1965a, *Bull. Astr. Inst. Neth.*, **18**, 5.
———. 1965b, *Sky and Telescope*, **29**, 219.
———. 1968a, *Ap. J.*, **151**, 765.
———. 1968b, *Bull. Astr. Inst. Neth.*, **19**, 537.
Underhill, A. B. and de Groot, M. 1964, *Bull. Astr. Inst. Neth.*, **17**, 453.
Wellmann, P. 1952, *Zs. f. Ap.*, **30**, 71.
Wright, K. O. and Odgers, G. J. 1962, *J. Roy. Astr. Soc. Can.*, **56**, 149; *Contr. Dom. Ap. Obs. Victoria* No. 77.
Wurm, K. 1937, *Zs. f. Ap.*, **14**, 321.

6. PECULIAR STELLAR SPECTRA

P. Swings and O. Struve: The Evolution of a Peculiar Stellar Spectrum:
Z Andromedae. *Astrophysical Journal*, **93**, 356, 1941.

P. Swings: Symbiotic Stars and Related Peculiar Objects.

THE EVOLUTION OF A PECULIAR STELLAR SPECTRUM
Z ANDROMEDAE

(With a Note on the Spectrum of IC 4997)

P. SWINGS AND O. STRUVE

ABSTRACT

The spectral lines of Z Andromedae have been measured on plates taken between May 12 and August 25, 1940, when the brightness had decreased to magnitude 9.7 after the outburst in 1939. The spectrum shows emission lines of a thin and dilute nebula and resembles the spectrum observed by H. H. Plaskett in 1923–1926. There is some evidence of a late-type companion, in addition to a hot nucleus, which resembles some of the slow novae, and a fairly pronounced P Cygni type spectrum of an expanding shell. The latter is a remnant of the shell which was formed in the outburst of 1939.

The nebular spectrum shows strong forbidden lines, but there are relatively few stages of ionization, which shows that the nebula consists of a fairly thin layer. The intensity of the auroral transition of [$O\ \text{III}$], λ 4363, is relatively strong—a phenomenon which is characteristic of other binaries of similar type, and of novae. Among the nebulae the stellar planetary IC 4997 shows the same feature. A list of lines of this object is given in Table 4.

I. INTRODUCTION

Z Andromedae is an exceptionally interesting object whose spectrum was first described in detail by H. H. Plaskett[1] on the basis of spectrograms taken during the period 1923–1926. At that time the spectrum showed no absorption lines but contained numerous bright lines, especially those of H, $He\ \text{I}$, $Mg\ \text{II}$, $Fe\ \text{II}$, $Ti\ \text{II}$, $He\ \text{II}$, $C\ \text{III}$, $N\ \text{III}$, $O\ \text{III}$, and the forbidden lines of [$O\ \text{III}$], [$Ne\ \text{III}$], [$Ne\ \text{V}$], [$Fe\ \text{III}$?] and [$S\ \text{II}$]. H. H. Plaskett divided the lines into two groups which he designated as the "stellar part" (lines of low excitation: H, $He\ \text{I}$, $Mg\ \text{II}$, $Fe\ \text{II}$, etc.) and the "nebular part" (lines of high excitation: $He\ \text{II}$, $N\ \text{III}$, [$O\ \text{III}$], [$Ne\ \text{III}$], etc.). The continuous spectrum was very strong in the red region, and on Plaskett's spectrograms Hogg[2] was able to see weak absorption band heads of TiO; these increased in intensity in the years following, when the star was fading in light. Mrs. Greenstein[3] determined the light-curve for the object until 1936; during the interval 1922–1936 the star slowly decreased in brightness with some irregular fluctuations. The light-curve has been extended by L. Campbell,[4] who recorded an increase in light from visual magnitude 10.7 in April, 1939, to 8.6 on August 21 and to 7.9 shortly afterward. Since the fall of 1939 the star has declined in brightness, reaching magnitude 9.7 on August 15, 1940.

It is tempting to consider Z Andromedae as a binary system consisting of a nova-like object, subject to moderate outbursts at somewhat irregular intervals of about fifteen years, and of a variable late-type companion whose period is probably not identical with the 650-day period suggested by Prager,[4] because the amplitude of the latter appears to be least when the nova-like object is faint.

This picture receives strong confirmation from a study of spectrograms taken at the McDonald Observatory during the past year. Spectrograms taken from September 14 to December 5, 1939, revealed a spectrum completely different from that described by

[1] Pub. Dom. Ap. Obs., Victoria, **4**, 119, 1928.

[2] Pub. A.S.P., **44**, 328, 1932; Pub. A.A.S., **8**, 14, 1934.

[3] Harvard Bull., No. 906, 1937.

[4] L. Campbell, Pop. Astr., **47**, 335, 447, 571, 1939; K. Himpel, Die Sterne, **19**, 210, 1939, and **20**, 14, 1940.

PLATE XVI

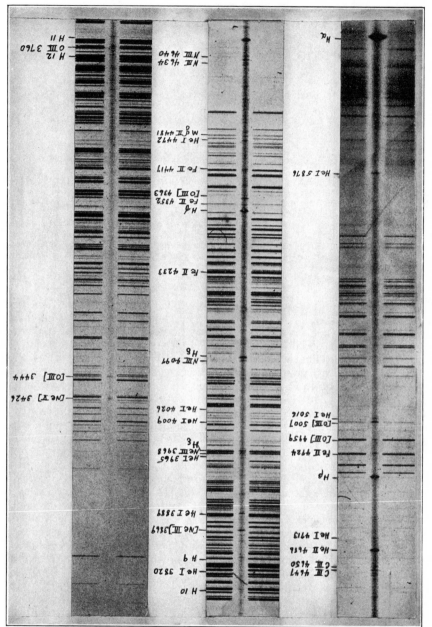

THE SPECTRUM OF Z ANDROMEDAE, AUGUST 15, 1940

Plaskett.[5] It was then of the P Cygni type with the "nebular part" of Plaskett's description completely missing. The spectrum showed many permitted lines of low excitation and was very similar to certain early stages of slow novae.

The transition from the high-excitation spectrum to one of P Cygni type must have occurred mainly between July 30 and September 14, 1939. On July 30, when the visual magnitude was about 8.9, the object was observed in Berlin-Babelsberg by P. Wellmann,[6] and the spectrum showed the lines of He II and N III; but there already seemed to appear weak absorption lines on the violet side of certain emissions, and traces of the TiO bands were still present at that time.[7]

The spectroscopic survey was resumed in the spring of 1940, when the star again became observable in the east; our first plate was taken on May 12. The entire aspect of the spectrum had changed. Again it resembled that described by Plaskett, although there were important differences. Strong nebular lines were again present; but, contrary to Plaskett's description, our spectrograms show many absorption lines. The development of the spectrum bears a strong resemblance to that of a slow nova.

This type of combination of a late-type star and an early-type companion—which, in this case, is nova-like—is by no means exceptional. Several other objects of the same type have recently been investigated, namely, T Coronae Borealis[8] (nova + M star), AX Persei and CI Cygni[9] (very high-excitation nebulosity + M star), RW Hydrae[10] (nebulosity of medium excitation + late K or early M), R Aquarii[11] (nebulosity + M7e + high-excitation companion), α Scorpii[12] (B4ne + M), o Ceti[13] (B8e + M), WY Geminorum[9] (B3e + M3e), W Cephei[9] (O? + Me), VV Cephei[14] (B9e + M2), etc.

The dynamical and physical processes involved in such binaries are not yet fully understood. It is probable that the emission lines are excited by the radiation of the early-type star outside its own reversing layer. There may be a departure from spherical symmetry in the excited layers, owing to the presence of the late-type companion. It is also probable that in some cases the radiation of the early-type component may excite the outer regions of the cool companion.

II. GENERAL DESCRIPTION OF THE SPECTRUM

The following description of the spectrum of Z Andromedae is based upon several spectrograms taken at the Cassegrain focus of the 82-inch reflector between May 12 and August 15, 1940, the best plate having been taken on August 15 with the quartz prisms and the 500-mm camera (dispersion 40 A/mm at λ 3933). The star was of about visual magnitude 9.7.

The visual region is quite strong, and a search for the TiO band heads found by Hogg (λλ 6159, 5448, 5167, and 4955) makes it appear probable that the red companion again begins to play an appreciable role in the visual region of the spectrum.

The results of the measurement of the spectrogram of August 15 are collected in Table 1, in which the late-type absorption features have not been included. The wave lengths have been corrected for the motion of the earth, but because of the complexity of the spectrum no attempt has been made to correct for the motion of the star.

Table 1 shows that the spectrum of Z Andromedae is very complex. We recognize at once the existence of the "nebula" whose spectrum was described by Plaskett. This

[5] Struve and Elvey, *Pub. A.S.P.*, **51**, 297, 1939; Swings and Struve, *Ap. J.*, **91**, 601, 1940.

[6] *Beobachtungs-Zirkular d. A.G.*, **21**, 105, 1939.

[7] *Vierteljahrsschrift d. A.G.*, **75**, 53, 1940.   [9] *Ap. J.*, **91**, 546, 1940.

[8] *Pub. A.S.P.*, **52**, 199, 1940.   [10] *Proc. Nat. Acad.*, **7**, 458, 1940.

[11] *Ap. J.*, **91**, 616, 1940; P. W. Merrill, *Spectra of Long-Period Variable Stars*, p. 82, 1940.

[12] *Ap. J.*, **92**, 316, 1940.

[13] P. W. Merrill, *op. cit.*, p. 75, 1940.   [14] C. and S. Gaposchkin, *Variable Stars*, 1938.

TABLE 1

LIST OF LINES IN Z ANDROMEDAE

λ	Int.	Identification* Elem.	Identification* λ	Identification* Int.
3121.5†.....	1E	O III	3121.71	5
3132.7†.....	6E	O III	3132.86	6
3202.9†.....	6E	He II	3203.16	....
3277.0†.....	1E	....	....	....
3281.1†.....	1E	....	....	....
3312.3†.....	4E	O III	3312.30	5
3323.6†.....	1E	....	....	....
3340.5†.....	5E	O III	3340.74	6
3345.8†.....	4E	$[Ne\ v]$	3345.8	....
3425.85.....	6E	$[Ne\ v]$	3425.8	....
3443.60.....	6E	O III	3444.10	5
3455.27.....	0E	O III	3455.12	5
3662.08.....	1A	H_{30}	3662.26	
3663.03.....	1A	H_{29}	3663.40	....
3664.25.....	1A }	H_{28}	3664.68	....
3664.92.....	0E }			
3665.70.....	1A }	H_{27}	3666.10	....
3666.53.....	0E }			
3667.47.....	1A }	H_{26}	3667.68	....
3668.26.....	0E }			
3669.12.....	1A }	H_{25}	3669.47	....
3669.94.....	1E }			
3671.21.....	1+A }	H_{24}	3671.48	....
3672.00.....	1E }			
3673.24.....	1+A }	H_{23}	3673.76	....
3674.42.....	1E }			
3676.07.....	2A }	H_{22}	3676.36	....
3676.94.....	1E }			
3679.10.....	2A }	H_{21}	3679.35	....
3679.75.....	1E }			
3682.37.....	2A }	H_{20}	3682.81	....
3683.27.....	1E }			
3686.19.....	1+A }	H_{19}	3686.83	....
3687.35.....	1E }			
3689.24.....	1+A }	....	....	....
3690.37.....	0E }			
3691.30.....	1+A }	H_{18}	3691.56	2
3692.21.....	1+E }			
3696.88.....	1+A }	H_{17}	3697.15	3
3697.79.....	1E }			
3704.18.....	1+A }	H_{16} / He I / Fe I / Ti II / Ca II	3703.85 / 3705.00 / 3705.57 / 3706.23 / 3706.03	4 / 30 / 700 / 125 / 40
3705.62.....	1+Enn }			
3711.56.....	1+A }	H_{15}	3711.98	5
3712.72.....	2E }			
3718.76.....	0A }	Fe I	3719.93	1000R
3719.91.....	0E }			
3721.40.....	1+A }	H_{14}	3721.95	6
3722.53.....	2E }			
3733.43.....	1A }	H_{13}	3734.37	8
3734.94.....	2E }			
3737.10.....	1E‡	Fe I / Ca II	3737.13 / 3736.90	1000r / 50

λ	Int.	Identification* Elem.	Identification* λ	Identification* Int.
3741.39.....	0E	Ti II	3741.64	200
3748.34.....	1E	Fe II / O II / Fe I / Fe I	3748.49 / 3749.47 / 3748.26 / 3749.49	8 / 125 / 500 / 1000r
3749.57.....	1+A }	H_{12}	3750.15	10
3750.86.....	2E }			
3755.00.....	2E	N III / O III / Fe II / He I	3754.62 / 3754.67 / 3755.56 / 3756.09	6 / 7 / 4 / 1
3757.81.....	1E	Ti II / Fe I / O III	3757.69 / 3758.23 / 3757.21	100 / 700 / 5
3760.08.....	2En	O III / Ti II / Fe II	3759.87 / 3759.30 / 3759.46	9 / 400 / 6
3762.35.....	0E	Fe II / Ti II / O II	3762.89 / 3761.32 / 3762.51	5 / 300r / 120
3770.52.....	1A }	H_{11} / N III	3770.63 / 3771.08	15 / 7
3771.25.....	2E }			
3795.27.....	1E	Fe I	3795.00	500
3797.30.....	1A }	H_{10}	3797.90	20
3798.69.....	3E }			
3819.35.....	1A }	He I / Fe I	3819.61 / 3820.43	50 / 800
3819.93.....	2E }			
3824.47.....	0E	Fe II	3824.91	4
3834.48.....	1A }	H_{9}	3835.40	40
3835.63.....	3E }			
3847.06.....	1E	....	....	......
3853.94.....	1E	Si II	3853.67	3
3855.92.....	3E	Si II	3856.09	8
3862.60.....	2E	Si II	3862.51	7
3868.75.....	7E	$[Ne\ III]$	3868.74	
3880.69.....	0A }	O II	3882.19	35
3882.22.....	0E }			
3887.84.....	5A }	He I / H_8	3888.65 / 3889.05	1000 / 60
3889.28.....	5E§ }			
3899.58.....	1E	Fe I	3899.71	500
3902.91.....	1E	Fe I	3902.95	500
3905.62.....	1Es	Si I / Fe I	3905.53 / 3906.04	20 / 5
3914.19.....	1En	Ti II / C II	3913.46 / 3918.98	70 / 80
3919.22.....	1En	C II / N II / O II	3920.68 / 3919.00 / 3919.28	200 / 35 / 35
3923.24.....	1E	Fe II / He II	3922.91 / 3923.51	600 / ....
3926.55.....	1E	He I	3926.53	7
3930.47.....	1E	Fe I	3930.30	600
3934.95.....	1E	Ca II / He I / Fe II / N III	3933.67 / 3935.91 / 3935.94 / 3934.41	600R / 4 / 6 / 3
3938.39.....	1+E	Fe II / N III	3938.97 / 3938.52	4 / 4
3945.35.....	1E	O II	3945.04	20
3961.77.....	1+E	O III / Al I	3961.59 / 3961.53	8 / 3000

* All the laboratory wave lengths and intensities have been taken from the M.I.T. wave-length table, except for *Fe* II (J. C. Dobbie, *Annals of the Solar Physics Observatory, Cambridge*, Vol. V, Part I) and the doubly or trebly ionized atoms (C. E. Moore, *Multiplet Table of Astrophysical Interest*). In the M.I.T. table the intensities in the arc or in the spark have been taken for the neutral or ionized elements. The lines of stellar intensity zero are uncertain.

† These lines were measured on a plate taken January 5, 1941.

‡ Probably absorption of P Cygni type.   § Violet wing weaker.

TABLE 1—*Continued*

λ	Int.	Elem.	λ	Int.
3964.84	2E	He I	3964.73	50
3967.33	2E	[Ne III]	3967.51	......
3970.34	4E	Hε	3970.07	80
3973.83	1E	O II	3973.27	125
4009.24	1E	He I	4009.27	10
4013.95	2A			
4014.85	1E	....	....	
4016.62	1+A			
4017.56	oE	....	....	
4025.56	2A			
4026.43	2E	He I	4026.19	70
4070.30	2En	C III	4068.94	10
		C III	4070.30	8
		O II	4069.90	125
		O II	4072.16	300
		[S II]	4068.50	
4073.89	2A			
4075.82	2E	O II	4075.87	800
4097.35	4Es	N III	4097.31	10
4101.97	6E||	Hδ	4101.73	100
4103.54	2E||	N III	4103.37	9
4121.31	1E	He I	4120.81	25
4123.65	oE||	Fe II	4122.64	4
4144.25	1En	He I	4143.76	15
4162.35	1E	C III	4162.80	4
4173.12	2En	Fe II	4173.45	8
4175.78	2A			
4178.17	2En	Fe II	4178.87	8
4225.15	oA	?Ca I	4226.73	500R
4226.21	oE	?Al II	4226.81	35
4230.95	1A			
4233.13	3E	Fe II	4233.17	11
4244.83	oEn	[Fe II]	4243.97	......
4264.55	1A	C II	4267.02	350
4267.17	1E	C II	4267.27	500
4277.23	oE	[Fe II]	4276.87	......
4288.70	oE	[Fe II]	4287.40	......
4340.67	8E	Hγ	4340.46	200
4351.83	3E	Fe II	4351.77	9
4363.46	6E	[O III]	4363.20	......
4366.92	oE	O II	4366.91	100
4369.24	1+E	?O II	4369.28	50
4379.66	oE	N III	4379.09	10
4384.16	1E	Fe I	4383.55	1000
		Fe II	4385.38	7
4387.63	2E	He I	4387.93	30
4395.19	1E	Ti II	4395.03	150
		O II	4395.95	80
4399.92	1E	Ti II	4399.77	100
4404.65	1E	Fe I	4404.75	1000
4413.86	oE	[Fe II]	4413.78	......
		Fe II	4416.82	7
4416.78	2En	[Fe II]	4416.28	......
		Ti II	4417.72	80
		O II	4416.97	150
4437.67	oE	He I	4437.55	10
4444.97	oE	?Ti II	4443.80	125
4447.41	oE	N II	4447.03	300
		O II	4448.20	70
4451.23	oE	Fe II	4451.54	4
4457.57	1E	?[Fe II]	4457.97	......
4470.08	1A			
4472.08	3E	He I	4471.48	100
4480.96	2E	Mg II	4481.33	100
4488.98	1E	Fe II	4489.19	4
4491.32	1E	Fe II	4491.41	5
4501.07	1E	Ti II	4501.27	100
4506.84	1A			
4508.26	1+E	Fe II	4508.28	8
4514.10	1A	Fe II	4515.34	7
4515.29	2E	N III	4514.89	7
4521.06	1E||	Fe II	4520.22	7
4522.89	1+E||	Fe II	4522.64	9
4528.75	1E	Fe I	4528.62	600
		?Al III	4529.18	6
4534.64	1E	N III	4534.57	3
		Ti II	4533.97	150
4540.92	1+E	He II	4541.61	5
		Fe II	4541.52	4
4549.04	1E‡	Fe II	4549.47	10
		Ti II	4549.63	200
4556.30	1E	Fe II	4555.89	8
4576.51	oE	Fe II	4576.33	4
4583.11	2E	Fe II	4583.83	11
4629.86	3E	Fe II	4629.34	7
		N II	4630.55	300
4631.69	1A?	N III	4634.16	8
4634.70	3En	Fe II	4635.33	5
4637.66	3A			
4640.30	4En	N III	4640.64	10
4645.74	3A			
4648.10	3En||	C III	4647.40	10
4651.05	2En||	C III	4650.16	9
		C III	4651.35	8
4653.87	1A?	C IV	4658.64	5
4657.92	1+E	[Fe III]	4658.18	......
4685.46	10E	He II	4685.75	300
4713.82	1+E	He I	4713.14	40
4861.59	10E	Hβ	4861.33	500
4920.08	2A	He I	4921.93	50
4922.59	2+En	Fe II	4923.92	12
4958.86	3E	[O III]	4958.91	......
5007.23	5Es	[O III]	5006.84	......
5012.21	2A			
5015.74	3E||	He I	5015.67	100
5018.93	2E||	Fe II	5018.43	12
5169	1E	Fe II	5169.03	12
5235	1E	Fe II	5234.62	7
5317	2E	Fe II	5316.61	8
5412	2E	He II	5411.55	50
5876	5Es	He I	5875.62	1000
6563	20E	Hα	6562.82	2000

\|\| Separation difficult, but the line seems definitely double.

spectrum consists of two groups of lines: the first includes forbidden and permitted lines usually associated with planetary nebulae; the other includes permitted lines, principally in the region λλ 4632–4658, of N III, C III, and C IV, bordered by absorption on the violet side. This second group—which exhibits some of the characteristics of lines in Wolf-Rayet stars, except that in Z Andromedae they are fairly sharp—we shall attribute somewhat arbitrarily to the hot nucleus. We do not wish to imply that this nucleus is a normal Wolf-Rayet star, and it is possible that what we have designated as the nuclear lines are really lines produced in the innermost layer of the nebular shell.

In addition to the "nebula" and the "nucleus" there is a spectrum of the P Cygni type. It consists of H, He I, and Fe II; and we are inclined to attribute to it also the bright lines of low excitation: Mg II, Si I, Si II, Ti II, Fe I, and extremely weak [Fe II]. This spectrum is the remnant of the P Cygni type spectrum which predominated in the fall of 1939, when the star was near maximum light. It has undergone large changes since December, 1939, having developed strong lines of He I and having lost most of its violet-absorption features. But it has not yet returned to the character of spectrum designated by Plaskett as "Z Andromedae-Star."

The strong continuous spectrum of Z Andromeda in the photographic region must come from the hot nucleus. It is too blue to be associated with the M star.

III. THE SPECTRUM OF THE NUCLEUS

Carbon is represented by very weak C II (λ 4267), strong C III ($\lambda\lambda$ 4648–4651), and C IV (λ 4658); all these bright lines are flanked by absorption components on the violet side. The carbon lines have the usual intensities of Wolf-Rayet stars and of pure re-

TABLE 2

VELOCITIES OF EJECTION

Element	V_{ej} in km/sec
Balmer lines of P Cygni type	83
λ 3889 He I	111
Other lines of He I	81
C III	174
N III	182
C II	186

combination spectra; in an O star of the 9 Sagittae type, λ 5696 C III would be an emission line instead of $\lambda\lambda$ 4648–4651, according to previous investigations by the authors.[9]

Nitrogen is characterized by the strong bright lines of N III at λ 4634 and λ 4641, with P Cygni absorption components. It is difficult to decide whether broader lines are superposed on the strong, sharp nebular N III lines measured at $\lambda\lambda$ 4097–4103.

Several weak O III lines which cannot be excited in the nebulosity by Bowen's fluorescence mechanism probably belong to the nucleus. The question arises whether the lines of H and He I also belong to the nucleus. Strong evidence against such an assumption is found in the measured differences between the radial velocities of the emission and absorption components of the lines. These are collected in Table 2.

The spectrum of the nucleus contains both nitrogen and carbon. Like the nucleus of NGC 6543,[15] it is an object intermediate between the usual carbon and nitrogen sequences. It would, indeed, be difficult to apply to it the usual classification criteria for Wolf-Rayet stars.[16] For example, we are not allowed to use the intensities of the He I lines which belong mainly to the P Cygni layer or those of the He II lines which belong to the nebula. This difficulty presents itself in all cases where a Wolf-Rayet star having relatively narrow lines is surrounded by an emission nebulosity. The absence of N v and the intensity ratio of the C IV and C III lines suggest that the nucleus may have a temperature of the order of 70,000°.

This would be in agreement with the lines observed in the nebula, where [Ne III] is stronger than [Ne v]. It should be remembered that the lines of the nucleus of Z Andromedae are much sharper than those of Wolf-Rayet stars. But the nucleus of HD 167362, which is an object resembling Campbell's hydrogen-envelope star, has also fairly narrow lines of Wolf-Rayet type. It is possible that the nucleus of Z Andromedae is of the same character.

[15] $Ap. J.$, **92**, 289, 1940.   [16] $Trans. I.A.U.$, **6**, 248, 1938.

IV. THE NEBULAR SPECTRUM

The following observed lines belong to a dilute nebulosity: (*a*) the transitions of nebular type of [*O* III] (int. 3–5), [*Ne* III] (int. 7–3), and [*Ne* v] (int. 6–4); (*b*) the transition of auroral type of [*O* III] (int. 6); and (*c*) the lines of *He* II (at least in part) and the fluorescence lines of *O* III and *N* III excited by the resonance line of *He* II. The emission lines of *H* and *He* I may also belong in part to the nebulosity. The excitation of this nebular spectrum may be attributed to the hot nucleus described in section III. The mean radial velocities of the pure emission lines are: forbidden lines, +5.8 km/sec; *N* III 4097–4103, +7.6 km/sec; Balmer lines (*Hε–Hβ*), +16.0 km/sec.

Since the nucleus continuously ejects nitrogen and since λλ 4097–4103 are fairly strong in the nebula, we should have expected to observe [*N* II] or [*N* I] in the nebulosity.

TABLE 3

COMPARISON OF INTENSITIES OF EMISSION LINES IN PLASKETT'S "NEBULAR PART"

λ	Elem.	INTENSITY		NOTES	λ	Elem.	INTENSITY		NOTES
		Plaskett	Swings-Struve				Plaskett	Swings-Struve	
3425.8.......	[*Ne* v]	4	6		4634.2.......	*N* III	3	3	
3444.1.......	*O* III	4	2		4640.6.......	*N* III	6	4	
3759.9.......	*O* III	3	2	1	4647.4.......	*C* III	2	3	
3868.7.......	[*Ne* III]	2	6	2	4650.2.......	*C* III	4	2	
3967.5.......	[*Ne* III]	2	2		4685.7.......	*He* II	9	10	
4068.9.......	*C* III	<1	2		4958.9.......	[*O* III]	2	3	
4097.3.......	*N* III	2	4		5006.8.......	[*O* III]	5	5	
4199.9.......	*He* II	<1	0		5411.6.......	*He* II	2	2	
4363.2.......	[*O* III]	6	6		5801.4.......	*C* IV	2	0	
4541.6.......	*He* II	3	1						

1. Blend.
2. Plaskett estimates intensity 6 on ultraviolet spectra.

In reality, neither their nebular nor their auroral transitions are observed.[17] This is certainly an effect of ionization, as [*O* II] and [*O* I] are also absent despite the high intensity of [*O* III]. This is quite different from what we normally observe in planetary nebulae and in the nebulosities of slow post-novae. For example, in Nova Herculis, 1934, we observe at the present time [*O* I], [*O* II], [*N* II], [*S* II], [*Fe* VII], and other elements,[18] in addition to strong [*O* III], [*Ne* III], and [*Ne* v]. Essentially similar elements are also present in the extremely slow nova RT Serpentis, 1909. The absence of other stages of ionization in the nebulosity of Z Andromedae suggests that this nebulosity consists of a rather thin layer.

The intensities of the nebular and nuclear lines, which we observed·in the summer of 1940, closely resemble those observed by Plaskett[1] in 1923–1926. Table 3 contains those lines which Plaskett attributed to the "nebular part" and which are not seriously blended with lines of the P Cygni spectrum. There are only minor differences in the intensities. We conclude that the nucleus and the nebula have not been appreciably altered by the outburst.

It is of great interest to compare Z Andromedae with the binaries of similar excitation. T Coronae Borealis, which, like Z Andromedae, consists of an M-type star and of a nova-like companion, has shown an irregular increase in brightness since 1936. The forbidden

[17] [*O* I], [*O* II], and [*N* II] were also absent on Plaskett's spectrograms.

[18] *Ap. J.*, **92**, 295, 1940.

lines now present in its spectrum belong to [O III], [O II], and [Ne III].[8] Hachenberg and Wellmann[19] have shown that the nebulosity of T Coronae was optically too thin to permit the application of the theory of Zanstra, which requires complete absorption of all the nuclear radiation beyond the Lyman limit. In RW Hydrae [O II] is extremely weak, whereas [O III] and [Ne III] have similar intensities.[10]

The striking intensity of the transition of auroral type, λ 4363, compared with the transitions of nebular type, N_1 and N_2, is well known in the novae at the beginning of their nebular stages. It is also a characteristic feature of all similar binaries, such as T Coronae, AX Persei, CI Cygni, RW Hydrae, etc. On the other hand, in only one planetary nebula observable in our latitude, IC 4997, is the auroral transition λ 4363 very strong compared with N_1 and N_2. In almost all other nebulae, N_1 and N_2 are much stronger than λ 4363.[20] Actually, IC 4997 is often considered as an object intermediate between nebulae and stars. The following section contains a description of its spectrum.

V. THE SPECTRUM OF IC 4997

Except for the high intensity of λ 4363, the spectrum of IC 4997 has the usual appearance of a planetary nebula, according to W. H. Wright[21] and R. H. Stoy.[22] It is certainly not a nebula of very high excitation,[23] because [Ne v] is absent. On the other hand, there is no reason to suspect an abnormally low abundance of oxygen, nitrogen, etc. Hence, the cooling effect[24] of the elements other than H upon the electrons, resulting from the photoionization of hydrogen, must be of the usual type. According to the work of Menzel and his collaborators,[25] the presence of a strong λ 4363 implies that the electron density of IC 4997 is much higher than in the other nebulae thus far investigated—say between 10^5 and 10^6 electrons per cubic centimeter, instead of 10^4. The following questions arise in connection with this stellar planetary: (a) Of what type is the exciting nucleus (which has not been observed previously)? (b) How do the elements other than [O III] behave? (c) Is there any way to observe "stratification-effects" in stellar planetaries of higher electron density and in which slitless spectra cannot reveal appreciable differences among the monochromatic images?

We secured in June and August, 1940, four spectrograms of IC 4997 with the 82-inch McDonald reflector. Three were taken with the quartz prisms and the F/2 Schmidt camera (exposures, 1^h20^m, 1^h57^m, 8^h44^m). The fourth was obtained with the glass prisms and the same camera (exposure, 8^h40^m). On the two long-exposure spectrograms the continuous spectrum is relatively strong and the Balmer continuum is very strong, extending at least to λ 3300. Superimposed on the strong Balmer continuum are eight lines of He I which we could measure only with difficulty;[26] these have not been included in Table 4, which contains all the other lines. The radial velocities given by the different lines are entirely consistent, and Table 4 gives the wave lengths corrected for the motion of the object.

[19] *Zs. f. Ap.*, **17**, 246, 1939.

[20] The majority of the brighter Magellanic nebulae are similar to IC 4997 (C. and S. Gaposchkin, *op. cit.*, p. 316, 1938).

[21] *Lick Obs. Pub.*, **13**, 193, 1918.

[22] *Lick Obs. Bull.*, **17**, 179, 1935.

[23] Even if the excitation were high, this would not have a great influence upon the electron temperature. Menzel and his collaborators have shown that the electron temperatures, which range from 6,000° to 10,000°, seem to be independent of the temperature of the central star.

[24] Because of the loss of energy by the electrons in the collisional excitation of the metastable levels of O III, Ne III, etc. (Menzel).

[25] Series of papers in *Ap. J.*, 1938–1940.

[26] Wave lengths: λλ 3634 (2), 3614 (2), 3599 (1), 3587 (1), 3554 (1), 3537 (1), 3530 (1), 3517 (1).

a) THE SPECTRUM OF THE NUCLEUS

The main characteristic of the nucleus is the group from λ 4637 to λ 4690. There is probably also a trace of *N* III 4634. Some *He* I lines seem to have weak P Cygni absorption components, which would indicate that *He* I plays some role in the nuclear spectrum. There is also a trace of *N* IV 4058 and of *C* III 4070, which belong to the nucleus.

It is immediately seen that the group of *N* III, *C* III, and *C* IV lines near λ 4650 is similar to that observed in the nucleus of Z Andromedae. *C* III and *C* IV are of the regular Wolf-Rayet type; *N* III is more nearly of the 9 Sagittae type. The nucleus must

TABLE 4

LIST OF LINES IN IC 4997

λ	Int.	Identification Elem.	Identification λ	Identification Int.	Note
3676.7......	1	H_{22}	3676.36	......	
3679.5......	1	H_{21}	3679.35	......	
3683.0......	1	H_{20}	3682.81	......	
3687.0......	2	H_{19}	3686.83	......	
3691.09.....	3	H_{18}	3691.56	2	
3696.98.....	3	H_{17}	3697.15	3	
3703.87.....	3	H_{16}	3703.85	4	W
3711.91.....	4	H_{15}	3711.97	5	W
3721.92.....	4	H_{14}	3721.94	6	
3726.1......	7	[O II]	3726.12	....}	W1
3728.7......	5	[O II]	3728.91	}	W
3734.25.....	4	H_{13}	3734.37	8	
3744.37.....	1	......	......	......	
3750.03.....	5	H_{12}	3750.15	10	W
3770.87.....	5	H_{11}	3770.63	15	W
3797.93.....	6	H_{10}	3797.90	20	W
3819.59.....	3	He I	3819.61	50	W
3835.53.....	6	H_9	3835.39	40	W
3868.82.....	15	{[Ne III] {He I	3868.74 3867.5	15}	W
3888.48*....	7	He I	3888.65	1000}	W1
3888.99*....	8	H_8	3889.05	60}	
3964.71.....	3	He I	3964.73	50}	
3967.40.....	10	[Ne III]	3967.51	....}	W1
3970.14.....	8	Hε	3970.07	80}	
4009.11.....	2	He I	4009.27	10	
4026.30.....	4	He I	4026.19	70	W
4058.2......	on	N IV	4057.8	2	
4068.70.....	3	[S II]	4068.62	......	W
4071.0......	1	C III	4070.43	8	
4076.37.....	2	[S II]	4076.22	......	W
4098.0†.....	0	N III	4097.31	10	
4101.70.....	10	Hδ	4101.75	100	W
4120.67.....	2	He I	4120.81	25	
4143.56.....	2	He I	4143.77	15	
4340.63.....	12	Hγ	4340.46	200	W
4363.25.....	15	[O III]	4363.2	......	W
4387.79.....	3	He I	4387.93	30	
4471.41.....	6	He I	4471.48	100	W
4637.06.....	2A}	N III	4640.64	10	
4640.64.....	2En}				
4645.22.....	2A}	C III	4647.40	10	
4649.11.....	2+En}	C III	4650.16	9	
		C III	4651.35	8	
4653.33.....	1+A}	C IV	4658.64	5	
4658.86.....	2En}				
4685.5......	1+n	He II	4685.81	300	I
4702.2......	2n				
4712.8......	3	{He I {[A IV]	4713.14 4711.4	40}	
4740.6......	1–2	[A IV]	4740.3	......	
4861.5......	18	Hβ	4861.34	500	W
4921.8......	2+s	He I	4921.93	50	
4958.8......	20	[O III]	4958.91	......	W
5006.4......	25	[O III]	5006.84	......	W
5016........	4	He I	5015.67	100	
5048........	1	He I	5047.74	15	
5755........	3	[N II]	5755.0	......	S
5876........	8	He I	5875.62	1000	S
6300........	6	[O I]	6300.2	......	S
6310........	4	[S III]	6310.2	......	S
6364........	3	[O I]	6363.9	......	S
6548........	5	[N II]	6548.4	......	S
6563........	25	Hα	6562.82	2000	S
6584........	7	[N II]	6583.9	......	S
6678........	3	He I	6678.15	100	S

* Separation difficult.
† Doubtful line.
W. Observed by W. H. Wright.

W. 1. Observed by Wright as a blend.
S. Observed by R. H. Stoy.
1. Also present in BD +30°3639.

be a Wolf-Rayet star containing both *N* III and *C* III and *C* IV. NGC 6543 is another planetary whose Wolf-Rayet nucleus also contains both nitrogen and carbon with similar intensities, but it is of earlier type than in IC 4997. We should classify the nucleus of IC 4997 as W7 or W8. The bright lines are abnormally narrow, but they are appreciably broader than the lines of the nebula.

The simultaneous presence of nitrogen and carbon in several planetary nuclei of Wolf-Rayet type is of great importance for the classification of the Wolf-Rayet stars. Other nuclei are typical members of the usual sequences; for example, the nuclei of Campbell's envelope star[27] and of HD 167362[28] are pure carbon stars containing no nitrogen, despite the high abundance of nitrogen in the surrounding nebulosities.

[27] *Proc. Nat. Acad.*, **26**, 548, 1940. [28] *Ibid.*, p. 454.

b) THE SPECTRUM OF THE NEBULA

The nebula is responsible for the continuous spectra[29] at the *H* limits, for the Balmer series up to H_{22}, for a rich spectrum of *He* I, and for a large number of forbidden lines. The latter are collected in Table 5.

There is no trace of the auroral transition of [*O* I] at λ 5577, although the transition probabilities are practically the same for [*O* I] and [*O* III].[30] This suggests stratification. We should expect the collisional cross-sections to be of the same order of magnitude for the corresponding metastable levels ^{1}S of *O* I (4.2 v.) and *O* III (5.3 v.). The different behavior of [*O* I] and [*O* III] may be due to the fact that the O^{++} ions are excited near the nucleus, whereas neutral oxygen occurs mostly at the outskirts of the nebula where the density is considerably reduced. [*S* III] also shows a strong line of the auroral type. It

TABLE 5

FORBIDDEN LINES IN IC 4997

NEBULAR TYPE		AURORAL TYPE		TRANSAURORAL TYPE	
Element	Intensity	Element	Intensity	Element	Intensity
O I..............	6–3	*O* III	15	*S* II	3–2
O II.............	7–5	*N* II	3	....	....
O III............	25–20	*S* III	4	....	....
N II.............	5–7	....	....	....	....
Ne III...........	15–9	....	....	....	....
A IV.............	3–2	....	....	....	....

would be interesting to search for the auroral transitions of [*O* II] at λλ 7319.0–7330.3 and for the nebular transitions of [*S* II] at λλ 6717.3–6731.5. This region is not covered by our spectrograms. [*Ne* III] is very strong: λ 3869 is certainly as strong as $H\gamma$. On our spectrograms we do not find the auroral transition of [*Ne* III] at λ 3342.8; it must be much weaker than the transitions of nebular type. [*Ne* III] is probably excited in the denser regions near the nucleus, together with [*O* III], [*A* IV], and, to a lesser extent, [*S* III]. The absence or weakness of the auroral transition of [*Ne* III] may be due either to the higher excitation potential required (6.9 v.) or to specific atomic properties affecting the collisional cross-section. The ionization potentials of Ne^+ and A^{++} being 40.9 v. and 40.7 v., respectively, we should expect the presence of [*A* IV] because we know that [*Ne* III] is very strong. The auroral transition of [*Ne* III] is also absent in Z Andromedae.

The ratio of intensity of the auroral to the nebular transition is much smaller for [*N* II] than for [*O* III]. This can also be understood if we assume that the collisional cross-sections for [*O* III] and [*N* II] are the same, because N^+ must extend much farther from the nucleus than O^{++}, but not as far as neutral oxygen.

VI. THE SPECTRUM OF THE P CYGNI TYPE OF Z ANDROMEDAE

In the Balmer lines the ratio of intensity of the emission components and the absorption components decreases steadily toward the higher members. From $H\epsilon$ to $H\alpha$ the

[29] A part of the continuous spectrum is probably due to the nucleus.

[30] The transition probabilities are shown in the accompanying table.

	Nebular	(^{3}P –^{1}D)	Auroral (^{1}D –^{1}S)
[*O* III]............	0.016	0.0055	2.7
[*O* I].............	0.013	0.0040	2.8

lines appear purely in emission. The Balmer continuum is still weakly present in absorption,[31] but it has decreased much since last fall, together with the absorption components of the lines of H, Ti II, Fe II, etc.

The radial velocities of the H and He I lines are collected in Table 6; all the Balmer lines with definite P Cygni characteristics have been included (from H_{15} to H_9), and they give very consistent results.

The lines of Ti II, Fe I, and $[Fe$ II$]$ are all very weak, and some identifications may be doubtful, but their presence may be regarded as very probable. The extreme weakness of $[Fe$ II$]$ and the absence of forbidden lines of any other element indicates that the P Cygni layer is rather dense or that its distance from the exciting nucleus is small.

In many respects this part of the spectrum is very similar to the present spectrum of BD$+11°4673$,[9] which also shows Fe I, Ca I, Si I, and Si II; but higher stages of excita-

TABLE 6

RADIAL VELOCITIES OF THE LINES OF P CYGNI TYPE

(In Km/Sec)

Lines Used	vEmission	vAbsorption	Differences
Balmer lines.............	$+46.2$	-37.2	83.4
λ 3889 He I.............	$+50.1$	-60.9	111
Other He I lines.........	$+27$	-53.8	80.8

tion are present in BD$+11°4673$. Of course, the comparison with a P Cygni type star which is single may be quite artificial, as the presence of the late-type companion may substantially distort the P Cygni layer. The dilution effect is apparent in the absorption component of He I 3889. This line, which arises from the metastable $2s^3S$ level, shows a fairly strong absorption component. The line λ 3965, which arises from the metastable $2s^1S$ level, is complicated by the strong neighboring line of $[Ne$ III$]$ 3967, but we believe that it has also a violet absorption line. In the series $(2p^3P^0 - nd^3D)$ of He I, λ 4026 shows an absorption line of intensity 2, while λ 4472 shows one of intensity 1. The fact that these triplet lines are present at all in absorption proves that the dilution is not excessive. The dilution factor can hardly be less than 0.01, and we doubt that it is more than 0.1. The P Cygni type shell should, therefore, have a radius $r \sim 5R$.

The selectivity observed in the Si II spectrum is striking. The presence of the group λλ 3853.7–3856.0–3862.6 $(3s\ 3p^2\ ^2D - 3s^2\ 4p\ ^2P^0)$ is certain, although the group λλ 4128–4130 $(3s^2\ 3d\ ^2D - 3s^2\ 4f\ ^2F^0)$ is absent; this had been noticed by H. H. Plaskett. It is also observed in P Cygni itself and in BD $+ 11°4673$.[9] It may be due to the fact that the lower level $3s\ 3p^2\ ^2D$, although not really metastable, is connected with the ground level $3s^2\ 3p\ ^2P^0$ by a weak transition (weak lines at λ 1817 and λ 1808). It should be noticed that the electron configuration of the ^2D level is $3s\ 3p^2$, whereas all the other terms giving strong lines are due to the addition of one excited electron to the closed subshell $3s^2$.

VII. CONCLUSIONS

During the outburst of Z Andromedae in 1939 the only spectrum which could be observed was that of the P Cygni type expanding shell. The M spectrum, the spectrum of the nebula, and that of the nucleus were not visible. At the time of our observations the visual magnitude of Z Andromedae was 8.2.

In the summer of 1940, when the visual magnitude was 9.7, the P Cygni shell was still present, but the violet-absorption components were weaker and the emission spectrum

---

[31] On Plaskett's spectrograms the Balmer continuum was in emission. On a spectrogram which we have taken on January 5, 1941, the Balmer continuum is clearly in emission and the P Cygni features are no longer visible.

had changed from that corresponding to an A star to one corresponding to a B star. Its evolution resembled that of a slow nova and evidently approached the kind of spectrum which Plaskett observed in 1923–1926. The forbidden lines of [Fe II], which were strong on Plaskett's plates, were still very weak on our 1940 plates.

The velocity of expansion of this shell has been approximately 100 km/sec. Since about 450 days have elapsed since the beginning of the outburst, the shell must have expanded over a distance of about 10^9 km. But the observable spectrum of the shell may not come from the same layer if material is being fed into the expanding mass over an appreciable period of time. Hence, the value of 10^9 km represents the upper limit of our estimate for the present distance between the P Cygni type shell and the surface of the nucleus.

When the star's brightness had declined to magnitude 9.7, the M star, the nebula, and the nucleus were again present in the spectrum. To all appearances they had not changed since before the outburst. It is possible that they never did change appreciably during the outburst and that their absence in the fall of 1939 was caused solely by the brightness of the P Cygni shell which required relatively short exposures. A search for the strong nebular emission line He II 4686 on our best plates of last fall shows that it was not visible at that time, also that it would not have been complicated by blending with strong lines of the P Cygni spectrum. The very strong nebular line should have shown quite readily with the shorter exposure required for Z Andromedae last fall, provided it was not superposed over a strong continuous spectrum. Whether it would still have been visible on top of the continuous spectrum is difficult to decide, but we have the impression that the line was really weaker last fall.

It is of interest to compute the relaxation time, τ, considered by Grotrian.[32] If we suppose that the ultraviolet exciting radiation of the nucleus was completely extinguished by the dense P Cygni shell in 1939 and that the nebular radiation had ceased completely, then τ would measure the length of time a pure hydrogen nebula would require to return to within 1.67 mag. of the final (normal) brightness. We follow Grotrian, but assume for the radius of the nucleus

$$R^* = 0.3R\odot ,$$

$$T^* = 70,000° .$$

We also assume, following Menzel, that the

$$\text{Number of electrons per cm}^3 = 10^6 .$$

Using Cillié's data for hydrogen, we find for the volume of the nebula

$$V = 10^{48} \text{cm}^3$$

and

$$\tau \sim 2 \times 10^6 \text{ sec} \sim 3 \text{ or } 4 \text{ weeks} .$$

This is quite consistent with the observations, if we assume that the P Cygni shell in August, 1940, was completely transparent to the ionizing radiation.

The remarkable similarity of the nebular spectrum before and after the outburst suggests that its distance from the nucleus is great compared with the radius of the P Cygni type shell. The best estimate we can get for the latter depends upon the dilution factor:

[32] Zs. f. Ap., **13**, 228, 1937.

$r = 5R^*$. The radius of the nucleus is not known. But all available evidence points to a radius R^* for a nova-like object which is considerably smaller than the radius of the sun.[33] If we adopt $R^* = 0.3R\, \odot$, the radius of the P Cygni type shell would be about $1.5R\, \odot$ or, roughly, 10^6 km. This is much smaller than the upper limit derived from the velocity of expansion. We are inclined to believe that the phenomenon cannot be treated as a single, thin layer which expands with a velocity of 100 km/sec. This is not surprising, since in all normal P Cygni type stars we are already accustomed to think of a continuous process of ejection. Once we accept the order of magnitude suggested for R^*, the small radius of the P Cygni shell follows from the dilution effect and from the weakness of the forbidden lines.

The radius of the nebula remains unknown. But the order of its size may be inferred from the estimate of the volume which we have made previously:

$$V = 10^{48} \text{cm}^3 .$$

If the entire sphere were occupied, we should find for the radius

$$r_{\text{nebula}} = 10^{16} \text{cm} = 10^{11} \text{km} .$$

This is of the same order of magnitude as the radius of the forbidden [Fe II] nebula which surrounds the B-type companion of α Scorpii.[34] This nebula, located at a distance of 100 parsecs, has a radius of $3''$.

The entire picture of the system agrees well with that which results from Kuiper's dynamical theory[35] of binaries, such as β Lyrae, WY Gem, etc. The late-type component of Z Andromedae must be a supergiant, and its outer atmosphere may have a radius which is considerably larger than that determined with the interferometer for α Orionis and α Scorpii. It is entirely possible that the nebular material of Z Andromedae is concentrated within the limiting surface computed by Kuiper.

We are indebted to Professor Leon Campbell of the Harvard Observatory for information concerning the light-curve of Z Andromedae, and to Dr. D. M. Popper for some of the spectrograms taken last winter.

McDonald Observatory
AND
Yerkes Observatory
September 1940

[33] We are indebted for this suggestion to Dr. G. P. Kuiper.

[34] $Ap.\ J.$, **92**, 316, 1940.   [35] $Ap.\ J.$, **93**, 133, 1941.

SYMBIOTIC STARS AND RELATED PECULIAR OBJECTS

P. SWINGS

Institut d'Astrophysique
Université de Liège

I. INTRODUCTION

Stellar classification in spectral types and absolute luminosities is based on the absolute or relative intensities of the absorption or emission lines of one or several atoms or molecules. The absorption or emission intensities, or more generally the profiles, result from the combined effects of absorption and emission in the stellar atmosphere. Certain geometrical, physical, kinematical, and dynamical factors may affect the intensities of emission or absorption of certain lines, rendering them anomalous in the sense that they differ from those of most stars having otherwise similar general characteristics. Since these factors may thus upset the spectral classification or even render it precarious or erroneous, it is important to gain a general understanding of them. Such a study will help in realizing the full meaning of normal intensities, hence of the normal classification. Actually one may wonder whether any star may be regarded as entirely free of peculiarities, or at least of a predisposition to develop them, either from time to time, or at certain evolutionary stages.

The factors introducing spectral anomalies are best studied in the case of conspicuously peculiar stars. In such cases, the mechanisms producing ionization, dissociation, and excitation must be examined individually. The present report is essentially devoted to the objects which received from Paul W. Merrill the name of "symbiotic stars"; in these objects high excitation emission lines are superposed on a low-temperature absorption spectrum, usually of type M. The symbiotic stars yield "combination spectra." In biology the word "symbiosis" indicates an actual dependence or interdependence of dissimilar organisms. This biological expression seems thus well adapted to the group of peculiar stars which we shall consider, since these stars are indeed characterized by a similar spectroscopic behavior of apparently inconsistent features. The symbiotic stars are extreme pathological cases.

In order to gain some understanding of the symbiotic stars (abbreviation: SS) I shall consider a wide range of other peculiar objects whose behavior will help in explaining various aspects of the SS. Obviously I shall not be able to describe in detail all the spectral and physical characteristics of the various peculiar classes, and so I shall confine myself to these characteristics which may help in explaining the phenomena observed in the SS. Occasionally priority will also be granted to phenomena observed at the McDonald Observatory, in the large program carried out under the leadership of Otto Struve. The emphasis will also be placed on individual objects, as our understanding of the nature and evolution of the astronomical bodies has been and still is conditioned to a considerable extent by the progress of individual spectroscopic investigations. Indeed we would have progressed neither far nor deeply in our physical knowledge of the stars and of their origin and evolution without the detailed spectroscopic investigations of the type carried out by Struve. To promote further progress, two factors which Struve stressed on numerous occasions remain as essential now as they were in 1939 when the 82-inch Otto Struve Memorial reflector became operational: widen the covered spectral range as well as the geometrical and the spectral resolution; accelerate the securing of observational material, especially by increasing the sensitivity of the receivers and adapting the newly discovered optical, photographic, or electronic equipment.

My task has been rendered easier by the fact that Miss Anne B. Underhill prepared the preceding chapter on the formation of lines in extended atmospheres, and was generous in letting me read her original text before mine was typed. I wish to acknowledge her help and courtesy. Her detailed discussions enable me to shorten several of my own.

In all peculiar stars, including the Wolf-Rayet, Of-, P Cygni-, Be-type and shell stars, and the novae and planetary nebulae, the observed spectra are superpositions of features that are formed in different regions of the atmosphere. In particular the effects of the geometrical and physical dilution factors[1] may be different for various absorption or emission lines. Many selectivities affecting the relative intensities of transitions in the same atom or ion disappear when the dilution decreases.

The prototype of the SS is Z And which we shall discuss in some detail, beyond the discussion in the paper "The Evolution of a Peculiar Shell Spectrum: Z Andromedae" (Swings and Struve 1941) reproduced in this volume. The brightness of this star fluctuates semi-regularly; the fluctuations are generally rather small, but at times they may amount to 3 magnitudes, the rises in brightness being steeper than the declines.

Most SS have been listed in Bidelman's "Catalogue and Bibliography of Emission Line Stars of Types later than B" (1954) and in Mrs. Gaposchkin's book *The Galactic Novae* (1957). A few recently discovered objects may be added. We shall consider here:

Z And (the prototype);
AG Peg, AX Per, CI Cyg, BF Cyg, RW Hya, Nova (T) CrB, R Aqr, Nova (RS) Oph, FR Sct, MWC 603.
The southern SS's WY Vel and AR Pav have been studied by Sahade (1952).

For comparison, the following objects which are related to the SS will also be considered:

17 Lep, AX Mon, RX Pup, MWC 17, RY Sct, CD $-27°$ 11944, MHα 328–116 MWC 349, B 1985, B 5481, WY Gem, W Cep, VV Cep, the companion of α Sco, W Ser, α Her, α Sco A, M1–2, and HD 45677.

Many of these objects are definitely binaries, others may be single.

In 1939 the spectroscopic group at the Yerkes and McDonald Observatories, under the leadership of Otto Struve, engaged in a large program of spectroscopic observations of all types of outer envelopes which are not accounted for by the usual theories of normal stellar atmospheres. We knew that the long and exceptionally successful period of development carried by the use of thermodynamic equilibrium was gradually being replaced by a new period in which emphasis was being placed upon departures from equilibrium conditions. On the other hand it was clear that our understanding of the physical, dynamical, and geometrical conditions prevailing in the SS would eventually be based on the spectroscopic variations in these stars which are actually all variable to some extent. The time was ripe, since a great deal of progress had been made in laboratory atomic spectroscopy. The Yerkes astronomers had at their disposal the new 82-inch reflector, equipped with an efficient Cassegrain prism-spectrograph, enabling observations in the near ultraviolet as well as in the usual region. The ultraviolet region of the peculiar stars had been very little explored. Soon

---

[1] The geometrical dilution factor at a point P is the ratio of the solid angle subtended by the photosphere at P to 4π:

$$W(r) = \tfrac{1}{2}\left[1 - \sqrt{1 - \left(\frac{r_*}{r}\right)^2}\right]$$

where r = radius of the "shell," and $r_* $ = radius of the underlying star. For $r/r_* > 2$, the approximation $\tfrac{1}{4}(r_*/r)^2$ suffices.

To this geometrical factor, a physical factor should be added, expressing the influence of absorption or emission features of the underlying radiation at the active wave length, taking into account the relative radial velocities.

it became also possible to do some work in the photographic red and infrared region. I had the great privilege to be a member of that spectroscopic group. On my very first night of observation, the ultraviolet spectrograms obtained revealed emission features which had never before been observed. So began an excitingly fruitful period of collaboration with Otto Struve, continuing my previous short periods of joint work with him in 1931 and 1936.

II. IMPORTANCE OF FURTHER EXPERIMENTAL AND ASTRONOMICAL SPECTROSCOPIC INVESTIGATIONS

Laboratory spectroscopy has lost some of the glamour that it used to have. Yet much experimental work of astrophysical importance remains to be done. The present report provides me with an opportunity to express a plea in favor of spectroscopy, experimental and theoretical as well as astronomical. We know that, even in a strategically located star, like the Sun, at least 30 percent of the absorption lines listed in the Revised Rowland Table are still unaccounted for, and, actually, this Table is still rather incomplete. Hundreds of lines remain unidentified in all kinds of astronomical objects. A few laboratory spectroscopists continue to render outstanding services to the astronomers. Among the latest examples, we should mention the long-awaited analysis of the Fe IV spectrum[2] which Edlén completed in March 1966. Progress is being made also in the field of transition probabilities.

At the invitation of the Joint Commission for Spectroscopy I published (Swings 1951) a general discussion "Spectroscopic Problems of Astronomical Interest"; another report "Problems of Astronomical Spectroscopy" was published ten years later (Swings 1960). In these compilations—which are outdated and should be repeated[3]—many unidentified absorption or emission lines and bands were listed. Of course several of these lines have now been assigned satisfactorily. But many puzzling cases remain, in practically all spectral types, including peculiar objects (Wolf-Rayet, Of-, P Cygni-type and symbiotic stars, and novae) especially in the red and near-infrared region. There remain hundreds of unassigned lines in the rare-earth stars, but good progress is being made in the analysis of the laboratory spectra of the neutral, singly ionized and doubly ionized rare earths: this problem should be re-examined, especially in α^2 CVn. There are also many unassigned absorptions in the spectra of hydrogen-poor stars, such as v Sgr; of R CrB stars, of phosphorus stars, etc. We must prepare ourselves for the moment which appears near when, thanks to orbiting astronomical observatories and to balloon-carried telescopes, we shall obtain spectra of many astronomical objects in the ultraviolet ($\lambda < 3000$), the X-region and the infrared.[4] A considerable amount of laboratory work is still lacking in the ultraviolet and the infrared.

---

[2] The Fe^{+++}-ion remained the last ionization stage of iron which had not been satisfactorily analyzed until 1966. The permitted Fe IV lines will probably be observed in the absorption spectra of hot stars (especially in the far ultra-violet). Certain forbidden [Fe IV] lines seem to be present is RR Tel (1953–54). The metastable levels of the $3d^5$ and $3d^44s$ configurations which give rise to [Fe IV] transitions in the region $\lambda > 3000$ are rather high (EP > 6.1 ev). Hence [Fe IV] will probably not play as important a role as [Fe II, III, V, VI, VII].

[3] The author began work on such a revision in 1968.

[4] The importance of the infrared will become more and more apparent. Very cool stars have been and will be discovered, including, probably, unknown cool companions of peculiar stars, which are presently supposed to be single. New forbidden lines will be found, including coronal and other forbidden lines. Even the near photographic infrared deserves scrutiny. For example, the observation of the auroral transition of [O II], $\lambda7319$–$\lambda7330$ (EP $= 5.0$ ev) which has a much higher probability than the nebular transition $\lambda3726$–$\lambda3729$ (EP $= 3.31$ ev), but a higher excitation potential would help in understanding peculiar stars. The other following forbidden lines in the photographic infrared are also interesting: [S II], $^4S^o$–$^2D^o$ ($\lambda6717$–$\lambda6731$) which have low transition probabilities (≈ 0.001 sec^{-1}) and low excitation potentials (1.8 ev); [Ar III], ^3P–$^1D_2$ ($\lambda7136$–$\lambda7751$) which have a high transition probability (0.32 and 0.085 sec^{-1}) and a low excitation potential (1.73 ev).

The extension of the covered spectral range into the ordinary ultraviolet proved most fruitful in the hands of the McDonald astronomers as soon as the 82-inch reflector became operational. I shall never forget the excitement which prevailed in our group when the ultraviolet region revealed to us the forbidden lines of Ne V and Fe VII and many other features: new forbidden lines of [Fe II], [Ni II], [Cr II], [V II], . . . The tremendous extension in spectral range which will be attained thanks to space vehicles will open entirely new possibilities for the study of the astronomical objects. In the SS, a wide range of ionization occurs; thus many lines may appear in the ultraviolet, as should also be the case in the richest planetaries, such as NGC 7027. Of course there will always remain obstructions, such as the region[5] from 912 Å down to the X-rays near 20 Å which is absorbed by the interstellar H, He, and He+.

There is still a great need for accurate measurements and the analyses of numerous atomic spectra, and for various transition probabilities. Garstang's computed probabilities for many astronomically important atoms and ions are of tremendous interest. Indeed, the data on [Fe II, III, V] recently computed by Garstang make imperative a revision of many old discussions of spectra of peculiar stars (including novae, long-period variables, etc. . . . .). This revision is at present under way in Liège.

III. EXAMPLE OF ASTRONOMICAL APPLICATIONS OF A SINGLE ANALYZED SPECTRUM: Fe III

In 1937 Edlén and I (1942) took the decision to analyze the Fe III spectrum, the original limited aim being to locate the metastable levels. At that time we had a vague hope that the forbidden lines of Fe III might possibly explain the coronal lines. Eventually we classified 1500 lines and established 320 energy levels. At the time we started, Bowen had only identified 3 multiplets of Fe III; later Green discovered 3 additional multiplets which we had independently found. Our spectrograms covered the region from approximately 500 Å to 6500 Å, with a dispersion of about 4 Å/mm shortward of 2000 Å.

The deepest electron configuration of Fe III is $3d^6$, the next higher is $3d^5 4s$; the lowest odd configuration is $3d^5 4p$. About 450 combinations belong to $3d^5 4p \rightarrow 3d^6$ (region 679 Å to 1143 Å); about 1000 connect $3d^5 4s$ and $3d^5 4p$ (region 1700–4000 Å); strong quintets and septets connect $3d^5 4p$, $3d^5 4d$, $3d^5 4f$, $3d^5 5p$ and $3d^5 5s$. The analysis took us two years of steady work. It has now been extended by S. Glad (1956).

Two types of strong permitted Fe III lines appear in the astronomical region. In the first group the lower levels are metastable or quasi-metastable even terms of the configuration $3d^5 4s$. In the second group the lower levels are the high terms associated with $4p$, $4d$, $4f$, $5s$ and $5p$ which are *not* metastable. In an extended atmosphere, excited by diluted radiation, the first group becomes considerably enhanced relative to the second. The enhancement gives an estimate of the dilution factor; in various early-type shells this is actually the easiest way to ascertain the presence of dilution. For example $\lambda4165$ and $\lambda4372$ (second group) are much stronger than $\lambda4419$ (first

---

[5] As a result, many important lines will be forever unobservable, such as the lines shortward of 912 Å which play a role in fluorescence mechanisms (He II, O III, N III). For the physics of the atmospheres of hot stars the region $\lambda < 912$ plays a major role; but this may be predicted on the basis of the observations $\lambda > 912$. For example, in high excitation nebulae, the Lα quanta of He II will provide an important source of electron kinetic energy by ionization of H. Such electron energy may excite resonance lines, such as $\lambda1551$–$\lambda1548$ of C IV. Actually the planetaries will not reveal many forbidden lines in the vacuum ultraviolet (for example [S III], [O II, III], [Ne III, IV, V], [Ar IV, V]), but many permitted lines—which may be excited by collision—will be found, including many resonance and nonresonance lines (atoms of C, N, O, Mg, Al, Si, Fe, S, Ar, . . . at various stages of ionization).

group) in ordinary absorption B stars, whereas the opposite is true in P Cyg, γ Cas, and ζ Tau.

The forbidden $^5D-^3F$ and $^5D-^3P$ transitions are very strong in many objects. Their strongest transitions are respectively those of J-values 4–4, $\lambda4658$ and 3–2, $\lambda5270$.

The transition probabilities of [Fe III] have been computed by Garstang (1957); $\lambda4658$ ($^5D_4-^3F_4$, EP = 2.66 ev) is intrinsically stronger than $\lambda5270$ ($^5D_3-^3P_2$, EP = 2.41 ev) and is observed as such when the star is not heavily reddened. Our analysis of the Fe III spectrum led to results which had nothing to do with the solar corona. Yet we were well rewarded for our labor.

Many unidentified absorption lines in B stars (example: γ Peg) turned out to be due to Fe III, as was shown as early as 1938–1939. The latest paper is one by B. Warner (1966), who used the Fe III lines to determine the abundance of iron in early-type stars. The region $\lambda < 3300$ of P Cyg contains one line of He I, four of Si III and nineteen of Fe III (lower metastable levels); certain multiplets, such as $a^5D-z^5P^0$ ($\lambda5127-\lambda5156$) which are weak in the laboratory are strong in P Cyg as a result of the dilution effect. γ Cas at the time of its sharp-line stage of 1939 was very rich in strong Fe III shell lines; except for H and He I, the Fe III lines constituted the most conspicuous features of the entire spectrum of γ Cas; the sharp Fe III lines remained very intense until late in 1940[6]; the multiplet $a^5D-z^5P^0$ was very characteristic in the visual region.

The [Fe III] lines are strong in MHα 328–116, BF Cyg, RY Sct, MWC 17, etc. They are found in many novae of the slow as well as the fast type. Of course they are more easily detected in the slow novae which, after the η Car stage ([Fe II] strong, [Fe III] absent), pass through stages showing [Fe III], the relative intensities of [Fe II] and [Fe III] giving an indication of the ionization conditions. Typical stellar comparison spectra are: η Car ([Fe II] only), MWC 17 ([Fe II] and [Fe III] present simultaneously), Nova (DO) Aql (1925) in 1931 ([Fe III] stronger than in MWC 17), Nova (RT) Ser (1909) in 1931 ([Fe III] strong), BF Cyg (very strong [Fe III] at certain phases), MHα 328–116 (extremely strong [Fe III], weaker [Fe II]) and RY Sct ([Fe III] only). [Fe III] is present in many gaseous nebulae, for example in the Orion Nebula.

In BF Cyg, a relatively unreddened star, $\lambda4658$ is stronger than $\lambda5270$, but in MWC 349, which is located in a dark region of the Milky Way and is reddened to a considerable extent, $\lambda5270$ is much stronger than $\lambda4658$.

Beside the strong $^5D-^3F$ and $^5D-^3P$ transitions, the line $^5D_4-^3H_4$ ($\lambda4881.14$) is found in certain nebulae.[7] The a^5D-a^3D and a^5D-a^7S multiplets (3200–3400 Å) did not appear on strongly exposed McDonald spectrograms of the Orion Nebula, taken for this purpose;[8] however traces of these multiplets have been found on strongly exposed spectra of BF Cyg taken at McDonald in 1947, 1949, and 1950) (Swings and Swensson 1953), and on Mt Wilson coudé spectrograms of planetary nebulae (Kaler, Aller, and Bowen 1965). No trace has been found of $^5D-^3G$ (EP = 3.8 ev) except a very weak line observed in NGC 7027 at 4008 Å ($^5D_4-^3G_4$). This is due to the lower probability of this transition (0.019 sec^{-1}), and possibly also to the higher excitation potential.[9]

The reason for reporting these astronomical applications of the analysis of an atomic spectrum is to encourage young astronomers to tackle research problems,

---

[6] In January 1941, the Fe III lines had disappeared; γ Cas had resumed its broad lined B-spectrum.

[7] The presence of $^5D_4-^3H_4$ has been interpreted by I. S. Bowen (1960).

[8] The weakness of a^5D-a^3D (EP = 3.81 ev) cannot be explained by low transition probabilities, the values for $^5D-^3D$ (0.23 sec^{-1}) being of the same order as for $^5D-^3F$ (0.44 sec^{-1}). Probably the weakness is due to the higher excitation potential.

[9] Several wavelengths of [Fe III] have been measured accurately on coudé spectrograms of planetary nebulae and of the Orion Nebula (Bowen 1960, Flather and Osterbrock 1960). Example: 4658.10 ± 0.05 Å.

especially the analysis of atomic spectra of astronomical interest, which may take a long time, patience, and precision. There are still quite a number of such problems.

It is gratifying to see that there is a real renewal of interest among young physicists in atomic and molecular spectra of astronomical importance. New excitation techniques such as the beam foil method are providing important new data. Convincing evidence of this renewed attention is provided by the great number of interesting contributions and discussions at the Symposium on Beam Foil Spectroscopy, organized by Professor Stanley Bashkin at the University of Arizona in November 1967 (Bashkin 1968).

IV. BEHAVIOR OF CERTAIN EMISSION LINES IN PECULIAR STARS

I shall place myself mainly at the point of view of the astrospectroscopist. The dilution effects[10] have been discussed theoretically in Miss Underhill's chapter. Generally we adopt as excitation mechanisms: the recombination of ions and electrons, the impacts by particles, the fluorescence excitation (either by discrete emissions or by continua) or the molecular processes. But other processes have also been occasionally considered, such as collisions of clouds in the atmosphere (conversion of kinetic energy of turbulence into excitation), ejections of subphotospheric layers (conversion of thermal energy) and shock waves.[11] I shall only consider spectroscopic effects given little or no discussion by Miss Underhill.

The selectivity effects observed among the atomic lines in Of stars (i.e., O-type stars with shells) have been studied by Miss Underhill (Liège 1957). The presence of the $\lambda 5696$ C III line ($2s\,3p\ ^1P_1{}^0-2s\,3d\ ^1D_2$) in emission while the other C III lines are found in absorption has been attributed to the formation of a C^{++} ion in the $2s\,3d\ ^1D_2$ level after ejection of a $2s$-electron from a C^+ ion by absorption of a $\lambda 304$ He II quantum. The energy match between $\lambda 304$ and the energy difference between $C^+\ 3d\ ^2D$ and $C^{++}\ 2s\,3d\ ^3D$ is not so close as in the case of $C^{++}\ 2s\,3d\ ^1D$. Another interpretation has been suggested by J. Gauzit (1966), on the basis of the coincidence between $\lambda 322.575$ C III (transition between the ground state and $2p\,3s\ ^1P_1{}^0$) and $\lambda 322.570$ N IV (transition $2s\,2p\ ^3P_1{}^0-2s\,3s\ ^3S_1$). The absorption of the N IV photon leads first to an unobservable emission $\lambda 2982.2$ C III, then to $\lambda 5696$. The observation of Of stars from space vehicles will possibly give us a convincing interpretation, although the region $\lambda < 912$ will not be reached. The observation of the photographic infrared, especially of $\lambda 8499.7$ C III ($3s\ ^1S-3p\ ^1P^0$) would be of interest.

As for the emission of $\lambda 4634-4640$ N III while the other N III lines are present in absorption, it was first thought by Struve and myself that it could not be due to the usual fluorescence effect described by I. S. Bowen (Lα of He II absorbed by O III, then absorption of resonance O III quantum by N III), since no O III line was observed in emission. However this point of view is not convincing. Indeed two of my collaborators, J. Humblet and G. Mannino (Liège 1957), stressed the point that the selectivity may indeed be due to Bowen's fluorescence mechanism. The absence of emission in the observed lines of O III does not imply that the resonance line $\lambda 374$ O III cannot be an emission line able to bring the ions N^{++} to the excited state $3d\ ^2D$.

---

[10] The dilution effects in the far ultraviolet would be profitably studied theoretically, in preparation for space observations.

[11] There may even be more "exotic" excitation mechanisms, of the type suggested by P. Swings and Y. Öhman (1939). Amorphous metals produced by condensation at low temperature seem to have a critical temperature at which the solid undergoes a transition from the amorphous into the crystalline state. Such a transition would be accompanied by a sudden emission around 200 or 300 Å. One may wonder if this mechanism could not explain emissions in regions of the sky where none of the usual energy sources seem to exist.

The Fe II spectrum behaves in a typical fashion in P Cygni-type stars. The a^6S–z^6P^0 multiplet ($\lambda\lambda4923.92$, 5018.43, 5169.03) differs strongly from the b^4P–z^4D^0 ($\lambda\lambda4233.7$, 4351.76, 4385.38) although the equivalent widths are very similar in the solar spectrum (sextet: 167, 210 and 154 mÅ; quartet: 139, 133 blended, 81 mÅ). In BD + 47°3487, HD 160529, 17 Lep, Z CMa, CD–27° 11944, RS Oph (at 1938 outburst), Nova Her 1963, the sextet is strong, while the quartet is very weak. Generally the a^6S–z^6P^0 transition consists of emission lines with shortward absorptions; the radial velocities differ appreciably. In Z CMa, the effect is striking. The multiplets with lower level b^4P, b^4F, a^4G are pure sharp emissions, without absorption component. On the contrary the a^6S–z^6P^0 transition has strong, violet-displaced absorption components. Yet the excitation potentials are very similar,[12] but a^6S has the same multiplicity as the ground level of Fe II. Actually the behavior of Fe II in P Cygni stars resembles those of Si III, N II, N III, N IV and C III in shells of higher excitation. In supergiants, some "multiplicity effect" is present, although all lines are present in absorption; at this point of view 3 Pup is closer to the P Cygni stars than to α Cyg.

The very different behavior of the O I triplets ($\lambda8446$) and quintets ($\lambda7772$) in peculiar stars is very striking. $\lambda8446$ sometimes appears in emission, while $\lambda7772$ is in absorption. It is generally assumed that Lβ may bring the oxygen atoms to the ^3D level, and thus give rise to emission in the triplets. But this is not necessarily required. B. Pagel (1960) has shown that oxygen atoms placed in a field of diluted radiation become overpopulated on level ^5S relative to ^3S. As a result $\lambda7772$ acquires a high opacity and appears in absorption. Moreover the ^5S level can be de-excited by collisions only, while the triplets may be de-excited by radiation; hence at very low densities, ^5S is less de-excited. More generally it should be clear that apparently abnormal relative intensities among emission lines may sometimes be simply due to the fact that certain emissions are more re-absorbed than others.

The behavior of the emission lines of He I is interesting. While the triplets are much stronger than the singlets in planetary nebulae, the opposite is true in certain peculiar stars. This important point will be discussed in detail later on.

Many peculiar stars show either Fe II or [Fe II] emissions or both. Actually the relative intensities of these emissions depend not only on the electron densities n_e, but also on the electron temperatures T_e. The observations indicate that the Fe^+ ions are excited by collisions to higher levels, with subsequent cascading to lower ones. The lower levels of the observed permitted Fe II transitions are metastable, and the forbidden [Fe II] lines correspond to jumps downward to the deepest electronic levels. If T_e is sufficiently high the permitted lines will be much stronger. A low T_e and a large extent of the radiating layers enhance [Fe II].

Certain lines vary rapidly in intensity with time. This is, for example, the case of [O III], [Fe III] and [Ne III] in BF Cyg. While [O III] may be the strongest transition at times, it may have become extremely weak a few days later. Such variations require a rather high n_e, say of the order of 10^6 cm^{-3} in the shell responsible for [Fe III]. We shall consider this point in relation to the "relaxation time" later on. Actually the minimum electron density 10^6 is more comparable with the densities found in nova shells than with the densities found in planetary nebulae.

Although the continuum of symbiotic stars is uncomfortably intense in the red and photographic infrared, the observation of emission lines in this region is helpful. The lines of H (Paschen series), He I, O I, N I, Ca II, C III, [S I], [Ar III], etc . . . may be strong (as in BF Cyg) and furnish useful information; for example the relative intensities of $\lambda7772$ (quintets) and $\lambda8446$ (triplets) of O I may vary considerably in a few

[12] The effect of the excitation potential on the bright Fe II lines is marked, the low-excitation lines being enhanced. For example: $\lambda3938.29$ (lab. int. 2; $a^4P - z^6D^0$, EP = 4.80 ev) is present in Be stars, while $\lambda3938.97$ (lab. int. 4; $d^2D - x^2F^0$, EP = 9.02 ev) is always absent.

days. Incidentally, many emissions of the peculiar stars in the red region are still unassigned. No [Fe I] has been observed;[13] the presumably strongest [Fe I] line is $^5D_3$–$^3P_2$, 5565.68 Å (EP = 2.27 ev).

Strong [Fe II] lines have been found in planetary and diffuse nebulae, in novae (slow and fast), and in a variety of stars, especially in systems composed of M- and B-type stars, such as VV Cep, B 5481, B 1985, WY Gem, the B-type companion of α Sco. Such observations are suggestive: possibly the association is necessary for the production of strong [Fe II] lines. It is true that HD 45677 shows strong [Fe II], whereas there is no evidence of a cool companion: but one may be found eventually (perhaps by infrared spectroscopic observations?).

The peculiar star η Car is especially rich in [Fe II]. Merrill identified in it the following transitions:

$$a\,^6D\text{–}a\,^6S,\ b\,^4P,\ b\,^4F;$$
$$a\,^4F\text{–}b\,^4F,\ a\,^4G,\ b\,^4P,\ a\,^4H.$$

Investigations of B 1985, WY Gem, W Cep, B 5481, HD 45677, etc. increased the number of observed [Fe II] transitions, some of them rather strong in the ultraviolet[14] and infrared:

$$a\,^6D\text{–}b\,^4D;$$
$$a\,^4F\text{–}b\,^4D,\ a\,^2G;$$
$$a\,^4D\text{–}b\,^2D.$$

B1985 and WY Gem are richest in [Fe II] in the ultraviolet.

The first ultraviolet spectrograms of B 1985, WY Gem, and HD 45677 revealed strong emission lines in addition to Fe II, [Fe II] and Cr II; they were later on assigned to [Ni II]: $a\,^2D\text{–}a\,^2P,\ a\,^2G$. Other transitions were found in the violet ($a\,^2D\text{–}b\,^2D$, $a\,^4P$) and in the infrared regions ($a\,^2D\text{–}a\,^2F$). The $a\,^2D\text{–}a\,^2F$ multiplet ($\lambda\lambda6667$, 7378, 7412, 8302) has a low excitation (1.92 ev) and is very strong in η Car.

In addition to the forbidden lines of singly ionized iron and nickel, only weak lines of a few other metallic ions have been observed (Swings 1951):

[Cr II]: $a\,^6S\text{–}a\,^4P,\ a\,^6D$ (found in η Car);

[Mn II]: $a\,^5D\text{–}a\,^5F$ (in B 1985, WY Gem, HD 45677);

[V II]: $a\,^5D_4\text{–}c\,^3F_4$ ($\lambda3334.66$ in HD 45677).

Thackeray has also considered [Cu II] and [P II] in η Car, but these assignments are not convincing. Among the forbidden lines of neutral metals one should mention:

[Mg I]: 4562.48 Å (a magnetic quadrupole transition observed in nebulae, but not yet in peculiar stars, not even in η Car); the permitted line Mg I $\lambda4571$ is often found in peculiar stars.

---

[13] One may hope to find [Fe I] in a peculiar star of low excitation or in a nova at a low-excitation stage.

[14] The identification of an infrared [Fe II] line of low excitation potential is permissible even when none of the usual violet [Fe II] lines are observed. As a general rule, any element will have stronger forbidden lines in the infrared than in the violet, because, in a statistical way, the infrared lines correspond to excitation energies lower than those of the violet lines. A [Fe II] $a\,^6D\text{–}a\,^4D$ multiplet near 12,000 Å may be expected to be very intense in all [Fe II] stars since the excitation potential of $a\,^4D$ is only about 1 ev. The $a\,^4F\text{–}a\,^2G$ transition in the infrared has been observed in η Car and other [Fe II] stars; it is present in υ Sgr while the violet [Fe II] lines are not found in that star.

In addition to [Fe III], the other doubly ionized metals do not play a major role. Some uncertain coincidences have been found (Swensson 1953):

[Mn III]: a ^{4}G–a ^{4}F, possible in Z And (1948), planetary nebulae;

[Co III]: doubtful: in Orion Nebula, η Car, WY Vel, planetary nebulae;

[Ni III]: rather convincing: a ^{3}F–a ^{3}P, $\lambda6001$ in Orion Nebula, Z And in 1949, Nova Ser 1948, novae;

[Cr III]: not excluded in RY Sct: a ^{5}D$_3$–a ^{3}P$_2$, $\lambda5714.6$;

[Ti III]: doubtful: ^{3}F–^{1}S, ^{1}G.

In many cases the predicted wavelengths are not accurate enough for convincing assignments. The forbidden lines of more highly ionized iron play an extremely important role in peculiar stars, planetary nebulae, novae, etc. Actually [Fe V, VI, VII] provide valuable information; the recently analyzed [Fe IV] may possibly be of help too.

[Fe V] had not been identified convincingly until 1942, despite Wyse's endeavors in nebulae. The first good observation was made in AX Per in January 1942, as a result of a systematic monitoring of the variable spectrum of this symbiotic object. The excitation in AX Per varies considerably. The McDonald observations during the period 1939–1941 always revealed [Fe VII, VI]. A phase of lower ionization produced [Fe V] in January 1942. Later the [Fe V] spectrum has been repeatedly found in AX Per, CI Cyg, Z And, etc., and on old spectrograms of Nova (RR) Pictoris and Nova (RT) Ser. Bowen's observations in planetary nebulae have provided accurate wavelengths: 3891.28 ± 0.08; 3895.52 ± 0.05; 4227.49 ± 0.11 Å. The strongest [Fe V] lines are a ^{5}D$_4$–a ^{3}F$_4$ (λ 3891.3) and a ^{5}D$_3$–a ^{3}P$_2$ (λ 3895.5); they fall near H8 + He I ($\lambda3889$) and hence are not easily separable in novae with broad Balmer emissions.

[Fe VI] has a characteristic line $\lambda3664$, ^{4}F$_{7/2}$–^{2}D$_{5/2}$ in a relatively clear region. The other transitions ^{4}F–^{2}G, ^{2}P, appear in the visual region and are not so easily observed in symbiotic objects on account of the presence of the late-type companion. [Fe VI] is found in many novae.

[Fe VII] has three characteristic transitions ^{3}F–^{3}P, ^{1}D, ^{1}G; the spectrum is observed in AX Per, CI Cyg, Z And, RX Pup, Nova (RR) Pic, Nova (RT) Ser, Nova (RS) Oph, and many other novae.

A new and systematic discussion of the behavior of the [Fe VII] lines in different objects would probably be fruitful. Certain "intensity anomalies" may have been caused by instrumental effects, for example to the "green dip" of the photographic emulsions. Image tube cathodes avoid this difficulty. A general study of the [Fe VII] lines is at present under way in Liège.

Thus forbidden lines of iron have been observed in many stages of ionization: II, III, IV, V, VI, VII, X, XI, XII, XIII, XIV, XV; many permitted lines have also been observed for Fe I, II, III, and higher stages (in the solar corona).

V. DESCRIPTION OF TYPICAL SYMBIOTIC STARS

1. Z Andromedae (HD 221650)

In these descriptions of typical symbiotic stars and closely related objects, the emphasis is placed on the spectroscopic observations which may help by intercomparisons to understand the physical, geometrical, dynamical, and evolutionary characteristics.

The article by Struve reproduced here[15] had been preceded by only one detailed spectroscopic description of Z And, by H. H. Plaskett (1928). Quite a number of papers followed. Z And is really a prototype of the family of symbiotic stars; for this reason it will be described and discussed in some detail. The history of Z And until 1941, as well as our geometrical, dynamical, and physical views up to that date have been given in the reproduced article, and will thus not be repeated here.

As will be discussed later on, the possible connections between symbiotic stars and somewhat similar objects (such as planetary nebulae, W-R stars, shell stars, novae, etc.) have been stressed repeatedly in recent years. Actually our present ideas of stellar evolution are still rather vague in the region of the H-R diagram between the red giant and the white-dwarf phases. It has often been suggested that planetary nebulae and (or) symbiotic objects may provide missing links. In 1940, Struve and I could of course not have guessed that there might be evolutionary connections between symbiotic stars and planetary nebulae. But we had been impressed by various spectroscopic similarities between the spectrum of Z And at certain phases and that of planetary nebulae, especially those having a fairly high electron density, such as IC 4997. It is for this reason that we had combined our spectroscopic descriptions of Z And and IC 4997 in a single paper; actually the spectroscopic similarities are very great indeed, although IC 4997 has a lower ionization ([Ne V] is absent). The "nebular spectrum" of Z And resembles that of a high-excitation planetary nebula with lines of [Ne III, V], [O III], etc. The chief spectroscopic differences lie in the intensity of [Fe V, VI, VII] and the high intensity of $\lambda4363$. On the basis of these differences, a higher density of the order of 10^6 ions cm^{-3} may be assigned to the nebular shell surrounding Z And, as against 10^3 or 10^4 ions cm^{-3} in the gaseous nebulae. The region $\lambda\lambda4632$–4658 of Z And, just as the nucleus of IC 4997, contains characteristic emissions of N III, C III and C IV, but they are sharper in Z And than in Wolf-Rayet stars. They were designated as the "nuclear lines," but they may possibly be produced in the innermost layer of the nebular shell.

Many light curves of Z And (Pickering, Shapley, Mrs. Greenstein, Parenago, Prager, Mrs. Gaposchkin, Whitney) have been published. There are small amplitude variations with a period of approximately 694 days, plus occasional sharp outbursts. The nova-like outburst of 1939 and the spectroscopic behavior during the period 1939–1942 were described by Struve and associates.[16] During the period August 1940–January 1941, $N_{1,2}$ increased in intensity relative to $\lambda4363$; the P Cyg absorptions disappeared; a Balmer continuum appeared in emission. In January 1941 (star at minimum) the spectrum of Z And was of the postnova type. The P Cyg absorptions had disappeared; the intensities of $N_{1,2}$ relative to $\lambda4363$ had increased. In 1941 there was a new maximum, but less pronounced than in 1939:[O III], [Ne III] and [Ne V] decreased in intensity; $\lambda4363$ increased relative to $N_{1,2}$; a new shell appeared, but no appreciable dilution was apparent. There was no variation in the structure of the "nuclear features" $\lambda\lambda4632$–4658, but these had weakened relative to the Fe II emission. Indeed there were increases in the following intensity ratios: Fe II/N III, C III, C IV; O III fluor./[Ne V]; He I/[Ne III]; Fe II/He I; Fe II/[O III]; Si II/continuum; Mg II/continuum.

----

[15] In Table I of Struve's paper, $\lambda\lambda3277.0$, 3281.1, and 3323.6 should be assigned to Fe II. In Section VI "The Spectrum of the P Cygni Type of Z Andromedae," a comparison is made with BD + 11°4673 (AG Peg), with the restriction that "the comparison with a P Cygni type star which is single may be quite artificial, etc. . . . ." Actually it turned out later on that AG Peg is a symbiotic object too!

[16] Struve and his associates were very fortunate during their work on symbiotic stars: a major outburst in Z And, followed by minor outbursts; important spectroscopic changes in AX Per, RW Hya, AG Peg. The fact that the near-ultraviolet region could be covered was also of great help.

Six months after this minor outburst the star had returned to its appearance nine months following the major outburst of 1939; the [Fe VII] lines were strong again. In 1942 a considerable increase of [Ne V] was observed in comparison with 1941. There were also very striking changes in the intensity of the fluorescent line $\lambda 3444$ O III (excited by $\lambda\,303$ He II). Indeed the intensity ratio of $\lambda 3444$ and $\lambda 3426$ [Ne V] appears to be very sensitive to changes in excitation, density gradient and velocity distribution in the shell. During the second half of 1942 [Fe VII] became very intense again: actually this was the first time in the history of Z And that such a high intensity of [Fe VII] was observed, making Z And similar to AX Per. [Fe V] was present, but weaker than [Fe VII]. There seems to have been a peak in excitation during the summer of 1942. Relative to 1941, there were strong variations in the relative intensities of the triplets and singlets of He I ($\lambda\lambda 4471/4388$; $4026/4009$; $4026/4144$), indicating that the relative contributions of the recombination and fluorescence mechanisms had varied. During the period November 1943–October 1944, the excitation increased; [Fe VII] and [Ne V] became very strong, while Fe II weakened (but [Fe II] increased in intensity). Late in 1946, when the star was again bright, Merrill observed a shell spectrum, resembling Pleione; the only previous observation of a shell in Z And had been made by Struve and collaborators in 1939. Merrill succeeded in obtaining high-dispersion coudé spectrograms (10.3 and 20.6 Å/mm). More than 200 shell absorption lines of early type were present; 60 emission lines were also observed; there were only traces of TiO bands. The Balmer series extended to H37 in absorption and was followed by a strong absorption continuum; emission was seen to Hη. In other words a steep decrement was present in emission, while the decrement was very slow in absorption. Struve and collaborators had found in the shell of 1939 that the continuous absorption shortward of $\lambda 3613$ was all due to the Balmer continuum, and that the ultraviolet continuum was absorbed by the overlying He I-atoms (absorption at $\lambda\lambda 3634, 3587, 3554$). The high-dispersion spectrograms showed that the bright lines were diffuse; the widths of the lines of different elements did not decrease with increasing atomic weights: the structure of the lines was thus not due to kinetic motion. Merrill (Liège 1957) suggested that the structure probably arises either (i) in gross motions of portions of the star's extended atmosphere (prominences?) or (ii) in motions of streams of gas coursing about in a system containing more than one star. Late in 1947, the shell had again disappeared; only emission was present in the Balmer series; there was no progression in radial velocity within the Balmer lines. The radial velocity of the He I triplets exceeded that of the singlets by +16 km/sec. The radial velocity fluctuations in the mean curve for [O III] and [Ne III] follow those for permitted lines (H and He I) by about 200 days. This time lag is probably a relaxation phenomenon of the type described by Grotrian for Nova (DQ) Her and applied in the accompanying paper by Struve. Z And increased in brightness in June 1959, then again in May 1961. During the period 1946–1959 a few spectroscopic observations had been made by Miss M. Bloch and Tcheng Mao Lin (in 1956 and 1957). In 1956–57 the spectrum was of the "minimum type": in addition to emissions of low excitation, nebular lines of high excitation including [Fe III–V–VII]; [Ca VII] and TiO absorption bands are observed.

Spectrograms taken in August 1960 by Bloch did not reveal appreciable changes since October 1959. They extended to $\lambda 3050$, and revealed the ultraviolet emissions of O III (fluor.), He II 3203, [Ne V], [Fe VII], [Ne III] 3343; the Balmer continuum appeared in emission.

Coudé spectrograms were taken at Mt. Wilson by Swings in September and October 1959 in the ultraviolet and the blue region. This high-dispersion material (10 and 20 Å/mm) is similar in quality to that obtained in 1946 by Merrill. It has been described in detail by F. Dossin (1964 and *Thesis*, unpublished). Many lines which previously had been observed as blends are resolved on these spectrograms; hence better identifications have been made possible. The H emissions are strong and broad, and have important wings, easily observable to H8; the Balmer series extends to H31 and the continuum is seen to $\lambda 3390$. The lines are asymmetrical, widened on the shortward side, with a sharper drop on the longward side; the profiles agree qualitatively with the occultation hypothesis. As previously observed, the He I singlets and triplets have different radial velocities (-15 and -10 km/sec respectively). Only the fluorescent O III and N III lines are present, with no trace of the recombination spectrum of O III. [O II] ($\lambda\lambda 3726, 3729$) is absent as usual (the density is too high). On the whole the spectra of 1959 resembled greatly those of 1942, a few months after an outburst, at the beginning of the decline. It would be most valuable to obtain accurate profiles on the basis of spectrograms of still higher dispersion; 2 Å/mm is technically possible.

Prager in 1939 suggested a period of 630 days for the variation in brightness; according to Merrill the radial velocities indicate a period of recurrence of approximately 680 days. The interstellar K line is weak, indicating that Z And is not very distant; its luminosity is considerably lower than that of an average star of class B0.

The general relation between the light curve and the spectroscopic behavior of Z And is as follows:

(*i*) When the object is near minimum brightness, emission lines of low and high excitation, generally including forbidden [Ne V] and [Fe VII], are present, together with an M-type absorption spectrum. The "stellar" emissions belong to H, He I, Fe II, Mg II, Ca II, Ti II, Cr II, C II, Si II, ... The "nebular" lines are due to He II, C III, N III, O III, [O III], [Ne III, V], [Fe III, VII], [S II], [Ca VII], ...

(*ii*) When the brightness increases the high excitation emissions weaken progressively; the other emissions (H) show intensity fluctuations. The TiO bands become less conspicuous.

(*iii*) Near maximum brightness a B-type shell develops, sometimes of P Cygni type. The remaining emissions are due to H, He I, Ca II, [O III], [Fe II]. The high members of the Balmer series are in absorption. The TiO bands have disappeared.

(*iv*) In the phase of brightness decline the metallic absorption lines disappear. The high members of the Balmer series and continuum appear in emission again. The TiO bands reappear progressively.

(*v*) Near minimum the nebular lines and the TiO bands are present again. There are considerable fluctuations in intensity ratios: for example $N_{1,2}/\lambda 4363$; Fe II/[Fe II]; emission lines/continuum; [Ne III]/[Ne V].

L. H. Aller estimated the "Zanstra temperature" on the basis of the He II-emission (during the nebular stage) and found approximately $T_Z = 90,000°$. The electron temperature based on the intensity ratio $\lambda 4363/N_{1,2}$ is estimated to be between 8500° and 10,500° near minimum brightness. At the time of an outburst the electron density in the expanding shell is so high that forbidden lines cannot appear. The "nebular" stage begins when the probability of the forbidden transition becomes approximately equal to the probability of collisional deexcitation.

2. AG Pegasi ($= BD + 11°4673 = HD\ 207757$)

AG Peg is a remarkable, bright, symbiotic object which possesses a variable magnetic field. The slow transformation of its Be-type spectrum into a combination spectrum has been observed; it takes a few decades, whereas Z And goes through the same evolution in a few months. The spectrum of AG Peg has been studied at various phases by Merrill (first description in 1929), Struve and associates, Aller, G. and M. Burbidge, B. Pagel, Miss Bloch, Dossin. The radial velocity has changed from -13 km/sec in 1915 to -200 km/sec in recent years, while the symbiotic character was becoming more apparent. This seems to indicate that a strong outflow of gas from a central region is involved in symbiotic phenomena (Merrill). The profiles and radial velocities of $\lambda 4363$ and of $N_{1,2}$ differ considerably: $\lambda 4363$ is rather narrow, while $N_{1,2}$ are broader and have much flatter profiles; the radial velocity of $\lambda 4363$ varies with a range of 65 km/sec and a period of 800 days, whereas $N_{1,2}$ are nearly stationary. These behaviors give some idea of the geometry and kinematics of the system: $\lambda 4363$ comes predominantly from an inner zone, while $N_{1,2}$ come probably from a shell so extended that motions in all directions radial from the center can be observed (although streams of the types considered by Struve, as for β Lyr, are not excluded).

The first descriptions based on McDonald spectrograms (1939) differed considerably from those of the Mt. Wilson plates (1929). N III was strong in 1939, absent in 1929. We found that the emission lines of P Cygni type showed the same selectivities as the typical stars of that type. For example, the Si II transitions $4p\ ^2P^o$–$4d\ ^2D\ (\lambda\lambda 5055, \ldots)$ and $4p\ ^2P^o$–$5s\ ^2S\ (\lambda 5958, \ldots)$ are present, while $3d\ ^2D$–$4f\ ^2F^o\ (\lambda\lambda 4128, 4131)$ is absent. Silicon in present in four stages, Si I, II, III, IV, probably indicating an effect of stratification. N II and N III are strong, while C II and C III are very weak or absent. AG Peg is thus related to the nitrogen sequence, rather than to the carbon sequence of Wolf-Rayet stars.

The trend toward higher excitation and more marked symbiotic character continued. By 1941, $\lambda 3905$ Si I disappeared, Ca II and Fe II decreased in intensity, He II, N III and Si IV increased appreciably.

In 1960 Miss Bloch observed intense N IV $(\lambda\lambda 3479$–$3485)$. Dossin examined the near infrared region, which is characterized by a strong emission $\lambda 8446$ O I, whereas $\lambda 7772$ O I is very weak.

The mechanism of excitation of the Balmer emissions has been discussed theoretically by G. R. and E. M. Burbidge (1953, 1954, 1955) and by B. Pagel (Liège 1957) in order to estimate the sizes of the emitting shells.

3. AX Persei and CI Cygni

These two typical symbiotic objects are treated together, because they both exhibit sharp emission lines of high excitation and very characteristic M-type spectra (AX Per: gM3e; CI Cyg: gM4e). There is however an appreciable difference: the spectrum of AX Per varies much more rapidly than CI Cyg.[17] Both stars were first studied by Merrill (1931–1932); later on they have been investigated by Merrill, Swings and Struve (1939–1943), Aller, J. Gauzit, Bloch, Dossin. The best measurements of CI Cyg, based on coudé spectrograms, have been published by Merrill (1950).

When Struve and his associates started their work on these two objects in 1939 it was immediately noticed that AX Per had changed greatly since 1931–32 (from Merrill's description) in the sense of an increased excitation, while the emission lines of CI Cyg had not changed appreciably. Actually there were even noticeable spectroscopic changes in AX Per between September 1939 and February 1940, while there was no appreciable variation of the emission lines of CI Cyg from 1931 to 1942. A few characteristics of both stars at the end of 1939 were:

[O II] $\lambda 3727$ extremely weak, and [O III] $\lambda 4363$ very strong (much stronger than $N_{1,2}$) as in the "stellar planetary" IC 4997.

O III fluor. and N III fluor. strong; [N II] fairly strong.

[Ne III, V] strong.

Fe II fairly strong, but [Fe II] uncertain; [Fe III] 4658 very weak.

---

[17] There seems to have been a photometric outburst of CI Cyg in 1911 (from $m = 12.1$ to 10.7, then back to 12.1 in 200 days).

[Fe V, VI] present (the first fairly convincing identification of [Fe V]).

[Fe VII] prominent (the only identification of [Fe VII] prior to 1939 was that of Bowen and Edlén in Nova (RR) Pic; [Fe VII] was very weak in 1931).

[Fe X] (coronal line) probable.

Si I $\lambda3905$ and Mg I $\lambda4571$ (characteristic of long period variables) present; a few other weak forbidden lines of ionized metals.

The spectrograms of AX Per taken during the period January–March 1941 exhibit very marked changes in one year; indeed the spectrum had resumed its aspect of 1931–32: [Fe VI, VII] had almost completely disappeared; there was a considerable increase in intensity of $N_{1,2}$ relative to $\lambda4363$ and of [Ne III] relative to H. New changes took place during the period Jan. 15 to May 30, 1941: [Fe VII] was absent, although [Ne V] was still present, as well as fluorescent O III; [O II] was absent. Then again the excitation increased, so that in August 1941 [Fe VII] and [Ne V] had recovered their intensities of 1939. The intensity ratios of the singlets and triplets of He I vary considerably. The spectra taken in January–February 1942 revealed striking new changes, compared to those of August 1941. [O II] had become stronger (lower densities in O^+-regions?), $\lambda4363$ had declined relative to $N_{1,2}$; [Fe VII] $\lambda3587$ had increased relative to fluorescent O III $\lambda3444$; [S II] was present for the first time. The ratio singlets/triplets of He I was intermediate between those in nebulae and in low-excitation objects like RW Hya (see section 5, below).

Essentially, from the spectroscopic point of view, the physical conditions in the nebular region had become such that [Fe V] and [Fe VI] could appear with considerable intensity, making AX Per the first striking example of a [Fe V, VI] object. A strong [Fe VI] feature at $\lambda3664$ ($^4F_{7/2}-^2D_{5/2}$) accompanied strong [Fe V] lines:[18]

$$^5D_4-^3F_4\ (\lambda3892),\ ^5D_3-^3F_3\ (\lambda3839);$$
$$^5D_3-^3P_2\ (\lambda3896),\ ^5D_2-^3P_1\ (\lambda4071).^{19}$$

Between February and November 1942 AX Per showed decreases in the intensity ratios [Ne V]/O III (fluor.) and [Fe VII]/[Ne III], indicating a decrease in excitation; [Fe V] was still present at the end of 1942. The last spectrograms taken by Struve and his associates in January 1943 revealed again differences, compared with the spectra of February and July 1942, in the sense of an increase in excitation: [Fe V]/[Fe VII] weaker, [Ne III] much weaker, [Ne V]/O III increased, He I singlets stronger.

In recent years AX Per has been placed on the observing program of Miss Bloch, who has found intensity variations similar to those observed by Struve and associates.

In a general way one could conclude that, between September 1939 and May 1941 the excitation had decreased (the disappearance of [Fe VII] may be due to a decrease in temperature of the exciting nucleus). T_e had not changed greatly, but n_e had decreased (by a factor 10?). Of course it is necessary to be careful in considering excitation criteria in an object such as AX Per, in which effects of stratification and asymmetry may be important: for example the emissions of [Fe VII] and of H or He I are presumably localized in different regions.[20] Moreover a change in the temperature of the exciting nucleus will influence the ionization of distant layers only after a certain lag which evidently varies with the distance from the nucleus, the density and the

---

[18] Fe III and Fe V have complementary electronic configurations $3d^6$ and $3d^4$ with regard to the half-closed shell $3d^5$. Hence similar forbidden transitions are found in both cases.

[19] This line, located between the two [S II] lines, had often been mistaken for [S II] in previous descriptions of nova spectra.

[20] In the case of a planetary nebula the "stratification effects" may be observed directly by taking slitless spectrograms or direct photographs with filters, or by scanning techniques, but this is not possible in the symbiotic stars!

physical characteristics of the different atoms. [Fe VII] is probably excited in the region closest to the nucleus and would be first to react to a nuclear variation. It is known that a post-nova nucleus fluctuates in brightness; such fluctuations affect the surrounding nebulosity after certain relaxation times which are functions of the nebular density and of other geometrical and spectroscopic characteristics (see the case of Z And in the accompanying article by Struve). A striking example of behavior of [Fe VII] in a postnova is given by Nova (RR) Pic. [Fe VII] attained its maximum intensity in 1932, then decreased in the following years. The "lag" may be short (a few days in BF Cyg, see next section) or may amount to months or years. The time lag in symbiotic stars has been studied by M. Johnson (1952).

The line emission of AX Per behaves in a way similar to that of the postnovae in their nebular stages, but the ejection process in AX Per is slower than in novae. In a nova shell the density decreases and, generally speaking, the excitation increases, but the fluctuations of the nova produce nebular variations, and in certain cases (as in symbiotic stars) the phenomena are recurrent.

The simultaneous presence of [Fe II] to [Fe VII] in AX Per and CI Cyg indicates a nebulous envelope exhibiting a wide range of excitations, akin perhaps to the planetary NGC 7027.

As for the radial velocities, whose complicated pattern has been studied in greatest detail by Merrill, the main results are: the displacement of the He I singlets differ greatly from the He I triplets (emissions in different regions of the radiating shell); [O III] and [Ne III] show the same displacements, indicating an origin in the same shell; H, He II, Fe II, O III and N III all show different shifts, as though the radiation of each ion tended to be concentrated in a different layer.

The He II and fluorescent N III lines are not closely correlated in intensity, but this is not strange on account of differences in ionization behavior.

4. BF Cygni

This object has been studied by Merrill, Struve and associates, Aller, Bloch, Dossin. Until 1965 BF Cyg was the object which showed the strongest known [Fe III] emission; recently another star, MHα 328–116, showing occasionally extremely strong [Fe III] has been described by Miss Bloch and by O'Dell; we shall consider this later on.

The apparition of very strong [Fe III] does not coincide with a maximum of [O III]. The underlying gM4 absorption spectrum is not always observable; most of the time it is masked by the continuous emission of the hot source. This contrasts with the case of AX Per and CI Cyg whose M-spectra always are prominent. BF Cyg has not reached a high excitation phase: the highest observed ionization is that of Ne III and O III.

The radial velocities deduced from Fe II, [Fe II], [Fe III], H, He I and the nebular lines differ considerably, indicating that they originate in different regions. The He I singlet/triplet ratio is much stronger than in planetary nebulae, as it is in most other symbiotic objects. The relative intensities of Fe II and [Fe II] vary in a complex manner, depending certainly upon T_e and probably on other factors. The excited Fe II and [Fe II] levels are presumably excited by collisions: a relative enhancement of [Fe II] relative to Fe II results from a decrease in T_e and (or) n_e. Aller's estimates are $T_e = 7500°$ to $15,000°$; $n_e \approx 10^6$ cm^{-3}. We are tempted to believe that T_e is even lower than 7500°. The enhancement of the Fe II lines of low excitation potential is such that identifications must proceed very cautiously. Sometimes a weak laboratory line, or even a predicted one of low excitation potential may play a greater role than a very strong laboratory line of higher excitation potential.

Extremely rapid changes in the intensities and velocities of [O III] and [Ne III] have been observed to take place in the course of one day. Additional theoretical work on the time lags in the case of BF Cyg is desirable.

Dossin has observed the photographic infrared region of BF Cyg. Certain intensity ratios, especially $\lambda 7772/\lambda 8446$ of O I vary strongly in an interval of two days. The identified infrared emissions are those of H, He I, Ca II, O I, NI, C III, [Ar III], but there remain several strong unassigned emissions which are not present in η Car.

In addition to the complex rapid variations of the intensities and displacements of the emission lines, BF Cyg shows slow variations. In recent years these have been followed by Miss Bloch. In the spring of 1955 the lines of [O III], [Ne III], [Fe III] were absent, whereas they were strong in October 1952 and in 1956.

5. RW Hydrae

This SS was first studied by Merrill in 1933, and later by Struve and associates in 1939–1943. It shows a "nebular" part having an excitation definitely lower than in AX Per and CI Cyg (no trace of [Ne V]); the excitation resembles that of R Aqr. The observed emission lines are those of H, He I, He II, [O III] (strong $\lambda 4363$, very weak N_1), very weak [O II] and relatively strong O III. Actually RW Hya is a unique case in which a strong, almost pure recombination spectrum of O III is present: no planetary nebula has been found showing such an intense and complete recombination spectrum of O III (see however sec. X). Fourteen lines of O III have been identified between $\lambda 3265$ and $\lambda 3962$. Bowen's fluorescence mechanism is not present to any appreciable extent: the lines of the singlet, triplet and quintet systems have the normal intensities of a recombination spectrum. The relative intensities of the O III lines are the same as in Campbell's object BD + 30°3639, but the latter belongs to the WC sequence, while RW Hya contains nitrogen and no carbon. New observations with higher dispersion are very desirable. One cannot exclude the possibility that the O III lines belong to a Wolf-Rayet nucleus of type WN with abnormally sharp lines. The WC nucleus of the stellar planetary HD 167362 (Swings and Struve 1940) has also relatively sharp lines, but the lines in RW Hya are still sharper. The estimated classifications of the late-type spectrum range from K5 to M0.

Further observational work is desirable because the spectrum of RW Hya varies slowly. Between 1939 and 1941 there was a slow gain in excitation, indicated by the ratios He II/He I and O III $\lambda 4363$/He I $\lambda 4388$. As in Z And, BF Cyg and RS Oph, the He I singlets are strong as indicated by the ratio of $2\,^1P^o-5\,^1D$ ($\lambda 4388$) and $2\,^3P^o-5\,^3D$ ($\lambda 4026$). In AX Per, CI Cyg, RX Pup and the planetary nebulae $\lambda 4026$ is much stronger than $\lambda 4388$. On general grounds one should expect that the singlets are favored by very low pressures. $\lambda 4686$ He II is present in RW Hya, suggesting that a recombination mechanism leading to the He I emission is also possible, but stratification effects must be present.

6. Nova (T) Coronae Borealis

The recurrent nova T CrB of small range, associated with a gM3-type star, may be a transition between the SS and the novae proper. Moreover T CrB is definitely a binary. In 1949 Sanford found that the M3 III component had a variable velocity with a period of about 230 days and a semi-amplitude of 21 km/sec. In 1958 Kraft demonstrated completely the binary character, the velocity of the hot component being obtained from the bright Hβ line (period $227^d.6$): Kraft concluded that the M3 giant of $M_v = -0.5$ and mass ≥ 3.7 ⊙ probably fills its lobe of the first critical equipotential surface in the restricted three-body problem. There may be transfer of matter from the M3-component toward the blue subdwarf companion, of $M_v = +4.4$ and mass ≥ 2.6 ⊙, through the Lagrange point L_1.

The most extraordinary spectroscopic event of T CrB was the appearance of the coronal [Fe X] and [Fe XIV] lines during the decline after the outburst in 1946 (Liège 1957, CNRS 1963). Similar observations of coronal lines have been made also in the composite object Nova (RS) Oph and in the probably-not-composite[21] object Nova (T) Pyx.

The spectrum of T CrB has been described by several observers, especially Struve and associates, R. Minkowski and Miss Bloch (CNRS 1963). After the 1936 outburst the excitation in the emission layers increased, and the intensity of the P Cygni-type absorption fringes of Hβ and Hα decreased as compared to the bright components. In 1940 the fluorescent lines of O III and N III were very conspicuous. The red companion appeared strongly in 1941, λ4686 He II and the fluorescent lines of O III, N III were strong, so that T CrB was approaching the appearance of RW Hya. However, in 1942 bright Fe II lines were present, and T CrB became somewhat similar to Z And, except that [Ne V] was absent from T CrB.[22] During the interval 1942–1943, considerable changes took place in the spectrum of the blue component. The H, He I, Ca II, Si I and Si II lines acquired a complex structure. Beside the previously observed He II λ4686, [O II], [O III], O III fluor., N III fluor., [Ne III] and Fe II, the spectrum showed a trace of [Ne V] λ3426, of [Fe VII] λ3586 and of the [S II] doublet. These changes are reminiscent of the analogous case of Z And.

The observations made by Miss Bloch and Tcheng Mao Lin (Liège 1957) in 1950 showed the presence of H, Ca II, He II, N III and [Ne III] in emission and of strong TiO and Ca I λ4226 in absorption. In 1955–1956, only the Balmer lines appeared in emission. Lawrence, Ostriker and Hesser (1967) have discovered very rapid oscillations (period from 98 to 112 seconds) similar to those found by Walker in Nova (DQ) Her (period 71 sec.).

7. R Aquarii

R Aqr is a complex assemblage of a hot source, a long period variable and an extended variable nebulosity. Here we shall not concern ourselves with the outer lenticular nebulosity. As for the long period variable its character seems quite regular.

This object has been studied by a number of observers, especially Merrill. When the McDonald group observed it in 1939, [O II] was strong while λ4363 was weak. This is an unusual behavior: [O II] must be produced in a nebular region of very low density. R Aqr deserves further scrutiny, especially more quantitative spectrophotometry. In 1939 there seemed to be a possible effect of occultation by the TiO bands: N_2, which falls close to a head of TiO seemed too weak relative to N_1. Old observations by W. H. Wright seemed to indicate that [Ne III] λ3968 was absent (perhaps absorbed by Ca II?) while [Ne III] λ3869 was present. A careful photometric study of the intensity ratio N_1/N_2 should be made, but this is a rather difficult investigation. Thus far the published results do not appear convincing.

[Fe III] has been prominent at times; this was the case in 1941, whereas [Fe III] was absent in 1931. At other times [Fe II] has been strong.

The most recent spectroscopic observations of R Aqr near minimum light are those by Herbig (1965) and by Ilovaisky and Spinrad (1966). The comparison of R Aqr near minimum with the Mira variable R Leo is very instructive. On Herbig's spectrograms, strong nebular emission lines appear due to the nebulosity extremely near the star. The Ilovaisky and Spinrad plates of 1966 (compared to those of Merrill obtained

[21] An infrared search for a possible cool component of T Pyxidis is desirable.

[22] A comparison between T CrB (1942) and Z And is especially interesting in the region λ < 3400. Like Z And, T CrB shows O III λ3344 (fluor.), Fe II λ3323, O III λ3133 (fluor.), Fe II λ3277, but there is no trace of [Ne V].

in 1919–1949) did not reveal any trace of the so-called "blue companion spectrum." Emissions of [Fe II], [S II], [O II], Mg I, Mn I, Fe I, Si I, Fe II, Sr II, and Balmer lines to H15 were present. The spectrum of R Aqr was rather similar to that of R Leo in October 1964. The intensity ratio in the [O II] doublet corresponded to $n_e \approx 10^3$ cm^{-3}. The "nebular component" had decreased considerably in intensity: [O III] $\lambda 4363$ and [Ne III] were present.[23]

8. Nova (RS) Ophiuchi

Four major outbursts have been observed: the first in 1898, a second in 1933 which was well-studied spectrographically, a third in 1958 also well investigated (CNRS 1963), and another in 1967 (Code 1968). In addition there have been minor outbursts. The astronomers around 1983 may have a chance to observe a spectacular new display! Since they will be able to cover the far ultraviolet ($\lambda > 912$ Å) and the X- and γ-ray regions from orbiting telescopes, we may imagine the exciting discussions which will take place. Actually it would be very interesting to have the far-ultraviolet spectrum of this object now, during minimum.

Nova (RS) Oph was the first instance in which the coronal lines were observed in a nova, soon after the outburst of Aug. 12, 1933. Although RS Oph had the characteristics of a fast nova, comparatively sharp nebular lines appeared soon: $\lambda 4363$ on Aug. 18; $N_{1,2}$, $\lambda 4640$ N III and λ 4686 He II on August 29. The [Fe X] line $\lambda 6374$ was probably present as early as September 7, enhancing the strength of $\lambda 6371$ Si II. At the end of October, $\lambda 5303$ [Fe XIV] had approximately the intensity of Hβ and $\lambda 6374$ was twice as strong as D$_3$ (He I). At the end of the observing season (Nov 13, 1933) the coronal lines were strong. By the beginning of the next season (March 1934) the coronal lines had completely disappeared.

The excitation in the [Fe X, XIV] regions of RS Oph is definitely lower than in most regions of the solar corona. RS Oph reveals several stages of ionization of Fe and other atoms, lower than X: the presence of [Fe VII] in RS Oph, and its absence in the corona may be explained by the transition probabilities (Bowen and Swings 1947):

$$\text{[Fe VII]: } 0.49 \text{ and } 0.37 \text{ sec}^{-1};$$
$$\text{[Fe X]: } \quad 69 \text{ sec}^{-1};$$
$$\text{[Fe XIV]: } 60 \text{ sec}^{-1};$$
$$\text{[Ar X]: } \quad 106 \text{ sec}^{-1}.$$

The identifications of the post-maximum lines were discussed by A. H. Joy and P. Swings (1945).

A minor outburst may have taken place on April 25, 1942. The red coronal line $\lambda 6374$ [Fe X] was observed on July 19, 1942, whereas there was no trace of the green line $\lambda 5303$ [Fe XIV].

The outburst of July 14, 1958 was the object of considerable attention (CNRS 1963); the parallelism with the 1933 outburst is extraordinarily close. As in 1933 coronal lines appeared in September 1958, especially [Fe X, XI, XIV], [Ar X, XI], [Ni XII, XV]. Very strong variations have been observed in the intensity ratios of the singlets and triplets of He I, of the auroral and nebular transitions of O III, of the $\lambda 8446$ triplet and the $\lambda 7772$ quintet of O I. The relative behavior of the quartets and sextets of Fe II was similar to that in the P Cygni stars. [O II] was absent (on account of the too high electron density: $n_e \gg 10^7$ cm^{-3}, probably $> 10^9$). An estimation of the mass of the gas cloud given off in the outburst is of the order of 10^{-6} ⊙.

---

[23] Sanford (1949) discovered that the long-period variable UV Aur also shows emission lines of H, [O III], and [Ne III].

As in 1933 the color temperature was low and the Balmer decrement rapid. The red color was not necessarily due exclusively to the interstellar scattering, although the interstellar absorption bands were strong; there may also be a reddening by the Kosirev-Chandrasekhar effect.

A strong emission line at λ6827 has been observed on numerous occasions in RS Oph and in Nova (RT) Ser. A tentative assignment to [Kr III] remains somewhat doubtful. The recent outbursts (March–April 1965 and October–December 1967) seem not to have been studied in detail. However, in January 1968 (three months after the maximum of 1967 October 26) Rosino and Mammano (1968) observed that the nova had reached a stage of extremely high excitation. Many coronal and nebular lines were present, including [Fe X], [Fe XIV], [Ar X], [O III], [Ni XII], He II, N III; there were also many Fe II lines.

9. FR Scuti

This object, first mentioned by W. P. Bidelman and C. B. Stephenson (1956), is definitely of symbiotic character with strong TiO bands. It has been described by Miss Bloch and Tcheng Mao Lin (Liège 1957). It is characterized by the emissions of H, Fe II, [Fe II, III], [O III].

10. MWC 603

This remarkable variable SS has been described by W. G. Tifft and J. L. Greenstein (Liège 1957) on the basis of 18 Å/mm spectrograms. Two hundred emission lines are listed: [O III], [Ne III], [Fe V], [Fe II], H, He I, Mg I, Si I, Si II, Fe II, He II, C III, O II, O III, N III, C III and probably [Ne IV] and [Fe VI]. He I shows variations in the singlet-triplet ratio. There is a strong dependence of line sharpness on excitation, and presumably on location in the nebulous envelope. Most diffuse of all are the forbidden lines of high excitation ([Fe V]).

VI. SPECTROSCOPIC DESCRIPTION OF PECULIAR BINARIES OR SUSPECTED BINARIES RELATED TO SYMBIOTIC OBJECTS

1. 17 Leporis and AX Monocerotis

These objects are related to the SS, but differ greatly from typical objects such as AX Per or Z And. 17 Lep and AX Mon display emission lines of low excitation; they have combination spectra (late- and early-types) and they suffer outbursts which are not strictly periodic (Cowley 1964, 1967).

17 Lep shows the TiO band at λ7054 (Slettebak); it combines a gM1 and an early-type (B9) companion; the orbital period is 260 days. The most probable masses are 1.4 and 4.6 ⊙, respectively. The B9 primary has a shell, but various broad features (He I λ4471, Mg II λ4481, Si II λλ4128, 4130) belonging to the underlying star are observable. 17 Lep displays similarities to T CrB. As for AX Mon, it combines a gM- or a gK-star with a rapidly rotating B3nn component. The period according to Mrs. Cowley is 232 days.

2. RX Puppis

Besides strong bright lines of H, He I, and He II, RX Pup shows lines of [O III], [Fe VII] (very strong), [Ne V], [Fe VI], [Ca VII] and C IV, which make it similar to CI Cyg as far as emissions are concerned. But the evidence for a late-type component is not quite definite. The low-dispersion spectra which have been obtained in the red region reveal certain absorption features which are similar to those of MWC 17 (see next paragraph), but no spectral type could be assigned to the possible red component. Spectra of higher dispersion in the red and near infrared would be most valuable, not only to confirm or deny the symbiotic character, but also to check on the possible presence of [Fe X] (close to $\lambda 6363$ [O I]) and of the auroral transition $\lambda 7319$–$\lambda 7330$ of [O II]. $\lambda 4363$ is very strong, yet the ratio $\lambda 4363/N_{1,2}$ is smaller than in CI Cyg or AX Per. Nebular [O II] $\lambda\lambda 3726$–29 is absent (density too high), but nebular [N II] is present. Nebular [O I] is present, but not the auroral transition: [O I] is excited on the outskirts. This star, unfortunately rather far south, deserves further observational attention.

3. MWC 17

The spectrum of this object is intermediate between RY Sct ([Fe III] only) and η Car (an [Fe II] star) as [Fe III] and [Fe II] appear with similar intensities. Beside the ^5D–^3F and ^5D–^3P multiplets of [Fe III] and the usual [Fe II] transitions, MWC 17 shows H, He I, Si II, [O I], [N II] and [S II], whereas [O III], [Ne III], [O II] and the permitted Fe III lines are absent. There is some suspicion of a red component, but better spectra should be obtained in the red. MWC 17 is possibly a SS of lower excitation (absences of O^{++}, Ne^{++}, Fe^{+++}, but presence of Fe^{++}, N^+ and S^+). As in RX Pup, the nebular [O I] lines must be emitted in regions where the ionization and the electron density are lowest. On the other hand the absence of nebular [O II] indicates that the electron density prevailing in the regions where oxygen is ionized is too high. $\lambda 5270$ ($^5D_3$–$^3P_2$) and $\lambda 4658$ ($^5D_4$–$^3F_4$) of [Fe III] have about the same intensity (an effect of reddening). The Si II lines show the same selectivity as in Z And, AG Peg and P Cyg.

4. RY Scuti

The early-type "component" of this binary occupies a special place among the P Cygni stars. Its spectrum has been described by Merrill and later by Struve and associates. Emissions of H, He I, [N II] (nebular and auroral) and [Fe III] are present. RY Sct represents a definite stage in the evolution of certain novae, which follows the η Car stage ([Fe II]). There is no trace of Fe II, Fe III or [Fe II]. $\lambda 5270$ is approximately of the same intensity as $\lambda 4658$. There remain several unidentified emissions in the visual region. There is a late-type component.

5. CD − 27°11944

This peculiar star of P Cygni type (with weak [Fe II]) may have a late component, possibly of R-type but not of M-type (no TiO bands). It should be observed in the red region with a higher resolution. The P Cygni emission features are unusually broad; they may extend over 600 or 700 km/sec. They are indeed broader than the emissions of some Wolf-Rayet stars (BD + 30°3639 [Campbell's object], HD 167362, the nucleus of IC 4997) and of slow novae. Ca II and Na I are present in emission.

CD–27°11944 belongs to the group of P Cygni stars in which the expanding shell is observed but not the exciting star, while other P Cygni- or Of-objects show the spectrum of the stellar reversing layer.

6. V1016 Cyg = MHα 328-116

This object, which rose abruptly in brightness from $m_{pg} = 15.3$ in 1963 to 11.9 in 1964, has been studied by Miss Bloch (1966), by Fitzgerald, Houk and McCuskey (1966), and by O'Dell (1967). The pre-outburst type was late M. On October 5, 7, and

13, 1965, strong emission lines of H, He I–II, C II–III–IV, N II–III, [N II], O I, [O I, III], [S II, III], [Ne III], [Ar III], Fe II, [Fe II, III] were present. H was represented by the Paschen series to P23 and the Balmer series to H24. O I λ8446 was strong, while λ7772 was weak. The red doublet of [O I] was intense, but the green line weak. The [O II] lines λ7320–λ7330 were intense, but $\lambda\lambda$3726–29 faint. The [O III] lines $N_{1,2}$ and λ4363 were among the strongest emissions, together with [S II] $\lambda\lambda$4068, 6730 and [Ne III]. The lines [S III] $\lambda\lambda$3721, 6312 and [Ar III] $\lambda\lambda$7136, 7751 were also present. About sixty [Fe II] lines were observed, as well as several of [Ni II]. [Fe III] was very well marked; indeed it seems that V1016 Cyg is the richest object in [Fe III] thus far observed. The relative line strengths of [O III] were found to vary as well as the Balmer lines, whose decrement indicated that the envelope was optically thick in those lines.

The symbiotic character of V1016 Cyg, which was doubtful for a time, is now well established.

7. MWC 349

This star which bears a striking spectral analogy to MWC 17 has been observed by Merrill in 1932, then by Struve and associates in 1941. [Fe III] was strong in 1932, as well as in 1941. λ5270 ($^5D_3$–$^3P_2$) is much stronger than λ4658 ($^5D_4$–$^3F_4$). This is certainly due to interstellar reddening. No late-type absorption features appear clearly, but better spectrograms should be obtained in the red.

8. Boss 1985 = HD 60414-5 and Boss 5481 = HD 203338-9

B 1985 consists of a M2 Iab and an early B-type star (B2V, or perhaps a hot subdwarf). It is characterized by strong [Fe II] and [Ni II]. Mrs. Cowley (1964) has obtained the period (about 27 years). In all probability the late-type component fills its lobe of the first critical equipotential surface. Certain emission features ([Fe II], [Ni II]) may arise in a very extensive nebulosity enveloping the system.

While no spectacular change has been observed during the period 1939–1947 except a slow intensity increase of the H emission from 1942 to 1946, very striking changes have been observed in the ultraviolet region beginning October 18, 1947. While previous spectrograms had mostly revealed emission lines of H, Fe II, Cr II, [Fe II] and [Ni II], the autumn 1947 McDonald material (Swings 1950) showed intense absorption lines of the shell type, similar to those at an outburst of Z And or an eclipse of VV Cep. Similar observations have been made by R. F. Sanford and by D. B. McLaughlin. The [Ni II] emission remained strong. The Balmer continuum appeared in absorption. Compared with the shell of Pleione or with the supergiant 3 Puppis, B 1985 exhibited stronger Sc II and Ti II absorptions (of low excitation), and much weaker Cr II and Fe II. The Balmer lines had the appearance of reversed P Cygni lines, similar to the lines of the eclipsing binary VV Cep. McDonald coudé spectrograms obtained during 1963–64 revealed the red [O I] emission λ6300, but not the auroral green line. Additional observations are needed. It would be interesting to measure the ultraviolet magnitude regularly.

In the triple system B 5481, the spectroscopic behavior of the primary (M1 I + B2 V) resembles that of B 1985, including the temporary production of a shell.

9. WY Geminorum and W Cephei

Both of these binaries consist of M- and Be-type components, and are rich in [Fe II], especially WY Gem which strongly resembles B 1985. Several ultraviolet emission lines are still unassigned. The ultraviolet spectrum of WY Gem is entirely free of the late-type component.

10. VV Cephei

This is an eclipsing binary consisting of an M supergiant (radius $\approx 1200 \odot$) and a Be-type main-sequence star. It has bright lines of H, Fe II, Ca II, [Fe II], [Ni II]. Struve and associates have described the early-type spectrum for $\lambda > 3100$ Å. The low-excitation transitions of Fe II are considerably enhanced. These Fe II lines do not share in the orbital motion of either of the components. The Fe II and [Fe II] emission originates most probably in an extended envelope surrounding the whole system, while the H emission must be closely associated with the Be star.

Investigations covering more than one orbital cycle of 20.4 years have been recently carried out in the blue region by Peery (1966) and in the photographic infrared by Glebocki and Keenan (1967). In the blue, a few unexplained peculiarities and discordances remain. The region 7000–8000 Å is particularly interesting. The transient appearance of the O I blends at $\lambda\lambda 7772$, 8446 which in 1944 had been found to last for less than 55 days, was again observed in 1964. No strong streaming motions are evident in the O I surrounding the early-type component; this absorption spectrum appears when the material is projected upon the M-type star. In the region occupied by the O atoms there probably is no appreciable Lβ flux, because the $\lambda 8446$ absorption line is relatively deep.

It would be interesting to examine the photographic infrared spectra of other similar systems, such as 31 and 32 Cyg and ζ Aur, at the times of their secondary minima.

11. The Companion of α Scorpii

α Sco B is of mag. 5.2, at a distance $3''$ from the M0 supergiant. Struve and associates have tried to determine the extension of the [Fe II] lines in the nebulosity surrounding the Be-type companion. The observation is rather difficult, but deserves being repeated with a longer focal length, a higher dispersion and perfect seeing. The radius of this nebulosity is approximately $2''$ or $2''.5$ (≈ 200 astronomical units).

The Be star itself has a [Fe II]-rich spectrum resembling MWC 17 and η Car. There is no trace of Fe I, [Fe I], Fe II, [Fe III]. If the H lines are bright, they are certainly weaker than [Fe II]. The absence of Fe II indicates that T_e is low. The abundance of H may also be abnormally low.

12. W Serpentis, α Herculis, and α Scorpii

The eclipsing variable W Ser has been the object of many spectroscopic investigations (C. A. Bauer (1945), Struve, J. Sahade, M. Hack, A. Beer, A. Fresa, etc . . . Liége 1957). While certain emission lines, for example [Fe II], are seen at all phases, other broad emissions appear only during eclipses and border the shell lines. The [Fe II] emissions must be produced in the outermost layers of the envelope surrounding the system.

Herzberg observed Fe II emissions (EP up to 5.6 ev) in the region $\lambda < 3300$ Å in the supergiants α Her and α Sco. These could not be due to the hotter companions. Herzberg assigns the emission lines to a "corona-like nebulosity."

13. $M1 - 2 = VV\,8$

This object was originally found by Minkowski (1946), and identified by him as a stellar planetary nebula. Attention was called specifically to it by Razmadze (1960). It has been studied by O'Dell (1966) who detected the absorption spectrum of a G2 supergiant. The object is either a G supergiant plus a hot companion, or a single hot star surrounded by a very extensive outer shell that is ionized only in the innermost region.

14. CH Cygni

The type M6 semiregular variable CH Cyg had an outburst in June 1967. The emission lines in the ultraviolet (3140–3500 Å) have been described by Swings and Swings (1967), and compared with those in VV Cep, B 1985, BF Cyg and η Car. These lines are mainly due to Ti II, Mn II, Cr II, and Fe II; lines of [Fe II] and [Ni II] are very weak or absent.

The 3680–5050 Å region has been described by Faraggiana (1968), and compared to the same region in 30 Her and VV Cep. She found sharp emission lines of Fe II, [Fe II], [S II], broad He I emissions, P Cygni-like structure at the Balmer lines and the H, K lines of Ca II. She lists the radial velocities from absorption lines, from the P Cygni structure, and from the emission lines of different elements.

15. HD 45677 (= MWC 142)

This star, which is one of the richest objects in [Fe II], is not known to be a binary. Yet the intensity of [Fe II], [Ni II] and other emissions, and also the spectroscopic variability of HD 45677 lead one to believe that a cool companion may be found.

At the time of the first McDonald observations (1939), the spectrum seemed to be identical to that described by Merrill (1923–1927). There were bright lines of H, [O I], Fe II, Cr II, [Fe II], [Ni II], and exceedingly sharp absorption cores in the Balmer lines to H29 or H30. The radial velocities from the lines of the shell agreed well with those from the reversing layer. The McDonald spectrograms revealed new [Fe II] transitions (EP up to 4.72 ev, by parallel studies of B 1985 and WY Gem), and led to the discovery of [Ni II] and [Cr II]. There remain some still unassigned emissions.

The spectrum changed appreciably during the period 1939–1943. Ca II $\lambda 3968$ (H) acquired a complex profile, similar to that in HD 190073. On the other hand the sharp emissions of Fe II, [Fe II] and [Ni II] remained the same. The profiles of the permitted and forbidden lines differ considerably.

This star is at present being investigated at Liège.

VII. COMPARISON WITH SPECTROSCOPIC OBSERVATIONS OF NOVAE

The SS display a bewildering variety of spectroscopic phenomena. In order to acquire general views regarding the physical characteristics of these objects it seems instructive to compare the behavior of the typical SS, not only with individual peculiar stars as we have done in the preceding section, but also to related classes of objects: novae, planetary nebulae (including their nuclei), Wolf-Rayet, Of and P Cygni stars. Obviously, in all these classes, we have retained and summarized only the spectroscopic phenomena which may help directly in understanding the SS.

In all the peculiar objects the continuum—on the assumption of elementary models—arises at depths which depend primarily on the wavelength and which differ completely from the layers giving rise to the discrete emissions. A light curve has a physical meaning only if it concerns a stated wavelength, hence a defined region of the star. The maximum of a light curve in the ultraviolet does not necessarily occur at the same time as the maximum in the visual or the infrared regions. In an expanding atmosphere the effective photospheric surface is a purely optical notion. In other words, at any given wavelength λ the luminosity (if due to a continuum) is determined by the diameter D and the temperature T of the non-static photosphere corresponding to this λ. The maximum of λ results from the combined effects of D and T. On the other hand the emitted energy may originate essentially in discrete lines and not in a continuum: this would be the case of coronae in the far ultraviolet. These considerations are especially applicable to novae and SS at the time of outbursts.

Actually what we need most in the photometric field is a set of monochromatic light curves, such as in Hα, in the pure late-type spectral region, in specific forbidden lines such as $\lambda 4363$, $N_{1,2}$, [Ne V], [Fe VII], and in the pure continuum of the hot component.[24] The establishment of a program of this kind for SS, novae and representative peculiar stars is imperative. It should include specially interesting phases of the nova evolution: apparition and evolution of the coronal lines, nitrogen outbursts, phases with molecular absorption. Once orbiting telescopes become operational it will become of great interest to record light curves in Lα and other characteristic ultraviolet emissions.

Two symbiotic objects, T CrB and RS Oph have had genuine recurrent nova outbursts. Indeed all SS suffer, at certain phases, variations of an explosive character. Z And has shown nova-like outbursts. All SS exhibit light variations, sometimes of small amplitude, always resembling certain phases of novae. In all cases the envelopes of SS as well as of novae depart from local thermodynamic equilibrium. There are of course differences in ejection velocities; the dilution effects may be quantitatively different. But essentially the mechanisms have close similarities. At the time of an outburst of either an SS or a nova, the expanding, probably asymmetric circumstellar shell has an electron density which is too high for forbidden lines. The latter appear when the density in certain regions of the expanding shell decreases to an extent such that the collisional de-excitation of the excited metastable levels is less efficient than the radiative de-excitation by emission of the forbidden lines. Each forbidden transition of each atom or ion has its specific requirements for appearance. Consideration of the required "time lags" is also essential. As for T_e, it is kept at a specific value, fairly constant on account of the thermostatic effect of the forbidden lines.

The similarity of the spectroscopic history of the SS with typical large scale nova outbursts is striking, but there is as yet no evidence that the physical causes of the outbursts are similar. Merely descriptive similarities of behavior may be deceptive. The total mass of a nova shell is of the order of 10^{-4} ☉, while that of an SS shell is definitely smaller.

The presence of molecular bands at certain phases of novae is probably not due to the same mechanisms as the presence of a cool component in the SS. It is known that the CN and C_2 bands have been observed in several novae, especially in Nova (DQ) Her 1934. The molecules may be much more abundant in a nova layer than in thermodynamic equilibrium at the "photospheric" temperature (whatever this may mean) as the radiation reaching the molecular layers may have been previously depleted in the spectral regions required to photodissociate the molecules. Indeed the departures from equilibrium may even be such that absorption bands of ionized molecules, such as CH^+ or CO^+ may someday appear in a nova!

The SS seem to possess abnormal abundances, and so do the novae. [Fe VII] seemed abnormally strong in Nova (RR) Pic (1925), [Ne III] in Nova (GK) Per (1901) and Nova (RT) Ser (1909), [Fe III] in Nova Ser and Nova (DO) Aql (1925). Nova Ser is rich in N III and poor in C III. But such anomalies may be purely apparent, in the novae as well as the SS, since the spectra integrate over a wide range of layers.

The observations of changing radial velocities of specific lines are similar in SS and novae. There is actually no need to envisage phenomena of acceleration or deceleration while the atoms are in flight. There is presumably an original spread in the velocity of ejection. "At different epochs, the material moving with one velocity may be more readily observed than that moving with another velocity, owing to the changing

---

[24] C. R. O'Dell has in progress at the McDonald observatory a program devoted to systematic observations of a number of SS. Certain emission-line ratios and the flux distribution in the continua are measured with a spectrum scanner. Large changes in these quantities have already been observed.

density and excitation which produce or destroy measurable spectral features"
(A. B. Underhill).

Most SS appear to have an individuality. So do the novae. The differences in
spectroscopic appearance result from the wide possible varieties of density and
velocity. Actually there are certain SS which look spectroscopically very similar,
more in fact than novae: this is the case of AX Per and CI Cyg at certain phases.

We have seen that strong coronal lines have been observed in two symbiotic novae:
RS Oph and T CrB; coronal lines have been found in other novae, such as T Pyx,
and very probably [Fe X] was present in two SS, AX Per and CI Cyg at certain phases.
The behavior of the coronal lines is in accord with the properties of the ions concerned.
The apparent anomalies are actually caused by the operation of the primary mechanisms of ionization and excitation, attention being paid to the transition probabilities,
the electron densities, the departures from equilibrium and the fact that the ionizing
underlying radiation may differ considerably from a black body.

The Zanstra mechanism has been applied to SS as well as to novae and planetary
nebulae. However it is clear that the usually adopted hypothesis of ultraviolet
ionizing radiation of the black body type is only a very crude approximation. In
fact atoms which have approximately the same ionization potential compete for the
ultraviolet quanta (continuum, modified by discrete emission and absorption features)
(Swings 1942). For example the ionization of O^{++} may be reduced by the fact that the
ionization potentials of He^+ and O^{++} are nearly equal (this explains observations in
NGC 6543). The ionization of C^{++} and N^{++} may be favored by the presence of emission lines of He II superimposed on the continuum. Indeed the far ultraviolet continuum—if there is any—differs considerably from a black body. If the underlying
radiation is absorbed by He II for $\lambda < 228$ there will result a higher abundance of
C^{+++}, N^{+++}, O^{++}, Ne^{++} and Ar^{+++} in the higher layers. Similarly a strong continuous
Lyman absorption increases the population of Si^+, Fe^+, N, and Ar, and the He I
absorption continuum affects the ionization of C^+, N^+, O^+, Ar^+ and Si^{++}. All those
phenomena will become clearer once we obtain ultraviolet spectra ($\lambda > 912$ and
$\lambda < 20$ Å) with orbiting telescopes (CNRS 1963).

VIII. SPECTROSCOPIC REMARKS ON INDIVIDUAL NOVAE

Certain post-novae reveal spectroscopic information which are of interest in
relation to the SS.

Nova (DQ) Her which was studied extensively at the time of its outburst in 1934 is
a typical representative of the slowly developing type, hence may be a source of
instructive comparison with SS. It has been re-observed at McDonald on numerous
occasions since 1940. At that time the N III spectrum was strong in emission; it was
not excited only by Bowen's mechanism because $\lambda4379$ was present. On the other
hand the strong O III emission lines may have been mainly excited by fluorescence.
[O I], [O II], [O III], [N II], etc. . . were present. $\lambda4363$ was fairly strong. In July
1942 the relative intensities differed greatly from 1940; [O II] was stronger, while the
ratio $\lambda4363/N_{1,2}$ had not decreased significantly. The range and variations in expansion and in radial velocity are rather complex, or even appear almost contradictory. The spectra taken in 1947 and 1949 reveal striking changes in the spectrum
of the nebulosity since 1942. The apparently contradictory spectroscopic changes may
be understood if we assume that the far-ultraviolet exciting radiation of the nucleus
differs considerably from a black body. This departure could bring about an irregular
distribution among the stages of ionization. It is also possible that fluorescent excitation by the underlying radiation plays an important role in the emission. Such a
mechanism would be affected considerably by the far-ultraviolet absorption or
emission features of the exciting nuclear spectrum. On the other hand the evolution

of regions where different physical conditions prevail must also differ and the geometry of the system is certainly far from spherical symmetry.

Observations made in 1950 showed that the striking spectroscopic changes observed between 1942 and 1949 continue while the mean velocity of expansion remains practically the same as in previous years. The nebular spectrum progresses toward a stage characterized by very strong [O II], strong He II, N III, H; the other usual emissions [O III], [Ne III], [Ne V] and [Fe VII] have practically disappeared in 1950.

Following Merle Walker's discovery that Nova (DQ) Her is an eclipsing binary of period 4^h39^m, various attempts have been made to determine a radial velocity curve. The most complete results based on wonderful prime-focus spectrograms taken at the 200-inch reflector have been described jointly by J. L. Greenstein and R. P. Kraft (1959), and by R. P. Kraft alone (1959). These two remarkable investigations deserve a detailed review. Outside eclipse the spectrum is qualitatively the same as it was in 1950 (strong [O II] and [S II], very weak [O III]). Complex phenomena which differ for different emissions take place in the course of a period. Greenstein and Kraft suggest that He II and the higher members of the Balmer series are produced in a rotating disk of gaseous material that follows the motion of the nova in its orbit; the material undergoes eclipse along with the star. However, Hβ together with the forbidden emission lines is produced mostly in the expanding nebular envelope seen on direct photographs. In 1956–1958 DQ Her consisted probably of a hot, rather bright, white dwarf of approximate mass $0.25\odot$ coupled with a "dark star", possibly dM3.

It is desirable that DQ Her continue to be investigated spectroscopically and photometrically from time to time; Greenstein's study of 1959 is a model for such work.

Nova (RT) Ser had an outburst in 1909 and has developed abnormally slowly. Its spectrum was taken by A. H. Joy 22 years after the outburst: [Fe III] was very strong. Spectra taken at McDonald in May 1940 revealed the following: [Fe III] had become very weak, while [Fe VI], [Ne III, V] were strong. Beside Hα the strongest line was [Ne III] $\lambda3868$, as in Nova Per 1901 and Nova Sgr 1936 at certain phases. The N III emission was strong, while O III was absent. In July 1942 [Fe V] and [Fe VI] had developed more completely than in 1940. Actually RT Ser was an ideal object for [Fe VI] in 1942. The McDonald spectrograms obtained in May 1950 revealed tremendous changes since 1942: [Fe VII] and [Ne V] had become very intense, while [Fe V, VI] and [Ne III] had become weak. $\lambda4363$ was still much stronger than $N_{1,2}$, 41 years after the outburst: the density in the nebular envelope was still high, compared with most planetary nebulae including IC 4997, but resembled that in AX Per and CI Cyg. Since the ionization has been increasing steadily during the period 1909–1950 it would not be surprising if RT Ser would acquire coronal lines in the not-too-distant future.[25] The spectrum of this object should be observed periodically. Herbig has discovered late-type absorption features in the red; thus RT Ser is a symbiotic nova.

The fourth outburst of the recurrent nova T Pyx in 1945 revealed spectroscopic phenomena similar to those in Nova (RS) Oph, including the appearance of coronal lines. However the ionization in T Pyx was higher than in RS Oph, while the density was lower in the layers which emit [Fe VII]. In fact [Fe VII] is stronger relative to [Fe X] and [Fe XIV] in T Pyx than in RS Oph. It would be interesting to search for a possible red component.

Another excellent representative of [Fe VII] emission, beside T Pyx and RT Ser, was Nova (RR) Pic (1925). [Fe VII] attained its maximum intensity relative to

[25] Two spectrograms obtained by G. H. Herbig in 1963 and 1964 do not yet reveal convincing evidence of [Fe X] or [Fe XIV].

$\lambda 4686$ and Hα in 1932, and then decreased in the following years. The behavior of RR Pic showed similarities to that of AX Per. Actually RR Pic is an important example of the slow nova type. It was described in considerable detail by H. Spencer Jones during the period 1931–1934, but many identifications, especially those of [Fe III, V, VI, VII] could be made only several years later: [Fe VI, VII] by Bowen and Edlén (in 1939), [Fe III] by Edlén and Swings (in 1939) and [Fe V] by Edlén (in 1939) and by Swings and Struve (in 1942). The broad [Fe V] emissions in RR Pic could be discussed only after [Fe V] had been clearly observed as sharp lines in AX Per. The successive appearance of [Fe II], [Fe III] in 1925, [Fe V] in 1926, [Fe VI, VII] in 1934 resembles that in RT Ser and in various SS and other peculiar stars. Actually a similar succession has also been observed in Nova (RR) Tel, in which Edlén has moreover made tentative identifications of [Fe IV].

Among recently observed novae, Novae Her 1960 and 1963 showed spectroscopic phenomena which resemble those in SS: appearance of coronal lines; variations in the intensity ratios $\lambda 8446/\lambda 7772$ of O I; sextets/quartets of Fe II; $\lambda 5812/\lambda 4658$ of C IV. In both novae the usual Wolf-Rayet emissions in the infrared are present: He I, He II, C II and C IV, but the usually strong C III infrared transitions which are prominent in WC stars are absent in the novae (CNRS 1963).

IX. GENERAL COMPARATIVE REMARKS ON THE EXCITATION MECHANISMS AND THE PHYSICAL PHENOMENA

We have mentioned earlier that coronal lines have been observed in several novae and probably in SS; all these emissions are actually forbidden transitions. In the far ultraviolet and in the soft X-region the permitted lines of the highly ionized atoms must be prominent at the phases when the coronal lines of the usual spectral region are intense. Actually high excitation permitted lines may also appear in the ordinary region (CNRS 1963). The high ionization is probably due to inelastic collisions with electrons; so may arise the forbidden and permitted lines of the highly ionized atoms. This would indicate that there may be "coronal regions" in nova envelopes where $T_e > 500,000°$ as in the solar corona. The ionization and excitation in the "coronal regions" of novae would thus differ radically from those in planetary nebulae. Actually it has generally been assumed that the excitation and ionization in the SS, novae, Wolf-Rayet and other peculiar objects of early type are due to electromagnetic radiation. But electrons of high energy may also play a major role. Magneto-hydro-dynamic methods must be applied.

The typical novae which show strong coronal lines are recurrent objects; in the case of RS Oph and T CrB a cool companion exists, and it definitely seems that recurrent novae and the U Geminorum stars have companions (CNRS 1963). According to Kraft the period of recurrence would be the time required by an unstable component to accrete enough matter coming from the other star; the critical instability would give rise to an outburst. From the spectacular photometric and spectrographic studies of Nova (DQ) Her and other post-novae one may indeed wonder if all post-novae, not only the recurrent of fairly short period, are not close binaries.

What causes the occasional ejections in the evolution of SS and at certain phases of novae? Should the radiation pressure be considered? In the course of the evolution of a shell a slight change in density or velocity distribution or in the "quality" of the exciting radiation may appear, which enhances the fluorescence excitation. As a consequence, the selective radiation pressure in the resonance lines will increase, and a more or less sudden ejective outburst may result owing to the fact that the increased velocity of the shell, relative to the reversing layer, displaces the shell absorption lines outside the depleted regions of the stellar spectrum. This explains the more or less sudden appearance of bright lines excited by fluorescence.

Turbulence and axial rotation may play a role in the fluorescence excitation. Axial rotation should be considered not only as a dynamical agent in the formation of a shell, but also a mechanism influencing the fluorescence excitation and the expansion which might result from the corresponding radiation pressure.

An extreme case may be the "nitrogen flaring stage" of novae, such as was described by W. H. Wright for Nova Gem (1912). When an adequate variation of the velocity gradient or stratification in the nova shells or of the quality of the underlying radiation appears, Bowen's fluorescence mechanism of He II, O III and N III may become prominent. The selective radiation pressure due to the accumulating quanta in N III $\lambda 452$ may expel suddenly the N^{++} atoms (Bowen 1935), thus giving rise to very broad N III and N IV bands. Since this abnormal ejection destroys the conditions required for Bowen's fluorescence mechanism, the outburst is only temporary. Its spectroscopic effects depend on the location of the atoms considered. The variation in the exciting nucleus reacts on the intensities and profiles of the nebular lines arising in the outer nebulosity. The resulting line modifications depend on the distribution of matter and physical conditions in the nebulosity and on the various relaxation times.

The continuous absorption of the flaring shell may affect considerably the lines arising in the deepest layers: a He II emission line arising in the inner layers may even be replaced by an He II absorption line arising in the outskirts of the "nucleus."

Eventually this mechanism may possibly help in explaining the formation of peculiar associations, such as have been observed in BD $+$ 30°3639, NGC 40, and other planetaries in which a pure carbon nucleus WC is surrounded by a nebulosity rich in N (characterized by strong [N II] lines). Other planetary nuclei containing both C and N are also surrounded by a nebulosity showing strong [N II] (e.g., IC 418, IC 4997, NGC 6572, NGC 6210).

The general analogy of the N-flaring stages of novae with the less spectacular variations observed in the Of, P Cyg or Be stars is obvious.

In the early stages of a nova or SS outburst there is only a minor influence of dilution, as the "shell" is not very distant yet from the region emitting the continuum. Later dilution effects appear, and give a possibility to estimate the distances of the shells from the photosphere. We refer to Miss Underhill's chapter for a thorough critical review of these dilution effects, including their influence on the spectral classification criteria. This chapter contains also a study of the effects of the strong emission lines, such as Lα of H ($\lambda 1215$) and He II ($\lambda 304$).

The paper by Struve on Z And stresses the importance of time lag in the sense considered by W. Grotrian. A change in the temperature of the exciting radiation will influence the ionization of distant layers only after a certain delay, which evidently varies with the distance to the nucleus, with the density, and with the atomic properties. [Fe VII] is probably excited in a region close to the nucleus and would be first to react to a nuclear variation. It is well known that a post-nova nucleus fluctuates in brightness, and in case of a surrounding nebulosity this will obviously give rise to a variation of the excited nebular spectrum after certain relaxation times. These may be short (of the order of a few days in BF Cyg) or long (months in other SS or novae). Since the fast particles reach greater distances the corresponding lag may be longer. The time lag has been reconsidered recently by J. S. Pecker (CNRS 1963), who suggested a theoretical treatment which is more complete than that of Grotrian.

X. COMPARISON WITH SPECTROSCOPIC OBSERVATIONS OF PLANETARY NEBULAE AND THEIR NUCLEI

The spectroscopic characteristic of planetary nebulae is the presence of strong forbidden lines; this is true also for the SS.

In the accompanying paper, Struve and I described the spectrum of IC 4997 on account of its resemblance to Z And at a certain phase (see the section on Z And).

Like the SS, the planetary nebulae are impermanent structures. Indeed L. H. Aller and W. Liller (in 1957) found an intensity variation of $\lambda 4363$ in IC 4997 in the course of the last 40 years. One may wonder if spectroscopic variations have not occurred also in IC 3568 (Swings and Fehrenbach 1958). Spectrograms taken in 1958 at the Haute Provence Observatory reveal a strong He II line which was not mentioned by W. H. Wright in 1918. Aller's description of IC 3568 does not mention [Ar IV] which is present on the Haute Provence spectrograms.[26] The very great variety of planetary nebulae is of course due partly to evolutionary effects. Slight differences of central star (including its age), T_e, n_e, and relative abundances, geometry and velocities give rise to very different spectra of the planetaries. Photographs and slitless spectrograms of planetary nebulae reveal their complex geometry, including stratification and a filamentary structure; different regions produce different spectra; in particular different filaments may have different T_e and n_e. Unfortunately in the peculiar stars we do not have such direct evidence of stratification effects although stratification is certainly present, and we have to use integrated intensity ratios of different emissions, essentially the intensity ratios of the nebular, auroral, and transauroral transitions of various ions and the [O II] ratios $\lambda 3726/\lambda 3729$. There are planetaries of low excitation (H, strong [O II], weak [O III]); of higher excitation (strong [O III], weak [O II], strong [Ne III]); of highest excitation (strong He II and [Ne V], strong fluorescence of O III and N III). We have encountered similar stages in the SS.

The recent spectroscopic descriptions of planetary nebulae based especially on Mt. Wilson, Lick and Haute Provence coudé spectrograms have brought a wealth of information which has been and will be in the future of great help in interpreting the SS spectra. For example, nebulae reveal $\lambda 4562$ [Mg I] and the $^5D_4$–$^3H_4$ transition of [Fe III] ($\lambda 4881.14$) and of [Fe V] ($\lambda 4227.49$).

The intensity ratio $\lambda 3726/\lambda 3729$ of [O II], $^4S_{3/2}$–$^2D_{5/2,3/2}$ depends on T_e and n_e in a theoretically well-established way and constitutes a valuable criterion of electron density in certain ranges of n_e (of the order of 10^3 cm^{-3}) and T_e. The transauroral doublet of [S II], $\lambda 4068$–$\lambda 4076$ is also valuable for assessing density fluctuations.

The Orion Nebula which is the brightest and one of the nearest of the H II regions (Kaler, Aller, Bowen 1965) has been observed spectrophotometrically with high dispersion (16 and 20 Å/mm); we thus have reliable intensities for [Fe III] and many permitted lines (He I, C II, N II-III, O II, Ne II, Si II, S III).

The richest nebular spectrum is that of NGC 7027 which has been studied in detail recently (Aller, Bowen, Wilson 1963). Many recombination lines (O II-IV, C II-III-IV, Ne II, Mg II, etc . . .) are observed together with many forbidden lines, revealing a great range in excitation (from Mg I and Fe II to recombination lines of O IV, N IV, N V and forbidden [Fe VII] among the filaments and other condensations. Among the observed emissions the following are of special interest: [Na IV-V], [Cl III], [N I], [Fe III-V-VII], [K IV-V], [Mg I].

L. H. Aller and W. Liller have prepared (1968) a new detailed report on planetary nebulae, with main emphasis on the physical mechanisms. Important theoretical contributions, especially by Seaton's group (Seaton 1963–1966), have recently interpreted many observational data on planetaries, and deduced the radii and masses of the nuclei, and the masses and luminosities of the nebulosities; many considerations may sometimes be applied readily to the SS.

As in the case of the SS, planetaries may present Bowen's fluorescence mechanism, or it may be absent. For example, Bowen's mechanism is absent in NGC 6543 (no trace of He II and O III emission in the nebula) and the strong lines of C II, C III and N III are due to a recombination process. On the other hand the SS RW Hya has a

---

[26] The [N II] doublet near Hα is absent. As in NGC 6543, IC 4997, and NGC 6826 the nucleus of IC 3568 contains lines of carbon and nitrogen.

very complete O III spectrum, excited by a recombination process, while N III is very weak or uncertain.

The great range in excitation observed in nebulae such as NGC 7027 is found also in SS, for example AX Per and CI Cyg.

The chemical composition of a planetary nebula (or of the nebulosity of a SS) is that of the outer skin of the star from which it originated. There seem to be real differences in chemical composition of the nebulae; these must be due to the nuclear processes and the amount of mixing in the pre-planetary object. We shall return to this point later on.

The photoionization processes (Zanstra mechanism) are, in principle, similar in SS and in planetary nebulae. In particular the effects of the overlapping regions of photoelectric absorption and of the far-ultraviolet emission or absorption features are of importance, as we have stated previously; the case of NGC 6543 is especially convincing (Swings 1940). The discrete far-ultraviolet features play an essential role. It is possible (but dangerous!) to draw synthetic far ultraviolet spectra of the nuclei, then to deduce the most active features (Swings 1942). In high excitation nebulae, H is ionized to an appreciable extent by strong emission lines of the exciting radiation, especially Lα of He II (λ 304). The photoelectrons are thus provided with a high kinetic energy and are able to excite various permitted lines of high excitation. The same consideration applies to the SS: many lines, including far ultraviolet lines, may be excited by collisions. It will be extremely interesting to obtain far ultraviolet spectra of SS stars by telescopes on space vehicles.

There is a considerable range in the spectra of the nuclei of planetary nebulae (Liège 1957), and resemblances to parts of the spectra of SS are common. We are only at the beginning of the construction of theoretical models for the atmospheres of nuclei (Böhm and Deinzer 1966); the adopted effective temperatures vary from 45,000 to 150,000°K. The central stars must have masses of the order of 0.6 $\odot$, *i.e.*, of the same order as the white dwarfs.[27] The masses of the ejected nebulosities are a few tenths of the solar mass.[28] The central stars of NGC 1514 and 2392 "flicker" like old novae (Lawrence, Ostriker, and Hesser 1967).

XI. COMPARISON WITH SPECTROSCOPIC OBSERVATIONS OF WOLF-RAYET STARS

At least half of the W-R stars are binaries, and various astronomers have suggested that the W-R stars are all binary systems. Different orientations of the orbital planes would explain why not all W-R stars show velocity variations. However there is no truly convincing evidence that all W-R stars are double. The close binary V444 Cyg (WN5 + O6) has been the object of numerous photoelectric and spectrographic observations and theoretical discussions, involving, as is necessary in most close binaries, the *problème restreint* and the Lagrangian points.

The spectral classification of the W-R stars has been re-examined recently (Hiltner and Schild 1966). Two parallel sequences within the WN group are recognized. They differ principally in emission band width and strength of the continuum. The sequence with the relatively weak narrow lines contains many binaries. In addition to the well-recognized carbon sequence that shows a regular increase of band width with excitation, a group of WC stars with relatively stronger emission (relative to the continuum) has been observed; two of them are central stars of planetaries. Their identity as a physical group is uncertain. A W-R star with only emission bands of He and Hα has been found. A class WC5 should be added to Beals' classification (very strong C IV, weaker lines of C III, O V, Hα).

---

[27] This was recognized first by Menzel in 1926.
[28] One of the queerest nuclei is an Of star with a strong O VI emission at $\lambda\lambda$3810–3835.

Actually C IV is found in the hottest WN stars, and it has been suggested that N lines are present in WC stars. Among the planetary nuclei there are Wolf-Rayet stars containing both N and C. There are also pure WC nuclei surrounded by a nitrogen-rich nebulosity. The subdivision into a WN- and a WC-sequence has generally been assigned to a difference in the relative abundances of N and C. However A. B. Underhill has suggested that the subdivision may be due to different physical conditions (different temperatures and departures from black-body radiation in the vacuum ultraviolet) rather than to different chemical abundances. There are strong arguments in favor of this view.

The spectra of the W-R stars which appear to be single have not changed appreciably over a few decades. We have to deal with atmospheres in a stationary state, although the motions are similar to those observed in novae. The shells of W-R stars are small (order of a few solar radii), much smaller indeed than those of novae.

The W-R stars do not show forbidden lines; reasonable values for n_e are of the order of 10^{12}–10^{13} cm^{-3}. Some are nebulous (Smith 1967). Further effort should be made to discover extended shells with the observational techniques developed to detect very weak nebulosities, such as that of G. Courtès who uses a focal reducer and a narrow interference filter at Hα.

The production of the broad, rounded W-R emission bands has been assigned either to ejection (by C. S. Beals, as in novae), or to electron scattering (by G. Münch), or to turbulent motion (by R. N. Thomas). As for the excitation itself, it is usually attributed to recombination,[29] but collisions of turbulent clouds, ejection of "subphotospheric" hot bubbles and shock waves have also been considered by E. R. Mustel (Liège 1957).

Great care has to be exercised in the interpretation of the Zanstra temperatures obtained for the W-R stars because of stratification effects and of the photoionization by emission lines, as was also the case for novae and SS. Lines from levels of high excitation (for example, the 4-n series of He II) are formed in deeper layers where the temperatures and pressures are higher than in the outer regions where lines of lower excitation (the Balmer lines) are formed.

XII. COMPARISON WITH SPECTROSCOPIC OBSERVATIONS OF Of STARS

The spectra of Of stars are superpositions of spectra of typical absorption O-stars and of shell emission lines of high excitation, principally N III and He II; sometimes H, He I, C III, N IV, N V and Si IV are also present. There is a wide range in intensity of these lines. All varieties are observed between pure absorption-O stars and pure W-R stars. There is also a wide variety in the relative intensities of the N III and C III lines, going from Of shells with strong N III and no C III, to Of shells with strong C III and weak N III. There remain several fairly strong unidentified emissions (for example, $\lambda\lambda 4485$, 4503 in 9 Sge); a detailed study of high-dispersion spectrograms might reveal more weak emission lines.

We have described earlier the selectivities observed among the lines of N III, C III, etc. These effects decrease when the intensity of the emission increases, and actually disappear in the W-R spectra.

No forbidden lines are observed; actually the electron density in the shell is of the order of 10^{13} cm^{-3}. The shell is rather close to the star, and the dilution factor is only of the order of 0.1. No dilution effects are observed, but it should be noticed that the usual dilution effects, such as those in He I, are theoretically much weaker around 50,000° than they are at 15,000° or 20,000° for the same geometrical dilution factor of the order of 0.1.

---

[29] The intensities of the stronger lines such as He II $\lambda\lambda 3203$, 4686 may be affected by self-absorption.

The absorption He I triplets are much stronger than the He I singlets as in normal B-type stars. Contrary to the case of emission He I in many SS, $\lambda 4686$ He II is a weak emission in 9 Sge while $\lambda 3203$ He II is a very strong absorption line. This is probably due to a fluorescence effect excited in He II by Lβ.[30]

A complete understanding of the Of stars and of the possible intensity variation of their emission lines will require accurate photometric data, and ultraviolet spectrograms taken from space vehicles.

XIII. COMPARISON WITH SPECTROSCOPIC OBSERVATIONS OF
STARS OF THE P CYGNI TYPE

As described earlier, SS occasionally go through a P Cygni-like phase; Z And provides an excellent example, described in the paper by Struve. The prototype P Cyg has changed in brightness in a nova-like manner and is related to the extremely slow novae such as η Car and RR Tel. The spectrum of P Cyg continues to vary definitely, although only very slowly; most other stars of P Cygni type such as HD 45910 and HD 218393 do likewise (CNRS 1963), imitating the recurrent novae on a much smaller scale. According to Pagel, the electron density in the shells of P Cygni stars is some 10^7 to 10^8 times higher than in planetary nebulae (Liège 1957).

The dilution effects observed in P Cyg (and in Z And), especially of Fe III, Si II, Si III and He I are very conspicuous (Swings 1944), but dilution alone does not seem able to explain all the striking selectivities observed.[31] P Cyg showed an interesting demonstration of the help gained by extending the covered spectral range. The infrared transition of C II, $3p\ ^2P^o$–$3d\ ^2D$ ($\lambda\lambda 7231$, $\lambda 7236$) connecting the upper level of $\lambda\lambda 6578$, 6583 (in emission) and the lower level of $\lambda 4267$ (in absorption) is intermediate (weak emission and absorption) between the two usual transitions of C II. The O I quintet ($\lambda 7772$) is much stronger in absorption than the triplet $\lambda 8446$. The ultraviolet region $\lambda 3000$–$\lambda 3300$ revealed one strong line of He I, 4 lines of Si III and 19 lines of Fe III, all with metastable lower level. All the observed transitions of Si III for which the term $4p\ ^{1,3}P^o$ is the upper level are present in absorption, whereas those transitions for which this term $4p\ ^{1,3}P^o$ is the lower level are in emission. These selectivities are not yet clearly understood.

The other P Cygni stars exhibit a similar behavior. An example is Z CMa, investigated by Merrill in 1927 and by Struve and associates in 1942. All the spectrograms revealed the striking behavior of Fe II stressed in a preceding section: the Fe II multiplets with lower level a^6S have very strong, violet-displaced absorption components, whereas the multiplets with lower level b^4P, b^4F and a^4G (of EP similar to a^6S) appear in pure emission. Many P Cygni stars behave likewise (17 Lep, HD 160529, CD$-27°11944$, BD$+47°3487$, etc.); a similar, but weaker effect is found in supergiants such as 3 Puppis (which however is related to P Cyg, as it shows bright Hα and [O I]).

Recombination, collisional excitation and fluorescence are all active for all elements, but their relative importance varies widely. In P Cyg, the emission lines of O I are produced by recombination, complicated by dilution effects. An ion of higher ionization potential such as Fe^{++} may be excited mainly by fluorescence. In shells of

---

[30] This cycle phenomenon renders the use of $\lambda 4686$ as a classification criterion rather dangerous. A valid classification must be based on features which originate approximately in the same regions of the atmosphere. This is not the case for the He I and He II lines.

[31] The selectivity in Si II observed in P Cyg, Z And, AG Peg, etc... is striking: the group $3s\ 3p^2\ ^2D$–$3s\ 4p\ ^2P^0$ ($\lambda\lambda 3854$, 3856, 3863) is present while $3s^2\ 3d\ ^2D$–$3s^2\ 4f\ ^2F$ ($\lambda\lambda 4128$, 4130), strong in normal B8 to B5 stars, is absent. This may be due to the fact that the lower level $3s\ 3p^2\ ^2D$, although not really metastable, is connected with the ground level of different electron configuration $3s^2\ 3p\ ^2P^0$ by a weak transition ($\lambda\lambda 1817$, 1808). Moreover the only way to reach $4f\ ^2F^0$ by absorption from the ground state is through $5d\ ^2D$, and the radiation thus required, of energy 13.87 ev, is only slightly higher than the ionization potential of H (13.54 ev).

lower excitation the predominance of low-level Fe II lines is very pronounced, which is probably a result of the collisional excitation at low T_e. These considerations serve to illustrate again the great caution that one should use when trying to estimate abundances of elements in peculiar stars, on the basis of either emission or absorption lines. Apparent anomalies may be due to the effects of dilution, fluorescence, or mechanical agents.

XIV. ADDITIONAL CONSIDERATIONS ON IONIZATION AND EXCITATION IN SYMBIOTIC STARS AND RELATED PECULIAR OBJECTS

All the essential mechanisms are discussed in the chapter by Miss Underhill. Only a few minor comments need to be added here.

The three usual excitation mechanisms—recombination, collisions and fluorescence—account satisfactorily for the spectra of the planetary nebulae. The situation is somewhat different in stellar envelopes of much smaller extension such as novae, Of, P Cygni and Be-type shells, and consequently in the SS at similar phases. At other phases, the SS should rather be related to planetary nebulae. The selectivities observed in He I, N II-III-IV, C III, Si II-III, Fe II, etc. may probably be assigned to recombination and fluorescence, coupled with the effects of the geometrical and physical dilution, the stratification or more complex geometry of the layers, the velocity and density distributions within the atmosphere. The details of these selectivities are not all clearly understood yet, nevertheless they are of great help in classifying the types of shells during the evolution of the SS, novae and other peculiar objects. A more satisfactory interpretation of the selectivities will require a detailed treatment along the lines of Rosseland cycles. The dilution effects may be quite different from the classical cases in which only the geometrical dilution of a continuous black-body radiation is considered. The fluorescence mechanism may become active when the mean interval of time between two collisions becomes longer than the average lifetime of an atom in the excited state. The density and temperature conditions are favorable in both the shells and nebulae.

The exciting radiation reaching an atom A at a specific location in a shell differs always from the black-body radiation at the effective temperature T_e reduced by a geometrical dilution factor independent of λ. Even if the photospheric radiation is that of a black body at T_e, the radiation reaching A will normally be depleted by discrete or continuous absorption features or will be enhanced by emission features. Whenever the stellar exciting radiation possesses absorption or emission features the resulting fluorescence should be considered as a superposition of monochromatically excited patterns, each of these having a specific intensity.[32]

The absorption coefficient for the fluorescence excited by the underlying radiation is much greater than the coefficient of photoelectric absorption. But since the photoionization is produced by a wide range of wavelengths beyond the ionization limit, the recombination mechanism will usually be more efficient than fluorescence, unless the ultraviolet ionizing radiation of the underlying layers is depleted.

The case of the emission lines of He I is especially interesting. On account of the high EP's in He I, collisional excitation is unlikely in objects of relatively low excitation such as T CrB, RW Hya, AG Peg, P Cyg, etc. If the He atoms are mostly neutral, electron captures are not frequent. The absorption of the underlying radiation enhances the singlet system relative to the triplets since the ground level is $1s^2\ {}^1S_0$, because the lowest triplet $1s\ 2s\ {}^3S$ corresponds to a high excitation potential (19.74 ev), and since there is no intercombination between singlets and triplets. The exciting wavelength range is essentially $\lambda500$–$\lambda580$; this region may contain absorption or

---

[32] The peculiar structure of the molecular bands in comets has been explained in this manner by taking into account the influence of the Fraunhofer lines of the exciting solar radiation.

emission features. Moreover the excitation depends on radial velocity effects. In objects of high excitation (CI Cyg, AX Per, RX Pup, nebulae) most He is in the He^+ or He^{++} state; the He I singlets are much fainter than the triplets. A convenient comparison may be based on $2p \, ^1P^o\text{--}nd \, ^1D$ ($n = 5$, $\lambda 4388$) and $2p \, ^3P^o\text{--}nd \, ^3D$ ($n = 5$, $\lambda 4026$). But one should keep in mind that stratification occurs and that neutral He may be expected on the outskirts of certain high-excitation shells.

The lowest singlet term of O III, $2p^2 \, ^1D$, is only 2.5 ev higher than the ground term $2p^2 \, ^3P$, and there are strong intercombinations between singlets and triplets. Hence the fluorescence of O III excited by the underlying star may contain triplets as well as singlets; this fluorescence may in fact be similar to a complete recombination spectrum if the exciting radiation does not contain strong discrete features coinciding with the wavelengths of the O III absorption from low levels. In RW Hya the He I singlets are much stronger than the triplets, and are to a considerable extent excited by fluorescence. But this does not mean necessarily that the O III spectrum is also excited by fluorescence, on account of stratification effects.

In P Cyg, the Si III transitions behave in a way similar to N III in Of stars. The absence of $3d \, ^2D\text{--}4f \, ^2F^o$ of Si II has been considered previously.

In the shells of lowest excitation, the Balmer emission lines themselves may be excited by fluorescence. The efficiency of this mechanism depends on the velocity and density distributions of the H atoms of the shell and on the profiles of the absorption lines of the exciting radiation. Any variation in these factors will give rise to changes in intensity and profile of the Balmer emission lines. If the underlying star has strong deep Lyman absorption lines and if the shell is stationary or expanding or contracting slowly, only a small proportion of the H atoms in the shell will be raised to the levels $n = 3, 4, \ldots$. If the shell is expanding or contracting rapidly, the Lyman absorption lines of the atoms in the shell may be shifted by the Doppler effect outside the deep parts of the Lyman absorption lines of the star, and fluorescence may operate. In such a case emission lines must be expected.

Why does Bowen's mechanism not affect O III in the Wolf-Rayet stars? This is due to the velocities of the O^{++} ions relative to the layer in which $L\alpha$ of He II is emitted. The O^{++} ions located at a specific place in the W-R shell are able to absorb only a small fraction of the resonance radiation of the ejected He^+ ions on account of the relative radial velocities. There is no selectivity in O III in the planetary nebula BD $+ 30°3639$ or in RW Hya (O III $\lambda 3444$, 3760 are not enhanced). In other SS (T CrB, AX Per, CI Cyg, RX Pup) the incomplete multiplets of O III and N III are prominent: in all cases the relative velocities are small and the situation resembles that in planetaries. In novae the role of the O III, N III fluorescence varies considerably with the phase.

Merrill has mentioned the possibility that the emission line Mg II $\lambda 3848$ which is anomalously strong in Z And may be excited by $L\beta$.

Since most of the atoms in the inner parts of the shells are either H or He^+, the stellar radiation will generally be strongly depleted for $\lambda < 912$ and $\lambda < 228$. As a result any ion whose ionization limit falls just shortward of $\lambda 912$ or $\lambda 228$ will not be further ionized as rapidly as ions whose limits fall on the longward sides. The absorption below $\lambda 912$ should cause an increase in the abundance of O I, Si II, Fe II; similarly the absorption shortward of $\lambda 228$ will enrich C IV, N IV, O III, Ne III, Fe IV. One should therefore expect low excitation potential lines of these ions (either forbidden or permitted) to be abnormally strong. In particular the lines of Fe II must be stronger than those of Ca II. High-excitation potential lines (usually permitted) are normally produced by recombination of the ion with an electron. This type of line should therefore appear with enhanced intensity for the next lower stage of ionization in each atom. This effect will be still further enhanced by the presence of strong emission lines of H and He II in the spectral regions just longward of $\lambda 912$ and

$\lambda 228$. For example the population of Fe IV should be increased with respect to Fe III. This may explain why RS Oph is rich in [Fe II] and [Fe VII] and poor in [Fe III] and [Fe V]. Ca^+ may be ionized from the metastable $3d$ ^2D level by absorption of $L\alpha$.

The third ionization potentials of C and N are almost identical (47.64 or 260 Å and 47.20 ev), while the corresponding value for O is higher (54.62 ev), approximately the same as the ionization potential of He^+ (54.14 ev or 228 Å). C^{+++} and N^{+++} are probably abundant in the same regions of high-excitation atmospheres; the same is true for O^{+++} and He. The recombination spectra of C III and N III should appear simultaneously, while those of O III and He II may be absent. Not only is the continuum stronger near $\lambda 260$ than shortward of $\lambda 228$; there may also be a crowding of strong bright emission lines of He II ($\lambda\lambda 256, 243, \ldots$), O IV, N IV and N V appearing between $\lambda 260$ and $\lambda 228$; the region $\lambda < 228$ is poorer in such emissions. The simultaneous appearance of [Ne III] (IP of $Ne^{++} = 40.9$ ev) and [Ar IV] (IP of $Ar^{+++} = 40.7$ ev) is readily understandable.

The same tendency to favor certain ions has been evident to varying degree in the spectra of AX Per, CI Cyg, RX Pup which show a wide range of stages of ionization. The fact that [Fe II] and [Fe VII] are both strong at the same time, in comparison to the intervening stages, has long been a puzzle of these objects. The interpretation has often been sought in the binary character or in stratification phenomena. The above interpretation may be correct; it should however be treated in greater detail on the basis of the transition probabilities of Fe II to VII which are now known. This work is at present under way in Liège.

We refer to Miss Underhill's chapter for other considerations and details on all these departures from the usual equilibria and on the excitation mechanisms.

XV. GENERAL CONSIDERATIONS AND CONCLUSIONS

A number of ordinary light curves of SS have been published but, as mentioned before, a great effort should be made to obtain more monochromatic and more accurate curves. We shall mention here only the light curves of Z And (visual and photographic) and AX Per (photographic only) discussed by Mrs. C. H. Payne Gaposchkin (1946). She has tried to separate two sources of light, variable in brightness, but constant in color (color index = 1.3 mag. for the M component, 0.0 mag. for the blue component). In Z And both "components" vary together with the 690-day cycle: the blue "component" has an observed range of approximately 6 mag., while the red "component" varies by 2.5 mag. at most. At the highest maxima of the blue "component" of Z And the spectrum is that of a minor nova outburst, which is succeeded as the star declines by the typical spectral development of a very slow nova. AX Per behaves in a similar way. Such photometric studies do not furnish an entirely convincing argument for or against the binary character.

Several SS have been found by H. W. Babcock (1958) to have magnetic fields; this is the case for AG Peg, WY Gem, HD 4174, VV Cep. Merrill (Liège 1957) has suggested that this magnetic field may give rise to driving forces by interaction of the charged particles with the field.

Before attempting to discuss the nature of the SS, let us collect a few last geometric, energetic or kinematical data. The theoretical intensity ratio of $N_{1,2}$ is known to be very nearly 3. If N_2 appears much weaker than it should relative to N_1, one might be tempted to assign the anomalous intensity ratio to a different absorption of N_1 and N_2 by TiO, since N_2 is nearer the strong TiO band head $\lambda 4954$ (degraded to the red) than is N_1. We have mentioned such a possible observation in the case of R Aquarii; a similar indication exists also for AX Per. However this is a very delicate type of photometric measurement which should be repeated with higher resolution and more accurate photometry (CNRS 1963).

As far as the range of variation is concerned we may distinguish the SS which are real recurrent novae (RS Oph, T CrB) and those which have more moderate and slightly more frequent outbursts (Z And, AX Per).

We have mentioned that the radial velocities of different elements, or even of different lines of the same element may differ appreciably in SS. In particular, the radial velocities of the He I singlets may differ greatly from the triplets: the latter show systematically greater displacements than do the singlet lines. [O III] $\lambda4363$ yields a velocity curve which differs appreciably from that of N_1. Similarly, Fe II and [Fe II] behave differently. The variations in brightness and in velocities appear to have the same periods (from P = 200 to 900 days). What do these variations mean? A volume pulsation in a large mass of gas (Merrill)? Or are they due to orbital motion? Or to the rotation of an unsymmetrical body? Or to a revolving stream or jet which discharges continuously into an outer zone? The observed velocity curves point strongly to specific localizations of the zones of emission of various lines. Remarkable time lags have been observed.

The first and essential question that arises is: are the SS queer associations of two very different stars, or exceedingly peculiar single stars? The next question is: why do outbursts occur, and where does the energy required to give rise to the high excitation features come from? The observational study of a wide variety of close binaries, combined with numerous theoretical investigations—beginning especially with Kuiper's dynamical theory of 1941—leaves little doubt that binary nature may stimulate the process of formation of shells or layers or streams, giving rise to line emission in the integrated spectra. These shells or layers may very well be stratified and unsymmetrical in nature. There is observational evidence in support of the hypothesis that the matter which forms the shells, layers, streams or rings comes from the secondary component through the Lagrangian point L_1. Matter may also leave the system through the external Lagrangian points (Sahade 1963, 1965). Prominence action may be the mechanism of loss of mass in the vicinity of L_1. Struve has described eight possible configurations; four relatively wide pairs and four close pairs (Liège 1957).

The close binaries present an enormous range of bright-line phenomena. One or both components may be near the main sequence; others may belong to the subgiant, giant or supergiant classes. Typical early-type binaries are the Wolf-Rayet binaries and the post-novae. Examples are found in practically all spectral types. On the other hand there are many close binaries which show no tendency toward emission.

Let us first consider the single star hypothesis. The first suggestion was given by D. H. Menzel in 1946: a hot star could be covered only partially, either toward the poles or at the equator, by a cool, M-type atmosphere; an outer envelope would be excited by the radiation from the stellar region. Sobolev in 1960 also suggested a hot nucleus surrounded by a very large envelope.[33] Aller in 1954 took the opposite point of view: a late-type star would be surrounded by a corona in which emission lines are excited by the dissipation of shock waves, much like the solar coronal lines in Schwarzschild's theory. A somewhat similar view had been expressed by F. J. M. Stratton for R Aqr. The "coronal" hypothesis was also adopted and extended by J. Gauzit (1955): the stellar corona surrounding the M star would be a hundred or a thousand times denser than the solar corona; the physical conditions of the SS would vary chiefly through the influence of strong prominences. In the period 1956–1957 various astronomers: A. J. Deutsch (1956), Aller (1959) and Swings (1959), independently envisaged that the SS may represent the turning-point in the evolution of the red giants. At this stage the nucleus contracts, becomes hotter, and tends to separate from the relatively cool atmosphere characterized by TiO bands. This

[33] H. L. Johnson (1967) has found that several known shell stars (ϕ Per, κ Dra, P Cyg) are abnormally red in K − L. This he interprets as due to infrared emission by circumstellar shells.

M-type atmosphere instead of contracting begins gradually to dissipate into space; the emission lines would originate below it. An SS would thus be a transition between a red giant and a blue predegenerate phase.

The hypothesis of a binary system is the oldest and simplest. It was proposed by F. Hogg (in 1934) who described Z And as a normal, possibly somewhat variable M giant, and a variable, very hot dwarf which excites a nebular envelope. In the binary hypothesis the nebular shell may arise by ejection of matter in the nova-like outbursts of the hot star. This hypothesis was considered by L. Berman (in 1932), Merrill (in 1935–40–57), C. H. Payne-Gaposchkin (in 1938–46–57), Struve and Swings (in 1939–1945), G. P. Kuiper (in 1941), Aller (in 1954) and defended especially by J. Sahade (1965). According to this hypothesis, a SS would be a binary consisting of a giant or supergiant red star and of a hot subdwarf; the red component would fill its lobe of the first critical equipotential surface. One or the two stars would be immersed in a small variable planetary nebula. Our current ideas on stellar evolution do not exclude the association of a giant late-type star and a hot companion; such a combination may produce a highly peculiar, explosive binary system. Kraft in 1958 showed that T CrB is a double-lined binary formed by a giant M3 and an underluminous, blue, emission-line companion. The red component of T CrB probably fills its lobe of the inner contact surface. Matter thus escaping from the M component is accreted by the blue companion.

Let us remember that certain ordinary novae, such as Nova (DQ) Her, are binary systems. Other binaries which have more frequent but less spectacular (no forbidden line!) outbursts are the U Gem stars; in the latter case the primary is a cool dwarf or subdwarf.

A solution intermediate between a single or a binary star has been suggested by Mrs. C. H. Payne-Gaposchkin (Liège 1957). In the course of its evolution a binary may become an essentially binucleate star. If the common pattern is developed, the more massive nucleus may outrun the other. "Perhaps such a developing binary . . . might show the characteristic spectrum and behavior of a *symbiotic variable*. Whether such an object would be a single or a binary star at this time is almost an academic question. . . . The greater the disparity of rate of development between the two components, the greater would be the range of excitation displayed."

Where are the SS located in the H-R diagram? Our ideas on stellar evolution are extremely uncertain after the red giant phase. There is a great temptation to consider the planetary nebulae as intermediate links, somewhere between the red giants and the white dwarfs, although we have no proof that all stars possessing a mass in the proper range pass through the stage of planetary nebula. But what are the immediate parents and descendants of the planetaries? Deeming (1965), among others, considers that "infrared supergiants" may be the predecessors of the planetary nebulae, and the predecessors of these infrared supergiants might be either the long period variables or the SS. What the evolutionary sequence really is we do not know. At any rate, there are many points in common between the planetary nebulae and the SS. Indeed, about ten years ago, Merrill (Liège 1957), Sobolev and others suggested that the SS may be the parents of the planetaries. Deeming separates them by the "infrared supergiants." Considering our very poor knowledge of the region between the horizontal branch and the white dwarfs, a greater effort in the study of the planetary nebulae and the SS is well worth while. The central stars of the planetaries are evolving toward the white dwarfs; the SS are somewhere in the same region, but their evolutionary track is most uncertain.

There is no doubt that, as Merrill stated: "Persistent observations, both spectroscopic and photometric, . . . of the brighter symbiotic stars would surely help us understand their mysterious behavior and might develop ideas of considerable general interest." Such investigations would certainly lead to a better understanding of evolutionary trends in stars. In particular the cataclysmic phenomena in stars would become clearer.

I hope that a few young astronomers who have access to a fairly large instrument, such as the Otto Struve Memorial Telescope, will endeavor to continue the investigations which Otto Struve and Paul W. Merrill conducted so brilliantly and which are now rather neglected. What should such a program on SS comprise? We would stress essentially the following points:

(i) light curves in specific wavelengths, either by slitless spectrograms or preferably by narrow-band interference filters or a scanning spectrograph and photoelectric receivers;

(ii) high resolution spectrograms, in order to obtain more accurate profiles, radial velocities and identifications;

(iii) attempts to determine abundances, especially the ratio He/H;

(iv) theoretical and observational study of time lags;

(v) attempt to measure accurately the intensity ratio of $N_{1,2}$;

(vi) make a more detailed observational and theoretical study of the line selectivities.

I do hope that we shall in a few years have a few doctoral theses in this field.

REFERENCES

All the papers contained in "Etoiles à raies d'émission" (8th Liège International Astrophysical Symposium, 1957) are simply referred to as (Liège 1957). All the papers contained in "Colloque International sur les novae, novoides et supernovae" (CNRS, Observatoire de Haute Provence, 1963, and *Ann. d'Ap.*, **27**) are referred to as (CNRS 1963).

Aller, L. H. 1957, *Ap. J.*, **125**, 84.
Aller, L. H., Bowen, I. S., and Wilson, O. C. 1963, *Ap. J.*, **138**, 1013.
Aller, L. H. and Liller, W. 1968, *Nebulae and Interstellar Matter*, ed. B. M. Middlehurst and L. H. Aller (Chicago: University of Chicago Press), Ch. 9.
Bashkin, S. (ed.) 1968, *Beam-foil Spectroscopy* (London: Gordon and Breach).
Bauer, C. A. 1945, *Ap. J.*, **101**, 208.
Bidelman, W. P. 1954, *Ap. J. Suppl.*, **1**, 175.
Bidelman, W. P. and Stephenson, C. B. 1956, *Pub. A.S.P.*, **68**, 152.
Bloch, M. 1966, *Comptes Rendus Acad. Sci. Paris*, **262**, 148.
Böhm, K.-H. and Deinzer, W. 1966, *Zs. f. Ap.*, **63**, 177.
Bowen, I. S. 1935, *Ap. J.*, **81**, 1.
——. 1960, *Ap. J.*, **132**, 1.
Bowen, I. S. and Swings, P. 1947, *Ap. J.*, **105**, 92.
Burbidge, G. R. and Burbidge, E. M. 1953, *Ap. J.*, **117**, 407.
——. 1954, *Ap. J.*, **120**, 76.
——. 1955, *Ap. J.*, **122**, 89.
Code, A. D. 1968, *Ap. J.*, **151**, L145.
Cowley, A. P. 1964, *Ap. J.*, **139**, 817.
——. 1967, *Ap. J.*, **147**, 609.
Deeming, T. J. 1965, *Pub. A.S.P.*, **77**, 443.
Deutsch, A. J. 1956, *Ap. J.*, **123**, 210.
Dossin, F. 1964, *Astr. J.*, **69**, 137.
Edlén, B. and Swings, P. 1942, *Ap. J.*, **95**, 532.
Farraggiana, R. 1968, *Mem. Soc. Astr. Ital.*, **39**, 291.
Fitzgerald, M. P., Houk, N., and McCuskey, S. W. 1966, *Ap. J.*, **144**, 1135.
Flather, E. and Osterbrock, D. E. 1960, *Ap. J.*, **132**, 18.
Garstang, R. H. 1957, *M.N.R.A.S.*, **117**, 393.
Gauzit, J. 1955, *Ann. d'Ap.*, **18**, 354.
——. 1966, *Comptes Rendus Acad. Sci. Paris*, **262**, 1309.
Glad, S. 1956, *Ark. f. Fys.*, **10**, 291.
Glebocki, R. and Keenan, P. C. 1967, *Ap. J.*, **150**, 529.
Greenstein, J. L. and Kraft, R. P. 1959, *Ap. J.*, **130**, 99.
Herbig, G. H. 1965, *Kl. Veröff. Remeis-Sternw. Bamberg*, **4**, Nr. 40, 164.
Hiltner, W. A. and Schild, R. E. 1966, *Ap. J.*, **143**, 770.

Ilovaisky, S. A. and Spinrad, H. 1966, *Pub. A.S.P.*, **78**, 527.
Johnson, H. L. 1967, *Ap. J.*, **150**, L39.
Johnson, M. 1952, *Trans. I.A.U.*, **8**, 839.
Kaler, J. B., Aller, L. H., and Bowen, I. S. 1965, *Ap. J.*, **141**, 912.
Kraft, R. P. 1959, *Ap. J.*, **130**, 110.
Lawrence, G. M., Ostriker, J. P., and Hesser, J. E. 1967, *Ap. J.*, **148**, L161.
Merrill, P. W. 1950, *Ap. J.*, **111**, 484.
Minkowski, R. 1946, *Pub. A.S.P.*, **58**, 305.
O'Dell, C. R. 1966, *Ap. J.*, **145**, 487.
———. 1967, *Ap. J.*, **149**, 373.
Pagel, B. 1960, *Ann. d'Ap.*, **23**, 850.
Payne-Gaposchkin, C. H. 1946, *Ap. J.*, **104**, 362.
———. 1957, *The Galactic Novae* (New York: Wiley).
Peery, B. F. 1966, *Ap. J.*, **144**, 672.
Plaskett, H. H. 1928, *Pub. Dom. Ap. Obs. Victoria*, **4**, 119. A few later papers on Z And beside the
    Liège 1957 and CNRS 1963 references are: Merrill, P. W.: 1944, *Ap. J.*, **99**, 15; 1947, **105**, 120;
    1948, **107**, 317; Struve, O.: 1944, *Ap. J.*, **99**, 205; Swings, P. and Struve, O.: 1945, *Ap. J.*,
    101, 224.
Razmadze, N. 1960, *Astr. J. U.S.S.R.*, **37**, 342.
Rosino, L. and Mammano, A. 1968, *I.A.U. Circular* No. 2052.
Sahade, J. 1952, *Trans. I.A.U.*, **8**, 842.
———. 1961, *Stellar Atmospheres*, ed. J. L. Greenstein, (Chicago: University of Chicago Press),
    p. 494.
———. 1965, *Kl. Veröff. Remeis-Sternw. Bamberg*, **4**, Nr. 40, 140.
Sanford, R. F. 1949, *Pub. A.S.P.*, **61**, 261.
Seaton, M. J. 1963–1966, a series of papers in *M.N.R.A.S.*: **125**, 437; **125**, 461; **127**, 217; **132**, 15;
    132, 347.
Smith, L. F. 1967, *Astr. J.*, **72**, 829.
Swensson, J. W. 1953, *Bull. Acad. Roy. Belg. (cl. Sci.)*, ser. 5, **39**, 405; *Liège Reprint* No. 353.
Swings, P. 1940, *Ap. J.*, **92**, 289.
———. 1942, *Ap. J.*, **95**, 112.
———. 1944, *Pub. A.S.P.*, **56**, 238.
———. 1950, *Ann. d'Ap.*, **13**, 134.
———. 1951, *J. Opt. Soc. Am.*, **41**, 153.
———. 1959, *L'Astronomie*, **71**, 114.
———. 1960, *Sitzungs. Heidelberger Akad. Wiss.*, p. 47.
Swings, P. and Fehrenbach, Ch. 1958, *J. des Observateurs*, **41**, 161.
Swings, P. and Öhman, Y. 1939, *Observatory*, **62**, 150.
Swings, P. and Struve, O. 1941, *Ap. J.*, **93**, 356.
Swings, P. and Swensson, J. W. 1953, *Liège Reprint* No. 370.
Swings, J. P. and Swings, P. 1967, *Ap. Letters*, **1**, 54.
Warner, B. 1966, *Observatory*, **86**, 15.

7. T TAURI STARS AND THE ORION POPULATION

O. STRUVE and M. RUDKJØBING: Stellar Spectra with Emission Lines in the Obscuring Clouds of Ophiuchus and Scorpius. *Astrophysical Journal*, **109**, 92, 1949.

G. H. HERBIG: Early Stellar Evolution at Intermediate Masses.

STELLAR SPECTRA WITH EMISSION LINES IN THE OBSCURING CLOUDS OF OPHIUCHUS AND SCORPIUS*

OTTO STRUVE AND MOGENS RUDKJØBING
Yerkes and McDonald Observatories
Received September 8, 1948

ABSTRACT

Six dwarf stars situated in the densest part of the obscuring clouds in Ophiuchus and Scorpius are found to have spectra with strong emission lines of hydrogen and calcium. The great intensity of Ca II H seems to indicate that the excitation of the emission spectrum is due to collisions, the principal source of energy perhaps being the potential energy of the surrounding nebula. In addition, three faint early-type stars were found.

Joy's[1] remarkable discovery of forty stars having strong emission lines in the vicinity of the great obscuring cloud of Taurus has inspired a study of the faint stars in the region of the dark nebulae in Ophiuchus and Scorpius. With the 82-inch telescope of the McDonald Observatory the authors have observed the spectra of twenty-six stars in this region. The Cassegrain spectrograph with quartz prisms and the $f/1$ Schmidt camera was used for all the stars with the exception of the two brightest, for which the glass prisms and the $f/2$ camera were used. On the chart (Fig. 1), a print made from a Ross *Atlas* plate, the stars are marked by arrows and by numbers indicating the order in which their spectra were taken. Of these stars, those having the numbers 5, 6, 7, 14, 16, 17, 18, 19 are situated outside the densest part of the nebula. Of the eighteen that are seen on the background of the optically thick part of the cloud, six were found to have emission lines. These are all of photographic magnitudes between 12 and 15. Nearly all stars in this densest part of the cloud and in this range of magnitudes have been observed. One of the emission-line stars (No. 13 in our list) is a variable star with a range of at least 1.5 mag. It is brighter than stars 9, 11, and 12 on Plate 2 in the Ross *Atlas of the Milky Way*, fainter than 9 on Plate 3, and fainter than 9, 11, and 12 on Plate 5.

These six stars range in spectral type, as estimated from the absorption features, between A and M. With the exception of one, No. 22, which is of early type, they are K and M stars and are thus similar to the majority of Joy's stars in Taurus, presumably belonging to the main sequence and having, in addition, a superimposed spectrum of strong emission lines, which make them stand out among the ordinary dMe and dKe stars which are occasionally found among ordinary stellar populations. A short description of these stars is given in Table 1.

An interesting feature of stars 9, 12, and 13 is the great strength of the blend of Ca II H and $H\epsilon$, which seems to be abnormal, since it exceeds the sum of the intensities to be expected in the two blended components. This may be explained as an effect of excitation of the Ca^+ ions by transfer of excitation from colliding hydrogen atoms, which in the process of collision make the transition corresponding to the $H\epsilon$ line. Because the excited Ca^+ ions have to send out more than 90 per cent of the excitation energy in the H line,[2] while the hydrogen atoms before the collision process had the approximately equal probabilities of emitting light both in a Lyman line, in the Balmer line in question, and in some infrared lines, the result is an increase in intensity of the H blend. The opposite process, which in thermodynamical equilibrium would balance the one mentioned, will in a diluted radiation field occur less frequently because it involves the collision of two atoms, both of which must be in excited states.

---

* *Contributions from the McDonald Observatory, University of Texas*, No. 158.

[1] *Pub. A.S.P.*, **58**, 244, 1946.

[2] D. R. Hartree and W. Hartree, *Proc. Roy. Soc. London*, A, **164**, 167, 1938.

232

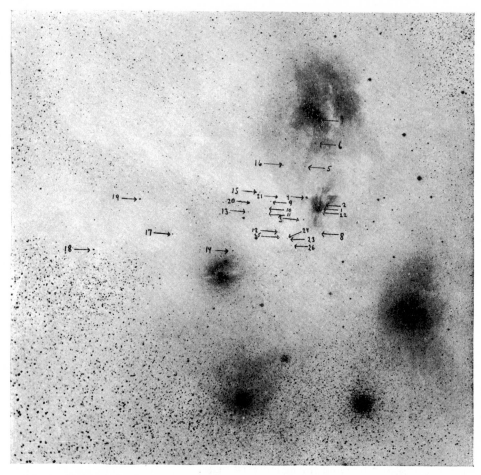

Fig. 1.—Region of the obscuring nebulosity in Scorpius and Ophiuchus. The numbers of the stars observed correspond to those given in Tables 1 and 2.

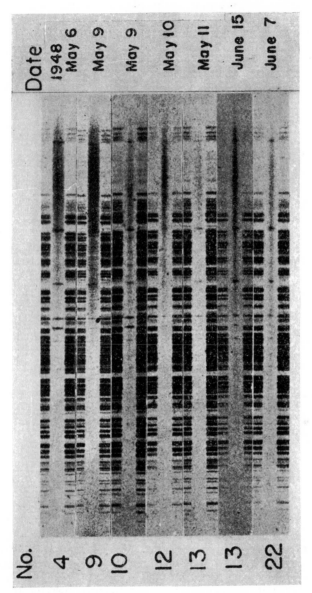

FIG. 2.—The spectra of six emission-line stars in Scorpius and Ophiuchus

That the extra intensity is really found in the *Ca* II line and not in the hydrogen line is suggested by one of Joy's[3] spectra, which has a greater dispersion than ours and which shows the same effect (the spectrum *c* on his Pl. XIV), being the only one of his spectra in which the *Ca* II lines are not too strong or too faint in comparison with the hydrogen lines to show the effect. It looks as if a strong *Ca* II H line is accompanied by a fainter hydrogen line.

The occurrence of dwarf stars with strong emission lines in cosmic dust clouds again raises the problem of the origin of their peculiar spectra. A computation based upon Hoyle's[4] expression for the accretion of mass of a star which rests within a nebula leads

TABLE 1

EMISSION-LINE STARS

No.	α(1948)	δ(1948)	Mag.	Sp.	Description
4...........	16^{h}22^{m}8	−24°13′	13	K5	Very strong *H* and *Ca* II and perhaps weak *He* I; the absorption lines are systematically weaker than normal
9..........	16 24.5	−24 14	12.5	K2	Fairly strong and very broad *H*, strong and sharp *Ca* II; *Ca* II H very much stronger than *Ca* II K
10..........	16 24.7	−24 16	15	K?	Very strong *H* and *Ca* II, moderately strong *Fe* II; spectral type estimated from the continuous spectrum because the emission lines overlie the absorption features
12..........	16 24.1	−24 34	13.5	M	Moderate *H* and *Ca* II, perhaps weak *Fe* II; *Ca* II H very much stronger than *Ca* II K; has *TiO* absorption bands
13..........	16 25.6	−24 20	$\left\{\begin{matrix} 12.5 \\ -14.0 \end{matrix}\right\}$	K	Very strong *H* and weak *Ca* II; *H*+*Ca* II H is much stronger than *Hδ*
22..........	16 22.3	−24 22	15	A?	Strong *H* and *Ca* II
					Faint Stars of Early Type without Emission Lines
3..........	16 23.0	−24 28	13	A0	Fairly sharp absorption of *H*
5..........	16 22.9	−23 50	13	A5	Diffuse lines of *H*
16..........	16 23.8	−23 50	13.5	A	Very weak and shallow absorption lines of *H*

to the following ratio of the energy released by the infall of particles to the radiation energy emitted by the star:

$$a = 16\pi \frac{(GM)^3}{RL} N m_H \left(\frac{3\bar{R}T}{\mu}\right)^{-3/2}.$$

Here *R*, *M*, and *L* denote the radius, mass, and luminosity of the star in question, *N* the number of atoms per unit volume, m_H the mass of the hydrogen atom, $\bar{R}$ the gas constant, and *T* and *μ* the temperature and molecular weight of the gas surrounding the star. This expression was derived from Hoyle's formula for the accretion of mass by multiplying the amount of matter gained per second by the potential energy GM/R per unit mass released during the motion of the matter toward the star. (There seems to be a small error in Hoyle's formula [17].) With the values of *M*, *R*, and *L* which Kuiper has given for the dwarf K0 star, 70 Oph A (Table 34*a* in Chandrasekhar's *Stellar Structure*), we find the following value for *a* (*μ* has been put equal to 1):

$$a = 10^{-4.95} \cdot N \left(\frac{T}{10^3}\right)^{-3/2}.$$

[3] *Ap. J.*, **102**, 168, 1945.   [4] *M.N.*, **106**, 412, 1946.

Thus an ordinary dwarf star, like the one considered here, could produce enough emission to be observed only if the number of atoms per cubic centimeter of the cloud is of the order of 10^3 or 10^4 for a temperature of the gas of the order of 10^4 degrees. However, too much stress must not be laid on this formula, first, because it is derived as a very rough approximation by Hoyle and, second, because much of the interstellar matter in the cloud is probably in the form of dust to which the formula does not apply.[5]

We should, according to the above mechanism, have emission lines only in those cases in which the pressure of radiation from the stars is not strong enough to drive away the gas. Thus an early-type star will have this kind of emission lines only if it has a small radius, giving it a great gravitational potential at the surface. Our star 22 is perhaps

TABLE 2

STARS WITHOUT EMISSION LINES

Star No.	Spectral Type
1	B3 (as in *HD Catalogue*)
2	G8
3	A
5	A
6	K0
7	G5
8	K2
11	K5
14	G5
15	(Underexposed)
16	A
17	K2
18	K0
19	G (overexposed)
20	M
21	K0
23	K5
24	(Underexposed), a double star
25	K5
26	Early-type star, F0?

such a case. It is situated near the eighth-magnitude star, HD 147889, which is our No. 1 and whose spectral type is given by the *Henry Draper Catalogue* as B3. This star is imbedded in the nebula, as is indicated by the brightness of the nebular matter around it. Number 22 is 7 or 8 mag. fainter, and, being of early type, is almost certainly underluminous.

In addition to the six emission-line stars, three faint stars of type A were found within the nebula. Their data are given in Tables 1 and 2. These spectra have no emission lines, and only No. 16 looks abnormal because of the weakness of its hydrogen absorption lines. This may be an underluminous A star or possibly even a white dwarf. The rest of the stars have normal spectra of types G, K, and M. Most of them are probably foreground stars, while a few, outside the densest part of the nebula, may be distant objects shining through the thinner portions of the cloud. Table 2 gives the estimated spectral types of the stars without emission lines. The emission-line spectra are reproduced in Figure 2. Of the variable star, No. 13, two spectra are given, the first taken in the beginning of May and the second in the middle of June, 1948. They do not suggest that any spectral changes have taken place during this interval of time.

[5] Dr. Hoyle has very kindly verified our computations, in a slightly different manner. His results and ours are in essential agreement.

EARLY STELLAR EVOLUTION AT INTERMEDIATE MASSES

G. H. Herbig

Lick Observatory
University of California
Santa Cruz, California

I. INTRODUCTION

It is generally believed that at the present time in our part of the Galaxy, stars are formed by condensation of interstellar material. On this hypothesis, we may distinguish in principle four distinct phases in the early history of a star of intermediate mass, which terminate with its arrival on the main sequence:

(1) The gravitational isolation of a sub-region of a dense nebula, which thereafter evolves as an independent object. There is next:

(2) A period of collapse which is halted when and if the material becomes opaque, and is then succeeded by slower contraction at a rate governed by the surface radiative loss. This is followed by:

(3) A rapid dynamical collapse, due to the onset of H_2 dissociation through a major part of the star. This rapid collapse ends when H and He are essentially all ionized, and

(4) The effective temperature reaches $3000°$–$4000°$. The star can then be plotted on a conventional H-R diagram, and the general features of its subsequent contraction toward the main sequence are well known, as we shall see.

The term 'star formation' is usually applied to process 1. Despite considerable attention given to the question, it cannot be said that theoretical studies of this stage have to date led to any fruitful interfaces with direct observation. In fact, some of the issues that have received considerable theoretical attention do not appear to be very relevant to the questions raised by direct observation. For example, it has recently been reemphasized (Huang 1965; Huang and Wade 1966) that there is no evidence that the angular momenta of the stars are derived directly from differential galactic rotation, yet theoreticians have wrestled with that problem for many years. And another: there is direct evidence that stars of intermediate mass can form in small numbers—certainly in twos and threes, possibly one at a time—yet theoretical expectation is that star formation must take place in associations or clusters, by the fragmentation of very large masses. And still another: there is direct evidence also that stars of intermediate mass often, or perhaps in the majority of cases, form in purely H I regions. Yet disproportionate attention has been given to the effect of an H II region upon a cool cloud, as an initiator of collapse. If theoretical investigations in this area are to exert any positive influence, it is important that full attention be given to the observational information that we do possess, small in amount though it may be.

II. THE EARLY PHASES

If direct observations are required to provide some reality to theoretical investigation, it is obviously important to press the observations back toward the earliest evolutionary phases. In the past two decades, a great deal of information has become available on the properties of stars in stage 4: their physical properties, their systematics, their instabilities. The Struve-Rudkjøbing paper of 1949 that is reproduced here was among the pioneering investigations in this area. A promising opening toward direct observations of stage 3, the period of dynamical collapse, has emerged from a recent study of the remarkable object FU Orionis (Herbig 1966). A strong case can be made for the hypothesis that the 6-magnitude flareup of this star in 1936 represented the last stage of such an event, so that we in effect then observed the arrival of a star at the top of its Hayashi track. The physical reason for such a collapse, described initially by Cameron (1962), is that at temperatures near 1800°K, the potential energy released by further contraction of the star is directed into the performance of internal work (initially into the dissociation of H_2) rather than into heating the gas, so that contraction continues instead of being braked by a rise in temperature. It is clear, however, that the 1936 flareup of FU Ori cannot be identified with the full collapse from an initial radius of the order of 2×10^4 R☉, which according to Cameron is appropriate for the precollapse configuration for 1 solar mass, because of the existence of a slightly variable sixteenth-magnitude star at the position of FU Ori prior to 1936. On the assumption of free-fall and with the observed time scale of the flareup, this "precursor" must have had a radius of only about 300 R☉, which raises the question of the nature of this apparent hold-up in the collapse. From the evidence of the Harvard plates, this delay must have lasted for at least 40 years. One possible explanation, which is the purest conjecture, is that the 16th-magnitude image at FU Ori prior to 1936 was that of a Herbig-Haro Object, and not of a star. In that case the radiation would not have been thermal, and so its interpretation as a step in the light curve would be incorrect. The source of excitation in the H-H Objects is unknown, but Magnan (1965) and Magnan and Schatzman (1965) have shown that the spectroscopic properties of the Objects could be produced by collisional ionization of interstellar gas surrounding a source of a strong flux of protons of about 100 KeV energy. Some small support for this interpretation of FU Ori is provided by the existence of two faint diffuse spots in the dark cloud very near the position of FU which may indeed be H-H Objects.[1] An interpretation in terms of an invisible pre-stellar source of high-energy particles is attractive in that it also provides a natural seat for the production of the light elements such as lithium, although in that case protons of energy greater than 20–30 MeV are required. The original proposal of Fowler, Greenstein, and Hoyle (1962) that Li is produced by electromagnetic acitivity in the atmospheres of T Tauri stars is contradicted by the fact that in FU Ori, less than 30 years after its appearance at the top of its Hayashi track and still not a conventional T Tauri star, the Li abundance is already 80 times that of the sun.[2] Unless Li is this abundant in the material from which FU Ori was originally formed, it must be produced at some intermediate stage in the contraction, for which the Magnan–Schatzman idea provides a site.

In regard to the H-H Objects themselves, it is often supposed that they represent some early phase of star formation. This supposition is based on their occurrence only in the denser nebulae (both dark clouds and H II regions) and always in areas where T Tauri stars are also found. The fact that T Tau and HL Tau are both centered in small emission nebulae that in both spectra and structure resemble a bright H-H Object strengthens the connection. Extrapolating from previous considerations, one might speculate that the H-H Objects mark the sites of some type of prestellar activity which generates a flux of high-energy particles. The presence of [S II] and [O I] emission lines in the integrated spectra of many T Tauri stars suggests that such

---

[1] The brightest of these is visible on the published 120-inch reflector photograph of the field: Herbig 1966, Fig. 4, 11 mm south and 15 mm west of the image of FU Ori. But a recent attempt to obtain its spectrum was unsuccessful, so this suggestion is as yet unconfirmed.

[2] The production of this quality of Li throughout a largely convective star in such a short time by proton spallation would require an enormous degree of surface activity. The considerations of Bernas, Gradsztajn, Reeves, and Schatzman (1967) indicate that a surface proton flux 10^{11} to 10^{12} times the present solar level would be necessary.

small low-density envelopes are common, and that the emission nebula at T Tauri is unusual only in the presence of a star, and in its possession of strong [O II] lines. The well-substantiated changes in this close nebulosity at T Tauri[3] may have some connection with the variations observed in a H-H Object without central star, No. 2 near NGC 1999 (Herbig 1957, 1968*b*). One could regard these nebulous envelopes of T Tauri stars in which the forbidden lines originate either as the remains of the original H-H Object, or as simply the excited low-density fraction of the material which most T Tauri stars are observed to be ejecting at the present time. The first seems more probable, because (i) many such nebulae are not associated with detectable stars, and (ii) there are some T Tauri stars in which the rate of mass ejection is very high, yet there is no sign of the forbidden lines.

Observational information on protostars in stage 2 will be most difficult to secure. Possibly some indirect information might be obtained from the H-H Objects, but Gould (1964) has suggested that a protostar in stage 2 might be detectable from its high luminosity in the 28 μ rotational line of H_2, wherein most of the energy is radiated when the gas is still quite cool. Until such observations can be made, stages 1 and 2 would seem to be reserved to theoretical exploration. However the unexpected discovery by Becklin and Neugebauer (1967) of a strong infrared source in the Orion Nebula may already have proven this presumption wrong. The Becklin-Neugebauer source, which has not as yet been detected optically in the region $\lambda < 1.0 \mu$, consists of a sharp ($<2''$ diameter) intensity peak superimposed upon broad wings (extending about $30''$ from the peak). The sharp source is brightest at 3-4 μ, and if interstellar extinction can properly be neglected, can be fitted to a Planck distribution with $T = 700°K$ and $M_{bol} = -2$. Very near this same position, Kleinmann and Low (1967) found a diffuse source having a diameter of at least $30''$ whose surface brightness at 22μ suggests $T \approx 70°K$, again with the Planck assumption and no extinction. The relationship of the Kleinmann-Low nebula, which is not detectable at $\lambda \leq 5 \mu$, to the Becklin-Neugebauer diffuse source which was discovered at 2.2 μ is not clear. The assumption that the Kleinmann-Low nebula is indeed a black body of radius 0.05 pc at $T = 70°K$ leads to $M_{bol} = -8$, but some confirmation that the emission is truly thermal is desirable. A further interesting development is the detection by Raimond and Eliasson (1967) of an OH emission source very near the Becklin-Neugebauer position. It is surely premature to pronounce at this time upon the evolutionary significance of these infrared objects. Hartmann (1967) has indicated some of the conclusions that follow if the luminosity, temperature and dimensions of the Kleinmann-Low nebula are taken literally, and the object is assumed to be in gravitational contraction.

It is tempting to believe, but not yet demonstrable, that there is some kinship between these infrared radiators in the Orion Nebula and the large infrared excesses discovered by Mendoza (1966, 1968) in a number of T Tauri stars and related objects. Mendoza found that if the V − R colors (effective wavelengths 0.55 and 0.70 μ) were chosen as reference, then the T Tauri stars are abnormally bright in the infrared with respect to normal stars having the same V − R. The precise amount of this excess depends upon the correction for interstellar extinction, which is very uncertain because of doubt as to the true intrinsic colors. R Mon is the most outstanding example of the phenomenon, and survives as such following reasonable correction for extinction. Following Mendoza, R Mon was observed by Low and Smith (1966) at 20 μ. They found that the energy distribution corresponded roughly to a Planck distribution for $T = 850°K$ except that the 20 μ point was too high by a factor of 5-10. This

---

[3] The emission nebulosity very close to the star, discovered visually by Burham in 1890 and studied photographically by Hubble and by Baade, should not be confused with the more distant reflection nebulosity NGC 1555, which also varies in surface brightness and structure. For references to the history of these nebulae, see Herbig (1950, 1953).

breadth of the energy curve can of course be explained by a source having more than one T. Low and Smith constructed a model consisting of a deep, but optically thin envelope of solid particles that emitted rethermalized Planck radiation derived from a central G-type star. The maximum dust temperature in this model (1385°K) occurs at the inner boundary of the envelope, and falls off radially outward as $r^{-1/2}$. The model was adjusted so that its radiated energy fitted the observations of R Mon. The analogy to the zodiacal cloud surrounding the sun is interesting, but of course the interpretation is not unique. It may be significant that three of the stars having the largest infrared excesses (R Mon, V380 Ori, R CrA) all are closely involved in bright reflection nebulae, and that the infrared photometry has been performed through large diaphragms that admit, beside the star, a considerable area of this nebulosity. At the distance of V380 Ori and R Mon, the diaphragm diameter corresponds to about 10^4 a.u., compared to the Low-Smith model of an envelope around R Mon having an effective radius of < 1 a.u. In other words, it has not as yet been proven that the infrared excesses originate in the T Tauri stars themselves, or even in their immediate vicinity. The example of the diffuse infrared sources in Orion, if the analogy is valid, demonstrates that radiation of this kind and amount can somehow be generated over a very considerable volume.

In the case of R Mon, there are the additional complications that the "star" is in fact a peculiar non-stellar object of high surface brightness, and that there is no spectroscopic evidence for the presence of a normal G- or K-type star within this structure (Herbig 1968a).

Aside from low-temperature thermal emission by a dust envelope, several other explanations of the infrared excesses in T Tauri stars have been considered. In view of the similar excesses found by Johnson (1967) in Be stars and P Cyg, near which any dust concentrations seem most unlikely, it would appear that an atomic source of infrared emission should not be ruled out. Kolesnik and Frank-Kamenetskii (1968, and references therein) have discussed a mechanism whereby the free-free and bound-free emission of H I could be raised above the thermal equilibrium level by an excess of suprathermal electrons. The earlier proposals to the same result by I. M. Gordon, involving a radiatively induced over-population of the higher states of H, will be recalled. The simple presence of a sufficiently cool companion would also explain the infrared excess. For example, RW Aur has a reddish, photometrically unresolved companion which, if it had the energy distribution of an M6 dwarf but with the observed $M_v \approx +6$, would account for the infrared brightness of RW Aur A + B. However, the type of RW Aur B is only about K3, which is in accord with the fact that no TiO bands are observed in the near infrared. This would seem to rule out the hypothesis, if B is a normal star. Of course, there is no way to be sure, the infrared excess being a totally mysterious phenomenon, that in this example it must originate in component A (the T Tauri star) rather than in B.

Other speculations on the origin of the excess have made use of the mass ejection exhibited by the T Tauri stars. In those examples studied by Kuhi (1964, 1966), the kinetic energy of the outflowing material, if converted into radiation would be sufficient only, roughly, to double the luminosity of the star. This would be an adequate source of the extra luminosity in the less extreme cases, although the conversion process would have to be worked out. Poveda (1967) and Huang (1967) have discussed such mechanisms.

III. EVOLUTION IN THE H-R DIAGRAM

An example of a false initial start in the interpretation of observations is provided by the attempts *circa* 1948–50 to see in the T Tauri stars direct evidence of the gravitation accretion of interstellar material. Struve's views of the time were shared by other workers in the field, but the Struve-Rudkjøbing (1949) paper reproduced here is one of the better statements of that climate of opinion. It was written at about the same time as Struve's section on "The Formation of Stars in Interstellar Clouds" for his book *Stellar Evolution*, and as his review of *Centennial Symposia* (1949). They all show that at that time he had accepted Ambartsumian's arguments that associations containing early-type stars must be young and have their origin in the surrounding interstellar material. But he was still troubled by the possibility that the emission-line stars associated with the Taurus and the Scorpius–Ophiuchus dark clouds might be no more than field stars accreting nebular matter. From his articles in *Sky and Telescope*, it appears that by 1952 his position had begun to change and by 1956 Struve (along with others) had come over to the position that the T Tauri stars as well as their spectroscopically milder counterparts in the nebulae were indeed very young stars, owing their spectroscopic and photometric peculiarities to their extreme youth. It is not easy today to reconstruct the reasoning which led to the shift of opinion. There may have been some not entirely logical reasons; for example, the recognition at about this time that accretion was neither able nor necessary to rejuvenate the very massive early-type stars might have been a factor. But there were also specific problems with the accretion idea: it was discovered in 1952 that (i) the brighter Taurus cloud stars were 1–3 mag. above the main sequence, (ii) their spectra had anomalously broad absorption lines, and (iii) that these stars were ejecting material, not collecting it. These suggested some deeper origin than simple infall of material, and the discovery in 1954 of faint emission-Hα stars in the II Persei association whose expansion age was believed to be only a few million years, pointed strongly toward an evolutionary explanation. Homology arguments by Salpeter (1954) and model calculations by Henyey, LeLevier, and Levee (1955) first demonstrated in a modern context that contracting stars ought to lie above the main sequence, although the same point had been made by H. N. Russell 25 years before, apparently on intuitive grounds. And especially when the observations by Walker (1956) of the elevated main sequence in the young cluster NGC 2264 appeared, the evolutionary interpretation seemed to be clinched. Subsequent developments have shown that the apparently close correspondence between the observations and the theory of that time was somewhat illusory, but the position had been established, and there has been no incentive to return to the earlier point of view.

Looking back, one now realizes that the ingredients for an inspired leap to the same conclusion lay in the earlier observations by Joy (1945) and in the color-magnitude diagram of the Orion Nebula cluster by Parenago (1950). Except for Ambartsumian and his school, however, no one in the late 1940's and early 1950's was fully prepared for that step.

Struve's own direct contribution to this subject is represented by his paper with Rudkjøbing reprinted here, and by a short paper with Swings (1948) on the spectrum of DD Tauri, a striking example of those T Tauri stars having a strong blue-violet and ultraviolet continuum. Yet Struve's influence did not end with those publications. I remember very well talking with him in early 1949, after I had worked up my McDonald observations of the nebulosity at T Tauri and of the stars in the Orion Nebula. I had made a major effort to secure these observations with the 82-inch reflector, then the third largest telescope in the world, and was not at all optimistic about my ability to continue effectively in this area with the small Lick telescopes when I returned to Mt. Hamilton later that same year. Struve had an intuitive sense for promising fields of investigation, and he urged me strongly to go on in this subject, with whatever means I could find. I did so, stressing those aspects where the limitations of the Lick equipment of that era were avoidable. I know that Struve also encouraged others, particularly G. Haro and M. F. Walker, to continue their work in this field. So although his own published contributions to the study of very young emission-line stars are not extensive, he nonetheless played an important part in the early history of the subject.

Beside the firm establishment of the evolutionary point of view, progress since 1948 has been considerable in almost every direction. Surveys by slit spectrograph as that carried out by Struve and Rudkjøbing go very slowly. It is much more productive to use a large objective prism or slitless spectrograph. Their examination of the ρ Ophiuchi region, for example, was extended by Haro (1949), by Dolidze and Arekelyan (1959), and by Hidajat (1961). As a result of such surveys, the systematics of the occurrence of emission-Hα stars are now fairly well known. Given the point of view that these members of the Orion population[4] evolve down and across the H-R diagram to terminal points on the main sequence specified by their final masses, this flow can be followed once theoretical evolutionary tracks are available. The crucial importance of the outer convection zone in the early stages of the contraction process was first emphasized by Hayashi in 1961 (see the review by Hayashi 1966) and increasingly more refined tracks have been calculated on that basis, most recently by Iben (1965), Bodenheimer (1965) and Ezer (1967).

There is however no theoretical explanation of the surface activity that accompanies the initial phases of contraction in the H-R diagram, so it is not clear to which portion of these tracks the Orion-type activity is confined. Correlation of observed M_v, B-V's with the theoretical M_{bol}, log T_e's gives results that are only roughly indicative because of the contamination of the colors of the Orion population stars by line and continuous emission. A rough estimate of the duration of the Orion phenomenon can be obtained indirectly from consideration of the distribution of the emission-Hα stars in the Taurus-Auriga dark clouds, which is instructive in other ways as well.

The T Tauri stars in the Taurus-Auriga nebulae, because of their nearness (150 pc) and favorable declination, have been more thoroughly studied than the population of any other nearby H I region, although the search for non-emission variables therein is quite incomplete (Herbig 1962a). Eventually the data on the Scorpius-Ophiuchus clouds will be sufficient to permit similar studies there. The concentration of some 70 emission-line stars to the Taurus-Auriga obscuration, first recognized by Joy (1945) and now thoroughly confirmed by Götz (1961, 1965) is very striking. The evidence in Taurus-Auriga is overwhelming, not only that extensive star formation has taken place over a large volume of highly irregular shape, but that it has gone on without forming massive early-type stars in the process. There are several conspicuous clumpings of T Tauri stars in the clouds but none of them contain hot stars. Furthermore, there are large areas of these clouds where there are no emission-line stars at all (see Herbig and Peimbert 1966, Fig. 1). From these simple observations we conclude that (i) for some reason, star formation can take place much more readily in some regions of the same nebulae than in others; and (ii) contrary to theoretical expectation, the presence of an H II region to initiate collapse of an H I cloud is not necessary.

It is very interesting that the most extreme examples of these emission-line stars, namely those having the most intense Hα emission which we presume thereby to be the youngest, are projected upon the most opaque regions of the nebulae. This fact was discovered by Götz, who showed that the same dependence appears to hold in all nebulae rich in emission-line stars. Another aspect of the same phenomenon was discovered by Herbig and Peimbert, who found that the number of emission-Hα stars per unit projected area of the Taurus-Auriga clouds increases in strong dependence upon the amount of obscuration.

Some estimate of the velocity dispersion of these stars is required to interpret the

---

[4] This refers to the collection of T Tauri stars, irregular variables with little or no line emission and flash variables that are found between about $M_{pg} = +4$ and the main sequence in nebulae, young clusters, and associations.

observations. These remarkable correlations must exist because stars form pref-erentially in regions where the density of interstellar material is highest, or some other factor is optimized. But the random motions of the young stars will tend thereafter to carry them out of the regions where they were born, and thereby blur such relation-ships. It is known from the widths of interstellar absorption lines both in the optical and in the radio region that the velocity dispersion of the gas in H I clouds is of the order of 1 km/sec (Van Woerden 1966). Proper motion studies (Blaauw and Pels unpublished, and Wenzel 1961) of the T Tauri stars in Taurus-Auriga, and the few accurate radial velocities that are available for the same stars (Herbig 1962b) show that the velocity dispersion of these objects also is about 1–2 km/sec. (The agreement of these two values reinforces, incidentally, the view that these stars have formed in cool H I clouds, not in H II regions). The linear scale of the Taurus-Auriga clouds is such that if the present surface distribution of emission-line stars was reshuffled by adding cross-motions of 3 to 5 parsecs in random directions, the obscured clumpings and correlations would essentially be destroyed. That is, the present patterns could not survive more than 3–5 × 10⁶ years, so that the existence of the relationships found by Götz and by Herbig and Peimbert indicate either that the average age of these stars is less than 3–5 × 10⁶ years, or the cross-motions are not directed at random. This latter covers the possibility that the stars move in orbits within the cloud such that they remain in about the same place with respect to the boundary for long periods of time. There is one piece of information which shows that not all such stars are frozen near their place of origin: the fact that a few emission-Hα stars are found outside the conventional cloud boundaries. The question can ultimately be settled on the basis of proper motions. We here *assume* that newly-formed stars are free to move radially in the clouds, and therefore associate a time scale of 3 to 5 × 10⁶ years with the duration of a readily detectable Hα emission line.

The total contraction time to the main sequence at 1.0 solar mass is $t_c = 5 \times 10^7$ years, and at 0.5 ○, about 2×10^8 years. On the basis of masses inferred from their absolute magnitudes, a mean contraction time of 10^8 years can be associated with the sample of Tau-Aur objects under discussion. Therefore, the duration of the emission-Hα phase is no more than 3 to 5 percent of their total contraction time. The Orion population includes the non-emission irregular variables of nebular type as well, and these are about 1.5 times as frequent as the emission-line stars (Herbig 1962b). From this fact, it is concluded that the interval spent by one of these stars in the Orion phase is, on the average, about 10 percent of the total contraction time or less.

An independent estimate of this quantity can be obtained by calculating n_{Ori}, the number of stars above the main sequence in the solar neighborhood of which a fraction f exhibit Orion population characteristics, and comparing it with n_{ms}, the present number of main sequence stars in the same mass interval and volume. If $\dot{n}(t)$ is the rate of star formation at these masses at time t in the past, then at the present time

$$n_{ms}(0) = \int_{t_c}^{t_{ms}} \dot{n}(t)\, dt \tag{1}$$

and

$$n_{\mathrm{Ori}}(0) \approx \dot{n}(0) f t_c \tag{2}$$

where t_{ms} is the lifetime on the main sequence, for which we assume $t_{ms} = 100\, t_c$ for all masses of concern here. The statistics of the two groups of stars represented in n_{ms} and n_{Ori} have to be reduced to the same mass or *main sequence* absolute magnitude interval. The observed M_{pg}'s of the Orion population stars in the solar vicinity can be compared with main sequence M_{pg}'s in the following way. Iben's (1965) theoretical evolutionary track for 1.0 solar mass, converted to an M_{pg}, B − V scale, will yield the value of M_{pg} for an Orion population star moving along this track, which here is assumed to be the time average of M_{pg} over the first ft_c years of the contraction, beginning with Iben's first model. The proper value of f can be obtained by iteration with equation (3), below, and thus is an average of the f's appropriate to all the stars in the sample. On this basis, one compares numbers of Orion population stars with main sequence stars 0.8 mag. (pg.) fainter.

To $M_{pg} = +11.25$, the total number of Orion population stars is estimated to be about 4500 within the approximately 180° of galactic longitude that has been surveyed, out to a distance of 800 pc.[5] Therefore, over this volume we can write the ratio of equations (1) and (2) as

$$\frac{n_{\rm Ori}\,(+3.5 < M_{pg} < +11.25)}{n_{ms}\,(+4.3 < M_{pg} < +12.05)} \approx \frac{f}{g}\frac{1}{100}, \tag{3}$$

where we have assumed that the average rate of star formation in the past was g times that at the present time. Schmidt (1963) has tentatively concluded that, in effect, g is about 2. The data for main sequence stars in the solar vicinity have been taken from the compilations of Schmidt (1963) and of Limber (1960), and the result from equation (3) is that $f = 0.05$. In other words, if the census of Orion stars in this M_{pg} interval is complete,[6] then such stars are identifiable only for the first 5 percent of their contraction times. The result is in acceptable agreement with the previous argument that $f \leq 0.10$, but one should not thereby be too optimistic about the finality of the result, which must be regarded as only the roughest of estimates. Especially it should be recognized that these values of f are averaged over a mixture of Orion population stars having an appreciable range in mass, across which f may change substantially.

Nevertheless, these results indicate that the Orion population comprises only a minority of the stars still in contraction toward the main sequence. It is highly important to find some means of identifying at least a few stars that are in the post-Orion phase, in order to follow the process of transition to conventional main-sequence objects. By definition, such post-Orion stars cannot be discovered by conventional methods because by then, the strong Hα emission will have disappeared and the irregular variability become small or have vanished. A star which might be such an object is BD +28°637, a dK star with somewhat diffuse absorption lines and strong Ca II emission that was found by Joy (1945). Ordinarily it seems to be constant in light but on one occasion underwent a spectacular spectroscopic flare (Metreveli 1966). It is possible that the faint companions of T Tauri stars (such as RW Aur B, UX Tau B, HL Tau = XZ Tau B) represent objects more massive than their primaries that have already evolved through the Orion phase.

Post-Orion stars might best be discoverable by searching with an objective prism for stars having unusually strong Ca II emission lines. One might also examine non-emission stars that illuminate reflection nebulae, or perhaps even by measuring UBV colors at random for stars projected upon heavy obscuration.

Although this transition domain has yet to be explored, considerable effort has been made to detect in young main-sequence stars in the solar vicinity some fossil traces of their recent passage through the pre-main sequence phase. Two properties of T Tauri stars that might a priori be expected to survive for a time in some small degree have not been detected in normal F- and G-type stars: the irregular variability in light, and the abnormal colors. Two other properties of the T Tauri phase have however been linked directly to phenomena of the main sequence: the possession of strong chromospheric activity, and the high surface abundance of lithium.

---

[5] The survey originally (Herbig 1962b) was believed to be complete to 1 kpc, but there are several associations near that limit for which the data are probably inadequate, so they have been excluded and the survey volume reduced accordingly.

[6] If not, this value of f is too small.

A very practical index of the degree of chromospheric activity is the strength of the H, K emission lines of Ca II. There is a clear tendency for the stronger H, K emission to appear in the younger G-type stars. This result was foreshadowed by R. E. Wilson's (1949) discovery that the dwarfs of the Hyades cluster had systematically stronger Ca II reversals than did field stars, and by Miss Roman's (1949) observation to the same effect in the Ursa Major cluster. The phenomenon has since been studied in detail by O. C. Wilson (1963) and by Wilson and Skumanich (1964). Wilson's interpretation of the data is that in a given star the chromospheric activity decreases with time on the main sequence, and that the physical phenomenon responsible is the decay of surface magnetic fields; this latter explanation was pointed out by Unsöld (1964). In the present context, we interpret this decay after arrival on the main sequence as no more than a continuation of the process that began much earlier during the evolution of these stars through and out of the Orion population. Whether the surface rotational velocities of stars are reduced in a similar way is an interesting subject for speculation (Wilson 1966, and Chapter 10 of this volume).

The connection of main sequence F- and G-type stars to the Orion phase through their lithium abundances is a most interesting but increasingly complex subject. The basic situation is as follows. If we measure surface Li abundance by an index l, and take as unit of the l scale the abundance ratio $n(\text{Li})/n(\text{Si}) = 0.38 \times 10^{-6}$ which was at one time believed to represent the solar surface abundance, then on this scale the actual solar value is about $l = 0.16$, while the values for T Tauri stars range from about $l = 80$ to 400. In F5–G8 dwarfs of the solar neighborhood, l ranges from about 35 to less than 2, the lower limit being of instrumental origin. It is probable that, depending upon the mass, a star loses some of its surface Li during contraction to the main sequence by means of deep convection to levels where the temperature is high enough to destroy Li through (p, α) reactions. But the statistics of Li abundances in main sequence stars and the evidence from clusters of known age indicate that further destruction of Li must take place after the star reaches the main sequence, on a much longer time scale. The general dependence of Li abundance upon spectral type can be understood in terms of the deepening of the surface convection zone with decreasing stellar mass; in particular, this explains the fact that no Li has been observed in any dwarf later than G8. The principal problem with this picture is that the observations demand deeper convection than is provided by present stellar models, which has led to considerable reexamination of the theory. The general situation with special attention to theoretical considerations has been discussed in a recent I.A.U. Colloquium (Böhm 1968), while the problem with emphasis on the observational aspects has been reviewed by Herbig and Wolff (1966) and by Danziger (1967).

IV. BINARIES: A FEW REMARKS

In most speculations such as those of the preceding section, one ignores the fact that a large fraction of the stars occur in close binary systems, and that evolution must necessarily go on there as well. This subject is discussed in detail in Chapter 10 of this volume, but a few remarks in this particular context of stellar evolution are in order here. Struve was intensely interested in the evolution of double stars, and subsequent work by others has dealt profusely with the post-main sequence evolution of such systems. But the pre-main sequence phase has received much less attention. Attention should be called to an older paper by Huang (1957) on total angular momentum as the decisive parameter in determining whether the end product of a contracting protostar will be a double star, a star plus planetary system, or a single star with an arbitrary degree of rotation. The division of a single rapidly rotating mass into two, and its subsequent evolution has been studied by Roxburgh (1966). He applied his ideas particularly toward an explanation of the systems with undersize (i.e., smaller than their Roche lobes) secondaries; the fainter components were inferred to be sub-giants from their location in that region of the H-R diagram. From this location and the assumption of evolution toward the main sequence, Roxburgh derived ages of the order of 10^6 years for these systems. But other considerations do not support such short ages. Particularly, although this age is comparable with those of massive O- and early B-type stars, the 10 binaries of this type listed by Roxburgh have no evident connection with nebulosity, with young clusters or associations, or with any other circumstance that could be considered a direct index of extreme youth. Further-more, the spectra of those secondaries that have been observed directly show none of the characteristics of the T Tauri stars, which populate that region of the H-R diagram. The application of Roxburgh's basic ideas to this particular type of binary is therefore dubious. Of course, the ubiquitous W Ursae Majoris systems, on which Struve also worked extensively, have also been considered in this connection, notably by Eggen (1961; see also Faulkner, Griffiths, and Hoyle [1963]). The membership of W UMa stars in galactic clusters such as Coma and the Hyades suggest ages of the order of 10^8 years, while the location of the two components near the main sequence shows that they are not now in contraction. The location of W UMa systems in solar neighborhood, far from any seat of star formation, is also compatible with such an age. The point is that, whatever the W UMa binaries were or appeared to be before they reached the main sequence, their present peculiarities are not caused by con-tractional evolution. A similar conclusion has been reached by Kurochkin and Kukarkin (1966).

Thus there do not seem to be any convincing cases known of pre-main sequence binaries. The most promising observational approach to the question would be a close photometric and spectroscopic examination of the later-type stars above the main sequence of NGC 2264,[7] because spectroscopic binaries are known to exist among the early-type members that are now on the main sequence; similar cases are known in other young clusters (Eggen 1964). A discovery that may bear upon this question has been made by Thackeray and Wesselink (1965), namely that the composite system HD 101712, of type M2 I$b$pe + B?, is apparently a member of the very young cluster IC 2944.

V. FINAL REMARKS

The intention of this review has been to sketch some of the ramifications of a topic with which, in a primitive form, Otto Struve was concerned in 1948–49. His attention to matters of the T Tauri stars was not central, but only a facet of his omnivorous interest in all aspects of observational astrophysics. That interest was not so much in the observations themselves as in the understanding that came from direct observa-tion. Nevertheless his observational results remain as the more permanent memorial, for their interpretation tends to change with the ebb and flow of scientific opinion. Struve was philosophic about this. He fully appreciated the transient character of the current interpretation; he once said that "the history of previous . . . evolutionary hypotheses teaches us that most of them were wrong." Where this hunger for under-standing would have directed his attention today one cannot be sure. He probably would agree that a field of vast promise exists in the area between physical studies of meteorites and the solar system, on one hand, and astronomical investigations of early stellar evolution on the other. The feeling is strong today that these two subjects must fit together somehow, and that in the achievement of that fit there will be great but unpredictable profit for both. It was in the pursuit of such goals that Otto Struve found his largest fame, and probably also his greatest satisfaction.

[7] An astrometric investigation by Vasilevskis, Sanders and Balz (1965) shows that a number of stars plotted in this region of Walker's original color-magnitude diagram are non-members.

REFERENCES

Becklin, E. E., and Neugebauer, G. 1967, *Ap. J.*, **147**, 799.
Bernas, R., Gradsztajn, E., Reeves, H., and Schatzman, E. 1967, *Annals of Phys.*, **44**, 426.
Bodenheimer, P. 1965, *Ap. J.*, **142**, 451.
Böhm, K.-H. 1968, *Highights of Astronomy as presented at the XIIIth General Assembly of the IAU, Praha 1967*, ed. L. Perek, p. 229.
Cameron, A. G. W. 1962, *Icarus*, **1**, 13.
Danziger, I. J. 1967, *High Energy Nuclear Reactions in Astrophysics*, ed. B. S. P. Shen (New York: W. A. Benjamin, Inc.), p. 81.
Dolidze, M. V., and Arekelyan, M. A. 1959, *Astr. J. U.S.S.R.*, **36**, 444; *Sov. Astr.-A.J.*, **3**, 434.
Eggen, O. J. 1961, *Royal Obs. Bull.*, No. 31, E115.
———. 1964, *Royal Obs. Bull.*, No. 82, E74.
Ezer, D. 1966, *Colloquium on Late-Type Stars*, ed. M. Hack (Trieste: Osservatorio Astronomico), p. 357.
Faulkner, J., Griffiths, K., and Hoyle, F. 1963, *M.N.R.A.S.*, **126**, 1.
Fowler, W. A., Greenstein, J. L., and Hoyle, F. 1962, *Geophys. J., R.A.S.*, **6**, 148.
Götz, W. 1961, *Veröff. Sonneberg*, **5**, heft 2.
———. 1965, *ibid.*, **7**, heft 1.
Gould, R. J. 1964, *Ap. J.*, **140**, 638.
Haro, G. 1949, *Astr. J.*, **54**, 188.
Hartmann, W. K. 1967, *Ap. J.*, **149**, L87.
Hayashi, C. 1966, *Ann. Rev. Astr. and Ap.*, **4**, 171.
Henyey, L. G., LeLevier, R., and Levee, R. D. 1955, *Pub. A.S.P.*, **67**, 154.
Herbig, G. H. 1950, *Ap. J.*, **111**, 11.
———. 1953, *A.S.P. Leaflet* No. 293.
———. 1957, *Non-Stable Stars*, ed. G. H. Herbig (London and New York: Cambridge Univ. Press), p. 3.
Herbig, G. H. 1962a, *Trans. I.A.U.*, **11B**, 299.
———. 1962b, *Adv. Astr. and Ap.*, **1**, 47.
———. 1966, *Vistas in Astr.*, **8**, 109.
———. 1968a, *Ap. J.*, **152**, 439.
———. 1968b, *Mitt. Budapest*, in press.
Herbig, G. H., and Peimbert, M. 1966, *Trans. I.A.U.*, **12B**, 412.
Herbig, G. H., and Wolff, R. J. 1966, *Ann. d'Ap.*, **29**, 593.
Hidajat, B. 1961, *Contr. Bosscha Obs.*, No. 11.
Huang, S.-S. 1957, *Pub. A.S.P.*, **69**, 427.
———. 1965, *Ap. J.*, **141**, 985.
———. 1967, *Astr. J.*, **72**, 804.
Huang, S.-S., and Wade, C. 1966, *Ap. J.*, **143**, 146.
Iben, I. 1965, *Ap. J.*, **142**, 421.
Johnson, H. L. 1967, *Ap. J.*, **150**, L39.
Joy, A. H. 1945, *Ap. J.*, **102**, 168.
Kleinmann, D. E., and Low, F. J. 1967, *Ap. J.*, **149**, L1.
Kolesnik, I. G., and Frank-Kamenetskii, D. A. 1967, *Astr. J. U.S.S.R.*, **44**, 905; *Sov. Astr.-A.J.*, **11**, 726.
Kuhi, L. V. 1964, *Ap. J.*, **140**, 1409.
———. 1966, *Ap. J.*, **143**, 991.
Kurochkin, N. E., and Kukarkin, B. V. 1966, *Astr. J. U.S.S.R.*, **43**, 83; *Sov. Astr.-A.J.*, **10**, 64.
Limber, D. N. 1960, *Ap. J.*, **131**, 168.
Low, F. J., and Smith, B. J. 1966, *Nature*, **212**, 675.
Magnan, C. 1965, *Comptes Rendus, Acad. Sci. Paris*, **261**, 3294.
Magnan, C., and Schatzman, E. 1965, *Comptes Rendus, Acad. Sci. Paris*, **260**, 6289.
Mendoza, E. E. 1966, *Ap. J.*, **143**, 1010.
———. 1968, *ibid.*, **151**, 977.
Metreveli, M. 1966, *Astr. Circ. U.S.S.R.*, No. 352.
Parenago, P. 1950, *Peremennye Zvezdy*, **7**, 169.
Poveda, A. 1967, *Astr. J.*, **72**, 824.
Raimond, E., and Eliasson, B. 1967, *Ap. J.*, **150**, L171.
Roman, N. G. 1949, *Ap. J.*, **110**, 219.
Roxburgh, I. W. 1966, *Astr. J.*, **71**, 133.
Salpeter, E. E. 1954, *Mem. Soc. Sci. Liège* [4°] **14**, 116.
Schmidt, M. 1963, *Ap. J.*, **137**, 859.
Struve, O. 1949, *Ap. J.*, **109**, 180.
Struve, O., and Swings, P. 1948, *Pub. A.S.P.*, **60**, 61.
Struve, O., and Rudkjøbing, M. 1949, *Ap. J.*, **109**, 92.
Thackeray, A.D., and Wesselink, A. J. 1965, *M.N.R.A.S.*, **131**, 121.

Unsöld, A. 1964, *Observatory*, **84**, 152.
Van Woerden, H. 1966, *Trans. I.A.U.*, **12B**, 391.
Vasilevskis, S., Sanders, W. L., and Balz, A. G. A. 1965, *Astr. J.*, **70**, 797.
Walker, M. F. 1956, *Ap. J. Suppl.*, **2**, 365.
Wenzel, W. 1961, *Veröff. Sonneberg*, **5**, heft 1.
Wilson, O. C. 1963, *Ap. J.*, **138**, 832.
————. 1966, *ibid.*, **144**, 695.
Wilson, O. C., and Skumanich, A. 1964, *Ap. J.*, **140**, 1401.
Wilson, R. E. 1949, in Joy, A. H., and Wilson, R. E.: *Ap. J.*, **109**, 243.

8. INTERSTELLAR MATERIAL

O. STRUVE: The Constitution of Diffuse Matter in Interstellar Space. *Journal of the Washington Academy of Sciences*, **31**, 217, 1941.

GUIDO MÜNCH: Interstellar Gas.

H. C. VAN DE HULST: Interstellar Grains.

ASTROPHYSICS.—*The constitution of diffuse matter in interstellar space.*[1] OTTO STRUVE, Yerkes Observatory. (Communicated by EDWARD TELLER.)

THE EMPTINESS OF SPACE

On a clear moonless night we can see with the unaided eye somewhere between 2,000 and 3,000 stars. With an average pair of binoculars the number of visible stars is increased to about 10,000, and on long exposures taken with the largest existing telescope the number would be 2 or 3 billion. If we recall that the entire celestial sphere contains 41,253 square degrees and that the visible area of the full moon is about one-fifth of a square degree, we find that there are at least 10,000 stars within the reach of our most powerful instruments for every area of the sky equal to the full moon. If we remember also that the number of invisible stars—too faint to be recorded even with the 100-inch Mount Wilson reflector—is at least thirty times greater, and that near the Milky Way the concentration of stars is roughly one hundred times greater than at the poles of the galaxy, we find that in many regions of the Milky Way the apparent star density must be of the order of 30 million for an area equal to the disk of the moon. It is not surprising that on the best photographs of the Milky Way the images of the stars are so densely crowded together that they flow into one another and give the appearance of an almost continuous mass of stars.

But this impression is misleading. In reality the individual stars are separated by distances of several light years, and if we liken the stars to raindrops their average distances would have to be 40 miles to give us the right idea of the density within our galaxy of matter in the form of stars. For every cubic centimeter of stellar matter there are 10^{22} cubic centimeters of transparent space. Our galaxy is a relatively dense object. Since the average distances between neighboring galaxies are of the order of 10^6 light years, while their diameters are about 10^4 light years, it is easy to compute that within the diameter of the explorable universe—some 600 million light years—with its 10^8 separate

[1] The Eleventh Joseph Henry Lecture of the Philosophical Society of Washington, delivered on March 29, 1941. Received March 29, 1941.

galaxies and 10^{52} cubic centimeters of stellar matter, there are approximately 10^{28} cubic centimeters of transparent space for every cubic centimeter of stellar matter. The average density of a star like the sun is a little greater than that of water. Hence the density of stellar matter in the universe is only about 10^{-28} g/cm^3. A density of the order of 10^{-15} g/cm^3 is considered a high vacuum in ordinary laboratory technique.

But the question arises whether all matter in the universe is concentrated in the form of stars. It is possible that free atoms and molecules or small particles of dust float in interstellar space without completely obstructing the light of distant stars and galaxies. Eddington once remarked that although astronomers do not know much about interstellar matter they talk a great deal about it; they are like the guest who refused to sleep in a "haunted" room and who, when asked whether he believed in ghosts, replied: "I do not *believe* in ghosts, but I am *afraid* of them." It is probably no exaggeration to say that interstellar matter was the ghost that has haunted astronomers for the past hundred years. Until about 15 years ago they steadfastly refused to believe that there existed any such matter, even though direct photographs of the Milky Way gave unmistakable evidence of large regions in space where the light of distant stars is more or less completely cut off by the screening effect of cosmic dust clouds. They were afraid of the ghost because they thought it would play havoc with their elaborate theories of the structure of the Milky Way. These theories all depended upon an application of the inverse square law for the brightnesses of the stars. If two stars are of the same intrinsic luminosity, for example, if both have spectra that exactly match the spectrum of the sun, but one star is of apparent magnitude 5 while the other is of apparent magnitude 10, then the astronomers reasoned that since each step in magnitude corresponds to a ratio of 2.5 in the brightnesses of the stars, the fainter star sends us one hundredth as much light as the brighter and that, consequently, its distance must be ten times greater than that of the brighter star. It is obvious that if a part of a star's light is intercepted by a screen of absorbing material, this computation would lead to erroneous results: the real distance of the faint star would be smaller than the one computed by means of the inverse-square law.

DISPERSION OF LIGHT IN SPACE

The first intimation of a possible effect of interstellar matter upon the propagation of light through cosmic space occurs in a letter by

Newton to Flamsteed, dated August 10, 1691. In the last sentence of this letter Newton, who was not an observer, asks the Astronomer Royal at Greenwich: "When you observe the eclipses of Jupiter's satellites I should be glad to know if in long telescopes the light of the satellite immediately before it disappears inclines either to red or blue, or becomes more ruddy or pale than before." The finite velocity of light had been measured in 1676 by Römer at Paris. He had used the predicted eclipses of the satellites of Jupiter and had taken advantage of the fact that Jupiter is at certain times much closer to the earth than at others. It was quite natural that Newton should try to find whether the velocity of blue and of red light is equal through interplanetary space, or whether an appreciable dispersion of the light takes place between Jupiter and the earth.

We have no record of Flamsteed's reply, and we do not even know whether observations were made to answer Newton's question. But astronomers gradually concluded that any possible dispersion was much too small to produce measurable effects in the satellites of Jupiter.

In 1855 Arago, in a course of public lectures at the Paris Observatory, applied the idea of Newton to the eclipses of distant binary stars whose orbit planes pass through the earth: "Let us then investigate what ought to be the density of this (hypothetical) interstellar gas in order that two rays, one red, the other blue, emitted at the same instant from a variable star, should arrive almost simultaneously at the earth notwithstanding the prodigious thickness of the matter traversed, notwithstanding the time of transmission which cannot be under three years; the solution of this simple problem of physics will astonish the imagination by its smallness." Arago gives no numerical results. He and others had "frequently examined periodic white stars in their different stages of brightness without remarking any appreciable coloring."

But these observations were made visually and were not very accurate. After the introduction of accurate photometric methods into astronomy, in 1908, it seemed for several years that a real positive effect of interstellar dispersion had been discovered independently by the Russian astronomer Tikhoff and the French astronomer Nordmann. These scientists found that when certain eclipsing variables are observed in red light the phase of central eclipse, or minimum light, occurs earlier than when the observations are made in blue light. For Algol (β Persei) the observed lag was 16 minutes $\pm$ 3 minutes. For λ Tauri it was 50 minutes and for RT Persei it was 4 minutes. This apparent lag of blue light with respect to red light has been verified in many instances, and it is now known as the Tikhoff-Nordmann phenomenon. But, as Lebedeff had pointed out almost immediately after the announcement of Tikhoff's and Nordmann's discoveries, the distance of Algol is about 60 light years while that of RT Persei is 740 light years. Yet the nearer star has the longer lag of 16 minutes, so that if interstellar dispersion were responsible for the phenomenon the dispersion constant would have to be $16/4 \times 740/60 = 40$ times larger in the direction of Algol than in the direction of RT Persei. Modern determinations give somewhat different distances for the two stars, but the conclusion of Lebedeff has been shown to be true. Unless we make the absurd assumption that the dispersion constant is entirely different for different stars—even if they are located in the same part of the sky—the Tikhoff-Nordmann phenomenon must be due to some other cause.

The final word in the matter came from Shapley. For the stars of the globular cluster M5, whose distance is about 30,000 light years, or 3×10^{22} cm, blue light and yellow light arrived on the earth within an interval of -10 seconds $\pm$ 60 seconds. This corresponds to a difference of less than 0.3 cm/sec between blue and red light, and shows that the velocities are the same to at least one part in 10^{11}. We conclude that there is no measurable dispersion of light in interstellar space.

GENERAL ABSORPTION

Apparently the first astronomer to worry about the dimming of star light by intervening clouds of diffuse matter was Halley, around 1720. The argument was revived by Chéseaux in Switzerland, about 1744. Both astronomers pointed out that an infinite universe with an infinite number of self-luminous stars should cause the entire heavens to be ablaze with light—for no matter in which direction we should look, our line of vision would always ultimately reach the surface of some distant star. It was tacitly assumed that the distribution of the stars in space is uniform and that there are no dark stars. In 1823 the famous German astronomer Olbers expressed a similar view in the following words: "God has made the transparency of space imperfect in order to enable the inhabitants of the earth to study astronomy in its details . . . Without this, we should have no knowledge of the starry heavens; our own sun would be discovered only with difficulty by its spots; the moon and the planets would not be distinguishable, except as obscure discs upon a bright background, like the sun . . . " Fortunately all these dreadful consequences had been removed by the foresight of the Creator who had introduced into interstellar space enough absorbing material to dim the light of the most distant stars so that the background is dark and not so brilliantly luminous as the sun.

It is futile now to speculate upon the scientific logic and the philosophical insight of these early astronomers. It is easy to see that even an infinite universe need not necessarily lead to a sky completely covered with stars, and the early arguments in favor of interstellar absorption are now, to say the least, unconvincing.

The mathematical theory of interstellar absorption was first precisely formulated by F. G. W. Struve, in Russia. His book *Études d'astronomie stellaire*, published in 1847, is the first really scientific study of the whole problem of interstellar matter. It forms the connecting link between the earlier semiphilosophical speculations of Olbers, Chéseaux, and Halley, and the brilliant theoretical researches of the last decades of the nineteenth century and the prewar years of the present century, principally by Seeliger, Kapteyn, and Schwarzschild. The common property of all these investigations was the tendency to smooth out the local irregularities in the observed structure of the Milky Way and to study an idealized or "typical" stellar system which retained certain characteristics of the Milky Way, such as galactic concentration, but purposely avoided the discussion of individual star clouds and dark regions.

Struve's work was based upon the star counts that Sir William Herschel had made with his giant telescopes in many different parts of the sky. These counts gave the numbers of all stars visible for every step in brightness, over a uniform field of the sky. The problem was to derive the true distribution of the stars in space and to find, if possible, whether there was an effect of absorption in space.

To consider this problem it is convenient to use the functions:

(1) $D(r)$—the density function, which measures the number of stars per unit volume, as a function of the distance from the sun, r.

(2) $\phi(M)$—the luminosity function, which measures the distribution of stars of different intrinsic luminosities.

(3) $i = f(r)$—the intensity function, which measures the apparent brightness of a star as a function of its distance from the sun, r.

The simplest assumptions that we can make and that we can test are

$$D(r) = \text{const.}; \ \phi(M) = \text{const.}; f(r) = \frac{i_0}{r^2} \cdot \tag{1}$$

This is, essentially, what Olbers and his predecessors had assumed, and it is fairly obvious that the appearance of the sky at once suggests one of two conclusions:

(a) the universe is finite, or

(b) there must be an appreciable absorption in space, in which case $f(r) \neq i_0/r^2$.

If we are not willing to abandon the inverse-square law, then it is easy to see that there must exist two simple, but important, statistical relations:

(1) The theoretical distances of stars of successive magnitude (apparent brightness) classes form a geometrical series whose coefficient is $(2.5)^{1/2}$

(2) The theoretical numbers of stars down to successive magnitude (apparent brightness) classes form a geometrical series whose coefficient is $(2.5)^{3/2}$

Since by definition the ratio of the observed intensities of two stars whose stellar magnitudes are m_1 and m_2 is

$$\frac{i_1}{i_2} = 2.5^{(m_1 - m_2)}$$

and since by (1)

$$\frac{i_1}{i_2} = \frac{r_2{}^2}{r_1{}^2}$$

we have the first relation; when $m_1 - m_2 = 1$

$$\frac{r_2}{r_1} = \sqrt{2.5} \cdot \tag{2}$$

Since, next, for uniform distribution of stars in space ($D(r) = $ const.) the number of all stars down to each magnitude step is proportional to the volume of the spheres occupied by those stars which have the required apparent brightnesses, and since these volumes are proportional to r^3, we have, when $m_1 - m_2 = 1$:

$$\frac{N_2}{N_1} = (2.5)^{3/2} \tag{3}$$

or, for any value of $m - m_0$:

$$\log N_m = \log N_{m0} + 0.6\,(m - m_0). \tag{4}$$

Struve tested relation (4) by means of Herschel's star counts, which gave directly the values of N_m. There were large systematic departures in the sense that the observed N_m were smaller than those predicted by the formula. Two possible conclusions presented themselves: (*a*) The assumption $D(r) =$ const. is wrong, and the density of the stars in space must decrease in all directions from the sun, or (*b*) the assumption $i = i_0/r^2$ is wrong, and there is an effect of interstellar absorption which makes i smaller than it would be without absorption. With regard to hypothesis (*a*), Struve writes: "Perhaps someone will say that there might be a gradual diminution in the star density in the principal plane, toward the outer boundaries of the Milky Way. But how much do we know concerning these boundaries? The Milky Way is for us absolutely impenetrable. What, then, is the probability that the sun should be located near the center of a disc whose extent is for us completely unknown? Let us recall, furthermore, that our study of the Herschel stars has led to the same average decrease in density, at right angles to the principal plane, which occurs in the neighborhood of the sun, up to the distance of stars of the 8th and 9th magnitude. From all these considerations I wish to state that we have discovered a phenomenon in which the extinction of star light unquestionably manifests itself." In place of the inverse-square law Struve adopts the relation

$$i = \frac{i_0}{r^2}\, e^{-\lambda r} \tag{5}$$

where λ is the coefficient of absorption. This is equivalent, according to Struve, to a loss of one stellar magnitude per 3,000 light years, or $\lambda = 3 \times 10^{-4}$ if r is expressed in light years. This value, obtained almost 100 years ago, is amazingly accurate. Modern results give an average loss of light of between 0.7 and 0.8 stellar magnitude per kiloparsec (3,000 light years). The agreement is as good as between individual results of modern observers.

It is strange that Struve's results were not universally accepted. The tendency was to assume that $\lambda = 0$ and to derive the resulting function $D(r)$. This procedure was adopted by Seeliger, Kapteyn, Schwarzschild, and Charlier. It resulted in a badly distorted picture of the Milky Way, with the sun near the center and the star density decreasing in all directions. Halm, in 1917, attempted to find λ under the assumption $D(r) =$ const., but his results were not accepted. As late as in 1923, Kienle, after a careful review of all available evidence, concluded that the loss of star light through absorption must definitely be less than 2 mag/kiloparsec and that it is probably less than 0.1 mag/kiloparsec.[2]

---

[2] Mag is used in this paper as the abbreviation for stellar magnitude.—EDITOR.

The first definite break in astronomical opinion came in 1930 when R. Trumpler published his results on open star clusters, obtained at the Lick Observatory. His work was based upon measurements of the diameters of galactic star clusters—groups of tens or hundreds of stars forming compact systems in which the motions of the individual members are all alike. Typical among these formations are the Pleiades, the Hyades, the cluster in Coma Berenices, and others. By observing the spectra of the members of clusters and their brightnesses Trumpler was able to prepare for each cluster a Hertzsprung-Russell diagram, in which the brightness of each star appears as the ordinate and the spectral type as the abscissa. In a diagram of this kind, as Russell had found many years ago, the majority of the stars are arranged in a definite narrow band—the so-called main sequence. Physically speaking, the main sequence of one cluster should be rather similar to the main sequence of another. But since the distances of the two clusters are, in general, not the same, it is necessary to shift the diagrams along the vertical coordinate in order to make the two sequences coincide. This displacement, measured along the vertical coordinate in stellar magnitudes, provides a measure for the relative distances of the two clusters. Distances determined in this manner are affected by absorption. Suppose we find that the vertical shift corresponds to 3 magnitudes. This means that all stars of the brighter cluster are $(2.5)^3 = 15.6$ times as intense as those of the fainter. If there were no absorption we should conclude from the inverse-square law that the fainter cluster is $\sqrt{15.6} = 3.95$ times as far away as the brighter cluster. But if absorption is present the real distance would be smaller. Trumpler conceived the idea of measuring the diameters of the clusters. He first made sure that he was measuring physically similar objects. If the brighter cluster had a diameter of 15′ the fainter should have a diameter of $15/3.95 = 3′.8$ provided the distances inferred from the Hertzsprung-Russell diagram are correct. From a large amount of very homogeneous material Trumpler concluded that the diameters of the fainter clusters were systematically too large. He suggested that there is an appreciable amount of interstellar absorption and derived for it a value of 0.67 mag/kiloparsec.

Among the many modern determinations of the average amount of interstellar absorption per unit distance, one of the most interesting is due to Joy at the Mount Wilson Observatory. The method depends upon determinations of radial velocities of Cepheid variables and upon the theory of the rotation of our galaxy, which predicts that the rotational component of motion of a star in the line of sight must be proportional to the distance from the sun. This results from the nature of galactic rotation in a central field of force. The matter is complicated by the fact that the stars have their own individual motions. But these are probably distributed at random so that if we take average velocities for groups of Cepheid variables arranged according to apparent magnitude we should derive a series of values the ratios of which, after correction for foreshortening, give us directly the ratios of the true distances for the various groups. Now, the Cepheid variables, as everyone knows, have the remarkable property that their periods of light variation are exactly related to their intrinsic luminosities, so that if we know the periods of two such variables we can tell at once how much brighter one is than the other. If we use this criterion of absolute luminosity, together with the apparent brightness, we derive the distance. If there were no absorption the two procedures should give identical results. From the departures, which are conspicuous, Joy, and later R. E. Wilson, derived an absorption of about 0.6 mag/kiloparsec.

Another method depends upon counts of extragalactic objects in different parts of the sky. These distant galaxies are seen through the thickness of absorbing matter in our Milky Way, and their numbers are greatest near the two poles of the galactic circle, while near the plane of the galaxy the absorption is so great that no outer galaxies are seen through it. In intermediate galactic latitudes the absorptions are proportional to the cosecants of the latitudes. For example, at galactic latitude 10° the absorption is 1.4 mag, at 20° it is 0.7 mag, at 30° it is 0.50 mag, at 60° it is 0.29 mag, and at 90° it is 0.25 mag. The smooth manner in which these values progress has suggested to Seares the idea that we are here dealing with "a widely diffused absorbing stratum extending equally above and below the galactic plane." The zone of complete avoidance of galaxies, which is irregular in shape, is associated by Seares with the obscuring clouds in the Milky Way which give their distinctive irregular appearance to the star clouds.

SELECTIVE ABSORPTION

In 1895 Kapteyn discovered that the average color of stars in the Milky Way is bluer than outside of it. He suggested that this phenomenon might be caused by selective absorption, which should make those stars appear bluer for which the absorption was least. Since at that time there was no conclusive evidence of general absorption, Kapteyn could not know that the color observations would be in disagreement with other evidence by requiring the absorption to be greatest at the poles of our galaxy. In 1904 DeSitter made a careful study of star colors and concluded: "True differences in the colors of the stars, or general absorption in certain spectral regions, or selective absorption by intervening cosmical clouds or nebulous masses, these are questions that can be put, but not yet answered." We now know that the intrinsically blue, hot stars have a much greater tendency to concentrate toward the galactic equator than do the cool, red stars. Hence the phenomenon of Kapteyn has no bearing upon the question of selective scattering.

However, the question was revived some years later when Turner and others noticed that in order to obtain photographic star images of equal densities for successive stellar magnitudes, it was not sufficient to increase the exposure times in the ratio of 2.5 for each step of one magnitude. In fact, it was found that the ratio of the exposure times is much more nearly 3 than 2.5—in spite of the fact that by definition a step of one magnitude signifies an intensity ratio of 2.5. In other words, the photographic density B is not a function of the product $(i \times t)$ alone, where i is the light intensity and t is the exposure time, but may be written as

$$B = f(i \times t)^p$$

where $0 < p < 1$. The quantity p is now known as Schwarzschild's exponent.

Tikhoff and Turner suggested that p had a cosmic significance. They argued that selective absorption will make the stars appear redder the greater their distances. But distant stars are, on the average, faint. Hence, faint stars are red, and must require relatively longer exposures on the blue-sensitive photographic emulsions—the only ones then used for these observations. Turner summarized the matter in the following words: "The fact that when the photographic exposure is prolonged in a ratio which ought to give stars fainter by five magnitudes, we only get four visual magnitudes" is an argument in favor of "the scattering of light by small particles in space."

The obvious thing would have been to check the result by means of visual observations made with the help of violet filters. There is no reason why this could not have been done. The test would have shown at once that the fainter stars are not appreciably redder and that the Schwarzschild exponent is not a measure of interstellar reddening.

The correct interpretation of p was given in 1909 by Parkhurst. It represents a characteristic property of the photographic emulsion and determines what we now call the reciprocity failure of the emulsion. It is different for different brands of emulsions.

A new attack upon the problem became possible with the development of accurate methods of photographic photometry. Miss Maury at Harvard had remarked in her work on the spectra of the stars that among representatives of a single class of spectrum there were some which were weak in violet light while others were strong. Kapteyn examined the available data on star colors and found that within narrow groups of spectrum those stars which were rich in violet light had large angular proper motions, while those which had little violet light had, at the same time, small proper motions. Since the angular proper motion of a star is a good statistical measure of distance, he concluded that the distant stars are reddened by absorption. Unfortunately, this conclusion was incorrect. It turned out later that another interpretation could be given. The stars examined by Kapteyn within any given group of spectrum were, through observational selection, all more or less similar in apparent brightness. Hence, the distant stars of each group were giants, while the nearby stars were dwarfs. Kapteyn could not have known, or even suspected, that the surface gravities of the giants are very much smaller than those of the dwarfs and that, consequently, the pressures of the former are one hundredth as great as those of the latter. Equal spectral types imply equal average ionization. But in a giant with its low atmospheric pressure the same degree of ionization is attained at a lower temperature than in a dwarf. Kapteyn had discovered what is essentially the basis of spectroscopic luminosities, but he had not found evidence in favor of space reddening.

In 1923 Kienle could only conclude that the coefficient of selective absorption must be less than 0.1 mag/kiloparsec. As we shall see, this result was unnecessarily pessimistic. Even before Kienle's paper was printed, in 1919, Russell had called attention to the fact that "the three most abnormally yellow stars of type B (ζ, o and ξ Persei) lie within 5° of one another, in a region full of diffuse nebulosity," and he had suggested that this might be caused by local selective absorption in some of the dark clouds of the galaxy. Several years later Hertzsprung compiled a list of abnormally yellow B stars, whose color excesses were larger than can normally be explained as a result of differences in absolute magnitude. This list was greatly enlarged by Bottlinger, and in 1926 the present writer found fairly convincing evidence "that the effect of reddening . . . is produced by light scattering in dark nebulae or in calcium clouds."

The modern evidence concerning selective absorption in interstellar space rests upon three types of observation:

(1) Color excesses of B-type stars determined photoelectrically by Stebbins, Whitford, and Huffer, or photographically by Seyfert and Popper. These results show a progressive reddening for the more distant stars. In setting the 82-inch McDonald telescope upon a faint B-type star I have often been amazed at the redness of these stars: their continuous spectra show energy distributions corresponding to some 4,000° or 5,000°, while their absorption lines correspond to a temperature of 20,000°.

(2) Observations of color indices of members of open clusters by Trumpler and his associates.

(3) Determinations of the energy distributions and colors of stars involved in dense clouds of absorbing matter, by Seares and Hubble, Baade and Minkowski, Henyey, etc.

An important quantity is the ratio of selective absorption to total absorption. Seares concludes from a discussion of all available data that:

Total absorption $= 10 \times$ selective absorption, for the diffuse stratum of interstellar matter, and

Total absorption $= 5.7 \times$ selective absorption, for the dark nebulae of the zone of avoidance.

The latter law corresponds to an absorption coefficient which is proportional to λ^{-1}. The departure from Rayleigh's λ^{-4} law was first established by Trumpler, and was then confirmed by Struve, Keenan, and Hynek. It is probable that some stars, observed by Baade and Minkowski in the Orion nebula, present appreciable departures from the smooth λ^{-1} relation, but the exact nature of these departures has not yet been explored.

DIFFUSE RADIATION

The existence of reflection nebulae, whose spectra are continuous and whose absorption lines are identical with those of the associated stars was demonstrated many years ago by V. M. Slipher. Since that time the present writer, Henyey, Elvey, Greenstein, and others have investigated in detail the character of the diffuse light that is scattered by dark nebulae and by interstellar space. The light is nearly of the same color as that of the illuminating stars; in the Pleiades the nebulosity is slightly bluer, but the difference between nebula and star is much less than that between the blue sky and the sun. This is in agreement with the λ^{-1} relation found from the absorption effects. The diffuse light is very little polarized, which all goes to show that the particles are too large for Rayleigh scattering. On the other hand, there is some evidence to show that the phase function of the particles throws

Fig. 1.—Region in Taurus, showing obscuring nebulae that radiate feebly by reflecting the light of the stars. (Photograph by E. E. Barnard.)

most of the light forward and does not show the "phases of the moon," as one would expect if the particles were larger than about 0.01 cm in radius. This conclusion rests upon a comparison of the diffuse surface brightness of dark nebulae of absorbing matter. These nebulae are illuminated by the light of all stars around them. The situation is somewhat analogous to the case of a spherical body which is illuminated from all sides. If the intensity of the general starlight is assumed to be uniform and equal to $L_1 = 56$ stars of magnitude 10.0 per square degree, if γ is the albedo of the particles, α is the phase angle, i is the angle of incidence and ϵ the angle of emergence, then the surface brightness of the nebula is

$$I = \frac{\gamma}{2\pi} L_1 \int \phi(\alpha) \, \frac{\cos i}{\cos i + \cos \epsilon} \, d\omega. \tag{6}$$

This formula predicts that the rim of the nebula must be brighter than the center, but the amount of the difference depends upon the form of the phase function $\phi(\alpha)$. If we do not wish to make the albedo unreasonably large, we are compelled to adopt a forward throwing phase function. This suggests that the particles have radii smaller than 0.01 cm.

Sizes between 0.01 cm and 0.001 cm are excluded because of the absence of diffraction nebulosities surrounding stars that are seen through absorbing nebulosity. If the particles acted like water particles in clouds, of the required size, they should give rise to bright rings, which are often incorrectly called halos, the surface brightnesses of which may be computed. As a bright star seen through a thin cloud gives a halo, so the stars shining through dust clouds should produce halos. A search made by the writer some time ago yielded rather definite evidence that there is no halo formation. Hence the particles are smaller than 0.001 cm.

The interval of possible sizes for the majority of the particles has now been narrowed to between 10^{-3} and 10^{-5} cm. To go beyond this we must make use of the complicated theory of Mie for the scattering of light by small particles, and this has been done by a number of investigators, for example, by Shalén in Sweden and Greenstein in the United States. The most frequent sizes of the particles are slightly greater than 10^{-5} cm.

In connection with the problem of illumination of dense obscuring nebulosities, Struve and Elvey found in 1937 that the surface brightnesses of some of Barnard's dark nebulae are only about 0.03 mag fainter than the background of the sky between the stars in star

Fig. 2.—Dark nebula projected upon luminous gaseous nebula, in Orion.
(Photograph by J. C. Duncan.)

clouds. If the star clouds were free of absorbing material the dark nebulae should be considerably more luminous. Hence we conclude that
either the scattering efficiency (albedo) of the particles composing the
nebula is very low, or there exists a large amount of diffuse radiation
in the star clouds of the Milky Way. A direct observational test by

means of a photoelectric surface photometer was made by Elvey and Roach. They confirmed the correctness of the second alternative: there is a diffuse "galactic light" in the Milky Way, corresponding to about 57 stars of magnitude 10 per square degree, in the galactic plane. A careful study by Henyey and Greenstein has permitted them to subtract from the galactic light that part which comes from emission sources. In Cygnus, where the galactic light is strongest they find the diffuse radiation to be equivalent to 80 stars of magnitude 10 per square degree. In Taurus the galactic light is equivalent to only 35 stars of magnitude 10 per square degree.

Henyey and Greenstein consider the absorbing layer to be a slab of emitting and scattering matter, stratified along parallel planes. If the incident starlight and diffuse light has an intensity of L_1, each element of the absorbing material having an optical thickness $d\tau$, contributes to the observed galactic light the amount

$$\gamma d\tau \int L_1\phi(\alpha)d\omega \tag{7}$$

where γ is the albedo of the particles. The formula is analogous to (6). The total surface brightness of the galactic light in a given direction is obtained by multiplying (7) by $e^{-\tau}$ and integrating over τ from 0 to the limiting optical thickness of the absorbing layer τ_0:

$$I = \gamma \int_0^{\tau_0} e^{-\tau}d\tau \int L_1\phi(\alpha)d\omega = \gamma(1 - e^{-\tau_0}) \int L_1\phi(\alpha)d\omega. \tag{8}$$

The quantity L_1 can be computed from star counts, provided we demand no refinements such as the variation of L_1 with τ. The phase function is not known. But it is clear that an isotropic distribution of the scattered radiation would make I larger than a strongly forward or backward throwing phase function. Indeed, at any given point on the slab of absorbing material the area of sky contributing to its illumination is greatest for $\alpha \sim 90°$. Hence a strongly forward throwing phase function will throw less radiation into the line of sight than an isotropic phase function. This presupposes that L_1 is uniform over the sky. In reality it is concentrated toward the Milky Way, and the integral must be evaluated numerically. In the region of Cygnus τ_0 must be very great. For all practical purposes we may assume that $\tau_0 = \infty$, so that

$$I_c = \gamma \int L_1\phi(\alpha)d\omega. \tag{9}$$

Knowing L_1 we can compute I for phase functions having different tendencies of throwing the light forward. We need not consider backward throwing functions because we have already eliminated them.

For the Taurus region

$$I_T = \gamma(1 - e^{-\tau_0}) \int L_1\phi(\alpha)d\omega \cdot \tag{10}$$

Taking the ratio, we find

$$1 - e^{\tau_0} = \frac{I_T}{I_c} = \frac{35}{80} = 0.44 \cdot \tag{11}$$

Since the integral $\int L_1 \phi(\alpha) \, d\omega$ is not strictly the same for the two regions Henyey and Greenstein derive a slightly different value. But we shall continue to use (11). Of course, τ_0 here refers to the Taurus region, where we evidently have relatively little material. In the Cygnus region the integral $\int L_1\phi(\alpha) \, d\omega$ is about 72 stars of magnitude 10 in the case of an isotropic $\phi(\alpha)$; 80 stars of magnitude 10 for a moderately forward throwing ϕ; and about 135 stars of magnitude 10 for an extremely forward throwing ϕ. Since γ, the albedo, can not be greater than 1, we conclude that:

(1) the phase function must be at least moderately forward throwing, and

(2) the albedo must be at least $80/135 = 0.6$.

Indeed, when $\gamma = 1$, we have

$$I_c = \int L_1\phi(\alpha)d\omega$$

and this is true only when ϕ is moderately forward throwing. If $\gamma < 1$, then ϕ must be even more forward throwing. For an excessively forward throwing ϕ, when all radiation is thrown forward, the integral is 135, so that $I_c/\int L_1 \phi(\alpha) \, d\omega = \gamma = 80/135 = 0.6$. If allowance is made for errors in the observations and for various refinements in the theory, Henyey and Greenstein find that the albedo must be greater than 0.3 and that the phase function must be strongly forward throwing.

STAR COUNTS

Our knowledge of the absorptions in individual regions of the Milky Way rests now largely upon star counts. The method first successfully used by Wolf and later improved by Pannekoek has in recent years received a new impetus through the work of Bok and his star-counting bureau at Harvard. The method consists essentially in the comparison of star counts made in obscured regions with those made in open regions. If there were no dispersion in the luminosities of the stars the interpretation of the results would be simple. The existence of $\phi(M)$ causes considerable complications, but Bok has developed a numerical procedure which is particularly adapted to the study of individual dark nebulae. A list of these nebulae prepared by Greenstein and quoted by Bok is given in table 1.

TABLE 1.—ABSORBING CLOUDS

Region	Area in square degrees	Distance in parsecs	Diameter in parsecs	Absorption in magnitudes
Taurus–Orion–Auriga...	600	145	65	1.1
Cepheus–Cassiopeia.....	450	500	170	0.7
Cygnus..............	85	700	130	1.5
Ophiuchus–Scorpius....	1050	125	80	0.9
Vela................	105	600	120	1.0(?)

Shalén has used similar methods, but has limited the material to a narrow range of spectral type. Morgan and, quite recently, O'Keefe have made considerable progress by studying the spectroscopic luminosity criteria of B-type stars and by deriving the best possible individual distances for highly reddened stars.

DYNAMICAL CONSIDERATIONS

An indirect method of estimating the density of interstellar matter rests upon Oort's study of stellar motions at right angles to the plane of the Milky Way. The motions of the stars in the galactic plane are governed largely by the mass of the galactic nucleus. But the motions across the galactic plane are determined almost wholly by the distribution of mass in the vicinity of the sun. From the observed radial velocities Oort computes the space motions of the stars. The Z-components are then analyzed with regard to the distances of the stars from the galactic plane. This leads to a determination of the accelerations in the Z-coordinate, and these depend upon the density of matter in the vicinity of the sun. The data of observation are best satisfied for an average density of 0.092 solar mass per cubic parsec, which is equivalent to 6.3×10^{-24} g/cm^3. Oort finds that the luminous stars in the solar neighborhood account for 0.038 solar mass per cubic parsec. Hence, the difference, or 0.05 solar mass per cubic parsec, must be due to dark stars and to diffuse matter in interstellar space. This corresponds to a density of 3×10^{-24} g/cm^3—the famous Oort limit for the density of interstellar matter.

This leads to an estimate of the upper limit for the radius r of the particles. If the absorption were due to obscuration we would have for the absorption coefficient $n\pi r^2$, where n is the number of particles per cubic centimeter. With Trumpler's value of the absorption we write

$$2.5 \log_{10} e^{-n\pi r^2 l} = 0.7,$$ (12)

where l is equal to 3×10^{21} cm. Let the density of each particle be $d = 5$ g/cm^3. Then Oort's limit

$$3 \times 10^{-24} \text{ g/cm}^3 = n \frac{4}{3} \pi r^3 d. \qquad (13)$$

Combining (12) and (13), we find

$$r = 10^{-2} \text{ cm.}$$

This is an upper limit for the average radius of the particles and not an estimate of the sizes of those particles which are mostly responsible for the scattering of light.

INTERSTELLAR GASES

In 1904 Hartmann found that the absorption lines of Ca II in the spectrum of the double star δ Orionis fail to take part in the periodic oscillations of the other lines. He eliminated the obvious explanation that the stationary Ca II lines come from a very massive secondary component of the binary system and concluded that "at some point in space in the line of sight between the sun and δ Orionis there is a cloud which produces that absorption, and which recedes with a velocity of 16 km/sec." Since that time other atoms and molecules have been found to originate in interstellar space. Among them are Na I, Ca I, K I, Ti II, CN, and CH. Na I was found many years ago by Miss Heger at the Lick Observatory; Ca I, K I, and Ti II were found by Adams and Dunham; CH was measured by Dunham in 1937 and was identified by Swings and Rosenfeld; CN is due to Adams and McKellar. There are several unidentified absorption lines from interstellar matter. Among them are sharp lines at λλ 3957 and 4233.[3] These lines are very prominent in several B-type stars, for example ζ Ophiuchi, and are clearly of interstellar origin. There are also a number of diffuse absorption lines, or bands, discovered by Merrill and by Beals and Blanchet. They are as yet unidentified, but there is evidence that their intensities are closely correlated with color excess, so that they are perhaps produced by the dust particles. The wave lengths of these broad lines are λλ 4430, 5780, 5796, 6284, and 6614.

The remarkable thing about the identified lines is that they all originate from the lowest level of each atom. Even if the level is multiple, as in Ti II, only those numbers of the resonance multiplet appear which come from the lowest sublevel. Forbidden transitions depopu-

---

[3] The symbols λ and λλ are used as abbreviations for "wave length" and "wave lengths," respectively; these wave lengths are expressed in Ångstrom units.—EDITOR.

late the higher levels to such an extent that absorption lines originating in them are not seen.

One of the substances which might be expected to produce interstellar absorption lines is Fe I. Swings and I searched for it in the fall of 1939 on our spectrograms of early-type stars. The lowest level is a^5D and there are several lines of different multiplets which arise from it. We searched in B-type stars for the strongest expected lines, $\lambda\lambda$ 3719.94, 3440.63, and 3859.92. The first is in the wing of the hydrogen line H_{14} 3721.94. A search on Process plates taken with the quartz spectrograph of the McDonald Observatory for several stars in which interstellar Ca II is strong, for example ξ Persei, χ Aurigae, χ^2 Orionis, α Camelopardi, strongly suggests that λ 3720 is present. But the line is weak and we are not prepared to make a positive identification with the material now available to us.

The intensities and radial velocities of the interstellar lines of Ca II and Na I have been measured in many stars. There is a pronounced correlation between distance and intensity, first found by the present writer, which has served to give us an independent method for finding the distances of hot stars. The method has been discussed by a number of astronomers; some believe that the density of interstellar calcium or sodium is not sufficiently uniform throughout space to give reliable information regarding the distances. Dunham has even announced from measurements in α Virginis (distance 53 parsecs) and η Ursae Majoris (distance 66 parsecs) that "there is probably a region of lower than average density close to the sun." Dunham also thought that there was a difference in the average density of 2.4×10^{-10} ionized calcium atoms in the direction of α Virginis and that of 5.2×10^{-11} atoms in the direction of η Ursae Majoris. This latter conclusion depends upon the adopted distances of the stars, which may be somewhat in error, as Morgan has recently demonstrated. On the whole, it may be said that distances derived from the intensities of interstellar lines are fairly reliable, provided a correction is applied to take care of the concentration of the gas toward the galactic plane.

There is also a tolerably good correlation between line intensity and color excess. But each of these quantities is much better correlated with distance, so that we are certain that condensations of reddening particles are not necessarily accompanied by increased densities of interstellar atoms.

The radial velocities of the interstellar lines show (a) a relatively small tendency toward peculiar motions and (b) a conspicuous relation with galactic rotation, the line of sight component of the latter being exactly one-half of its value derived from the corresponding stars. In view of the small peculiar motions the galactic rotation effect may be used for deriving the distances of the stars. However, there are numerous cases of stars showing double interstellar lines. These have been explained as being caused by two separate clouds of atoms, each having its own motion.

Whenever the star lines are broad and diffuse the attribution of a sharp line to interstellar matter is quite unambiguous. But when the star lines are sharp the distinction is not always easy, except in spectroscopic binaries. There has been in recent years too much of a tendency to take it for granted that all sharp Ca II lines in spectral types B3 and earlier are interstellar. I have recently measured the Ca II line λ 3933 in the luminous B1 star β Canis Majoris and have found on Texas Coudé plates that the line shares the oscillation of the star lines. Morgan has remarked that especially in supergiant stars the stellar calcium lines may sometimes persist as far as B0.

A problem of great interest and one that has not yet been completely solved is that of the line contours of the interstellar atomic lines. These contours are deep and become appreciably broadened in the more distant stars. It was at first suggested, by Unsöld, Struve, and Elvey, that the broadening may be caused by the effect of galactic rotation, and Eddington attempted to explain the relative intensities of Ca II to Na I as a result of the corresponding curves of growth. However, observations, principally by Merrill, Wilson, Sanford, and others at Mount Wilson, have shown that there are no striking differences in line contours of distant stars located (in galactic longitude) near the nodes and near the maxima or minima of the curve of galactic rotation. Clearly, near the nodes the broadening can not be due to galactic rotation, and in the absence of other Doppler effects, it must be due to the natural widening of the line by radiation damping. It is probable that a small effect of this kind does exist, but there can be no doubt that the lines in the two groups of stars are much more nearly alike than the theory predicts. Hence it may be regarded as certain that the contour is determined largely by turbulence, which masks the galactic rotation broadening except, perhaps, in the most distant objects. The average turbulent velocities have not been accurately determined, but they are probably of the order of 10 to 20 km/sec, in the line of sight.

THE IONIZATION PARADOX

In 1926 Eddington investigated the ionization of the interstellar gas. Since it is reasonable to suppose that the gas is in a steady state, we write for the equilibrium of Ca I and Ca II:

$$\text{No. of ionizations} = \text{No. of recombinations} \tag{14}$$

or

$$\frac{\text{No. of bullets of quanta}}{\text{disrupting atoms}} = \frac{\text{No. of ions colliding with electrons}}{\text{and forming atoms.}}$$

This is equivalent to:

$$\frac{N_1 \times \text{No. of quanta of}}{\text{appropriate power}} = \frac{N_2 N_e \times \text{function of cross sections for col-}}{\text{lisions and of velocity or temperature.}} \tag{14a}$$

In other words,

$$N_1\phi(\text{Ioniz. pot.}, T) = N_2 N_e f(T, \sigma)$$

or

$$\frac{N_2}{N_1} N_e = F\ (\text{Ion. Pot.}, T).$$

This is in effect Saha's ionization equation. In thermodynamic equilibrium it is

$$\log \frac{N_2}{N_1} N_e = -\chi \frac{5040}{T} + 1.5 \log T + 15.38 + \log \frac{2u_2}{u_1} \tag{15}$$

where χ is the ionization potential and u_2, u_1 are the statistical weights of the ground states of the ion and the atom, respectively. In the interstellar gas the radiation is very greatly diluted, and its spectral composition corresponds to the integrated effect of all the stars. In (14) or (14a) the left side is proportional to the density of the radiation and this is equal to Planck's function for the appropriate average temperature and to the dilution factor W, which measures the departure from thermodynamic equilibrium. W is equal to the ratio of the available density of radiation to that which would be available if for the appropriate temperature T there existed thermodynamic equilibrium. The right side of (14) or (14a) is not altered. We can, therefore, write the ionization equation for interstellar matter in the following manner:

$$\frac{N_2}{N_1} N_e = W \times F\ (\text{Ion. Pot.}, T),$$

or

$$\log \frac{N_2}{N_1} N_e = -\chi \frac{5040}{T} + 1.5 \log T + 15.38 + \log \frac{2u_2}{u_1} + \log W. \tag{16}$$

This, however, presupposes that the distribution of velocities of the electrons in the gas is Maxwellian and corresponds to the same temperature T, as the one which results from the summation of the energy curves of all stars. This will not, in general, be the case. A more elaborate treatment for an electron temperature T_e gives

$$\log \frac{N_2}{N_1} N_e = - \chi \frac{5040}{T} + 1.5 \, \log \, T + 0.5 \, \log \, \frac{T_e}{T} + \log W. \quad (17)$$

Finally, in some applications of the theory it is necessary to allow for the gradual diminution of the ionizing radiation by absorption in the gas. This introduces an extra factor of $e^{-\tau}$ in the left side of (14) and (14a), so that our final expression is

$$\log \frac{N_2}{N_1} N_e = - \chi \frac{5040}{T} + 1.5 \log T + 0.5 \log \frac{T_e}{T} + \log W + \log e^{-\tau}. \quad (18)$$

All logarithms are to the base of 10. The dilution factor can be determined from the observed distribution of the stars. Eddington proceeds in the following manner. He determines the appropriate temperature by summing the energy curves of all stars in the approximate wave length range where the ionization of calcium is produced. This turns out to be about 15,000°. In thermodynamic equilibrium Planck's formula gives an energy density of 387 erg/cm³ for 15,000°. In interstellar space conditions are different. We know that the sun radiates 3.8×10^{33} erg/sec and that a star of bolometric absolute magnitude 1.0 radiates 36 times as much, or 1.4×10^{35} erg/sec. By definition at a distance of 10 parsecs the absolute magnitude of a star is equal to its apparent magnitude. If we spread the energy radiated per second by our first magnitude star over a sphere of 10 parsecs in radius we have

$$\frac{1.4 \times 10^{35}}{4\pi \times 100 \times 9 \times 10^{36}} = 10^{-5} \text{ erg/cm}^2 \text{ sec.}$$

The energy density is obtained by dividing this by c. But the total of all star light is equivalent to 2,000 stars of the first apparent magnitude. Hence the radiation density in interstellar space is

$$\frac{2000 \times 10^{-5}}{3 \times 10^{10}} = 7 \times 10^{-13} \text{ erg/cm}^3.$$

The dilution factor is

$$W = \frac{7 \times 10^{-13}}{387} = 2 \times 10^{-15}.$$

More accurate values have been derived by Gerasimovič and Struve, by Greenstein, and by Dunham.

We can now proceed and compute N_2/N_1 for Ca II/Ca I by substituting numerical values into (18). The question is what to use for N_e. Eddington determines a very rough value for the density of interstellar matter by assuming that the gaseous nebulae are condensations of the interstellar medium and that the theory of isothermal gas spheres may be applied to these objects. This gives $\rho = 10^{-24}$ g/cm^3. Since the electron pressure is not very different from the total gas pressure this estimate gives N_e and hence also N_2/N_1.

Gerasimovič and Struve proceeded in a different manner. In the absence of any information regarding N_e they assumed that all elements are equally effective in producing electrons in interstellar space, so that N_e could be determined from the equation

$$pN_e = N_2(\text{Ca}^+) + 2N_3(\text{Ca}^{++}), \tag{19}$$

where $p = 0.015$ is the percentage abundance of calcium by atoms in the crust of the earth. This equation led to a very small value of $N_e = 10^{-3}$, which in turn gave a very high degree of ionization. The ionization of calcium and sodium derived by Gerasimovič and Struve is shown in table 2.

TABLE 2.—IONIZATION OF CALCIUM AND SODIUM
($N_e = 10^{-3}$)

Calcium		Sodium	
Ca............	3×10^{-11}	Na........	2×10^{-7}
Ca$^+$..........	7×10^{-5}	Na$^+$......	1
Ca^{++}..........	1	Na^{++}.....	6×10^{-3}
Ca^{+++}........	2×10^{-8}	—	—

Since the observed line intensities of Na I and Ca II are very similar, the table suggests that in cosmic clouds sodium atoms must be 300 times more abundant than atoms of calcium. On the earth the abundance of sodium is only about 1.3 times that of calcium. This ratio is probably true of other cosmic sources, and it is strange that in interstellar space the abundance of sodium should be several hundred times greater. This result is similar to that of Eddington. It formed a serious barrier to further work.

SOLUTION OF THE PARADOX

In 1926, when Eddington's work was published, and in 1929, when Gerasimovič and Struve published their computations, there was no reason to doubt that calcium was a very abundant substance. The fact that calcium and sodium were the only elements then known in the spectrum of interstellar matter, combined with the great intensity of the calcium lines in the solar chromosphere and in prominences, led quite naturally to the idea of "calcium clouds," which were believed to consist largely of calcium. The idea advanced by Gerasimovič and Struve that the abundance of calcium might be only 1 or 2 percent was revolutionary in 1929.

However, later work on the composition of the sun by Russell a short time afterward established beyond doubt the tremendous preponderance of hydrogen over all other elements. Hence it became necessary to review the problem of the ionization of interstellar matter with the idea that hydrogen might supply the overwhelming majority of the free electrons.

The basis of the new discussion consists of Dunham's recent measurements of the intensities of the lines of Ca I 4226 and Ca II 3933 in several stars. From these intensities it is possible to derive the numbers of atoms per cm^2 of Ca$^+$ and of Ca, and thus determine by observation the ratio N_2/N_1. This was done independently by Struve and by Dunham. The ionization equation then leads directly to a determination of N_e. Struve finds 30; Dunham obtains values of 14.4 and 7.3, depending upon which of two assumptions he uses for the integrated ultraviolet radiation of all stars. Within the past few days we have received a communication from Bates and Massey, of Northern Ireland, who have corrected the ionization formula (18) for the fact that the ionization processes take place almost exclusively from the ground level of the Ca atom, while in recombination many atoms first find themselves in higher energy states, from which they rapidly cascade downward to the ground level. This correction effectively reduces the ionization so that the same observed ratio N_2/N_1 can now occur at a lower electron density N_e. The factor is approximately equal to 6. Taking Dunham's upper value, 14.4, we find:

$$N_e = 2.4 \text{ cm}^{-3}. \tag{20}$$

We shall see in the next section that there is a good confirmation of this value from observations of interstellar hydrogen emission lines. The uncorrected value of N_e is sufficient, as I pointed out in 1939, to greatly improve the Na/Ca paradox. Using this N_e, and adjusting W and T for the required ionization potentials, we compute:

$$\left.\begin{array}{ll}
\dfrac{N_2(\text{Ca}^+)}{N_1(\text{Ca})} = 100 \quad \text{and} & \dfrac{N_3(\text{Ca}^{++})}{N_2(\text{Ca}^+)} = 10 \\[3ex]
\dfrac{N_2(\text{Na}^+)}{N_1(\text{Na})} = 270, & \dfrac{N_3(\text{Na}^{++})}{N_2(\text{Na}^+)} = 0
\end{array}\right\}. \tag{21}$$

From the equivalent widths of Ca K and Na D and the corresponding distances, Wilson and Merrill found

$$N_1(\text{Na}) = 3 \times 10^{-9} \text{ cm}^{-3}, \tag{22}$$

while for Ca K, Merrill and Sanford found

$$N_2(\text{Ca}^+) = 9 \times 10^{-9} \text{ cm}^{-3}. \tag{23}$$

Combining (21) with (22) and (23) we obtain the total numbers of sodium and calcium atoms:

$$\left.\begin{array}{l} N_1(\text{Na}) + N_2(\text{Na}^+) = 8 \times 10^{-7} \text{ cm}^{-3} \\ N_1(\text{Ca}) + N_2(\text{Ca}^+) + N_3(\text{Ca}^{++}) = 1.2 \times 10^{-7} \text{ cm}^{-3} \end{array}\right\}. \tag{24}$$

Sodium is about 7 times as abundant as calcium. Dunham finds a larger ratio,

$$\frac{\text{Na}}{\text{Ca}} = 25, \tag{25}$$

but even this is very much better than the discrepancy of 300 found in the earlier work. Strömgren suggests that the small remaining inconsistency of (25) with terrestrial data will disappear when the correction of Bates and Massey is applied to Na, as well as to Ca. Since the theory is difficult, expecially in not being able to give us very reliable values of T and W, we may consider the paradox of the ionizations as having been eliminated. Moreover, we must remember that the determinations in (22) and (23) rest upon the curves of growth for Ca and Na, and these are still incompletely known. Finally, the distances of the B stars are not accurately known.

INTERSTELLAR HYDROGEN EMISSION

We have already expressed the suspicion that the large value of N_e may come from the ionization of interstellar hydrogen. Consider the ionization of hydrogen. If we disregard the factor $e^{-\tau}$ in (18) and set $T_e = T$, we find with the appropriate values at $\lambda = 900$ A, viz. $T = 25,000°$, and $W = 10^{-17}$

$$\frac{N_2(\text{H}^+)}{N_1(\text{H})} \approx 10. \tag{26}$$

Since W and T are uncertain, this value is only a rough approximation.

The only lines of hydrogen which are accessible to observation are those of the Balmer series. Let us see how many H atoms we must have in order to account for the value of N_e which we have derived from the calcium absorption lines. For simplicity let us use not the final result (20) but a value which is close to that derived by Dunham, namely

$$N_e = 10 \text{ cm}^{-3}.$$

Evidently, since $N_e = N_2(\mathrm{H}^+)$ we conclude from (26) that we have in space one neutral atom of H per cm³. The question arises whether there is any chance of observing the bright H line in interstellar space.

We first compute the numbers of atoms in the third quantum level. Since this level is not metastable the population is proportional to W, so that

$$\frac{n_3}{n_1} = W e^{-(h\nu/kT)} \tag{27}$$

With $W = 10^{-17}$ and $e^{-h\nu/kT} = 10^{-2}$ we have

$$n_3 = 10^{-19} \text{ cm}^{-3}. \tag{28}$$

Consider now the depths of space. Our Milky Way extends to tens of thousands of parsecs from the sun, but in effect the more distant regions are cut off by obscuring nebulosities. We can make only a rough guess as to the effective thickness of the visible layer:

$$D = 1,000 \text{ light years} = 10^{21} \text{ cm}. \tag{29}$$

Hence we shall have

$$n_3 D = 10^2 \text{ atoms/cm}^2. \tag{30}$$

The question arises: Can the emission of 100 hydrogen atoms per cm² be observed? The energy emitted by the layer is

$$n_3 D h\nu A_{31}$$

where $h\nu = 0.5 \times 10^{-11}$, $A_{31} = 0.5 \times 10^8$, and $n_3 D = 100$. This is

$$n_3 D h\nu A_{31} \approx 10^{-2} \text{ erg/cm}^2 \text{ sec}.$$

This is considerably brighter than the limit of human vision for surface brightnesses. The latter is roughly equivalent to one star of magnitude 9 per square degree. Since the apparent magnitude of the sun is -26, and since it covers about one-fifth of a square degree this means that we can detect surface brightnesses which are 35 magnitudes or about 10^{14} times as faint. The total radiation of the sun is about 10^{11} erg/cm² sec. Our eyes are sensitive to only a fraction of this, let us say to $\frac{1}{5}$. Hence the minimum surface brightness we can see is

$$\frac{2 \times 10^{10}}{10^{14}} = 2 \times 10^{-4} \text{ erg/cm}^2 \text{ sec.}$$

This is a fiftieth as faint as the expected brightness of the interstellar hydrogen emission. There is all reason to search for it. This was done a few years ago at the McDonald Observatory, with a specially constructed nebular spectrograph of great efficiency. The instrument consists of a narrow plane mirror, which acts as the slit. The light from the sky is reflected by the slit-mirror along the direction of the polar axis. At a distance of 75 feet it is intercepted by a plane mirror and returned to the prism box and f/1 Schmidt camera. Each spectrum is accompanied by a comparison spectrum from some other part of the sky.

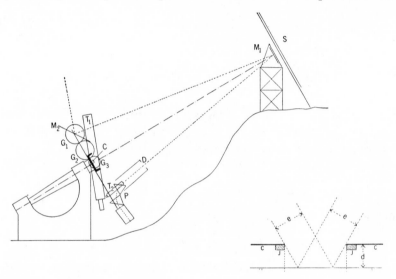

Fig. 3.—Nebular spectrograph of the McDonald Observatory. M_2 = narrow plane mirror acting as slit; M_1 = stationary plane mirror; P = prisms over camera; T = guiding telescope.

The results of the observations numbering nearly 80 long exposures show that—

(1) There are large regions in the Milky Way where H, [O II] 3727, [N II] 6548, 6484, and occasionally [O III] N_1 and N_2 appear in emission.

(2) These nebulae show but little concentration toward individual early-type stars, thereby differing conspicuously from ordinary gaseous nebulae.

(3) There is no emission at high galactic latitudes.

(4) The emission regions of H in Cygnus, Cepheus, and Monoceros are fairly sharply bounded on the outside. They are roughly circular

Fig. 4.—Nebular spectrograph of the McDonald Observatory.

in appearance. The intensity drops somewhat from the inner parts of each region toward the outer boundary, but this variation is relatively small.

(5) Whenever [O III] is seen in emission, it occurs in the central parts of the H regions, without any sharp boundaries.

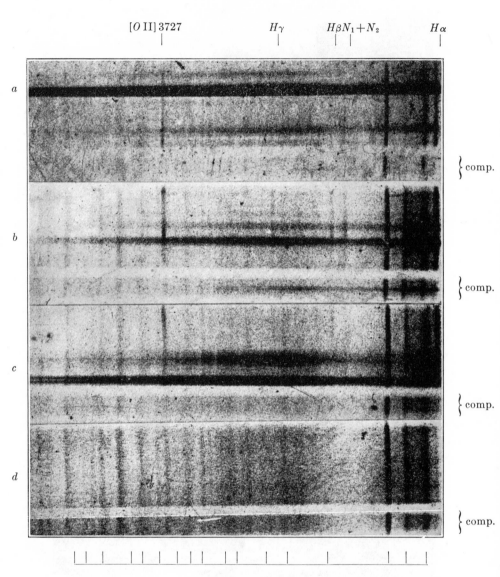

Fig. 5.—Spectra of Milky Way regions showing emission of H, [O II] and [O III] superimposed over spectrum of night sky.

Auroral Lines

(a) Guiding star: 56° 2604. Nebula IC 1396
(b) Guiding star: 60° 504. Loose cluster
(c) Guiding star: 59° 559. Loose cluster
(d) Guided 6′ south of 29° 741

(6) The H regions are probably associated with groups of O-type stars. This is demonstrated in Figs. 7 and 8 where the spectroscopic results are shown at the top, the distribution of the O stars at the bottom. The emission regions are shown by solid circles. Absence of emission is denoted by open circles.

Fig. 6.—Milky Way in Cygnus. A large part of this area covering nearly 400 square degrees shows emission lines of H and [O II].

(7) The ratio in intensity [O II]/(H+[N II]) is large in Monoceros and Canis Major and small in Cygnus, Cepheus, and Sagittarius. This effect demonstrates a conspicuous difference between the physical conditions of the emission regions in two different parts of the sky.

(8) Some of the emission regions show a slightly milky background on the direct photographs of Ross and Barnard, but there are other regions of similar milky appearance which shine by reflected light.

(9) The hydrogen emission regions sometimes cover dark markings while in other instances the emission does not seem to extend over the

dark nebulosities. In the Taurus region there is no emission in the dark clouds, except in the immediate vicinity of bright O and early B stars. In the Ophiuchus nebulosities emission is seen only near the B1 star σ Scorpii. The other bright stars, of type B3, produce only reflection nebulae.

(10) Slipher reports that in the region $\alpha = 18^h\ 8^m$, $\delta = -18°\ 16'$ the emission of H covers not only the star cloud, but also the dark marking B92.

The sizes of the emission regions are of the order of $s_0 = 7°$ for the one around λ Orionis and $s_0 = 5°$ for the one in Cygnus. Assuming reasonable absolute magnitudes for the associated O stars we find the following linear dimensions:

Region	Radius
near λ Orionis	40 parsecs
Cygnus	130 parsecs
Monoceros	85 parsecs

The intensities of the observed hydrogen lines may be used to compute the number of atoms per cm² in the third energy level. In principle this is simple and can be done by reversing the procedure of the first part of this section. In practice there are many difficulties because we are dealing with an extremely faint light source. A fairly good average result from two independent series of observations is

$$n_3 = 5\ \text{cm}^{-2}.$$

Assuming $s_0 = 300$ parsecs (for which we shall see the justification later)

$$N_3 = \frac{5}{3 \times 10^2 \times 3 \times 10^{18}} = 5 \times 10^{-21}\ \text{cm}^{-3}. \tag{31}$$

We next apply the formula

$$\frac{N_3}{N_1} = \frac{g_3}{g_1} W e^{-(h\nu/kT)},$$

which gives

$$N_1 = 3 \times 10^{-2}\ \text{cm}^{-3}.$$

Allowing for the ionization, as in (26):

$$\frac{N_2(H^+)}{N_1(H)} = 6,$$

we have

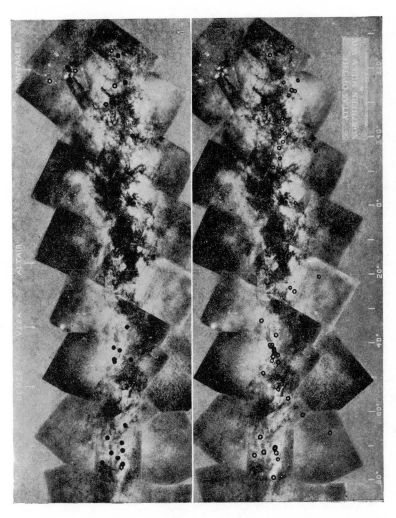

283

Fig. 7.—Summer Milky Way. At the top, regions showing emission of H and [O II] are plotted as solid circles. At the bottom, the known stars of spectral type O are plotted as open circles. The composite map was made by Miss Mary R. Calvert from the Atlas of the Milky Way by F. E. Ross and Mary R. Calvert.

284

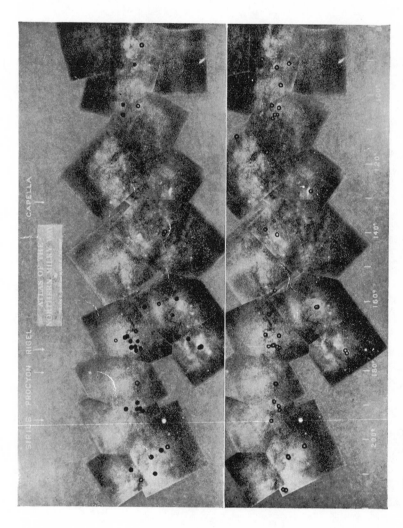

Fig. 8.—Winter Milky Way. At the top, regions showing emission of H and [O II] are plotted as solid circles; regions without emission are plotted as open circles. At the bottom the known stars of spectral type O are plotted as open circles.

$$N_2(H^+) = 0.2 \text{ cm}^{-3}.$$

But we should not have used the three-state problem. Evidently several other H levels will contain approximately the same populations as the third level. In fact, the second level may even be overpopulated. Strömgren has allowed for this effect accurately, but we shall simply estimate that the error due to the neglect of other excited levels corresponds to a factor of 10. Then the total number of hydrogen atoms is:

$$N_2 \approx N_e \approx N(H) = 2 \text{ cm}^{-3}.$$

From his detailed analysis which is based upon my and Elvey's observational data Strömgren finds

$$N(H) = N_e = 3 \text{ cm}^{-3}. \tag{32}$$

This agrees closely with the value derived from calcium.

STRÖMGREN'S THEORY

The significant element of this theory is the retention of the term $e^{-\tau}$ in the ionization formula (18):

$$\frac{N_2(H^+)}{N_1(H)} N_e = C \times \frac{1}{s^2} e^{-\tau}, \tag{33}$$

because $W = R^2/4s^2$. The constant C involves the ionization potential of H, the temperatures T and T_e and the radius of the exciting star R. Since presumably almost all free electrons came from H

$$N_e = N_2(H^+). \tag{34}$$

The optical depth τ is measured near the limit of the Lyman series:

$$d\tau = N_1(H)a_u ds, \tag{35}$$

where a_u is the continuous absorption coefficient near λ 900 per neutral hydrogen atom. This quantity is known. In our ionization equation

$$\frac{[N_2(H^+)]^2}{N_1(H)} = C \times \frac{1}{s^2} e^{-\tau} \tag{36}$$

the ionization decreases with increasing distance from the ionizing star. Since a_u is small this decrease is at first mainly caused by $1/s^2$, because for small s, the factor $e^{-\tau}$ is close to 1. But as the distance increases $e^{-\tau}$ becomes more important. Because of the exponential it causes a very abrupt change in ionization, producing a sharp boundary beyond which H is almost completely neutral and inside of which it is almost entirely ionized. The radius of this boundary, s_0, can be computed. It depends upon the temperature and radius of the ionizing star and upon the density of hydrogen, $N(H) = 3$ cm^{-3}. Physically what happens is this: "In the immediate neighborhood of a star, interstellar hydrogen will be ionized. With increasing distance from the star the proportion of neutral hydrogen atoms increases, and hence the absorption of the ionizing radiation increases. Ultimately, the ionizing radiation is so much reduced that the interstellar hydrogen is un-ionized." Table 3 gives the quantity $N_2/N_1 + N_2$ for several values of the ratio s/s_0.

TABLE 3.—IONIZATION OF HYDROGEN AS A FUNCTION OF DISTANCE

s/s_0	$\dfrac{N_2}{N_1 + N_2}$
0.00	1.00
0.58	1.00
0.74	0.99
0.84	0.98
0.93	0.96
0.97	0.94
1.00	0.85
1.03	0.33

The very rapid decrease of $N_2/N_1 + N_2$ near $s/s_0 = 1$ is of fundamental importance. It shows that each star is surrounded by a sharply limited volume of space where H is almost wholly ionized, while outside of this volume H is un-ionized.

Table 4 gives Strömgren's computations for the radius s_0 of the ionized hydrogen region, as a function of spectral type. The computation is based upon values of the radii corresponding to main-sequence stars.

TABLE 4.—RADII OF REGIONS OF IONIZED HYDROGEN

Spectrum	T	Vis. Abs. Mag.	s_0 in parsecs	s_0 in parsecs for main sequence and $N=3$ cm^{-3}
	$^\circ$			
O5	79,000	−4.2	$54 \times R^{2/3} N^{-2/3}$	67
O6	63,000	−4.1	40	52
O7	50,000	−4.0	29	41
O8	40,000	−3.9	20	31
O9	32,000	−3.6	13	22
B0	25,000	−3.1	7.2	12
B1	23,000	−2.5	5.6	8
B2	20,000	−1.8	4.2	5
B3	18,600	−1.2	3.1	3
B4	17,000	−1.0	2.2	2
B5	15,500	−0.8	1.6	1.8
A0	10,700	+0.9	1.	0.2

The table illustrates the great preponderance of the hot stars in creating volumes of ionized hydrogen. In the regions of ionized hydrogen the Balmer lines will be excited by several mechanisms and will give rise to emission lines. Strömgren identifies these regions with the extended areas of hydrogen emission found in the McDonald Observatory survey of the Milky Way. In these areas O stars are abundant. For a group of n stars the radius s_0 increases as $n^{1/3}$, so that from Table 4 we infer that a group of 10 or 15 late O stars would create an ionized sphere of about 300 parsecs in diameter. This is of same order of size as the observed region in Cygnus (p. 248). In the un-ionized regions nearly all the radiation beyond the Lyman limit, and also that in the Lyman lines L_β, L_γ, etc. has been converted into L_α and low-frequency lines. There will be almost no excitation of the third and higher levels of hydrogen, but the second level may be superexcited by a factor of 10^3.

It is of interest to consider the manner in which the radius of the ionized region depends upon the various quantities involved. In this connection it is important to remember that the discussion applies only to an element which furnishes the vast majority of free electrons. In the ionized regions this is hydrogen, and within these regions the ionization of elements of higher ionization potential is not influenced by a term of the form $e^{-\tau}$. Strömgren derives for s_0, when $T_e = T$:

$$\log_{10} s_0 = -5.85 - \tfrac{1}{3} \log a_u - \tfrac{1}{3}\theta I + \tfrac{1}{2} \log T + \tfrac{2}{3} \log R - \tfrac{2}{3} \log N,$$

where

$$I = \text{ionization potential}, \quad \theta = \frac{5040}{T}, \quad a_u = 6.3 \times 10^{-18}\ \text{cm}^{-2}.$$

But

$$\log R = \frac{5900}{T} - 0.20\ M_v - 0.02.$$

Hence

$$\log_{10} s_0 = -5.85 - \tfrac{1}{3} \log a_u + (3900 - 1660\ I)\,\frac{1}{T} + \tfrac{1}{2} \log T - 0.13\ M_v - \tfrac{2}{3} \log N.$$

Clearly s_0 is very sensitive to T, because of the term $(3900 - 1660 I)$ $1/T$. Except when I is small, $1660\ I/T$ predominates. For example,

when $I = 10$ volts,	$T = 8,300$;	$s_0 = 1$
" $I = 10$ "	$T = 16,600$	$s_0 = 10$
" $I = 50$ "	$T = 8,300$	$s_0 = 1$
" $I = 50$ "	$T = 16,600$	$s_0 = 10^5$

The conclusion is that when I is large hot stars alone are important.

The effect of M_v is relatively slight, because of the coefficient 0.13. For a main-sequence star and a supergiant of class A we have

$$\text{main sequence} \quad A0: M = +0.9$$
$$cA0: M = -6.0$$

The value of $\log_{10} s_0$ changes by $0.13 \times 6.9 = 0.9$. If for a main-sequence A0 star Table 4 gives $s_0 = 0.2$ parsec, the supergiant will give $s_0 = 16$ parsecs. This is of the order of size required for α Cygni which may be responsible for the luminosity of the North America nebula and the surrounding emission region. The distance is about 200 parsecs and the apparent radius $s_0 = 4°$. This gives $s_0 = 14$ parsecs, which is close to the value derived from Strömgren's theory.

In the case of γ Cygni, which is of type F8 and which is believed to be responsible for some faint hydrogen emission showing a symmetric arrangement around this star, the agreement is poor. The distance is about 140 parsecs and $s_0 = 4°$. Hence the radius should be $s_0 = 10$ parsecs. But the temperature is much lower than that of α Cygni. It is difficult to see how γ Cygni can be responsible for the emission region unless it radiates at $\lambda 900$ like a late B star. The importance of supergiants may become appreciable in the case of early class B.

Let us next apply the theory to two relatively close O stars, γ Velorum and ζ Puppis. For these stars

$$M_v \approx -4$$

and

$$D \approx 200 \text{ parsecs.}$$

For type O8 Table 4 gives $s_0 = 31$ parsecs. Hence we are well outside the ionized regions surrounding these two stars. Observations confirm this: there is no emission at high galactic latitudes.

EMISSION OF FORBIDDEN OXYGEN

The McDonald Observatory results show the following:

(1) [O II] is nearly always present in the regions of ionized H.

(2) There is a relative strengthening of [O II] in the winter Milky Way and a weakening in the summer Milky Way.

(3) The regions of [O II] coincide with those of H and have the same sharp boundaries.

(4) [O III] is rarely observed, but when it does occur it is limited to the inner parts of the H regions, and it shows no sharp boundaries.

(5) [O I] was never observed, although a special effort was made to distinguish it from the strong auroral lines in the spectrum of the night sky.

These results are in good agreement with Strömgren's theory. Since N_e depends on H, the decrease of [O III]-intensity is caused by the term $1/s^2$. The same is true for [O II]; at the boundary of H the ionization of O will also suddenly stop. Since all exciting stellar radiation is cut off at s_0, elements of ionization potential higher than that of H, 13.54 volts, are un-ionized. This applies to the following:

O	I.P. = 13.56 volts	
O^+	34.94	
O^{++}	54.88	
He	24.48	
C^+	24.28	
N	14.49	

Elements of lower ionization potential may be ionized in the regions where H is neutral. This applies to

C	I.P. = 11.22 volts	
Ca	6.09	
Ca^+	11.82	
Na	5.11	

Carbon is probably the most abundant of these elements and it must be the source of the free electrons in these regions. Allowing for the low abundance of C with respect to H, we estimate that in the non-hydrogen regions N_e is between 10^{-2} and 10^{-3} cm^{-3}. It will be recalled that this is very similar to the value originally inferred by Gerasi-movič and Struve for interstellar space. Hence conditions of ionization for Ca and Na in the nonhydrogen regions must be approximately those which they had derived. This means that Ca is nearly all doubly ionized, so that if we observe a star through a series of hydrogen and nonhydrogen regions, it is the former that give most of the absorption within the stationary lines of Ca II.

Na I is even more reduced in the nonhydrogen regions and Ca I originates almost entirely in the regions of ionized hydrogen.

It is not at once obvious why [O I] is not observed outside the regions of ionized hydrogen. It is certainly not ionized, because the ionization potential of O almost coincides with that of H. The forbidden lines of [O I] are also weak in planetaries. There we could make the plausible assumption that the gas ceases to exist in the ring in which we should otherwise expect [O I] to be strong. But this does not help in the case of interstellar space. Provisionally it seems possible that because of the low electron density in the nonhydrogen

regions there are not enough collisions to excite the metastable levels
of O I. Since the high levels are probably not appreciably excited by
radiation and since there is no ionization, recombination and cascading can not help to populate the metastable levels. We are compelled
to assume that excitations by electron collisions are not sufficiently
numerous to produce the forbidden lines.

It is of some interest to compute the abundance of O from the observed intensities of the line λ3727. The observational data give us
n_2, the number of atoms per cm² in the upper, metastable level:

$$\log n_2 = 11.97.$$

This is very large compared to hydrogen, because of the metastability
of the term. We can now apply two methods of reasoning in order to
determine n_1: (1) excitation by pure radiation and (2) excitation by
collisions.

Method (1) depends upon the formula for the three-state problem,
with the second level assumed to be metastable:

$$\frac{n_2}{n_1} = W \frac{A_{32}}{2A_{21}} \rho_{13}$$

where

$$\rho_{13} = e^{-(h\nu_{13}/kT)}.$$

Using $T = 25{,}000°$, $W = 10^{-17}$ we have $\rho_{13} = 10^{-4}$. Since $A_{32} = 10^8$ sec⁻¹;
$A_{21} = 2.4 \times 10^{-5}$ sec⁻¹, we have

$$n_2/n_1 = 2 \times 10^{-9}.$$

$$n_1 = 5 \times 10^{20}.$$

If $s_0 = 300$ parsecs $= 10^{21}$ cm, we find

$$N_1 \approx 10^{-1} \text{ cm}^{-3}.$$

Since nearly all O is ionized we infer that

$$N(O) = 10^{-1} \text{ cm}^{-3}.$$

Method (2) makes use of the equation

$$\frac{n_2}{n_1} = \frac{N_e a_{21}}{A_{21} + N_e a_{21}} e^{-(h\nu/kT)},$$

where a_{21} is the collisional probability:

$$a_{21} = \sigma^2 2 \left(\frac{2\pi kT}{m}\right)^{1/2} = 3 \times 10^{-10}$$

For the low metastable level

$$e^{-(h\nu/kT)} = 0.4.$$

If we assume $N_e = 1$ cm^{-3}, we get approximately

$$\frac{n_2}{n_1} = \frac{N_2}{N_1} = 5 \times 10^{-6}.$$

Since $n_2 = 10^{12}$ we find

$$n_1 = 2 \times 10^{17}.$$

Adopting again $s_0 = 10^{21}$ cm, we have

$$N_1 = 2 \times 10^{-4}.$$

Evidently, the collisional mechanism is much more efficient than the radiation mechanism. The abundance of O may be somewhere between the two limits. Strömgren has independently estimated from the same observational data that the interstellar abundance of oxygen is 10^{-2} or 10^{-3} atoms to one atom of hydrogen. The method upon which these computations are based is very rough, but as a preliminary result we shall adopt

$$N(O) = 10^{-3} \text{ cm}^{-3}.$$

The heterogeneity of the Milky Way in regard to the relative intensities of [O II] and H leads to interesting speculations. It may be due to real differences in abundance. But it may also be due to different conditions of excitation.

INTERSTELLAR EMISSION OR ABSORPTION

In the past astronomers have sometimes been searching for interstellar emission lines of Ca II and Na I, and it is of some interest to explain why these lines are observed in absorption, while H is observed in emission. Let us compare the emission of H and Ca II. From the absorption intensities we know that

$$N_2(\text{Ca}^+) = 10^{-8} \text{ cm}^{-3}.$$

The total emissions in H$_\alpha$ and Ca K are

$$4\pi E_\alpha = n_3 h\nu_{32} A_{32} \text{ for H}$$
$$4\pi E_K = n_K h\nu_K A_K \text{ for Ca}^+$$

Since we are only concerned here with orders of magnitude we may put

$$h\nu_{32} = h\nu_K$$

$$A_{32} = A_K$$

Accordingly,

$$\frac{E_\alpha}{E_K} = \frac{n_3}{n_K} = \frac{5}{n_K}$$

because $n_3 = 5$ cm^{-2} (page 248). But

$$n_K = 10^{-8} \times 10^{21} \times W \frac{g_K}{g_1} e^{-(h\nu/kT)}.$$

For $T = 15,000°$ and $W = 10^{-16}$,

$$\frac{E_\alpha}{E_K} \approx 10^3 \text{ or } 10^4,$$

so that the emission of Ca K can not be observed.

In a similar manner we can show that no interstellar absorption lines of H are expected in the spectra of distant stars, for example in a Nova which provides an emission background on which an interstellar line could easily be seen. The absence of any such line in Nova Lacertae, at a distance of about 900 parsecs, shows that

$$n_2 < 10^{12} \text{ cm}^{-2},$$

because if it were stronger our spectrograms would show it. We also know that for the third level, which is not metastable,

$$n_3 = n_1 W e^{-(h\nu_{13}/kT)} = 5,$$

while for the second, which is metastable:

$$n_2 = n_1 W \frac{A_{31}}{A_{21}} e^{-(h\nu_{13}/kT)},$$

Since $A_{31} = 10^8$ sec $^{-1}$ the inequality becomes

$$\frac{1}{A_{21}} = \tau_2 < 2 \times 10^3 \text{ sec.}$$

The question arises whether this is in accord with the theory. Breit and Teller find that in the absence of collisions the mean life of the $2s$ state of hydrogen is about 1/7 second. Hence the Balmer absorption lines should not be observable.

COMPOSITION OF INTERSTELLAR GAS

Table 5 summarizes the results. I have combined the results of Dunham and Strömgren with my own, without making an attempt to avoid slight inconsistencies. In my opinion the theory, as well as the observations, permits only a very rough orientation. In Dunham's work the atoms refer to the space between the earth and χ^2 Orionis; those for the molecules represent the means for several stars.

TABLE 5.—COMPOSITION OF INTERSTELLAR GAS
LOGARITHMS OF NUMBERS OF PARTICLES PER CM^3

Element	This paper	Dunham	Sun	Nebulae
Electrons.....	0.2	1		
Hydrogen	0.2	1	0	0
Oxygen......	−3		−1.5	−2
Sodium......	−6	−4	−3.3	−4
Potassium....		−5	−3.7	−5
Calcium......	−7	−5	−3.8	−4
Titanium.....		−7	−5.3	−4
CH..........		−6		
CN..........		−6		

For comparison the table gives also the relative abundances in the sun from Russell and in the nebula NGC 7027 from Bowen and Wyse. These values were adjusted for one hydrogen atom. The discrepancies between the results of this paper and Dunham for Ca and Na are attributable to differences between the determinations of Dunham and of Merrill, Wilson, and Sanford. However, there are real differences in different parts of the sky. Dunham found for the neighborhood of the sun a density of 10^{-10} Ca ions per cm^3; for the space between the earth and χ^2 Orionis he finds 10^{-7}, while Merrill and Sanford find 10^{-8} for the average of many stars. There are also serious difficulties with the curve of growth for interstellar absorption lines. Accordingly, we need feel no concern about the differences. It seems that the composition of the gas is similar to that of the sun and of the nebulae.

It seems to me that the most important task now is to study in detail the heterogeneity of the galaxy. We have definite observational indications that the relative intensities of different atoms are different in various parts of the sky. It is surprising, for example, that some of the newer interstellar lines are strong in ζ Ophiuchi. Adams has shown that λ 4232.6 and CH 4300.3 are relatively strong in this star. But Ca K is not particularly strong. The star is of very early B type (Morgan finds it may even be an O) and probably creates a moderate volume of ionized H. But it lies quite far from the Milky Way and from other O stars. The data now available are not sufficient for a detailed study, and we can only give a few hints as to the topics such a study might cover:

(1) The relative weakness of K and the strength of molecular lines in ζ Ophiuchi, far from the other O-type stars and from the Milky Way, suggest that the ionized volume is small and that molecular lines are strengthened in the un-ionized regions. We have as yet no accurate theory of the equilibrium of CH and NH and it is difficult to predict the outcome. Swings has made some computations and has shown that certain diatomic molecules, like CH and NH, must be quite frequent. The ionization of H may well impede the formation of the molecules, and it is quite possible that they will be relatively more numerous in the un-ionized regions.

(2) Adams states that Ca I 4227 does not appear in ζ Ophiuchi. This agrees with the prediction of Strömgren's theory (page 249) for a relatively un-ionized region. This is also true of the correlation between Ca I and Ca II, noticed by Adams.

(3) Because of the fact that interstellar Ca II and Na I originate principally in the ionized H regions, we should expect that there would be appreciable departures from the simple one-half relation in the galactic rotation term, which was discovered by Oort, and confirmed by Struve and by Plaskett and Pearce. However, since the regions are large it is important that the comparison of radial velocities be confined to those stars and their Ca II lines which are principally responsible for the creation of a volume of ionized H.

(4) In order to aid in the study of the heterogeneity of the galaxy, more material is required with the nebular spectrographs.

(5) An application of Strömgren's theory to diffuse gaseous nebulae, such as the Orion nebula, and to planetaries, should be made, with special attention to the theoretical and observational study of absorption lines, such as the lines of He I discovered by O. C. Wilson.

(6) A regional study of interstellar Ti II should be relatively easy. From the theory we should expect a behavior which is essentially similar to that of Ca II and Na I.

(7) The observational problem of determining N_3 for interstellar H has not been solved with sufficient precision.

(8) A study of the effects of stellar lines in early B stars and the determination of good spectroscopic parallaxes for these stars will improve our knowledge of N_2 (Ca^+) and N_1 (Na).

(9) The nature of the contours of the interstellar lines must be cleared up by extending the determination of line contours to very distant stars located near the nodes and near the maxima or minima of the curve of galactic rotation as a function of galactic longitude.

INTERSTELLAR GAS

GUIDO MÜNCH

California Institute of Technology and
Mount Wilson and Palomar Observatories

I. HISTORICAL INTRODUCTION

The interest of Otto Struve in problems related to the interstellar medium probably originated when, as a graduate student, he was measuring spectrograms of B-type stars taken at the Yerkes Observatory. His determination of radial velocities for the "stationary" Ca II lines, as they were then called, and his estimates of their strength provided the basis for his first studies on the subject, which appeared summarized in 1925 (Struve 1925a, 1925b). These papers analyzed the various correlations existing between the intensities and radial velocities of the Ca II lines, and the positions, distances, and spectral types of the stars in which they were observed. Somewhat earlier J. S. Plaskett (1924a, 1924b) had advanced the hypothesis that the narrow Ca II lines observed in early-type stars were produced by "vast clouds of calcium which surround the stars and which are excited and ionized by some sort of intensive radiation from the stars," a suggestion similar to those made independently by Ludendorff (1920) and Henroteau (1921). In his summarizing articles, Struve pointed out a number of observed facts which could not be accounted for in terms of Plaskett's hypothesis, such as: (1) the stationary Ca II lines are very faint or entirely absent in many stars at high galactic latitude, but they do occur in some such stars; (2) their radial velocities do not show the K-effect and individually are considerably smaller than those of the B-type stars; (3) these individual velocities are not distributed at random, but indicate group motion for certain well-defined regions of the sky; (4) these regions are associated with some of the more distinct Milky Way star clouds or with regions of greatest star density; (5) the intensities of the stationary lines are high in certain regions of the sky and low in others; (6) the stars which have the strongest stationary lines are appreciably redder than those in which the lines are faint. It is of importance to realize that Struve arrived at these conclusions on the basis of observational material severely limited by quality and resolving power, if judged by the standards reached fifteen years later with the large coudé spectrographs. Nevertheless, every one of the points he raised became, in time, a building block for the concepts held today about the structure of the interstellar medium.

In his Bakerian Lecture, Eddington (1926) formulated an alternative hypothesis to Plaskett's, *i.e.*, that the stationary lines were truly interstellar in nature, originating in gaseous masses unrelated to the stars in which they were observed. The analysis of Eddington fundamentally changed the course of further advancement on the subject. The paper published by Struve (1926) following the appearance of Eddington's lecture considers the alternative hypotheses in the light of the observed facts. He found that the data related to the velocities of the lines, their intensities, and regional variations, supported the interstellar nature of the clouds. On the other hand, a decrease in the mean intensities of the lines for stars at distances larger than about 600 pc, which appeared in his data, was considered by Struve to be an objection to Eddington's hypothesis, although the reality of the effect was not well established. The strongest argument raised by Struve against Eddington's hypothesis was based on the "sudden disappearance of the detached lines beyond spectral type B3." In this respect Struve was misled by insufficient resolving power and the limited material available on distant stars of late B-type. On the basis of the observations then available, Struve considered that it was not possible to accept definitely either of the two hypotheses. In the next paper on the subject published by Struve (1928), only two years later, it appears that his thoughts had become more like those expressed by Eddington. This was chiefly the result of further observations, which removed the apparent decrease of the K-line intensities for stellar distances larger than 600 pc, and the only obstacle considered by Struve still remaining in the way of Eddington's hypothesis was the nonoccurrence of detached lines in spectral types later than B3, but he expressed the view that additional observations of faint stars would "doubtless settle this point before long." It is of some interest to recall further that Struve in this paper (1928) raised as a strong argument in favor of the hypothesis of interstellar calcium the observation of the K-line in the spectra of novae.

A last paper on the same subject published by Struve in collaboration with Gerasimovich (1929) deserves special comment, because it contains an attempt "to derive from the observational data all possible information concerning the density and distribution of diffuse matter in interstellar space and to investigate the resulting physical properties of this hypothetical gaseous substratum of our galaxy." In this paper the problem of the relative interstellar abundance of sodium and calcium is primarily considered. Since at the time there was no information regarding the interstellar density of electrons, which is needed to evaluate the degree of ionization from a Saha equation as modified by a geometrical dilution factor, Struve and Gerasimovich assumed that all elements were equally effective in producing interstellar electrons. From the terrestrial abundance of calcium they obtained thus an electron density of 10^{-3} cm^{-3}, implying a very high degree of ionization for both calcium and sodium. Based on the observed fact that the intensities of the strong Ca II and Na I lines are nearly equal, they were forced then to infer that the interstellar abundance of sodium was about 300 times larger than that of calcium. A similar result had been obtained earlier by Eddington. In the light of current knowledge regarding the cosmic abundance of elements, and the origin of the interstellar electrons in regions where hydrogen is neutral, we realize today that the value of the electron density adopted by Struve and Gerasimovich was nearly the correct one, but for the wrong reasons. It happened that the relative cosmic abundance of calcium with respect to carbon, the element which provides most of the interstellar electrons when hydrogen is neutral, is about the same as that of calcium in the earth's crust. The Na/Ca abundance paradox thus raised was very real and, to a small extent, has not been completely explained up to this date, as will be discussed later.

The discovery by Dunham (1937) of the Ca I resonance lines as an interstellar absorption feature led Struve to reconsider the Na/Ca paradox, since the existence of Ca I to a detectable extent required an electron density higher than had been previously assumed. The question to be decided was whether most of the electrons were provided by hydrogen or by the less abundant elements. In the former case the Balmer lines would appear in emission. On the basis of approximate calculations, Struve (1939) showed that if the electrons derived from hydrogen, the recombination line spectrum of hydrogen should be detectable over a path length of about 300 pc. In order to search for the interstellar hydrogen emission, a spectrograph of great efficiency was designed and constructed at the McDonald Observatory, and with this instrument the extended emission regions around groups of O- and B-type stars were discovered. These observations had a transcendental importance for the development of the ideas that led to our present understanding of the nature of the interstellar medium. Most important of all, they established that the abundance of interstellar hydrogen relatively to other elements is as high as that found in the stars. They also provided the basis for the theoretical analysis of the state of ionization in interstellar space carried out by Strömgren (1939).

The number of papers published by Struve on the subjects related to interstellar matter sharply declined after the publication of the summarizing paper reproduced in this volume (Struve 1941). When the 82-inch telescope of the McDonald Observatory became operational, the attention of Struve was directed mainly to problems of stellar astronomy. His interest in problems related to the interstellar medium remained, nevertheless, undiminished. The tasks remaining to be done in the field, outlined with prophetic vision in the closing paragraphs of that summarizing paper, had become an objective of the most powerful telescopes. When, in 1951, this writer left the Yerkes Observatory to join the staff of the Mount Wilson and Palomar Observatories, Struve commented on and emphasized the importance of following further the work on interstellar absorption lines with the highest spectral resolutions possible. He also expressed to this writer his opinion regarding the value of high-dispersion observation of interstellar emission lines. To some extent, thus, the advice of Otto Struve provided the starting point for some of the work to be described in the following sections.

II. THE CLOUD STRUCTURE OF THE INTERSTELLAR MEDIUM

Structure of the interstellar absorption lines.—The observation of 300 O- and B-type stars with the coudé spectrograph of the 100-inch telescope enabled Adams (1949) to show that the interstellar Ca II lines are generally complex, in the sense that they are formed by the superposition of a number of narrow components. Among the stars observed by Adams, 87 have lines that clearly can be measured as double, 17 as triple, and 4 as quadruple, and in many of the remaining stars the lines appear sufficiently broad as to indicate almost certain complexity. The resolving power of Adams material is about 0.06 Å, but because of photographic effects and the shape of individual components, only components separated by no less than 0.1 Å appear effectively resolved. There is every indication to show that as the resolving power is increased the complexity of the lines increases markedly. For example in ε Ori the Ca II lines were found by Adams (1949) to have components at $+1.2$, $+10.6$, and $+26.5$ km/sec, on plates with 2.9 Å/mm dispersion. Plates at dispersion 1.1 Å/mm later obtained at Mt. Wilson show components at $+3.0$, $+11.8$, $+18.3$, $+25.5$, and $+28.2$ km/sec (see Fig. 2). Under these conditions it is natural to inquire how far the resolving power has to be increased in order that the true structure of the interstellar lines can be observed.

The observation of the interstellar D lines by Livingston and Lynds (1964) with the Czerny spectrograph of the McMath solar telescope at the Kitt Peak National Observatory, using an image intensifier tube, has shown in a striking fashion how the lines become more and more complex as the spectral dispersion is increased. A spectrum of α Cyg in the first order of the grating, with a resolution of around 0.05 Å, shows only one broad component separated from a narrower one by about 0.23 Å. In the fourth and fifth-order spectra, the broad component appears clearly resolved into four components, while a fainter one is suspected between the two noticed at the lower dispersion (see Fig. 1). In the spectra with the highest resolution, it appears that the intrinsic profiles of the lines are no longer seriously affected by the instrumental resolution.

Observations of interstellar absorption lines with extremely high spectral resolving powers have also been carried out by interferometric techniques. Using a stack of three Fabry–Perot pressure-scanned etalons, Hobbs (1965) has studied the interstellar D_2 line in the spectrum of α Cyg, with the 120-inch reflector of the Lick Observatory. The effective resolving power reached with this instrument is close to 500,000. In addition to the features reported by Livingston and Lynds (1964), Hobbs finds an apparent doubling of the longest and shortest wavelength components of the entire line complex with a spacing close to that expected from hyperfine structure, namely

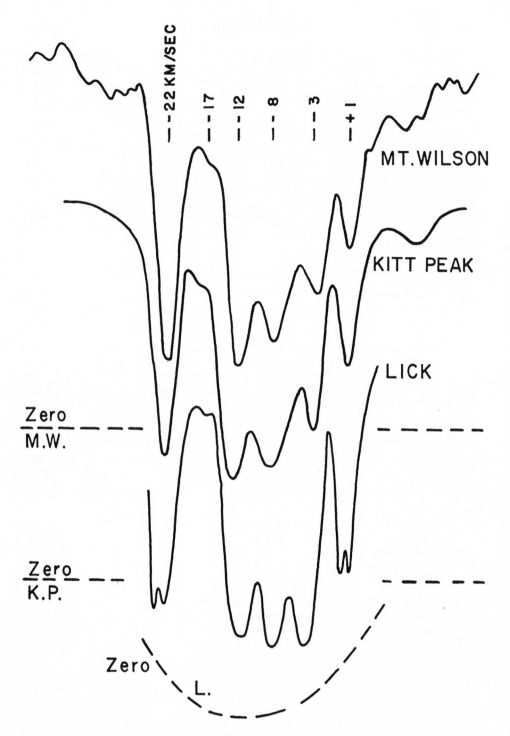

Fig. 1.—High resolution profiles of the interstellar D_2-line in α Cygni obtained with diverse instruments. The relative intensities of the various components present in the tracing obtained with the PEPSIOS spectrometer (bottom) do not agree with those observed by means of an image tube at Kitt Peak (middle) or interferometrically at Mt. Wilson (top). The origin of the doubling shown by the two strongest lines in the PEPSIOS tracing is not understood yet.

0.020 Å. It is not altogether clear whether hyperfine structure in the D lines has actually been detected, because of the apparent relative intensity of the two components and the saturation effects that would be implied. The possibility of an instrumental effect arising from imperfect alignment of the Fabry-Perot etalons cannot yet be discounted. It appears in any case that the internal turbulent velocities of individual interstellar clouds may be lower than 1 km/sec.

Interferometric observations of interstellar lines in a number of stars have been carried out recently also by Münch and Vaughan with a single Fabry-Perot etalon, with the coudé spectrograph of the 100-inch telescope used as pre-disperser. The resolving power reached is about 300,000 with an entrance aperture of 0.5 mm, which is sufficiently large to admit all the light of a star under good seeing conditions. A comparison of the results obtained by this technique with those of conventional photographic methods indicates an overall gain in efficiency by a factor of ten in favor of the interferometer. The linearity of the photomultiplier detectors makes possible the measurement of the true intensities of interstellar lines in stars brighter than visual magnitude 7.0 in times short enough to make practicable an extended survey. As an illustration of the results obtained with this instrument, the profile of the D_2 line in α Cyg, together with the profile obtained by Hobbs (1965) with the PEPSIOS spectrometer, and a microphotometer tracing of a plate obtained with the image intensifier tube at Kitt Peak by Livingston and Lynds (1964) is reproduced in Figure 1. The interstellar D lines in two other stars, obtained at Mount Wilson with the Fabry-Perot scanner are shown in Figure 2.

On the basis of the results being obtained with these instruments, it appears that the resolution of the interstellar lines into components will no longer be limited by the finite instrumental resolving power, but by the intrinsic width of the lines. Evidently, the observation of a few dozen stars at the highest resolution will be most informative.

The Discrete Cloud Model.—The existence of discrete components in the interstellar absorption lines raises the question of how large are the masses or clouds producing them and how often do they occur in space. Attempts to delineate in space the boundaries of the region where a particular component arises have been unsuccessful, because it is in practice impossible to obtain observations in stars sufficiently close to each other. Under these conditions the structure of the interstellar medium as a whole is described in a statistical sense. The discrete cloud model, originally introduced by Ambartsumian and collaborators (Ambartsumian 1940, Ambartsumian and Gordeladse 1938, Markarian 1946) to account for observed properties of the extinction produced by the interstellar solid particles, has proved to be useful also to characterize the distribution of the interstellar gas. In this model it is assumed that the interstellar clouds are identical, simply connected masses distributed at random in the galactic plane, with peculiar velocities given by some probability density function Φ(V). Over restricted regions near the large complexes of interstellar matter, as the Orion Nebula or the Taurus dark cloud, the assumption of randomness is questionable. On the other hand, over regions with large linear dimensions—say greater than 1 kpc— the structural features of the galactic system or spiral arms become apparent (Münch 1957) and the randomness in the distribution of the interstellar clouds disappears. The discrete cloud model, thus, applies only to regions within a spiral arm, with a few hundred pc extension, and it is characterized by the following parameters:

(1) The mean number of clouds ν (kpc^{-1}) intersected per unit distance by a line of sight in the galactic plane. The probability of intersecting n clouds in a distance r (pc) is then given by Poisson's law

$$P_n(r) = (\nu r)^n \frac{e^{-\nu r}}{n!} . \tag{1}$$

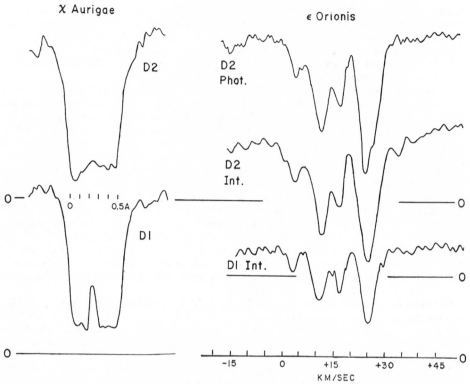

Fig. 2.—Profiles of the D lines obtained with the Fabry-Perot photoelectric scanner at the Mt. Wilson coudé with a resolution of 0.0206 Å. The profiles in χ Aurigae ($m_V = 4.9$) represent counts of 700 pulses per spectral element, in intervals of about 50 seconds, with a total of 40 minutes per line. The profile of the D_2 line in ε Orionis ($m_V = 1.7$) (top right) has been obtained photographically with a dispersion of 1.6 Å/mm, with an exposure of 2 hours. Each of the interferometric profiles (middle and bottom right) has been scanned in about 10 minutes.

From the statistics of the separation of the interstellar Ca II components, Blaauw (1952) finds that ν is around 8 kpc^{-1}. The mean number of clouds ν per unit distance can be also estimated from observational data related to the interstellar solid particles. From the statistics of color excesses of early-type stars, for example, Münch has obtained $\nu = 7$ kpc^{-1}, a value nearly the same as that indicated by the interstellar line components. It is suggestive to suppose, then, that the clouds producing the interstellar reddening are identical with the clouds where the interstellar absorption line components originate. Clearly a more detailed proof of the identity of the two kinds of clouds was highly desirable, especially since some doubts have been cast in this respect (Evans 1941, Schildt 1947). To obtain such a proof was not an easy task, however, because interstellar lines are strongly saturated while color excesses are not. A thorough analysis of the correlation between the color excesses E_1 in the system of Stebbins, Huffer, and Whitford (1940) and the intensities of interstellar D lines, carried out by Spitzer (1948), has shown that the clouds producing the interstellar extinction in a statistical sense are indeed identical to those producing the interstellar absorption lines. The color excess produced by the average cloud was found by Spitzer to be $E_1 = 0.024$ mag.

(2) The equivalent width of the absorption line of a certain ion or atom produced by a typical cloud. In principle this quantity can be related to the total number of atoms or mass per unit cross section in the line of sight and, depending on an assumed cloud geometry, the total mass can be introduced as characteristic parameter. According to Spitzer (1948) the average intensity of the absorption D lines produced by the typical cloud is $(D_1 + D_2)/2 = 0.15$ Å. The corresponding total mass of a cloud with radius R, evaluated immediately below, is near 10 solar masses.

(3) The number density of clouds in the galactic plane, N_c, or the characteristic "radius" of a cloud defined by

$$\nu = \pi R^2 N_c.$$ (2)

From the statistics of reflection nebulae illuminated by stars of various luminosities, Ambartsumian and Gordeladse (1938) have estimated that $N_c = 1.2 \times 10^{-3}$. With $\nu = 8$ kpc^{-1}, equation (2) gives $R = 5$ pc. This is a lower limit, since only a small part of a cloud need lie within the predicted zone of illumination about a given star to be seen as a reflection nebula. Oort and van de Hulst (1946), on the other hand, have pointed out that only 14 percent of all stars earlier than B2 are associated with emission nebulae. If the distribution of early-type stars and that of interstellar matter are independent of each other (an assumption which may be open to question) it would follow that the fractional volume of space occupied by clouds is $4\pi R^3 N_c/3 = 0.14$, and the resulting cloud radius is again about 5 pc. It should be emphasized that these characteristic values for the mean spatial density and the radii of clouds are subject to large uncertainty, depending on the method followed to evaluate them. These uncertainties arise from idealizing all clouds to be of the same dimensions. There are some observational indications that the slow-moving clouds may be more massive than those forming well-separated optical components to which the statistics of interstellar clouds mainly refer. Observations of the 21-cm hydrogen line in absorption have provided values somewhat larger than those given above. For example, Clark, Radhakrishnan, and Wilson (1962) have determined a mean diameter of 13 pc for four clouds, which at the measured temperature of 60°K provide densities of 20 H atoms cm^{-3} and masses around 600 ⊙. The dense cloud in front of χ^2 Orionis has, according to Strömgren (1948), a density of 60 atoms cm^{-3} and a diameter of 15 pc, or a total mass of 2600 ⊙.

III. KINEMATICS OF THE INTERSTELLAR CLOUDS

The motion of the interstellar clouds in the discrete cloud model is described by a probability distribution function $\Phi(V)$ of the peculiar velocities V of the clouds. From statistics of the measured radial velocities, however, it is not possible to derive $\Phi(V)$ in a direct fashion. The effect of galactic rotation on the mean radial velocities of the interstellar lines is well established, and for distances not larger than about 1 kpc from the sun the radial component of the rotational velocities is nearly that of a point at half the distance of the star. This means only that the individual components forming the entire line are randomly distributed, but obviously there is no way of determining the distance to the particular cloud giving rise to certain components. The identification of the measured radial velocities, reduced to some fixed standard of rest, with the peculiar motions will lead, as a consequence of this uncertainty, to a broadening of the apparent velocity distribution of the observed components. In principle, by grouping the stars in regions of galactic longitude near the nulls and the maxima of galactic rotation the broadening effect could be established. It is not possible to do this at present, however, because of limitations in the size of the samples available. A statistical study of the observations by Adams (1949) has been carried out by Blaauw (1952) under the further assumption that the velocity distribution function is isotropic. This assumption is actually inconsistent with the observations, which show an obvious asymmetry between the numbers of observed high-velocity components of approach and of recession. Irrespectively of how the interstellar clouds are initially set in motion, moreover, we would expect an ellipsoidal distribution of velocity because of the asymmetry of the gravitational potential. The latter effect is small, as indicated by the peculiar motions of the early-type stars. The asymmetry of the high velocities is more conspicuous, as it is discussed below, but again, because of our inability to determine the position in space of individual clouds and of sampling limitations, it can not be accounted for in the derivation of $\Phi(V)$. The analysis by Blaauw considers that all clouds produce lines of the same strength, and on the basis of an assumed Gaussian form

$$\Phi(V) = \frac{1}{\sigma(2\pi)^{1/2}} e^{-V^2/2\sigma^2}, \tag{3}$$

or an exponential form

$$\Phi(V) = \frac{1}{2\eta} e^{-|V|/\eta}, \tag{4}$$

a fit to the observed frequency distribution of velocity components, in stars grouped according to distance, is reached by selecting values of ν and the velocity dispersion σ or $\eta\sqrt{2}$. For stars at a distance of 900 pc, Blaauw obtains $\sigma = 8.2$ km/sec and ν between 8 and 12 kpc^{-1}, while for distances up to 300 pc, $\eta = 5.0$ km/sec. For a Gaussian law the value of σ is between 8 and 10 km/sec. It appears that an exponential law like equation (4) provides a better fit to the observed frequency functions, although in either case a considerable excess of high velocities is observed. It is of importance to remark that the velocity dispersion of the interstellar clouds, as indicated by the profiles of the 21-cm emission line of hydrogen, agrees with the optical results.

The asymmetry in the numbers n_+ and n_- of the high velocity clouds of either sign present in Adams' material is very obvious and was noticed by Blaauw and by Schlüter, Schmidt, and Stumpff (1953). Table 1, taken from the latter authors, shows

TABLE 1

OBSERVED DISTRIBUTION OF INTERSTELLAR CLOUD VELOCITIES

Radial velocity (km/sec)	n_+	n_-	Radial velocity (km/sec)	n_+	n_-
0–10	143	164	30–70	7	10
10–20	21	37	>70	1	2
20–30	4	30			

the observed distribution after allowance for solar motion and for the mean galactic rotation. The preponderance of negative values at intermediate velocities is very well marked, and it is not the result of insufficient statistics. An explanation of this phenomenon has been offered by Oort and Spitzer (1955) on the basis of their theory for the origin of the motions present in the interstellar medium. According to these authors, the interstellar H I regions are accelerated by thermal effects following the sudden formation of an O-type star and its associated region of ionized hydrogen. The direction of their motions is thus pointed away from the O-B associations, where the O- and B-type stars are predominantly found. Since many of the stars observed by Adams belong to such associations, and also because of the apparent arrangement of emission regions within 1 kpc from the sun, we observe most frequently the interstellar clouds moving toward the sun.

IV. THE TEMPERATURE OF THE INTERSTELLAR CLOUDS

State of ionization.—It was first pointed out by Strömgren (1939) that the physical state of the interstellar matter is largely determined by the degree of ionization of hydrogen, because of the large abundance of this element. The photon flux emitted by a star beyond the Lyman limit (UV flux) determines the number of interstellar hydrogen atoms around the star which can be ionized. If the optical depth of the medium in the Lyman continuum is large, the thickness of the transition region between the ionized and neutral zones is of the order of the mean free path of a UV quantum in the neutral region, and accordingly is small compared with the dimensions of the ionized region. In the steady state, the number of ionizations or the flux of UV quanta F_{uv} emitted by the star is equal to the number of recombinations which give rise to radiation that may escape from the ionized region. If N_e and $N(H^+)$ are the number densities of free electrons and protons, respectively, and α_B is the recombination coefficient of hydrogen in all levels but the ground state, the volume $\mathscr{V}$ of the ionized region is given by

$$F_{\text{uv}} = \int_{(\mathscr{V})} N_e N(H^+) \alpha_B(T_e) \, d\mathscr{V}. \tag{5}$$

In general, the recombination coefficient depends on the electron temperature, T_e, which is not necessarily the same as the temperature, T_s, which characterizes the stellar radiation. On the assumption of constant density and temperature, for a star radiating like a black body of radius R_s, we obtain from equation (5) the radius $R(H^+)$ of the ionized region in the form

$$\frac{8\pi^2 k^3 T_s^{\;3}}{h^3 c^2} R_s^{\;2} F(y) = \tfrac{4}{3}\pi R^3(H^+) N_e N(H^+) \alpha_B(T_e), \tag{6}$$

where

$$y = \frac{I_H}{kT_s} \quad \text{and} \quad F(y) = \int_y^\infty \frac{x^2 \, dx}{e^x - 1} \tag{7}$$

and I_H is the ionization energy of hydrogen. In this manner Seaton (1960a) obtains the numerical relation

$$R(H^+) = 78[N_e N(H^+)]^{-1/3} \left(\frac{R_s}{R_\odot}\right)^{2/3} \frac{F^{1/3}(y)}{y} \text{ [pc]}, \tag{8}$$

where the value $\alpha_B = 2.58 \times 10^{-14} \text{ cm}^3 \text{ sec}^{-1}$, corresponding to $T_e = 10^4 \,^\circ K$, has been adopted. The values of $R(H^+)$ obtained from equation (8) agree closely with those derived by Strömgren (1948) by a different procedure.

The temperature of H II regions.—The kinetic temperature in the ionized region is determined by the conservation of electron kinetic energy. The net gain, G_{ep}, per electron is the difference between the average energy of a photoelectron and the average energy of a recombining electron. If the star is assumed to radiate as a black body, the net gain G_{ep} will depend on T_s and T_e. The energy loss, L_{ei}, occurs mainly by collisional excitation of abundant ions with low-lying levels, from which forbidden-line emission takes place. It can be calculated on basis of the abundances A_i of the various ions and the cross sections for the collisional excitation of their low-lying levels. The value of the electron temperature can then be obtained from the condition

$$G_{ep}(T_s, T_e) = L_{ei}(T_e, A_i). \tag{9}$$

When Spitzer (1949) made this calculation, only the forbidden-line emission in the photographic region (intersystem lines of O^+, O^{++}, N^+) was included. Actually the fine structure in the ground states of ions as Ne^+, O^{++}, Ne^{++}, etc., also radiate appreciably as pointed out by Burbidge, Gould, and Pottasch (1963). A reexamination by Osterbrock (1965) of the thermal balance of H II regions has shown that the predicted nebular temperatures are in the range 7000°–9000°, somewhat below the values obtained directly from the observations.

Temperatures in H II regions can be measured from the relative intensities of forbidden lines of the same ion arising from levels of different excitation energy. The intensity ratio between the line of O^{++} at $\lambda 4363$ ($^1S \rightarrow {}^1D$) and the N_1 and N_2 lines ($^1D \rightarrow {}^3P$) is a convenient indicator of temperature, because it is nearly independent of density. For a galactic emission region the temperature obtained by this method is close to $T_e = 9500°$, depending of course on the values adopted for the excitation cross sections. The correctness of this value is confirmed by a direct measurement of the thermal Doppler widths of the emission lines in the nebulae, as is described in a later section.

The temperature of H I regions.—In the regions where hydrogen is essentially neutral, the radiation field beyond 912 Å is missing and only elements with ionization energies less than 13.6 ev can provide free electrons. Among these, carbon has the highest abundance, namely around 3×10^{-4} with respect to hydrogen, and its photoionization provides the main energy input. The energy of the photoelectrons, as in the case of the H II regions, is partially lost in the excitation of low-lying levels of ions, among which the most abundant is C^+. For a given kinetic temperature, then, the energy losses are not drastically reduced with respect to those in an H II region, but the energy gains are cut down by the ratio between the carbon and hydrogen abundances. It follows that the balance between the photoionization energy gains and the losses by ionic emission is established at much lower temperatures than in H II regions, at around 50°K. This result is fairly independent of the density. If H_2 molecules exist in H I regions, as seems very likely, the collisional excitation of rotational levels followed by quadrupole radiative decay also provides an important energy loss at somewhat higher temperatures. The solid particles present in H I regions contribute an additional source of cooling, because they must radiate as black bodies and in the steady state their temperature T_g should be lower than T_e. Atoms colliding with the solid grains, then, will lose their kinetic energy and evaporate at a rate determined by T_g. Observationally the only means for measuring the kinetic temperature of H I regions is through emission profiles of the 21-cm line of hydrogen, which provides a mean value around 125°K, considerably in excess of the predicted value. It follows that the gas in H I regions must be gaining energy through processes other than photoionization by stellar radiation.

It has been argued by Kahn (1955) that collisions among H I clouds provide the dominant energy input. With the numbers given before as characteristic of the discrete cloud model, the mean lifetime of a cloud between collisions is $\tau_c = 10^7$ years. During a collision the interacting parts of the clouds will heat up to about 1300°K, and this heat will be radiated in a time of the order of magnitude of τ_c. The average energy input to the medium deriving from this mechanism amounts to 5×10^{-4} erg gm^{-1} sec^{-1}, according to Kahn, in comparison with the input 1.7×10^{-6} erg gm^{-1} sec^{-1} provided by photoionization to a cloud of density $N(H) = 10$ cm^{-3}. Calculations have been made by Kahn to show that the mean temperature over the time of a cloud heated by collisions in this fashion would be the observed value of 125°K, if only one percent of the hydrogen is in molecular form.

Encounters of cosmic rays with interstellar atoms will also produce ionization and provide kinetic energy. A hydrogen atom has a photoionization cross section around 10^{-19} cm^2 for a 100 Mev proton, and the photoelectron acquires an energy of about 50 ev. If the flux of primary cosmic rays in interstellar space is equal to that estimated for the neighborhood of the earth, namely 0.5 cm^{-2} sec^{-1}, the energy input to a cloud of density $N(H) = 10$ cm^{-3} would be only 2.4×10^{-7} erg gm^{-1} sec^{-1}, considerably smaller than the heat input due to collisions and to photoionization. However, the flux of cosmic ray particles with energy below the cut-off set by the solar magnetic field could be high enough to become the dominant source of heating (Spitzer 1949). In conclusion we can say that at present the energy balance in the interstellar H I regions cannot be established with any precision, the outstanding questions in this regard being the concentration of molecular hydrogen, the dissipation of kinetic energy of mass motion, and the density of low-energy cosmic rays. It appears fairly well established that mechanisms other than photoionization by starlight and ionic cooling are playing an important role in the thermal balance.

V. THE INTERNAL MOTIONS OF H II REGIONS

In the development of our understanding of the nature of emission nebulae, observations of the Orion Nebula have played a prominent role, although because of its high density, surface brightness, and extent, it hardly can be considered as typical. The first observations of the Orion Nebula with high spectral resolving power were carried out by Fabry and Buisson (1911) and Buisson, Fabry, and Bourget (1914) for the purpose of determining the widths of the then-unidentified nebular lines. These authors measured the widths of the hydrogen and nebular lines, essentially with what is known today as a Fabry–Perot interferometer, and from them they inferred that the mean atomic weight of the element producing the nebular lines had to be around 3, if the widths resulted only from thermal Doppler motions. These earlier observations showed also that variations in the wavelength of the lines existed, indicating differential mass motions between various parts of the Nebula. The work of Campbell and Moore (1918) extended our knowledge of the internal motions of the Orion Nebula very considerably, when they determined the radial velocities of the [O III] and Hβ lines in about 125 points on the Nebula. A further major effort in the subject was carried out by Baade, Goos, and Minkowski (1933), who employed a Fabry–Perot etalon crossed with a small quartz spectrograph. Through this work it became clear that the broadening of the emission lines in the Orion Nebula is not set purely by thermal motions, but that there are also mass motions with linear scale extending to the limits set by the angular resolving power of the observations.

Interpretations of these early observations of the Orion Nebula were advanced by Minnaert (1948) and von Hörner (1951). In a general way, it could be said that the aim of these authors was to relate the radial velocity variations across the Nebula with the non-thermal component of the line widths, through a geometrical model of the Nebula. The more thorough analysis of von Hörner attempted to describe the state of motion in the Nebula in terms of the equilibrium theory of homogenous and isotropic turbulence for an incompressible fluid, as proposed by Kolmogoroff (1941) and von Weizsäcker (1946). On the basis of this theory it would be expected that if V_1 and V_2 are the velocity components of mass motion along a fixed direction at two points in the fluid separated by a distance r, then the root-mean-square velocity difference, averaged over all points separated by the same distance r, satisfies a relation of the form

$$\langle (V_1 - V_2)^2 \rangle_{\text{av}}^{1/2} = cr^{1/3} \qquad (c = \text{const.}), \qquad (10)$$

for values of r within a range than can be defined from the nature of the problem. On the basis of the Lick measures (Campbell and Moore 1918), von Hörner actually found that a relation of the form (10) was satisfied in the Orion Nebula for $r < r_c$, with the distance r_c being of the order of one-tenth the linear dimensions of the bright part of the Nebula in the plane of the sky. If the turbulence is isotropic and homogeneous, however, the line widths expected from a nebula with thickness of the order of r_c are considerably smaller than the observed ones. With the purpose of finding out the origin of this discrepancy, Wilson, Münch, Flather, and Coffeen (1959) undertook the measurement of line widths and radial velocities in the Orion Nebula with spectral dispersion and angular resolving power higher than that of previous investigations. The analysis of the new data established that the width and structure of the lines vary a great deal over the Nebula. In some regions the lines are asymmetrical, sometimes quite broad and occasionally even divided into two separate components with splitting up to 25 km/sec. The root-mean-square velocities $\langle V^2 \rangle_t^{1/2}$ and $\langle V^2 \rangle_m^{1/2}$ of thermal and mass motions, respectively, have been determined by comparing the line widths of the H and [O III] lines. In those regions where the lines appear narrowest, the thermal motions correspond to temperatures near 10^4 °K, in agreement with independent estimates, while the values of $\langle V^2 \rangle_m^{1/2}$ range from 5 to 7 km/sec. The statistics of the radial velocities in the plane of the sky, on the other hand, obey a law of the form (10), indicating an effective depth for the formation of the lines incompatible with the line widths. The conclusion is that the mass motions in the Nebula are not homogeneous in the sense of Kolmogoroff's law. The observations directly reveal the reasons for the failure of the theory of incompressible turbulence to account for the state of motions in the nebula. The sudden appearance in certain regions of distinctly double lines suggests immediately the existence of discontinuities in the flow, which very likely have the nature of shock waves. The different behavior of the components of the [O II] and [O III] lines in the two masses giving origin to the double lines suggests the existence of varying physical conditions, but so far the corresponding densities and temperatures have not been measured.

The overall mean trend of the radial velocities observed in the Orion Nebula suggest that the Nebula is in a state of differential expansion around its center, which appears to be the Trapezium. In a frame of reference fixed in the Trapezium cluster, the [O III] lines in the mean show a negative velocity of 10 km/sec, exceeding the approach velocity of the zone where the [O II] lines are formed by about 3 km/sec. The surrounding H I gas, on the other hand, is essentially at rest, as revealed by the velocities of the 21-cm line of neutral hydrogen. The expansion velocity of the emission region thus appears to decrease outward and does not extend into the main mass of cool surrounding gas.

VI. NEBULAR ABSORPTION LINES

The presence of the He I $\lambda 3888$ line, arising from the metastable 2^3S level, in the absorption spectrum of the brighter stars imbedded in the Orion Nebula was discovered by Wilson (1939). Observations of Adams (1944) showed that this line is double in some stars and that considerable variations in the intensity and velocity appear from star to star. The profiles of the He I $\lambda 3888$ observed in the stars imbedded in the Orion Nebula derived from unpublished material obtained by Münch are shown in Figure 3. The lines of He I at $\lambda 10830$ and $\lambda 3188$, also arising from the 2^3S level, have been observed in absorption in the spectra of the brighter stars (Boyce and Ford 1966, Herbig unpublished).

The complex structure and the variations from star to star in the radial velocity of the He I $\lambda 3888$ line clearly suggest the existence of discrete masses within the Nebula moving with supersonic velocities. An analysis of the observed strengths provides, besides, important information regarding the physical state of the gas.

The number of helium atoms in the metastable level $N(2^3S)$ D along the line of sight is determined by the observed equivalent width of the absorption line W, through the appropriate curve of growth

$$N(2^3S)\ D = H(W, b_i),\tag{11}$$

where b_i is the internal Doppler width. On the assumption of thermal motions only, for the typical value $W = 0.1$ Å, we obtain

$$N(2^3S)\ D = 2 \times 10^{-5}\ \text{cm}^{-3}\ \text{pc}.\tag{12}$$

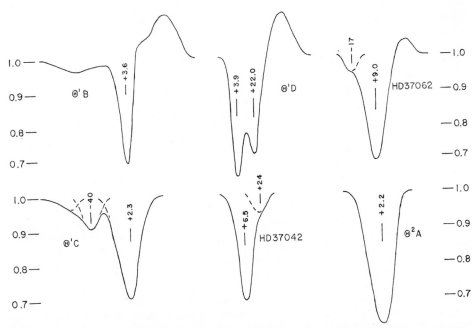

Fig. 3.—Profiles of the He I λ3888 absorption line in the bright stars imbedded in the Orion Nebula. The original spectrograms have a dispersion of 4.5 Å/mm, with the exception of that for HD 37062, which has a resolution twice as low as the rest. The nebular emission apparent in the tracings of $\theta^1 D$ and $\theta^1 B$ and HD 37062 may affect significantly the absorption profiles.

The number density $N(2^3S)$ may be related to the total density of helium, which is predominantly in the singly ionized state, by requiring that the recombination rate in the triplet system of helium equal the deactivation rate of the metastable 2^3S level, in the form

$$N(\mathrm{He^+})N_e\alpha(T) = N(2^3S)(A + fN_e + pr^{-2}),\qquad(13)$$

where $\alpha(T)$ is the recombination coefficient, $A = 2.2 \times 10^{-5}$ sec^{-1} is the probability for two-photon emission decay (Mathis 1957a), f is the probability for collisional deactivation and p is the probability of photoionization from the 2^3S level by radiation of a star at unit distance. With the values α $(T = 10^4$ °K$) = 2.1 \times 10^{-13}$ cm^3 sec^{-1}, as given by Burgess and Seaton (1960), $f = 9.7 \times 10^{-9}$ cm^3 sec^{-1} for the deactivation rate by electron superelastic collision (Meier-Leibnitz 1935) and collisional transitions to the 2^1S level (Phelps 1955), and the value of p corresponding to a star of spectral type O5, by combining equations (12) and (13) we find (Münch and Wilson 1962):

$$N(\mathrm{He^+})N_e\,D = 2100(1 + 4.4 \times 10^{-4}N_e + 0.077r^{-2})\ \mathrm{cm^{-6}\ pc},\qquad(14)$$

where r is measured in pc. For dense nebulae, $N_e = 10^3$ to 10^4 cm^{-3}, and we can see from equation (14) that the three processes of radiative decay, collisional deactivation, and photoionization may be competitive. For a hydrogen to helium ratio $N(\mathrm{He})/N(\mathrm{H}) = 0.1$, we then obtain

$$N_e^2\,D = 1.7 \times 10^4(1 + 4.4 \times 10^{-4}N_e + 0.077r^{-2})\ \mathrm{cm^{-6}\ pc}.\qquad(15)$$

This quantity represents the emission measure of one element capable of producing a λ3888 He I absorption line with strength $W = 0.10$ Å. In comparison, the emission measure of the Nebula derived from the surface brightness in Hβ, uncorrected for interstellar and internebular extinction, is $N_e^2 L = 10^6$ cm^{-6} pc. It follows thus that the emission measure of one element of Nebula producing observable $\lambda 3888$ absorption is at least a full order of magnitude less than the emission measure derived from the surface brightness, unless r is as small as 0.04 pc. Actually, the true emission measure with allowance for extinction is about 5×10^6 cm^{-6} pc, and to account for it around 50 elements, each capable of producing detectable $\lambda 3888$ absorption, are needed on the line of sight, if their distance from the Trapezium is of the order of the radius of the bright central region of the Nebula, around 0.2 pc. Because the stellar disks subtend a very small angle compared with the angular resolving power of the surface brightness measures, there is no obvious contradiction with the observed number of $\lambda 3888$ components. The picture that many elements contribute to the emission lines but only very few produce nebular absorption on the star is substantiated by the lack of correlation between the radial velocities of the $\lambda 3888$ absorption line and those of the nebular absorption lines. It is believed on this basis that the Nebula is not homogeneous, but that it must have density fluctuations of considerable magnitude.

The existence of density fluctuations in the Orion Nebula has been inferred by Osterbrock and Flather (1959) from comparison of the values of the electron density indicated by the intensity ratio of the [O II] lines at $\lambda\lambda 3726, 3729$, and by the surface brightness in the radio-frequency range. These authors estimate that the fractional volume of the Nebula occupied by the condensations is only about $\frac{1}{40}$.

From the previous discussion of the conditions under which absorption in $\lambda 3888$ may take place, it is evident that the surface brightness of many emission nebulae is sufficiently high to expect that nebular absorption should be observable in their exciting stars. The writer has undertaken a spectroscopic survey of the stars imbedded in emission nebulae, and has found that, indeed, many of them show $\lambda 3888$ in absorption. The large emission nebulae M8, M20, and M16 in Sagittarius contain stars with $\lambda 3888$ in absorption. The exciting stars of the nebulae NGC 7635, NGC 6822, IC 1396, and NGC 2467 also show detectable He I $\lambda 3888$ in absorption. However, a search for the line in the spectrum of the exciting stars of planetary nebulae, where it would be expected because of their high surface brightness, has not been successful. BD $+ 30°$ 3639 is the only planetary nebula with a central star showing with certainty $\lambda 3888$ absorption; not one, but actually two components, were discovered by Wilson (1950). The lack of absorption in the central stars of other planetary nebulae of high surface brightness has led this writer (Münch 1963) to suggest that there is another mechanism deactivating the 2^3S level of He I, different than those included in equation (13). Because of the large optical depth of all nebulae in the hydrogen Lα line, the energy density in this line may be built up considerably by resonance trapping over the value expected from pure recombination in an optically thin medium, and photoionization by Lα may become the dominant deactivation mechanism of He I 2^3S. A quantitative test of this hypothesis has been attempted by O'Dell (1964) from his observations of emission-line intensities in planetary nebulae. A large population of the metastable 2^3S level provides a source for the collisional excitation of the 2^3P level and thus the emission line $\lambda 10830$ (2^3P $\rightarrow 2^3$S) is enhanced with respect to other triplet He I lines with upper states not directly populated by collisional transitions from the metastable level, as is the case for the $\lambda 5876$ (3^3D $\rightarrow 2^3$P) line. The intensity ratio between $\lambda 10830$ and $\lambda 5876$ has thus been used by O'Dell to measure the metastability of the 2^3S level and hence the energy density in Lα.

The diffusion of resonance radiation in an optically thick medium with no motions other than thermal, has been often studied in relation to the possible dynamical effects that Lα radiation pressure may have on the structure and evolution of the planetary nebulae. Among the most recent treatments of the subject are those of Field (1959) and Osterbrock (1962). The energy densities required by O'Dell (1964) to explain the observed population of the 2^3S level in the planetary nebulae appear to be large in comparison with the values predicted by these simplified theories. The origin of the remaining discrepancy is not yet completely clear. It is believed by O'Dell that a shell of neutral hydrogen around the planetaries may scatter $L\alpha$ radiation back into the ionized region, in such a manner that the mean number of scatterings suffered by a $L\alpha$ photon before destruction or escape is increased considerably. In order to verify the correctness of this explanation, it appears that the problem of transfer of $L\alpha$ radiation through a nebula should be studied in more detail, paying special attention to the effects of differential Doppler motions.

VII. THE CHEMICAL COMPOSITION OF THE INTERSTELLAR GAS

Absorption lines.—The intensities W of the interstellar absorption lines provide a measure of the number of absorbing atoms NL along the line of sight, through a curve of growth which we denote by

$$W = G(NL, \Phi), \tag{16}$$

where Φ is the distribution function of the velocity components of the absorbing atoms or ions along the line of sight. When a line arises in a single cloud, Φ describes the internal motions; while if the line arises in many clouds, Φ also involves the relative motions of the clouds. For reasons of simplicity, the form of Φ is supposed to be Gaussian or exponential, in such a manner that it is completely specified by the velocity dispersion σ. For the case of atoms with two or more observable lines whose f-values are in fixed ratios arising from the ground level, more than one relation between NL, σ, and the equivalent widths can be written down, and both parameters can be determined. This procedure, known as the doublet-ratio method, was applied first by Wilson and Merrill (1937) to the resonance doublets of Na I and Ca II. The explicit dependence of the intensity ratio W_2/W_1 of the two lines of a doublet and one of the W's on the two parameters NL and σ has been evaluated by Strömgren (1948) for the case of a Gaussian Φ and by Münch (1957) for an exponential Φ. Applying the doublet-ratio method to the mean intensities of the Na I and Ca II lines, one finds that the mean densities of Ca II and Na I within 500 pc from the sun are

$$N(\text{Ca II}) = 2 \times 10^{-9} \text{ cm}^{-3} \quad \text{and} \quad N(\text{Na I}) = 2 \times 10^{-9} \text{ cm}^{-3}. \tag{17}$$

To derive the total densities of the species, it is necessary to specify the ionization equilibrium, which in general is described by an equation of the form

$$N''N_e\alpha = N'\Gamma \tag{18}$$

where N' is the density of parent atoms and Γ their photoionization rate, while N'' is the density of ions and α is the recombination coefficient, a function of the electron temperature T_e. The α's do not depend on the environment of the ions and can be evaluated in thermodynamic equilibrium. The Γ's depend on the photoionization coefficient of the parent ion and on the intensity of the radiation field. It can be shown that equation (18) can be rewritten in a form similar to Saha's equation,

$$\frac{N''N_e}{N'} = 2\frac{g_1''}{g_1'}\frac{(2\pi mkT_r)^{3/2}}{h^3} e^{-\chi/kT_r}\left(\frac{T_e}{T_r}\right)^{1/2} W \tag{19}$$

where g_1'' and g_1' are the statistical weights of the ground states, χ is the ionization potential of the parent ion, T_r is a temperature characterizing the radiation field, and W is a dilution factor, which is the ratio between the actual number of ionizations from the ground state and the total number of ionizations in thermodynamic equilibrium at the temperature T_e. The right-hand member of equation (19) has been evaluated by Strömgren (1948), Seaton (1951), and Weigert (1955) for the atoms and ions with observable interstellar lines and for the mean stellar radiation field described by Dunham (1939). The ionization equilibrium calculations are uncertain primarily because the stellar radiation field is unknown at the far ultraviolet wavelengths relevant for the photoionizing processes. The radiation field of Dunham represents stellar fluxes by black bodies and may, therefore, be badly in error. Improved evaluations of the mean energy density of starlight at the sun, based on calculated fluxes for model stellar atmospheres of early-type stars have been made recently by Zimmerman (1964). The extinction properties of the interstellar dust in the far ultraviolet are still unknown, however, and the stellar photoionizing flux remains uncertain on this account. Observations from orbiting telescopes no doubt will provide the necessary data in the near future. Let it be supposed that the values of α/Γ are known. Considering the ionization equilibria Na I $\rightleftarrows$ Na II and Ca II $\rightleftarrows$ Ca III then it follows that

$$\frac{N(\text{Ca})}{N(\text{Na})} = \frac{N(\text{Ca II})}{N(\text{Na I})} \frac{N_e + (\Gamma/\alpha)_{\text{Ca II}}}{N_e + (\Gamma/\alpha)_{\text{Na I}}}. \tag{20}$$

From the mean values of $N(\text{Ca II})$ and $N(\text{Na I})$ given by equation (17) and the ionization calculations of Seaton, it is found that for H I conditions

$$\frac{N(\text{Ca})}{N(\text{Na})} = \frac{N_e + 0.025}{N_e + 0.68}. \tag{21}$$

Since in H I regions $N_e = 3 \times 10^{-4} N(\text{H})$, it follows that $N(\text{Ca})/N(\text{Na}) = 0.036$, a value about one-twentieth of the abundance ratio found in the sun. This result is quite independent of the values of the temperature and electron density. The "calcium paradox" referred to by Struve has thus been reduced by an order of magnitude through the use of the correct ionization equation, but is not quite completely removed. The inaccuracies remaining in the atomic data at present are certainly too small to account for the discrepancy. To avoid an interstellar Ca/Na ratio a full order of magnitude less than the "cosmical" value, substantial changes may have to be introduced in the radiation field.

Clearly, in view of the unresolved discrepancies existing in the relative abundances of elements in the interstellar medium with respect to those in other cosmical sources, it is not possible to make definite statements regarding the absolute overall abundances of the species from the intensities of interstellar absorption lines. It is nevertheless meaningful to verify that the mean density of the interstellar gas, derived on the assumption of cosmical abundances, is about the same as that obtained from direct observations of hydrogen. For example, we consider the case of Ca II, for which ion we write

$$N(\text{Ca II}) = N(\text{Ca})N_e\left(\frac{\alpha}{\Gamma}\right)_{\text{Ca II}}, \tag{21}$$

since $N(\text{Ca III}) \approx N(\text{Ca})$ in an H I region. From the observed mean density of Ca II ions, setting $N_e = N(\text{C})$, equation (21) may be rewritten in the form

$$N_H{}^2 = N(\text{Ca II}) \frac{N_H}{N(\text{Ca})} \frac{N_H}{N(\text{C})} \left(\frac{\Gamma}{\alpha}\right)_{\text{Ca II}}. \tag{22}$$

The assumption of solar relative abundances of Ca and C relative to hydrogen, with the mean value of $N(\text{Ca II})$ derived from the line intensities and the appropriate Γ/α, equation (22) provides a value for the total density N_H. Using the value of $N(\text{Ca II})$ found in equation (17) and the ionization functions of Seaton, it is found $N_H = 0.3$ cm^{-3}, a value which should be compared with the mean density $N_H = 0.7$ cm^{-3} found from 21-cm line observations (Van de Hulst 1958). If the same procedure is applied to the Na I mean densities, $N_H = 2$ cm^{-3} is obtained. The analysis of the other atomic interstellar absorption lines leads to similar results, and on this basis it is inferred that the abundances of the elements in the interstellar gas are of the same order of magnitude as in the Sun and stars. This result refers directly to the metals Na, K, Ca, and Ti and indirectly to the elements supplying the free electrons in H I regions, mostly carbon.

Emission recombination lines.—The specific intensity of an emission line of frequency ν_{21} in an optically thin medium is given by the number and density N_2 in the upper level of the transition involved, in the general form

$$I_{21} = \frac{h\nu_{21}}{4\pi} \int_{(s)} N_2 A_{21} \, ds, \qquad (23)$$

where A_{21} is the probability for spontaneous emission and the integral is taken along the line of sight.

In a hydrogen-like pure recombination spectrum (collisions ignored) the populations of the energy levels are determined from the steady-state conditions. Any level of quantum numbers n and l is then populated by direct radiative capture into that state and by cascade transitions from higher levels. If $\alpha_{nl} = \alpha_{nl}(T_e)$ is the recombination coefficient in the level (n, l) and $A_{n'l',nl}$ is the probability for the spontaneous transition from (n', l') to (n, l), the condition of steady state is expressed by

$$N_e N(\mathrm{H}^+)\alpha_{nl} + \sum_{n'=n+1}^{\infty} \sum_{l'=l\pm1} N_{n'l'} A_{n'l',nl} = N_{nl} A_{nl}, \qquad (24)$$

where

$$A_{nl} = \sum_{n''=n_0}^{n-1} \sum_{l'=l\pm1} A_{nl,n''l''}. \qquad (25)$$

For the case of a medium optically thin in the Lyman lines $n_0 = 1$, since transitions $n \to 1$ produce radiation that escapes the medium (case A). When the medium is optically thick in the Lyman lines (case B) $n_0 = 2$, since radiation $n \to 1$ ($n > 2$) can excite hydrogen atoms in the ground state back to the level n, which eventually decays back to the ground state via $n = 2$, with the production of Lα. A general solution of equations (24) can be obtained by defining $P_{n'l',nl}$ as the probability that population of (n', l') is followed by direct transition to (n, l):

$$P_{nl',nl} = \frac{A_{n'l',nl}}{A_{n'l'}}. \qquad (26)$$

Let further $C_{n'l',nl}$ be the probability that population of (n', l') is followed by a transition to (n, l) via all possible cascade routes. These cascade coefficients $C_{n'l',nl}$ can be obtained by summing over all possible intermediate states that can be reached by direct radiative transitions. Starting with

$$C_{nl,nl} \equiv 1, \qquad (27)$$

successive coefficients of higher order are obtained from

$$C_{n'l',nl} = \sum_{n''=n}^{n'-1} \sum_{l''=l\pm1} P_{n'l',n''l''} C_{n''l'',nl} \qquad (28)$$

The solution of equation (24) is then written in the form given by Pengelly (1963):

$$N_{nl} = \frac{N_e N(\mathrm{H}^+)}{A_{nl}} \sum_{n'=n}^{\infty} \sum_{l'=0}^{n'-1} \alpha_{n'l'} C_{n'l',nl}. \qquad (29)$$

It is usual to express N_{nl} in terms of the population of the level (n, l) in thermodynamic equilibrium and a dilution factor b_{nl}, in the form

$$N_{nl} = N_e N(\mathrm{H}^+)(2l + 1) \left(\frac{2\pi m k T_e}{n^2}\right)^{-3/2} \exp\left(-\frac{\chi_{nl}}{kT}\right) b_{nl}, \qquad (30)$$

where χ_{nl} is the ionization energy of the level. Numerical values of $b_{nl} = b_{nl}(T_e)$ have been obtained by Pengelly on the basis of the recombination coefficients α_{nl} calculated by Burgess (1964). In conditions of thermodynamic equilibrium all quantum states of the same energy have equal populations and

$$N_{nl} = \frac{2l + 1}{n^2} \sum_{l'} N_{nl'}. \qquad (31)$$

In general, the solutions of equations (24) do not satisfy equation (31). Requiring that the populations of the angular momentum levels actually satisfy (31), we write the condition of steady state in the form

$$N_e N(\mathrm{H}^+)\alpha_n + \sum_{n'=n+1}^{\infty} N_{n'} A_{n',n} = N_n A_n, \qquad (32)$$

the general solution of which has been obtained by Seaton (1959).

Line intensities obtained with allowance for the non-degeneracy of the n-states differ considerably from those resulting when the population of the l-levels corresponds to equilibrium. We must therefore find to what extent the angular momentum levels are mixed by collisions in emission nebulae. The answer to this question has been given by Pengelly and Seaton (1963), who find that for nebular conditions ($N_e = 10^4$ cm^{-3}, $T_e \approx 10^4$ °K), collisional transitions $(n, l) \to (n, l')$ proceed faster than radiative transitions $(n, l) \to (n', l')$ $(n' < n)$ for $n > 22$. The population of highly excited states, taking into consideration collisional transitions to $n \pm 1$, collisional ionizations, and three-body recombinations, radiative capture, cascade, and collisional mixing of the l-levels has been obtained by Seaton (1963). It is only for $n > 40$ that the level populations differ significantly from those obtained in the purely radiative theory. Whereas the latter theory predicts a sharp increase of the b_n factors at sufficiently large n, the former predicts a more gradual rise in the b_n's, beginning at about $n = 50$. Accurate spectrophotometry of the Balmer lines with high upper levels n has not been made. The population of discrete levels very near the ionization limit has recently become a matter of great interest because observations of emission transitions between levels of very high n in the radio-frequency range have become possible (Höglund and Mezger 1965).

In order to compare observed emission-line intensities with theory it is necessary to account for the effects of interstellar absorption and reddening. Under the assumption that the reddening produced by the interstellar solids can be described by a universal function of wavelength, its effect on line-intensity ratios can be ascertained by comparing nebulae reddened to different degrees and from observations of stars. Actually, galactic emission nebulae have not yet been studied extensively from this point of view, and the method is best illustrated by considering the case of the planetary nebulae in the manner suggested by Osterbrock, Capriotti, and Bautz (1963). In Figure 4 is shown the relation between the Hα/Hβ and Hβ/Hγ intensity ratios for a number of planetary nebulae. The effect of interstellar reddening is shown by the lines constructed from the extinction curves determined by Whitford (1958) and by Wampler (1961) for the Cygnus region. The effect of self-absorption (Capriotti 1964) in the Balmer line ratios is shown by the dotted line. A nebula affected by interstellar reddening and self-absorption should fall in the wedge-shaped region defined by the two straight lines crossing at the point representing the radiative

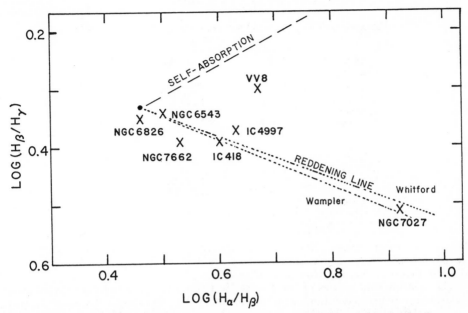

Fig. 4.—Relation between the intensity ratios between the first three Balmer lines in planetary nebulae (crosses). Interstellar reddening would displace the observed points along parallels to the dotted lines. Self-absorption would displace the observed points along parallels to the dashed line. The solid dot represents the prediction of hydrogen recombination theory for an electron temperature of $10^4°$K.

recombination theory. As can be seen in Figure 4, all points fall close to, but not exactly on the theoretical reddening line. The discrepancies between theory and observation may be 0.03 in log ($H\beta/H\gamma$), somewhat larger than the observational error. Inaccuracies in the numerical results of radiative recombination theory can not account for the discrepancy. The effects of self-absorption in the Balmer lines are apparent in the dense and small-diameter planetaries IC 4997 and VV8, but cannot help to bring the observations of less dense planetaries into agreement with the theory.

As far as the abundance problem is concerned, however, the small discrepancies remaining between theory and observation are of little relevance. What is required is the true absolute value of the surface brightness in particular radiations, and the interstellar extinction must be allowed for correctly. Departures from a constant value for the ratio between color excess and total absorption may introduce considerable uncertainties in the extinction corrections. By comparing the observed intensity ratio between Paschen and Balmer lines arising from the same upper state with the theoretical predictions, the scale of the reddening correction curve can be conveniently determined. The application of this procedure is straightforward for the case of the planetaries, because the absorbing matter certainly is in the foreground. However, galactic emission nebulae contain large amounts of solid particles, the absorbing and/or scattering by which may not be the same as those in the general field.

The theory of the recombination spectrum of helium has been developed by Mathis (1957a, 1957b), Seaton (1960b), and Pottasch (1960). Since the capture cross-sections except for the lowest levels are hydrogen-like, the results are similar to those for hydrogen, especially for the $2^1P–n^1D$ and $2^3P–n^3D$ transitions. Some lines of the triplet system, however, are significantly affected by the metastability of the 2^3S

TABLE 2

ABUNDANCE OF HELIUM (BY NUMBER) RELATIVE TO HYDROGEN
IN NEBULAE

Object	He/H	Object	He/H
Diffuse H II regions			
NGC 1976 (Orion)	0.117	M 20	0.094
NGC 604	0.102	M 8	0.11
Planetaries			
IC 418	0.10	NGC 2392	0.16
IC 2149	0.13	NGC 6543	0.18
NGC 6826	0.15	NGC 7027	0.18
NGC 7662	0.15	IC 4997	0.19
M15 (K648)	0.18		
B-type stars	0.16		

level discussed previously. Not only lines directly connected with 2^3S, as $\lambda 10830$ and $\lambda 3888$, are affected in this manner, but others such as $\lambda 7065$ (2^3P–3^3S) may be indirectly involved through Rosseland-type cycles.

The measured intensities of the H and He I recombination lines are the basis for determining the relative numbers of atoms along the line of sight. By selecting pairs of lines relatively close together in wavelength, as $\lambda 5876/H\beta$, $\lambda 4471/H\gamma$ and $\lambda 4026/H\delta$, the corrections for differential interstellar absorption are minimized and uncertainties arising from variations in the reddening or extinction law are made insignificant. In order to obtain the abundance ratio, however, we must know the relative distribution of He I and H atoms along the line of sight. For some planetary nebulae it is known that the regions where the two kinds of lines arise do not have the same dimensions (Wilson 1950), but the differences are small in most cases. In galactic emission nebulae it appears that the differences between the relative distribution of H and He I are even less than in the planetaries. The problem of the relative dimensions of the H II and He II regions has been studied by Swihart (1952) and by Hummer and Seaton (1964), who find that for He/H < 0.20 the regions emitting the H and He I lines are very nearly the same, provided the temperatures of the exciting stars are higher than about 50,000°K.

In galactic emission nebulae the concentration of He^{++} is small and the intensity ratios between the H and He I lines give directly the He/H abundance ratio. For planetary nebulae of high excitation the concentration of He^{++} is not negligible but the relative intensities of the He II lines at $\lambda 5412$ and $\lambda 4686$ to $H\beta$ provide the He^+/H ratio. The derived total He/H abundance ratios are in every case very insensitive to the assumed temperature, because all the intensities of the H, He I and He II lines vary with electron temperature in a manner that keeps their ratios nearly constant. The abundance ratios He/H obtained by Mathis (1962) for three of the brightest H II regions are given in Table 2 together with the value obtained for the giant nebula NGC 604 in the Sc spiral galaxy Messier 33. Considering that the probable error of determinations is about 20 percent, or 0.03 in He/H, it is apparent that the large emission nebulae have essentially a constant He/H abundance ratio. It is of interest to compare the values of the He/H ratio found for extended H II regions with those of the planetaries.

The results of the photoelectric observations of O'Dell (1963) are given in Table 2 for a number of planetaries covering a range in excitation and surface brightness. The mean value of He/H for the eight planetaries listed is 0.156, with an r.m.s. spread of

±0.03, nearly equal to the expected probable error resulting from the observational and theoretical uncertainties. The He/H abundance ratio in the planetaries, thus, appears to be essentially the same as in galactic extended diffuse nebulae, within the uncertainty of the determinations. It is interesting to notice that the helium abundance in the planetary nebula K648 in the "metal-poor" globular cluster M15 is 0.18 (O'Dell, Peimbert, and Kinman 1964), very close to the values found in galactic planetaries. Recalling that the mean He/H ratio found in B-type stars (Aller 1961) is 0.16, we conclude that the helium abundance in all these objects of diverse age and location is essentially the same. On this basis it is difficult to accept the hypothesis that the primordial galactic material was formed of pure hydrogen. It might be argued that the large helium abundance found in K648 and the disk planetaries reflects the hydrogen burning that has occurred in evolved stars, originally of low helium abundance. But then the fact that the helium abundance appears to be everywhere the same would have to be considered a matter of chance.

Forbidden emission lines.—The steady state population N_2 of the upper levels of forbidden lines in the photographic region of the spectrum is obtained from the collisional Q_{21} and radiative A_{21} deactivation rates, in the form

$$N_2 = N_2^{(0)} \frac{Q_{21}}{Q_{21} + A_{21}}, \tag{33}$$

where $N_2^{(0)}$ is the equilibrium population at the kinetic temperature, given by the Boltzmann formula. The corrections for the population of other excited upper levels, metastable or not, are in general small and can be carried out as perturbations. Radiative transfer effects and stimulated emissions are of importance only for infrared and microwave transitions.

From the surface brightness of a nebula in the radiation of some forbidden line, corrected for interstellar extinction, the number of ions along the line of sight producing it is then readily obtained. In order to derive the total abundance of the atomic species involved, it is necessary to specify the relative populations of the various stages of ionization and their concentration along the line of sight. A complete theoretical study of this aspect of the problem encounters great difficulties and has not been carried out. Accordingly, a semi-empirical procedure, first suggested by Bowen and Wyse (1939), is generally followed. This consists in the construction of a mean curve for the concentration of the various oxygen ions as a function of ionization potential and then the assumption that the corresponding curve for other ions has the same shape and differs from that of oxygen only by a scale factor, which in fact represents the relative abundance of the particular species with respect to oxygen. For example, in Figure 5 is reproduced the ionization curve for the various oxygen ions present in NGC 7027, as found by Seaton (1960a). The values of log (X^{+m}) corresponding to the N and Ne ions have been shifted by a constant amount for each species to obtain a "best" fit. Clearly, one cannot expect accurate results from this procedure, since the ionic concentrations under conditions removed from thermodynamic equilibrium depend not only on the respective ionization potentials, but also on detailed properties of the radiation field and of the photoionization cross sections. The existence of an empirical ionization curve in nebulae, as NGC 7027, covering ions with such widely different ionization energies as O^0(13.6 ev) and O^{+5}(135 ev) does not imply that the ionization conditions can be represented by a homogeneous mass at fixed density and temperature. Rather, there may be large-scale stratification, apparently not present in the particular case of NGC 7027, or inhomogeneities in density and temperature, which nevertheless are taken approximately into account by the Bowen and Wyse procedure.

In the case of extended nebulae the stratification effects in the ionization are obvious. In the Orion Nebula, for example, the intensity ratio of the [O II] to the [O III]

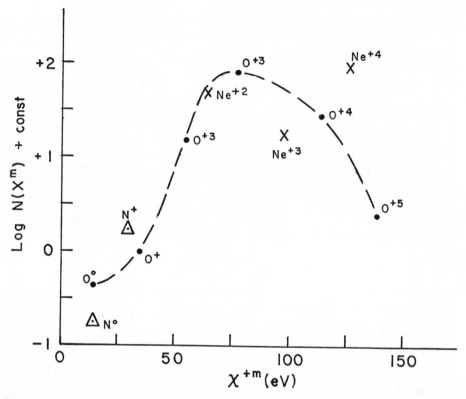

Fig. 5.—Empirical ionization function for the oxygen ions present in the spectrum of the planetary nebula NGC 7027. The points representing the neon ions (crosses) have been obtained by adding an arbitrarily chosen constant to the ionic densities in such a way as to fit the oxygen curve in a best possible way. A different constant has been added to the densities of nitrogen ions (triangles).

lines steadily increases as the distance from the exciting stars in the Trapezium increases, and the lines of [O I] are relatively strongest in the outermost and faintest regions. The degree of ionization and the electron temperature in diffuse nebulae are fortunately sufficiently low that the corrections to the abundances for unobserved ions are small, especially for the important cases of oxygen and neon. For nitrogen, on the other hand, represented only by [N I] and [N II], the correction for unobserved ions is large, and consequently the nitrogen abundance is poorly known. The important element carbon is represented in the nebular spectra accessible from ground only by a weak recombination line of C^+ at $\lambda 4267$, the atomic parameters for which are poorly known. The relative abundances of some of the more abundant ions present in the Orion Nebula are given in Table 3. For comparison, the mean abundance derived for the planetary nebulae and for stars are included. From the Table it appears that the nebular abundances derived by different authors show discrepancies by factors of two or three, resulting most probably from the ionization corrections. Comparing the nebular abundances with the stellar ones, a good overall agreement is found, especially in the case of oxygen. One exception is worth noticing. The O/H ratio in the planetary nebula of the globular cluster M15 is 1.5×10^{-5}, at least one order of magnitude less than in galactic planetaries, extended diffuse nebulae and in the stars (O'Dell et al. 1964).

TABLE 3

RELATIVE ABUNDANCES OF HEAVY ATOMS IN NEBULAE

Ratio	Orion Nebula		Planetaries			Stars		
	Aller	Mendez	Mean	NGC 7027	K648	10 Lac	γ Peg	Sun
10^4 O/H	4	3	6	11	0.15	6	4	
C/O	0.6	4	0.7			0.4	0.9	0.59
N/O	0.13	0.12	0.4	0.7		0.4	0.2	0.11
Ne/O	0.6	0.2	0.19	0.09	0.01	0.9	1.2	
S/O	0.25	0.07	0.09				0.15	0.22
Ar/O	0.01	0.01	0.01				0.02	
Fe/O		0.013						0.004

Further theoretical work on excitation cross sections and in the ionization factors should make possible a more accurate determination of the abundances in nebulae. Photoelectric line intensity measures and corrections for interstellar extinction are today reliable enough to provide higher accuracy than the present state of theory allows.

VIII. THE FUTURE

Our understanding of the nature and composition of the diffuse matter in the interstellar medium will advance fundamentally when observations are carried out with orbiting telescopes. The opening of the ultraviolet spectral region down to the Lyman limit will provide a wealth of information (and no doubt surprises!) that will help to remove some of the outstanding problems we face today. A preliminary study of the interstellar absorption lines expected to be observable below $\lambda 3000$ has been made by Spitzer and Zabriskie (1959). These authors have listed the ultimate ultraviolet lines produced by ions of the most abundant elements H, C, N, O, Mg, Si, S, Ar, and Fe. On the basis of assumed solar abundances and approximate ionization calculations for an H I region containing 2×10^{20} H atoms per cm^2, these authors have estimated equivalent widths W for the most important lines. For the O I line at $\lambda 1302$, for example, they obtain $W/\lambda = 4 \times 10^{-4}$, which value should be compared with the strength $W/\lambda = 2 \times 10^{-5}$ expected for the D$_2$ line in the same path length. The absorption lines of O I, C II, N I and Mg II thus will be an order of magnitude stronger than those available in the region accessible from the ground. The Lyman lines of neutral hydrogen will of course be extremely strong. For the pathlength taken by Spitzer and Zabriskie, $W = 9$ Å for Lα and $W = 0.063$ Å for L8. Of particular significance will be the determination of the concentration of H$_2$ molecules through the observation in absorption of the R(0) lines of the bands in the Lyman system. For a number of absorbers as small as 3×10^{15} H$_2$ molecules per cm^2, Spitzer, Dressler, and Upson (1964) estimate that the strongest H$_2$ lines would have equivalent widths as large as 0.02 Å. The problem of the concentration of H$_2$ molecules has often been discussed in view of the great importance that it might have in the net mass balance of the galaxy. Estimates of the relative concentration of H$_2$ to H, however, vary widely. Gould and Salpeter (1963), for example, estimate that the concentration of H$_2$, relative to H, is about unity, while Varsavsky (1966) believes that uncertainties existing in the actual formation and dissociation rates of H$_2$ may bring the abundance of H$_2$ as low as 10^{-4}. Under these conditions it is clear that the observation of H$_2$ will not only settle the question of the mass balance but also provide insight into the processes of formation of other interstellar molecules.

The study of emission nebulae will also advance greatly with the availability of the ultraviolet spectrum. The emission lines expected to be observable and their strengths have been discussed by Osterbrock (1963). It is also expected that a number of lines will appear in absorption, such as the resonance lines of N^+, N^{+2}, C^{+3}, Si^{+3}, and Si^{+4}.

Finally it may be mentioned that the operation of a large telescope in orbit should enable us to extend the study of diffuse matter to the intergalactic medium, through the observation of absorption features in the spectrum of extragalactic objects.

REFERENCES

Adams, W. S. 1944, *Pub. A.S.P.*, **56**, 119.
———. 1949, *Ap. J.*, **109**, 354.
Aller, L. H. 1961, *The Abundance of the Elements* (N.Y.: Interscience).
Ambartsumian, V. A. 1940, *Bull. Abastumani Obs.*, **4**, 17.
Ambartsumian, V. A. and Gordeladse, S. G. 1938, *Bull. Abastumani Obs.*, **2**, 37.
Baade, W., Goos, F., Koch, P. P., and Minkowski, R. 1933, *Zs. f. Ap.*, **6**, 355.
Blaauw, A. 1952, *B.A.N.*, **11**, 459.
Bowen, I. S. and Wyse, A. B. 1939, *Lick Obs. Bull.*, **19**, 1.
Buisson, H., Fabry, C., and Bourget, H. 1914, *Ap. J.*, **40**, 241.
Burbidge, G. R., Gould, R. J., and Pottasch, S. R. 1963, *Ap. J.* **138**, 945.
Burgess, A. 1964, *Mem. R.A.S.*, **69**, Part 1.
Burgess, A. and Seaton, M. J. 1960, *M.N.R.A.S.*, **121**, 472.
Campbell, W. W. and Moore, J. H. 1918, *Pub. Lick Obs.*, **13**, 96.
Capriotti, E. R. 1964, *Ap. J.*, **139**, 225; **140**, 632.
Clark, B. G., Radhakrishnam, V., and Wilson, R. W. 1962, *Ap. J.*, **135**, 151.
Dunham, T., Jr. 1937, *Publ. A.S.P.*, **49**, 26.
———. 1939, *Proc. Am. Phil. Soc.*, **81**, 277.
Eddington, A. S. 1926, *Proc. Roy. Soc. London*, **A.**, **111**, 424.
Evans, J. W. 1941, *Ap. J.*, **93**, 275.
Fabry, C. and Buisson, H. 1911, *Ap. J.*, **33**, 406.
Field, G. 1959, *Ap. J.*, **129**, 551.
Ford, K. W. and Boyce, P. B. 1966, *Sky and Telescope*, **32**, 12.
Gould, R. J. and Salpeter, E. E. 1963, *Ap. J.*, **138**, 393.
Henroteau, F. 1921, *J. Roy. Astr. Soc. Can.*, **15**, 62, 109.
Hobbs, L. M. 1965, *Ap. J.*, **142**, 160.
Höglund, B. and Mezger, P. G. 1965, *Science*, **150**, 339.
Hummer, D. G. and Seaton, M. J. 1964, *M.N.R.A.S.*, **127**, 217.
Kahn, F. D. 1955, in *Gas Dynamics of Cosmic Clouds*, ed. H. C. Van de Hulst and J. M. Burgers (Amsterdam: North-Holland Pub. Co.), p. 115.
Kolmogoroff, A. N. 1941, *Doklady Acad. Sci. USSR*, **30**, 301.
Livingston, W. C. and Lynds, C. R. 1964, *Ap. J.*, **140**, 818.
Ludendorff, H. 1920, *Astr. Nachr.*, **213**, 3.
Markarian, B. E. 1946, *Soobsc. Burakan Obs.*, No. 1.
Mathis, J. S. 1957*a*, *Ap. J.*, **125**, 318.
———. 1957*b*, *Ap. J.*, **126**, 493.
———. 1962, *Ap. J.*, **136**, 374.
Meier-Leibnitz, H. 1935, *Zs. f. Phys.*, **95**, 499.
Minnaert, M. 1948, *B.A.N.*, **10**, 405.
Münch, G. 1957, *Ap. J.*, **125**, 42.
———. 1963, *Ann. Rep. Mt. Wilson and Palomar Obs.: Carnegie Yearbook*, **62**, 63.
Münch, G., and Wilson, O. C. 1962, *Zs. f. Ap.*, **56**, 127.
O'Dell, C. R. 1963, *Ap. J.*, **138**, 1018.
———. 1964, *Ap. J.*, **142**, 1093.
O'Dell, C. R., Peimbert, C. R., and Kinman, T. D. 1964, *Ap. J.*, **140**, 119.
Oort, J. H. and Van de Hulst, H. C. 1946, *B.A.N.*, **10**, 197.
Oort, J. H. and Spitzer, L., Jr. 1955, *Ap. J.*, **121**, 6.
Osterbrock, D. E. 1962, *Ap. J.*, **135**, 195.
———. 1963, *Planet. Space Sci.*, **11**, 621.
———. 1965, *Ap. J.*, **142**, 1423.
Osterbrock, D. E. and Flather, E. 1959, *Ap. J.*, **129**, 26.
Osterbrock, D. E., Capriotti, E. R., and Bautz, L. P. 1963, *Ap. J.*, **138**, 62.
Pengelly, R. M. 1963, *M.N.R.A.S.*, **119**, 90.
Pengelly, R. M. and Seaton, M. J. 1963, *M.N.R.A.S.*, **127**, 165.
Phelps, A. V. 1955, *Phys. Rev.*, **99**, 1307.
Plaskett, J. S. 1924*a*, *Pub. Dom. Ap. Obs. Victoria*, **2**, 335.
———. 1924*b*, *M.N.R.A.S.*, **84**, 80.
Pottasch. S. R. 1960, *Ap. J.*, **131**, 202.
Schilt, J. 1947, *A. J.*, **52**, 209.

Schlüter, A., Schmidt, H., and Stumpff, P. 1953, *Zs. f. Ap.*, **33**, 194.
Seaton, M. J. 1951, *M.N.R.A.S.*, **111**, 368.
———. 1953, *M.N.R.A.S.*, **113**, 81.
———. 1959, *M.N.R.A.S.*, **119**, 90.
———. 1960a, *Rep. Prog. Phys.*, **23**, 313.
———. 1960b, *M.N.R.A.S.*, **120**, 326.
———. 1963, *M.N.R.A.S.*, **123**, 177.
Spitzer, L., Jr. 1948, *Ap. J.*, **108**, 276.
———. 1949, *Ap. J.*, **109**, 337.
Spitzer, L. and Zabriskie, F. R. 1959, *Pub. A.S.P.*, **71**, 412.
Spitzer, L., Dressler, K., and Upson, W. L. 1964, *Pub. A.S.P.*, **76**, 387.
Stebbins, J., Huffer, C. M., and Whitford, A. E. 1940, *Ap. J.*, **91**, 20.
Strömgren, B. 1939, *Ap. J.*, **89**, 526.
———. 1948, *Ap. J.*, **108**, 276.
Struve, O. 1925a, *Pop. Astr.*, **33**, 639.
———. 1925b, *Pop. Astr.*, **34**, 1.
———. 1926, *Ap. J.*, **65**, 163.
———. 1928, *Ap. J.*, **67**, 353.
———. 1939, *Proc. Nat. Acad. Sci.*, **25**, 36.
———. 1941, *J. Wash. Acad. Sci.*, **31**, 217.
Struve, O. and Gerasimovitch, B. P. 1929, *Ap. J.*, **69**, 7.
Swihart, T. L. 1952, M.S. Thesis, Univ. Indiana.
Van de Hulst, H. C. 1958, *Rev. Mod. Phys.*, **30**, 913.
Varsavsky, C. M. 1966, *Space Sci. Rev.*, **5**, 419.
Von Hörner, S. 1951, *Zs. f. Ap.*, **30**, 17.
Von Weiszäcker, C. F. 1946, *Zs. f. Phys.*, **124**, 614.
Wampler, E. J. 1961, *Ap. J.*, **134**, 861.
Weigert, A. 1955, *Wiss. Zs. Schiller U. Jena*, **4**, 435.
Whitford, A. E. 1958, *A. J.*, **63**, 201.
Wilson, O. C. 1939, *Pub. Amer. Astr. Soc.*, **9**, 274.
———. 1950, *Ap. J.*, **111**, 279.
Wilson, O. C., Münch, G., Flather, E. and Coffeen, M., 1959, *Ap. J. Suppl.*, **4**, 199.
Wilson, O. C. and Merrill, P. W. 1937, *Ap. J.*, **86**, 44.
Zimmermann, H. 1964, *Astr. Nachr.*, **288**, 95.

INTERSTELLAR GRAINS

H. C. van de Hulst

Leiden Observatory

The copy of Struve's Washington Academy lecture (Struve 1941) in our library bears the stamp: "Astronomische Instituten." It was one of a collection of American publications which still reached the Netherlands' observatories during the first war years. It was a pleasure to read and try to digest it from cover to cover.

The first part of Struve's review is devoted to the solid particles in interstellar space. Struve avoids the word dust, which physical chemists reserve for particles produced by a grinding process. The more appropriate word would be "smoke"; we shall use "grains," a noncommittal term chosen by Spitzer.

When Struve wrote this review, the interstellar grains formed a subject about ten years old which seemed already to have reached a deadlock. The interstellar extinction (then commonly called absorption) effectively blocked progress in the study of large-scale galactic structure. Thus its knowledge was, in a negative way, of tremendous importance. It was known to be roughly proportional to λ^{-1}. With preset assumptions about the chemical composition and size distribution of the grains, the predominant size could be determined rather well, as the fine work by Greenstein (1937) and by Schalén (many papers) had already shown by that time. The chemical composition itself could not be inferred with any degree of certainty from this comparison. Nor did the interesting calculations on the force of radiation pressure exerted on these grains by stars of various types give very conclusive results.

At that time I had read all of the available literature on the interstellar particles. By number, the papers dealing with metallic particles certainly outweighed those dealing with other hypotheses. There were two reasons for this preponderance, a fair one and an unfair one. The fair reason was a link, however vague, with meteors and meteorites. For the best evidence on meteor velocities still seemed to make a substantial part of them interstellar; only the radar measurements by Lovell's group about 1948 put a radical end to this. The unfair reason was that metallic particles require a smaller ratio of size to wavelength and hence shorter calculations to be fitted to the extinction curve.

What to do in such a deadlock? The situation is and was quite familiar in scientific research. And so is and was the answer: look for more and better observations, and for more and better theory.

Otto Struve thoroughly chose the side of observation. His reasoning, briefly, must have been as follows: If we cannot learn more about these grains from properties of the light which is taken away, then we must look at where it goes, *i.e.*, at the scattered light. Thus for many years he inspired a group of young astronomers at Yerkes Observatory to study reflection nebulae and the possible existing diffuse galactic light.

Before pursuing this subject, we shall review three other avenues which were explored about the same time and are still being pursued in order to obtain a more firmly founded knowledge of the interstellar grains.

The first, also observational, was the attempt, by Chalonge in Europe and by Stebbins and Whitford in the U.S., to obtain utmost precision in the extinction curve (extinction vs λ^{-1}) over a wide range of wavelengths. Among the recent work of high quality we should mention that of Johnson (1965) and of the Edinburgh group (Nandy 1965). The wavelength range is now being widened by studies in the far-infrared and in the rocket-ultraviolet. Perhaps we need further data with independent calibrations before we can put the same confidence in these extensions as in the data of the classical domain.

A closely allied problem is to find the ratio R of total to selective extinction or, in more familiar terms, of absorption to reddening. It is part of the same problem because the ratio determines where the zero-line of the extinction curve should be drawn, and this fixes the point at $\lambda^{-1} = 0$, provided there is no neutral absorption (which would require too much mass). Much has happened from the first determination of this ratio by Halm (1917) to the recent work of Johnson, but the principles of the determination have remained the same. The intricacies are nicely set forth by Oort (1938). Certainly, it came as a relief that R (and hence the extinction curve) is now found not to be the same in all regions of the sky. Exact equality would not go well with the concept of clouds having different dynamical histories, since the extinction curves depend rather sensitively on grain size.

I should like to skip over the polarization. Not because it is unimportant but because this surprise discovery, made in 1949, and the many subsequent observations have not basically changed our outlook on the grain composition. There is a second curve to be fitted now, the curve of polarization vs λ^{-1}. This extra restraint has not proved so severe as to permit a clear solution of the problem of grain composition (Greenberg and Shah 1966).

A second approach was based on the conviction that some day the unidentified interstellar lines must be identified. If they belong either to the solid grains or to molecules which are building stones of the grains, such identification must tell us something about the nature of the grains. As long as the solution has not been found, a note like that of Swings and Öhman (1939) may be as relevant as many papers in the following decades. Herzberg has in the course of years devoted much research, theoretical and experimental, to various uncommon combinations of common elements, but so far without success. Herbig (1967) has brought by careful observations the number of observed lines to 26. Yet the upshot is that at this moment we do not know if we are close to, or far away from an answer.

A third approach, first suggested by B. Lindblad (1934) and pursued in Holland 1943–1948 is to study by astrophysical (or rather astrochemical) theory how the grains behave in the medium of gas and radiation which exists in interstellar space. We know more about this medium (the second half of Struve's review paper) than about the grains. Of course, a laboratory physicist or colloid chemist would find the problem far too messy for a decent experiment. But with reasonable astrophysical certainty we did conclude that the grains will be cold enough to trap most of the colliding gas atoms at their surface. Only helium and the majority of hydrogen will evaporate again. The grain composition arising from this process (Van de Hulst 1949) has gone into the literature as "dirty ice."

I now return to the fourth method to learn more about the grains, namely by observing the light they scatter. This was the method on which Struve had put his hopes. In the 1941 paper Struve very briefly summarizes the elegant reasoning he had put forth elsewhere (Struve 1939) to exclude all but the size range of 10^{-5} cm to 10^{-3} cm for the interstellar grains, mainly by discussing the scattered or diffracted light. To obtain more precise results within this size range is a tougher problem. In his first big paper on reflection nebulae (Struve 1937) he concludes that "While, even now, it is not possible to derive the phase function accurately, a number of interesting features, common to many nebulae, can be satisfactorily explained by the theory of diffuse scattering."

This statement sounds rather optimistic in view of the following history; even now, 30 years later, we can at most repeat the statement. It is interesting to explore why—in a science in which some subjects are renewed beyond recognition in 10 years—this subject has remained stationary. Obviously this is not another example of a problem remaining untouched for years because it has wrongly been filed into the box "completely solved." I rather think that it came in the box "difficult but unattractive."

There is not a homogeneous class of objects to be studied here. At one end of the scale are the prominent reflection nebulae scattering the light from one star. At the other end is the diffuse galactic light consisting of light from "the general stellar background" scattered by "the general interstellar medium." The dark nebulae and lanes, which usually exhibit some scattered light, take an intermediate position.

A statistical view of the present status of observations on the first type of objects (about 200 nebulae) may be had from Dorschner and Gürtler (1965); for a modern high-quality study of three individual nebulae we refer to Elvius and Hall (1966). At least for these objects we now have reliable data on photometry, color, and polarization, which would merit a thorough theoretical study. For instance, the polarization in the famous nebula near Merope in the Pleiades is found to rise from about 2 percent at distances smaller than 5′ to about 10 percent at larger distances. In the same range of angles the color changes gradually from redder than the star to bluer than the star.

Let us pass over the problem whether the grains in reflection nebulae are representative for interstellar grains in general. The theory then offers no fundamental problems, at least if we wish to be content with models consisting of spherical particles arranged in a simple geometry (e.g., a slab). But no reasonably accurate and complete set of models has yet been calculated. The individual grains require Mie scattering computations, a once formidable task, which now has become a routine which has served on at least a dozen electronic computers as a practising example. For the entire nebula we need multiple scattering theory (or radiative transfer), again a problem which can be solved by brute computer-force and in many cases also by analytical theory. A full assessment of the properties of reflection nebulae using first-order Mie scattering has been presented by Greenberg and Roark (1967). Estimates of the second order have (in simple cases) already been made before this century. Struve (1937) mentions with some satisfaction that adding this second order by Schoenberg's theory did not change his conclusions. Sufficient results of machine computations are now available to say a great deal more. For instance, I find for isotropic scattering that in transmitted light (with the nebula as a slab between the star and the observer), for optical depth 1 and albedo 1 the first- and second-order scattering in typical cases form together between 55 and 61 percent of the total diffuse light. To me it is not obvious that the remaining 40 percent will have the same color- and light-distribution as the first and second order. Hence I feel that rather more complete studies of the multiple scattering are required before an accurate comparison with observations can be made. The situation is more favorable for particles which have a small albedo.

The difficulties in dealing with the diffuse galactic light are reversed. Here the theory is relatively easier, because the radiation field to which each part of the nebula is exposed simply is the general radiation of the starlit sky, including the diffuse radiation. Under some plausible assumptions multiple-scattering calculations may even be completely avoided. The observations, on the other hand, are notoriously difficult. Struve's four-page note (1933), or Dufay's more elaborate review (1939) already reveal that the crux of the problem is not so much in the photometry itself as in the separation, without arbitrary assumptions, of faint stars, zodiacal light, terrestrial night glow, and the diffuse radiation sought. Struve's 1941 lecture contains a rather full discussion of the then-just-completed study by Henyey and Greenstein (1941). These authors concluded that the grains must have a relatively high albedo combined with a forward-throwing phase function. Pleasing though this result may be in view of the "dirty-ice" theory, we must still regard the result with caution. The subsequent 25 years should have given ample time for independent confirmation by photoelectric photometry. Several such studies have indeed been made, partly with other chief aims: Roach, Pettit, Tandberg-Hanssen, and Davis (1954), Elsässer and Haug (1960), Roach and Smith (1964), Smith, Roach, and Owen (1965), Witt (1968). Detailed comparison of these and older studies reveals differences up to 25 percent. They can be explained away by calibration errors and color differences but it is difficult to become convinced that all correction factors are now final. Dufay (1957) and Smith et al. (1965) conclude that the existence of the diffuse galactic light is established; Van Houten (1967) is more skeptical. Van de Hulst and De Jong (1969) reinterpreted Witt's observations and reach conclusions rather similar to those of 1941.

We have now completed the main task of this paper, namely an updating of Struve's 1941 review regarding the solid particles to the present epoch, 25 years later. This is a different assignment from writing a complete present-day critical review. For one thing, I have been one-sided in my references as a comparison with e.g., the I.A.U. tri-annual reports will show. Also, I have not dealt with certain lines of development which start abruptly, either with a surprise discovery (like the polarization) or with an odd theoretical suggestion.

The most imaginative hypothesis of the latter class undoubtedly is that of Platt (1956) who proposes a type of particles made of light elements with a chemical structure such that in spite of their small size (about 10 Å) they still obey the observed extinction law. It is hard to say more about these particles, quantitatively, than Platt did, but the "Platt particles" remain an important alternative. At least one modern study of the growth process of grains from the interstellar gas (Schmidt 1963) favors this explanation.

Another alternative is graphite. First suggested because the anisotropy of graphite could help explain the polarization, graphite particles later became favored by some authors because of the stronger extinction in the rocket-ultraviolet (Nandy and Wickramasinghe 1965) or because they seemed helpful in the early stages of stellar evolution, or because they provided free condensation nuclei for ice grains (Wickramasinghe 1965; Donn, Wickramasinghe, Hudson, Stecher 1968). In spite of early efforts, the question of the condensation nuclei had not been solved; to get them from smoky stars would be a nice way out. A fair comparison of these ideas with observations requires the use of the same set of assumptions to explain polarization and extinction but also the solution of such tough optical problems as: what is the extinction of a graphite flake embedded in a nearby spherical ice mantle? I am afraid that in the literature so far available these problems may still have been weighed too lightly. The optical properties of graphite spheres (Friedemann and Schmidt 1966) or of small graphite flakes (Greenberg 1966) are better known.

At the time of Struve's review, interstellar space for a theorist still was one domain, without evolution and without structure, except for stationary, embedded regions of ionized hydrogen. We still have no clear concepts of the evolution of interstellar clouds; the wide scope of problems involved can be seen from Wentzel (1966). Yet we know a lot more about the evolution of stars and groups of stars associated with these clouds and their arrangement in spiral arms. Interstellar space now has, like everything in astronomy, a history. This forms the background of many of the papers cited and some entirely new processes had to be considered. One of these is Wickramasinghe's (1965) sputtering process: the gradual eating away of the outer layers of a grain when it passes near a B-type star. The observations have to be similarly directed toward new aims. The studies of solid particles inside diffuse emission nebulae (O'Dell, Hubbard, and Peimbert 1966; and earlier papers) are particularly relevant.

The work inspired by Struve in his pupils and his pupils' pupils has not yet brought the study of interstellar grains to the stage of tedious refinements but the subject is alive and full of fundamental questions: a fitting tribute to a great astronomer.

REFERENCES

Donn, B., Wickramasinghe, N. C., Hudson, J. P., and Stecher, T. P. 1968, *Ap. J.*, **153**, 451.
Dorschner, J. and Gürtler, J. 1965, *Astr. Nachr.*, **289**, 57.
Dufay, J. 1939, *Ann. d'Ap.*, **1**, 195.
————. 1954, *Nébuleuses Galactiques et Matière interstellaires* (Paris: Albin Michel).
Elsässer, H. and Haug, U. 1960, *Zs. f. Ap.*, **50**, 121.
Elvius, Aina and Hall, J. S. 1966, *Lowell Obs. Bull.*, **6**, 257.
Friedemann, C. and Schmidt, K. H. 1966, *Mitt. Univ. Sternw. Jena*, Nr. 72.
Greenberg, J. M. 1966, *Ap. J.*, **145**, 57.
Greenberg, J. M. and Shah, G. 1966, *Ap. J.*, **145**, 63.
Greenberg, J. M. and Roark, T. P. 1967, *Ap. J.*, **147**, 917.
Greenstein, J. L. 1937, *Harvard Obs. Circ.*, No. 422.
Halm, J. 1917, *M.N.R.A.S.*, **77**, 243.
Henyey, L. G. and Greenstein, J. L. 1941, *Ap. J.*, **93**, 70.
Herbig, G. 1967, *Radio Astronomy and the Galactic System*, ed. H. van Woerden (London: Academic Press; I. A. U. Symposium no. 31), p. 85.
Houten, C. J. van. 1967, *Bull. Astr. Inst. Neth.*, **19**, 303.
Hulst, H. C. van de. 1949, *Rech. Astr. Obs. Utrecht*, **11**, part 2.
Hulst, H. C. van de and Jong, T. de. 1969, *Physica*, **41**, 151.
Johnson, H. L. 1965, *Ap. J.*, **141**, 923.
Lindblad, B. 1934, *M.N.R.A.S.*, **95**, 12.
Nandy, K. 1965, *Pub. Roy. Obs. Edinburgh*, **5**, 13.
Nandy, K. and Wickramasinghe, N. C. 1965, *Pub. Roy. Obs. Edinburgh*, **5**, 29.
O'Dell, C. R., Hubbard, W. B., and Peimbert, M. 1966, *Ap. J.*, **143**, 743.
Oort, J. H. 1938, *Bull. Astr. Inst. Neth.*, **8**, 233.
Platt, J. R. 1956, *Ap. J.*, **123**, 486.
Roach, F. E. 1966, *Modern Astrophysics*, ed. M. Hack (Paris: Gauthier-Villars), p. 49.
Roach, F. E. and Smith, L. L. 1964, *NBS Tech. Note* No. 214.
Roach, F. E., Pettit, H. B., Tandberg-Hanssen, E., and Davis, D. N. 1954, *Ap. J.*, **119**, 253.
Schmidt, K. H. 1963, *Astr. Nachr.*, **287**, 215.
Smith, L. L., Roach, F. E., and Owen, R. W. 1965, *Plan. Space Sci.*, **13**, 207.
Struve, O. 1933, *A. J.*, **77**, 153.
————. 1937, *Ap. J.*, **85**, 194.
————. 1939, *Ann. d'Ap.*, **1**, 143.
————. 1941, *J. Wash. Acad. Sci.*, **31**, 217.
Swings, P. and Öhman, Y. 1939, *Observatory*, **62**, 150.
Wentzel, D. G. 1966, *Ap. J.*, **145**, 595.
Wickramasinghe, N. C. 1965, *M.N.R.A.S.*, **131**, 177.
Witt, A. N. 1968, *Ap. J.*, **152**, 59.

9. THE BETA CANIS MAJORIS VARIABLES

O. STRUVE: The Present State of Our Knowledge of the β Canis Majoris or β Cephei Stars. *Annales d'Astrophysique*, **15**, 157, 1952.

A. VAN HOOF: The Beta Canis Majoris Stars.

THE PRESENT STATE OF OUR KNOWLEDGE
OF THE β CANIS MAJORIS OR β CEPHEI STARS

by Otto STRUVE

Berkeley Astronomical Department
University of California

ABSTRACT. — *This paper summarizes the observational results obtained in recent years for seven stars of the β Canis Majoris and β Cephei types. The principal result is the discovery of the importance of axial rotation in explaining the wide variety of the observed variations in light and radial velocity. Those stars which have large rotations (sigma Sco) are usually characterized by a single short period of 4 to 6 1 /2 hours and irregular variations in the amplitudes. These same stars show very large periodic variations in the line widths, but not in spectral type or color. Those stars which have relatively sharp lines usually show two interfering periods, differing by a few minutes, and resulting in beat periods of the order of 7 to 49 days. In a few stars the beat phenomenon, though easily observed, is complicated by an irregular variation in the amplitudes that correspond to one of the two interfering oscillations.*

The last section of the paper reviews four hypotheses that have been advanced, and stresses the vibrational character of the two oscillations. It is probable that the frequencies of oscillations in the equatorial regions of a rapidly rotating star differ from the frequencies of oscillations in the vicinity of the poles. The variety of possible combinations of different frequencies is further augmented by the random distribution of the axes of rotation in different stars.

During the past quarter of a century most astronomers have carefully avoided the stars of the β Canis Majoris or β Cephei type. But during the twenties a number of investigations were devoted to them, and some results of great interest were obtained by F. HENROTEAU at Lick, and later at Ottawa ; by R. K. YOUNG at Victoria ; by P. GUTHNICK at Berlin-Babelsberg ; and by Joel STEBBINS at Wisconsin. However, except for the fact that the photometric and spectrographic periods of such stars as β Cephei and 12 Lacertae were found to be constant, or slowly but regularly varying, the observations were often quite disappointing. STEBBINS, for example, has commented upon the variability—apparently irregular in character — of the amplitudes and shapes of individual light curves, and HENROTEAU's efforts to find some regularity in the velocity curves were, on the whole, unsuccessful.

In recent years only one astronomer continued steadfastly to study one of these objects. I am referring to W. F. MEYER's observations of β Canis Majoris made with MILLS three-prism spectrograph at the Lick Observatory.

Beta Canis Majoris was not the first star of its kind to be discovered. The first was β Cephei, found in 1906 by E. B. FROST at the Yerkes Observatory, to

329

have a variable radial velocity with a period of only 4 1/2 hours. But, unfortunately, the spectrographic observations were continued only in a sporadic manner by C. C. CRUMP, by the present writer, and by several other astronomers. Hence, we do not now know nearly as much about the velocity curve of β Cephei as we do about that of β Canis Majoris. Yet, the photometric variability of β Cephei should have prompted the spectroscopists long ago to undertake an exhaustive study of its spectrum, as MEYER had carried out for β Canis Majoris.

In β Canis Majoris E. A. FATH also found a change in brightness, but his discussion is not very reliable because the star is low over the horizon at Lick, and it is so bright that there are few suitable comparison stars.

At McDonald observations of β Canis Majoris were made about ten years ago by SWINGS and STRUVE and some discrepancies were found when the results were compared to MEYER's predictions. This led to an exchange of information, which was interrupted by MEYER's death, and a year later by the death of J. H. MOORE.

As a MORRISON research associate at Lick in 1950, I completed the discussion of β Canis Majoris, and the results were published about a year ago in the *Astrophysical Journal*. Thereafter I have concentrated upon the study of other stars of the same class. The phenomena that have come to light in this work are of a surprising variety, and it is no exaggeration to say that a new and exceedingly fascinating field has been opened up for research. The present discussion is largely based upon the work of a group of astronomers at the Berkeley Astronomical Department in 1950-1952, and my thanks are due especially to Mr. Merle WALKER who was mainly responsible for the photoelectric observations, and to Messrs. D. H. NcNAMARA, R. LEVÉE, S. M. KUNG, R. P. KRAFT, and A. D. WILLIAMS, who collaborated with me in the spectrographic work. To the Mount Wilson and Lick Observatories we are indebted for the opportunity of securing the observational material.

SUMMARY OF OBSERVATIONAL RESULTS

1. The best observed star is 16 Lacertae, whose 4-hour period was discovered in 1950 by M. WALKER. Its velocity curve and light curve consist of two interfering harmonic oscillations :

$$P_2 = 0.17085 \text{ day} ; \qquad K_2 = 4.5 \; \frac{km}{sec} ; \qquad \Delta m_2 = 0.05 \text{ mag.}$$
$$(4^h \, 6^m \, 1^s 0)$$

$$P_1 = 0.16917 \text{ day} ; \qquad K_1 = 14.8 \frac{km}{sec} ; \qquad \Delta m_1 = 0.07 \text{ mag.}$$
$$(4^h \, 3^m \, 34^s 4).$$

The beat period $P_3 = 17.16$ days $\sim 100 \, P_2$.

The spectral lines vary in *width,* but their *equivalent* widths remain constant.

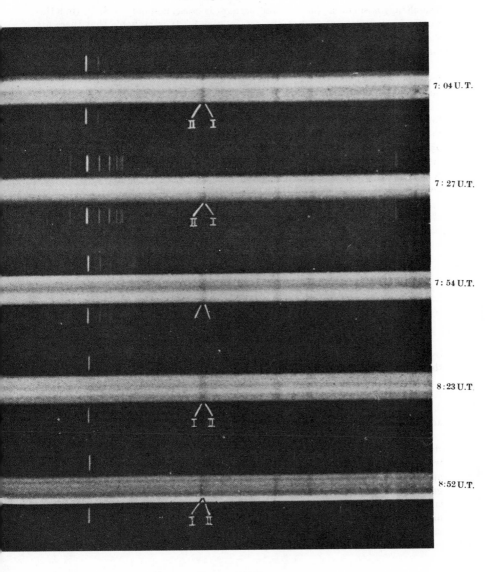

7: 04 U. T.

7 : 27 U.T.

7 : 54 U.T.

8 : 23 U.T.

8 : 52 U.T.

FIG. 1. — Spectra of 12 Lacertae.
(original linear dispersion 10 A/mm)

The change is small in this star and perhaps is associated only with P_2 (but this cannot be definitely proved at the present time). It is probable that there are slow changes in K_2, from one year to the next, perhaps also in Δm_2. These changes may be irregular. There is no change in color.

The star shows, in addition to the foregoing changes, a periodic variation in the mean (or gamma) velocity with $P_5 = 12.097$ days, $K_5 = 23.4$ km/sec. This may represent ordinary binary motion. The spectral type does not vary.

The star is *brightest* when the radial velocity is near γ, from positive values to negatives ones (phase of greatest compression on pulsation hypothesis). At the same time the lines are broadest. However, if the line widths change only with P_2 then they are *broadest* half way on the descending branch of the P_2, K_2 velocity oscillation. The lines are sharpest on the ascending branch of the P_2, K_2 oscillation of velocity. At this stage they (especially Ca II K) undergo rapid and (probably) irregular variations in intensity and width.

The spectrum shows the forbidden Stark components of λ 4 472 (HeI), etc. It is thus a B_2 main-sequence star — perhaps slightly under-luminous. (I reserve the designation P_4 for a possible periodicity in the slow variation of K_2 ; P_1 is is that oscillation which is not associated with a change in line profile.)

2. β *Canis Majoris* resembles 16 Lacertae but

$$P_1 = 0.25002246 \text{ day}, \qquad K_1 = 5.8 \text{ km/sec}$$
$$(6^h 0^m)$$

$$P_2 = 0.2513003 \text{ day}, \qquad \begin{cases} K_2 = 4.2 \text{ km/sec (1909-1931)}, \\ K_2 = 3.0 \text{ km/sec (1931-1948)}. \end{cases}$$
$$(6^h 2^m)$$

The beat period $P_3 = 49.1236$ days. The light changes, but the observations by FATH do not distinguish between the two periods. The star is brightest when the velocity is near γ, from positive to negative values.

The line widths change conspicuously and are associated *only* with the P_2, K_2 oscillation. They are broadest when the velocity passes through γ on the descending branch of this oscillation. Since $K_2 < K_1$ the phase, when the lines are broadest, appears to " slide along " the combined velocity curve.

LEDOUX has attempted to explain the observed oscillations, the one with a stationary vibration, the other with *one* of two theoretically possible traveling waves in a rotating star. He proceeds to show that the lines vary in width, approximately as in a normal Cepheid. But the observed change in line profile greatly exceeds the broadening that can be *directly* accounted for in a star whose surface elements have different radial components of δr. Thus, *this* part of his discussion cannot be maintened. Also it is an established fact that as a rule *maximum* line width occurs when the measured radial velocity is zero (maximum compression on old radial-type of pulsation hypothesis).

OTTO STRUVE

The spectrum is B_2, but the luminosity inferred from the absorption lines is much greater than for 16 Lacertae. Miss A. B. UNDERHILL thought that the luminosity is slightly greater when the lines are sharpest. There is no change in spectral type or in equivalent widths.

3. 12 *Lacertae* shows only *one pronounced* oscillation, with large, irregular changes, but there is no *definite* beat phenomenon.

$$P_2 = 0.193089 \text{ day}, \qquad K_2 = \begin{cases} 15 \text{ km/sec}, \\ 75 \text{ km/sec}. \end{cases}$$

I suspected that there may be a characteristic interval $I_3 = 39$ days in the variation of K_2, in place of a true beat period P_3. The lines change very greatly in width with P_2. There is a general tendency for this change to be greatest when K_2 is greatest (75 km/sec).

When K_2 is large the absorption lines become strikingly double (fig. 1), resembling a spectrospic binary with components of identical spectral type, the one having very sharp lines, the other having very broad lines. Thsee double lines have been measured on several dates. The result is shown schematically in figure 2.

The amplitude of the sharp component was about twice as large as the amplitude of the broad component.

The conclusion was drawn that the P_2, K_2 oscillation in this star *may* not represent a real motion of the center of mass of the star, or even the integrated effect of a complicated vibration, but the rotation of a turbulent spot producing *deep* lines on a rapidly rotating star having otherwise a uniform surface (fig. 3).

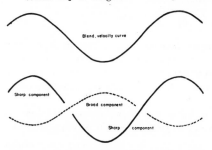

FIG. 2. — Schematic Velocity curve of 12 Lacertae.

The broad-line stage in 12 Lac is suggestive of fairly rapid axial rotation. The light varies with P_2 and $\Delta m_2 \sim 0.1$ mag. The star is brightest when the lines are broadest and when the velocity passes through gamma from positive values to negative values.

I believe that in this star K_2 is so large that a value of K_1 similar to that found in β Canis Majoris or even in 16 Lacertae would be overlooked. The P_2, K_2 oscillation dominates the picture. The irregular variation of K_2 is con-

FIG. 3. — Spot of turbulent gases produces sharp line displaced ot red ; rest of star produces broad line shifted slightly to violet.

sistent with similar, but smaller, variations of K_2 in β CMa and 16 Lac. The spectrum is B_2, main sequence, probably under-luminous.

From his own and J. J. RUIZ' photoelectric observations of 12 Lac, C. DE JAGER of Utrecht concludes that there *is* a beat period :

$$P_2 = 0.193089 \text{ day},$$
$$P_1 = 0.19735 \text{ day}.$$

and that the amplitude Δm associated with P_1 is much smaller than that associated with P_2. He finds that the same beat phenomenon can be discerned in my radial velocities. However, there is an irregular variation superposed over the P_1, Δm_1, K_1 oscillation, which accounts for my failure to establish the beats in the velocity curve. This irregular variation seems to be present in all stars, but it is most conspicous in 12 Lac, ν Eridani and σ Sco, where K_1 is large.

4. σ *Scorpii* shows only one oscillation of velocity with

$$P_2 = 0.246835 \text{ day} \qquad K_2 = \begin{cases} 40 \text{ km/sec} \\ 65 \quad \text{»} \end{cases}$$

K_2 varies in an irregular manner. There are no beats ; P_1, K_1 may exist, but then $K_1 \ll K_2$. The change in line width is fantastically large : at maximum width the lines suggest V sin i (rotation at equator) = 120 km/sec. At minimum we find V sin $i <$ 20 km/sec. The spectrum is B_2, main-sequence, probably under-luminous, as 16 Lac and 12 Lac.

The lines become double as in 12 Lac, but LEVÉE has pointed out that the maximum line widths are not the same in all cycles (as would be required on the turbulent spot hypotheses).

The mean velocity varies with P_5 = 33 days. As in 16 Lac, this may be ordinary binary motion.

Change in light was observed by HOGG in Australia ; $\Delta m = 0.08$ mag. HOGG believes that the star is bluer at maximum light, as in the case of Cepheids generally. According to LEVÉE the star is brightest when the velocity is near maximum of recession. This result contradicts the phase relation in the other stars.

5. β *Cephei.* — We have many radial velocity observations. At present only one period, P = 0.190 day, is known, but the corresponding K is variable. We do not yet know whether there are beats. The light changes with the same P, and $\Delta m \sim 0.05$ mag is probably variable.

There is no change in line profiles, even with the highest COUDÉ dispersion (2.9 A/mm at the Mc-Donald Observatory). Hence, I have previously designated the period as P_1. But it is possible that beats do exist, and the change in profile is too small to be observed. It was near the limit of the COUDÉ dispersion in 16 Lac.

6. *H. D.* 199 410. — R. M. Petrie has an extensive series of observations. Line profiles vary conspicuously, but Petrie finds only one period, $P_2 = 0.20103$ day (perhaps $K_1 \ll K_2$) $K_2 \sim 60$ km/sec and is thus very large. Walker finds $\Delta m \sim 0.2$ mag — larger than in the other stars.

7. ν *Eridani.* — We have many radial velocity observations. The value of K varies greatly and the line profiles vary as in σ Sco.

From Henroteau's early velocity observations Zessevich found a beat period of $P_3 = 39.P_2$ days, $P_2 = 0.174$ day. Thus the beat period is about 7 days. K_2 varies between 12 and 38 km/sec. NcNamara, Kung, Kraft, Williams and the writer find that the beat period of 7 days is approximately correct. But the curve which relates 2K and P_3 (beat) has a larger scatter than is consistent with the high precision of the individual values of K. Thus, there is an irregular variation of K_2 (and perhaps also K_1) superposed over the normal beat-curve. The star resembles in this respect 12 Lacertae.

The phase of broadest lines is always on the descending branch of the velocity curve, that of sharpest lines on the ascending branch. But is seems that these phases are somewhat earlier than is the case in several other stars. Thus, the sharpest lines invariably appear about half-way between the minimum of radial velocity and the succeeding phase of γ-velocity.

Walker finds that Δm varies between 0.03 and 0.06 mag. The phase relation is approximately the same as in other stars. We describe the principal period as P_2 because presumably it is the one that is associated with the change in line profiles. There is no change in color.

Review of Several Hypotheses

I. The first hypothesis was suggested by the observations of double lines in 12 Lac. Closely similar periods of certain vibrations are not an unusual occurrence in nature. Take, for example, the problem of the variation of latitude. We have here two periods : Chandler's of 14 months, which is essentially the Euler period of *free* oscillation of the earth in its rotation, and the annual period of 12 months. The latter is reasonably constant in amplitude and phase ; the former is subject to irregular changes in amplitude and phase. We associate the annual period with the transport of air mass to the continent of Asia during the winter, and thus must regard the 12-month period as corresponding to a *forced* oscillation. It appears reinforced because of resonance, the difference between its period and that of the free oscillation being relatively small.

As to the origin of the free oscillation, we can think of irregular variations in the meteorological conditions associated with the annual transport of air —

somewhat along the lines of G. Udny YULE's free pendulum bombarded by pellets at irregular intervals, and Harold JEFFREYS's extension of this idea to the determination of the damping constant of the earth's free oscillation.

In the β CMa stars we perhaps also have a forced oscillation caused by a very small, very dense companion. There would be just enough room to accommodate such a companion since its period would be

$$P = \frac{2 \pi a^{3/2}}{g \sqrt{\mathfrak{M}}}.$$

With $P = 0.19$ d and $\mathfrak{M} = 20 \mathfrak{M}_0$, this gives $a = 3 \times 10^6$ km $= 4$ R$_0$.

Already in 1926 EDDINGTON showed that for β Cephei ($\mathfrak{M} = 5.1 \mathfrak{M}_0$), $P = 0.190$ d, $R = 1.52 \times 10_2$ km) an infinitesimal mass point would be at a distance of 1.7×10^6 km from the star's center. This might be the cause of the *forced* oscillation — the latter's period being exactly equal to that of the motion of the center of mass of the bright star, and to that of the motion of the turbulent spot which I though to be responsible for the sharp component of the double lines.

The motion of the satellite would be uniform, but since the physical conditions pointed to variable turbulence of an irregular nature, it seemed reasonable to associate them with irregular explosions of prominences — and to YULE's pellets bombarding his pendulum and setting up a free oscillation with changes in phase and amplitude.

There is, however, no reason why the free oscillation should be often so very close in P to the forced oscillation, or orbital motion. As long as β CMa was the only case with the two oscillations, this was reasonable, but with three well established cases out of a total of less than ten stars this does not seem reasonable.

There are also other difficulties:

a) The absence of typical turbulent line contours (except in β CMa, where they are *always* present).

b) The restriction produced by orbital motion in the size of the star (β Ceph, 12 Lac, ν Eri, 16 Lac are main-sequence stars but β CMa is not).

c) Similarity in spectral class: all are B$_2$ or nearly so.

It seems to me now that the satellite hypothesis must be dropped. But I still adhere to the physical explanation (as distinct from dynamical hypotheses) of the double lines and changes in line profiles, in general. These are undoubtedly associated in some way with the star's axial rotation.

II. The second hypothesis is based upon the — at present rather distant — resemblance of the β CMa stars and the RR Lyrae or cluster-type Cepheids. HALF a dozen of these latter stars are known to have two periods each, which are very similar. Thus, in RR Lyrae, DETRÉ found $P_1 = 0.5668$ d, $P_2 = 0.5590$ d. The beat period is thus 40.6 days. In RS Bootis Oosterhoff found $P_1 = 0.377$ d, beat period $P_3 = 537$ days. The ratios P_3/P_1 range between 64 and 1 424. The ratios in β CMa stars are between 39 and about 250, which is not too different from the RR Lyrae stars.

For the RR Lyrae stars we showed some years ago at the McDonald Observatory that the velocity curves also show the two interfering periods.

Mrs. KLUYVER-PELS has suggested that there are two modes of radial pulsations in a standard model star with ratio of specific heats $\gamma = 1.54$, the fundamental one and the third, whose periods are in the ratio of two to one: $P_1/P_0 = 0.515$. These modes will be coupled, as in a coupled pendulum. If the separate frequencies are ν and σ, when coupled they produce a variation with $\sigma - \nu$ and an amplitude greatly increased by resonance. The frequencies σ and $\sigma - \nu$ are similar, so that the periods $2\pi/\sigma$ and $2\pi/(\nu - \sigma)$ are also similar.

But ROSSELAND has shown that there are reasons for believing that $\gamma = 1.40$. Thus he suggests that the fundamental mode with the second give roughly a $2:1$ commensurability.

Even more important are the large differences in the two groups of stars. They may be summarized as follows :

	RR LYRAE STARS	β CANIS MAJORIS STARS
Spectrum	A — F	B_2
Velocity variation	K_1 about the same as......	K_1
	K_2 much smaller than	K_2
Light variation	Δm large	Δm small
Line profile variation	none or in accordance with pulsation hypothesis	large with broadest lines at γ vel on asc. br. of vel. curve
Phase relation of velocity and Δm..	mirror image (or very small shift from it)	shiflted by $\pi/2$
Shape of velocity curves	unsymmetrical	harmonic or nearly so
Shape of Δm curve	unsymmetrical	harmonic or nearly so
Bright lines	on rising branch of Δm curve (desc. branch of velocity curve)	none (or very rare)
Double absorption lines	on rising branch of Δm curve (desc. branch of velocity	at max. and min. of velocity curve
Rotation	none	often large (σ Sco)
Abs. Mag.	0	— 1 to — 3
Population	II	I

Thus, the physical differences are very large, and while similar physical causes may operate to cause the pulsations they certainly assume very different characteristics in the two groups.

III. The pulsations considered by Miss KLUYVER represent different modes of the usual radial oscillations of ordinary Cepheid variables. These different modes would have approximately the same influence upon the broadening of the lines. In reality, the one oscillation (P_1) is not associated with any perceptible broadening, even though sometimes $K_1 > K_2$. On the other hand, P_2 is always associated with such a broadening — and the latter can sometimes become very large, indeed.

Hence, LEDOUX has proposed a new theory of non-radial oscillations. This third hypothesis bears an analogy to a famous problem of the theory of tides, which was first discussed by Lord KELVIN in 1879, and later by Sir Horace LAMB. Consider, first, a circular basin of water with uniform depth and with a vertical, cylindrical boundary. If an initial disturbance causes the water to oscillate, but if no outside forces, other than gravity, act upon the water the differential equations of the motion of an element of water on the surface admit of many different solutions — symmetrical as well as unsymmetrical. The symmetrical solutions, obtained with the boundary conditions ξ = finite at $r = o$ and $\dfrac{d\xi}{dr} = 0$ at the boundary $r = 0$ (where ξ is the elevation of the free water surface) correspond to systems of annular ridges and furrows in the manner of stationary waves.

The unsymmetrical solutions correspond to a system of waves traveling around the origin with an angular velocity σ/s in the positive and also in the negative direction. The letter s corresponds 1, 2, 3 ...., while for the symmetrical solutions $s = 0$.

But if the water in our basin rotates around a vertical axis, the solutions are all changed. There is again a system of symmetrical stationary oscillations with annular ridges and furrows but the frequencies are no longer the same as in the stationary vessel. There are also two traveling waves, in opposite directions, whose frequencies σ_1 and σ_2 differ from each other.

The case of a spherical star is more complicated. It has been treated by P. LEDOUX ($Ap. J.$, **114**, 373, 1951). He finds, for the simplest kinds of oscillations three frequencies ;

$$\sigma, \qquad \sigma - 2\,\beta\,\omega, \qquad \sigma + 2\,\beta\,\omega,$$

where approximately $2\,\beta = 1.6$. The frequency σ represents the stationary wave, the frequencies $\sigma - 1.6\,\omega$ and $\sigma + 1.6\,\omega$ are the two traveling waves. Because the latter should have similar properties, while the observed oscillations are different with regard to light variations and change in line profile, LEDOUX identifies the observed periods with σ and one of the two traveling waves, giving us in the case of β CMa

$$1.6\,\omega = \frac{1}{49} \quad \text{days} = \frac{1}{\text{beat period}}.$$

This would give as the period of rotation of β CMa

$$P_{rot} = 80 \text{ days}.$$

The linear velocity at the equator would be $\dfrac{2\pi R}{80 \text{ days}} = 5 \dfrac{km}{sec}$, if $R = 5R_{\bigcirc} =$

$5 \times 7 \times 10^5$ km. This is very slow — too slow, in fact, for the observed line widths.

LEDOUX has computed the expected broadening from the three oscillations (see his Table 2, p. 381) and has found that if τ_0 is the period of the velocity curve (P_1 or P_2 in my designations), the following relations must hold :

Stationary oscillation	line of sight 54 degrees	gives a periodic broadening with $\tau_0/2$
Positive traveling wave	equatorial line of sight	$\tau_0 - 1.6 \omega$ amplitude $\pm$ 10 km/sec
Negative traveling wave....	equatorial line of sight	$\tau_0 + 1.6$ amplitude $\pm$ 10 km/sec.

Since the longer observed period is associated with changes in line profiles, this must be the negative wave. But then the maximum broadening turns out to be advanced by $\pi/2$ against the velocity curve. In other words, the lines should be broadest on the ascending branch. This is opposite to the observations. We must thus follow LEDOUX and at least temporarily abandon this hypothesis.

The observations of other stars (σ Sco and ν Eri) prove beyond question that the broadening is not simply due to the integrated effect of the oscillations δr over the visible star disc. Consider σ Sco. When the lines are broadest they resemble the dish-shaped lines of rapidly rotating stars. Hence, I had originally suggested that this is the *normal* line profile. If this is true, then the narrow lines of σ Sco must represent a disturbance of some sort on the star. In other words, there is an area producing strong and deep lines. If this interpretation is incorrect and the narrow lines represent the normal line profile, then the broad lines must be the result of the pulsation. The extreme edges of the broad line profiles would then represent the velocity of at least a limited area on the star's surface. This velocity persists for at least $1/2 \, P_2 = 3$ hours. Thus, the total distance traveled by this area tis about $\pm 120 \text{ km} \times 3 \text{ h} \times 60 \text{ m} \times 60 \text{ s} = 10^6$ km. The radius of σ Sco is about 1.5×10^6 km. (This is again EDDINGTON's estimate for β Cep ; it must also be approximately correct for σ Sco). It is at once clear that an oscillation with $\delta R = 2/3 \, R$ incompatible with the fact that the brightness changes less than 0.1 mag, and the temperature less than about 1 000°.

It would also be difficult to understand, on the latter assumption, why the lines should be broadest when the average velocity is zero, and when the volume of the star (according to the pulsation hypothesis) should be a minimum.

These inconsistencies lead us back to the original idea : the broad lines are principally produced by rotation. The narrow lines must then be explained in terms of an additional mechanism — prominences, turbulence, etc. over a limited area.

The situation in the case of β CMa is, of course, less extreme, and were it alone we would perhaps not be able to advance this line of reasoning. But there can be no doubt that the physical processes in all these stars are quite similar.

It must, however, be recognized that the rotational interpretation of the broad lines has, in the case of some O—type stars, suffered a serious set-back (see Publ. A. S. P., **64**, 118, 1952) and we should retain an open mind with respect to the broad lines of σ Sco, 12 Lac., etc.

IV. The last hypothesis is one that has been recently put forward by D. H. MENZEL, and is as yet unpublished. He supposes that these stars have magnetic fields — not large enough to show in Horace BABCOCK's observations but large enough to influence the pulsations if they are present. He assumes a circular ring, with an electric current, around the equator, at an appropriate depth below the surface of the star. Then the pulsations at the pole and at the equator will have different frequencies (fig. 4).

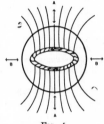

FIG. 4.

SUMMARY OF HYPOTHESIS

It is probable that the final solution of our problem will recognize the following observational facts :

1. The variation in line broadening is related to the rotation of the star.

2. When this broadening is greatest, as in σ Sco, we observe only P_2, K_2. Presumably P_1, K_1 is then lost in the variation of P_2, K_2.

3. There is a correlation between K_2 and the amplitude of line broadening.

4. Δm varies in both oscillations, and there is a correlation, but not a one — to —one relationship of Δm and K.

5. There is no relation between the $K's$ or $\Delta m's$ and luminosity.

6. The phenomenon is restricted to a small range in the spectrum near the main sequence.

7. The phase relations previously discussed between velocity and Δm, and between velocity and variation in line broadening are important properties of these stars.

8. There are irregular variations in K_2 and in the line broadening ; and the two are correlated.

9. There is no relation between P and the $K's$, or the $\Delta m's$, or the changes in line profiles.

10. There are no changes in temperature and perhaps only very slight ones in luminosity criteria.

11. The constant of the P $\sqrt{\rho}$ relation differs markedly between Cepheids and RR Lyrae stars.

12. The sharp-lined stars, or sharp-lined stages of stars, show forbidden He I and thus indicate a small amount of under-luminosity. But β CMa shows rather high luminosity from its spectral lines.

13. Although β CMa and σ Sco are very different in spectrum (high and low luminosity, respectively), their periods are nearly alike. This would seem to violate the P $\sqrt{\rho}$ relation which holds so well for the Cepheids and long-period variables, as well as the RR Lyrae stars.

In order to account for these results we propose the following :

a) There are two oscillations — one mainly polar (P_1, K_1) the other mainly equatorial (P_2, K_2).

b) Differences arise from 1) orientation of the line of sight (β CMa, 16 Lac are seen from $i \sim 45°$; σ Sco from $i \sim 0$; β Ceph from $i \sim 90°$) and 2) velocity of equatorial rotation (large in σ Sco ; intermediate in 12 Lac, ν Eri ; small in β Ceph).

c) The variations in line profile are due to indirect causes, and are not simply produced by the δr of the pulsations. We see them most conspicuously when $i \sim 0°$ (σ Sco).

d) The doubling of the lines is really an unsymmetrical broadening associated with the rotation. (Perhaps we should recognize that in 12 Lac the duplicity sometimes persists on the entire descending branch of the velocity curve : the line when broadest resembles two diffuse components of equal intensity).

e) The light variation affects both oscillations, but the (P_2, K_2) more so than the (P_1, K_1).

The facts which we have outlined favor pulsations of either the LEDOUX or the MENZEL type. The former has one disadvantage : it is not clear why one traveling wave is suppressed. The latter, while meeting this obstacle, is not yet mathematically sufficiently well developed to permit a detailed application. No magnetic fields have been observed by BABCOCK in the β CMa stars. On the other hand, true magnetic stars show phenomena of the spectrum-variable kind. They do not resemble the β CMa phenomena.

It is clear that we require more facts. One of the first tasks should be a study of the velocity curve of β Cep. We shall know a great deal more when we can definitely state whether or not it displays the beat phenomenon. It would also be of interest to determine whether *all* stars at B_2, main sequence, are oscillating in the maner of β CMa. The RR Lyrae stars are now known to occupy a well-defined region in the H-R diagram, and there is ample evidence supporting M. SCHWARZSCHILD's conclusion that all stars within this region do, in fact, vibrate.

Manuscrit reçu le 10 *mai* 1952.

THE BETA CANIS MAJORIS STARS

A. van Hoof

University of Louvain

I. INTRODUCTION

All astronomers who spent some time in Otto Struve's company in the early nineteen-fifties will remember the keen interest he then took in the β CMa stars, and the real happiness he felt about every new fact that became known, or discovery that was made—often under his stimulus—in the "new and fascinating field" that he had done so much to open up for research.

The article reprinted here is one of the many manifestations of this enthusiasm. Struve may well have written it mainly for himself, for the purpose of getting a clear picture of what he knew already about the subject, and of the knowledge he still lacked. There is in the article a sober report of the observationally known facts, an enumeration of the hypotheses already advanced to account for them, and an exposition of the reasons for discarding these hypotheses as erroneous. Finding that none of the explanations is adequate, Struve—as for himself again—then makes the inventory of the systematics and correlations that can be distilled from the observations, and which the final interpretation will have to explain. Behind the great variety of observed phenomena he discerns two agents: (1) pulsations (which cause the variations in light, velocity and line width) and (2) rotation (which complicates these variations) and he weaves them into an interpretation of his own. It is however doomed to remain tentative so long as the observational base on which it rests has not been extended, whence the final conclusion: "More observations!".

That recommendation has inspired the planning and the carrying through of numerous observing campaigns aimed at finding new members of the group, and extending our knowledge of the known members.

Under the first, there may be mentioned the searches made by Walker (1952) who investigated photoelectrically a total of 44 B stars and found only one new β CMa variable. Pagel (1956) was more successful and discovered three new members among the six stars examined on account of reported velocity variations. Lynds (1959) added two more members to the list of stars known for certain to be β CMa variables, while Petrie and Pearce (1962) noticed possible β CMa characteristics in eleven out of 570 B-type stars on their velocity program. A further extensive search by Milone is reported to be in progress at Cordoba.

Under the observing campaigns with objective (2) we must first mention the many photoelectric and spectroscopic series secured by Struve himself and his Berkeley co-workers, the intensive photoelectric observations by Van Hoof of southern β CMa stars, the world-wide observation campaign on 12 (DD)Lac, organized by De Jager (1962), observations on individual objects made by Milone (1963b), Hogg (1957), Rodgers and Bell (1962), Petrie (1954), and Miczaika (1952), and by various astronomers in the U.S.S.R., Japan, India, Germany, Italy and other countries.

One of the results of these campaigns is that they have outlined the position of the β CMa stars in the sky and on the H-R diagram. In the sky these objects seem to avoid the vicinity of the galactic equator. With only one exception, β Cru, they all have $|b| > 8°$, a peculiarity which was first noticed by De Jager (1953) and which probably has a cosmological significance. They further prefer the neighborhood of nebulosities, and many of them belong to open clusters (α Lup, β Cru, θ Oph to the Sco-Cen cluster; 12 and 16 Lac to the Lac cluster).

TABLE 1

The β CMa Variables

Star	P_2	P_1	MK type	M_v	$(\Delta m)_2$	$(\Delta m)_1$	$2K_2$ (km/sec)	$2K_1$ (km/sec)	Remarks
V 986 Oph	6^h56^m*	?	B0.5 III	-4^m5	—	—	—	—	
α Lup	6 14^m*	?	B2 II	-4.6	0^m025	?	15	2	Slow changes in Δm, P
β CMa	6 02	6^h00^m	B1 II-III	-4.7	0.004	0.021	7	12	
σ Sco	5 55	5 44	B1 III	-4.6	0.040	0.021	110	15	No color variation in P_1
β Cru	5 40	5 45:	B0 III	-4.5	0.020	0.014:	5	2.5:	Δm is U
ξ^1 CMa	—	5 02	B1 IV	-4.2	—	0.040	—	36	
BW Vul	4 49	—	B2 III	-4.1	$\{0.19, 0.26\}$	—	150	—	
KP Per	4 48*	?	B2 IV	-3.5?	0.09:	?	20	5	Δm variable
12 Lac	4 38	4 44	B2 III	-4.1	0.074	0.042	36	15	
β Cep	—	4 34	B2 III	-4.1	—	$\{0.02, 0.005\}$	—	18–46	
15 CMa	4 26*	?	B1 IV	-3.5?	?	?	?	?	$\langle\Delta m\rangle = 0.02$,
τ^1 Lup	4 16	4 21	B2 IV	-3.0?	0.030	0.005	?	?	$\langle 2K\rangle = 7$
ν Eri	4 10	4 16	B2 III	-4.1	0.106	0.060	29	17	$\langle 2K\rangle = 11$
16 Lac	4 04	4 06	B2 IV	-3.3	0.055	0.035	30	9	
δ Cet	—	3 52	B2 IV	-3.3	—	0.025	—	13	
53 Ari	—	3 40	B2 IV	-3.0	—	?	—	6	Var. ampl.
γ Peg	—	3 38	B2 IV	-3.0	—	0.015	—	7	
θ Oph	3 22*	3 18:	B2 IV	-3.0	0.025	0.005:	11	3:	Var. ampl.

* No information available.

On the H-R diagram they are confined to a narrow strip, somewhat above and to the right of the upper main sequence, with limits B0.5 and B2 in spectral type, classes II-III and IV in luminosity. They have however to share this territory with non-variable B stars, so that a negative answer has to be given to Struve's question at the end of his article.

Eighteen objects are at present known for certain to belong to the group. Their names and characteristics are collected in Table 1. For a few more B stars, of the correct spectral types and luminosity classes, variability has been reported by some observers, constancy by others. These reports are not necessarily conflicting; they may simply indicate that in those stars the amplitude of variation changes slowly between a well-observable maximum and a minimum below the limit of detectability. The case of α Lup is instructive in this respect. It is therefore important that those stars be kept under observation; they might indeed well be β CMa stars in the initial or final stages of variability, which would make them particularly interesting.

Table 1 shows clearly that many β CMa stars have more than one period. In some, the light- and velocity curves at times diverge so much from simple sine curves that representation by two slightly different periods (Struve's P_1 and P_2) is insufficient. Examples of those light curves are shown in Figure 1.

It is at present also well established that the light changes are accompanied by color changes. Stebbins and Kron's (1954) six color photometry of β Cep, Hogg's (1957) UBV observations of σ Sco and Van Hoof's UV lightcurves of several southern β CMa stars all give evidence that those stars are bluer at maximum light than at minimum light.

II. THE PERIOD-LUMINOSITY LAW

The most striking fact illustrated by Table 1 is that there exists a definite run of the absolute magnitude with the main period. The first to notice the existence of a period-luminosity law among the β CMa stars were Blaauw and Savedoff (1953), who, from six stars derived the relation

$$M_v = -10 - 9 \log P. \tag{1}$$

Shortly afterwards McNamara (1953) and Petrie (1954a) also arrived independently at the conclusion that a period-luminosity relation exists. To express it, Petrie proposed however the formula

$$M_v = +0.4 - (18.1 \pm 2.3)P \tag{2}$$

which gives absolute magnitudes that are systematically 0.4 mag. fainter than those of equation (1).

More than did the detection of color changes, the discovery of a period-luminosity law supported Struve's point of view that pulsations were the main cause of the observed β CMa type variability. But what kind of pulsations? Many astronomers were reluctant to accept the idea of purely radial pulsations like those in the cepheids, because of marked differences in the behavior of the two groups of stars. It was soon realized that the applicability of the Van Hoof-Wesselink method for the derivation of the stellar radius would provide a criterion.

The test was successfully applied by Stebbins and Kron (1954) to β Cep, a star with fairly simple light and velocity changes, but its application to stars with more complicated variations, especially to those showing strong variations in the spectral line profiles, failed completely—De Jager (1953) for 12 and 16 Lac and ν Eri, Hogg (1957) for σ Sco, Walker (1954) for 16 Lac. Even negative values for the radius were obtained, so that the existence of *radial* pulsations was seriously questioned. In Van Hoof's mind however, the failure of the method was due exclusively to the absence in the β CMa stars of the large phase-lag between light and radius changes which is present in cepheids, for which the method had been devised. He showed (1962a)

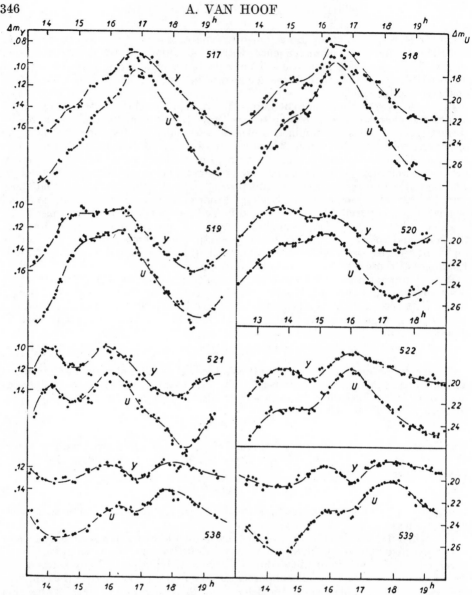

Fig. 1.—Yellow (Y) and ultraviolet (U) light curves of σ Sco, observed with the 60-inch Rockefeller telescope of the Boyden Observatory. The numbers are JD *minus* 2438000, the abscissae Bloemfontain sidereal time, the ordinates differences in photoelectric magnitude: σ Sco *minus* τ Sco.

that the existence of an inconspicuously small lag of the velocity curve and of the light curve behind the color curve must unavoidably lead to a negative value for the radius.

III. THE VAN HOOF EFFECT

The detection of small phase-lags showing the progressive run of the compression waves through the star's atmosphere could then be utilized as an argument which could outweigh the objection raised against the assumption of pulsations with spherical symmetry. Measurements by Van Hoof and Struve (1953) on large-scale tracings of spectrograms of β CMa taken by Struve with the Mt. Wilson 100-inch coudé spectrograph at a dispersion of 3 Å/mm disclosed that there was indeed a lag of about 0.04 period of the velocity curve derived from Hγ with respect to the curve derived from the O II lines $\lambda\lambda 4345$, 4349. A similar lag, termed the Van Hoof effect by Struve, was afterwards found to exist between the velocity curves derived from the hydrogen lines on the one hand and those derived from all the other lines on the other, in 16 Lac (Van Hoof, DeRidder, and Struve 1954), BW Vul (McNamara, Struve, and Bertiau 1955) and σ Sco (Struve, McNamara, and Zebergs 1955).

Van Hoof (1957) later expressed the opinion that the effect was not entirely due to the fact that the spherically symmetric compression waves attained the deep layers of the stellar atmosphere earlier than the upper ones, but that it was also caused by the changes which the turbulence suffered in the course of a pulsation. Turbulence will cause the hot rising gas elements to contribute more than will the cooler descending ones to the formation of spectral lines with high excitation potential. Consequently these lines will have their optical center displaced somewhat to the violet with respect to the lines of low excitation potential, in the formation of which the cooler descending gas masses took a relatively larger part. An increase in turbulence will thus augment the shift, while a decrease in turbulence will diminish it. Turbulence, varying inversely with the radius, will thus produce not only the observed line width changes, but it will bring about a Van Hoof effect as well.

The explanation implies that in the case of β CMa itself where $K_2 < K_1$, the Van Hoof effect can have the opposite sign when the two oscillations present are in opposite phases. This was indeed found to be the case. Moreover, if the explanation is correct, there will be no Van Hoof effect associated with those oscillations that do not show line broadening. McNamara (1956) found this to be the case in ξ^1 CMa, a star which shows only the $(P_1 K_1)$ oscillation.

IV. THE DETERMINATION OF THE RADIUS

Better photoelectric observations in different wave lengths have given good evidence that the β CMa stars are hottest and brightest at the phase of minimum volume (on the old pulsation hypothesis) and are coolest and faintest at maximum volume. Under these circumstances and for purely radial pulsations, the stellar radius R can be derived from

$$\left(1 + \frac{|\Delta R|}{R}\right)^2 = \frac{\left(1 + \frac{|\Delta T|}{T}\right)^4}{1 + |\Delta m_v| + |\Delta BC|} \tag{3}$$

where $\Delta R \approx KP/2$ is the amplitude of the radius variation, K being in km/sec and P in seconds;

$T =$ the effective temperature;

$\Delta T =$ the amplitude of the variation in effective temperature as deduced from the variation in color or in spectral type;

$\Delta m_v =$ the amplitude in visual light;

$\Delta BC =$ the amplitude of the bolometric correction, deduced from the known correlations between BC and spectral type or color.

If plausible values for R are obtained in this way, they will naturally support the hypothesis of purely radial pulsations on which the derivation is based.

Van Hoof (1961) applied the procedure to ν Eri and found concordant results from the main oscillation and from the many secondary oscillations which his analyses of the light and velocity curves had disclosed. They are shown in Table 2. β Cru (Van

Hoof 1962b), θ Oph (Van Hoof 1962c), and other members of the group were treated by him in the same way, and radii were obtained which agreed fairly well with the values to be expected from the temperature and luminosity criteria.

TABLE 2*

COMPUTED RADIUS OF ν ERI (IN 10^6 km)

R_0	5.25	R_1	5.3
R_{02}	4.74	R_{03}	6.0
R_{13}	5.60	R_{14}	5.5
R_{35}	5.67		
R_{46}	5.62	R_2	5.5

average: $\overline{R}_\nu = 5.46 \pm 0.07$

* In our notation a single index i refers to the order of the overtone, 0 stands for the fundamental mode. A double index ij refers to the difference oscillation between the modes i and j.

A reservation which must be made in appreciating these results is that they were obtained from photometric and spectrographic data that were far from being simultaneous. The combination of such data together in one formula is rather hazardous, since several β CMa stars are known to change their amplitudes slowly with time.

V. THE NATURE OF THE PULSATIONS

The discovery of a period-luminosity law leaves open the question which kind of pulsations are involved: radial, pseudo-radial or non-radial. The different effect of the P_1 and P_2 oscillations (in Struve's notation) on the line widths suggests a rather different nature or a different mode of excitation for them. Much thinking has been devoted to the elucidation of this enigma. Struve (1955) at one time thought of a shell which was ejected from the stellar surface at each expansion, rose to a certain height and then fell back on the star. One of the two P's would be the pulsation period proper, the other would be the "flight time" of the shell. The idea has been further developed by Odgers (1955) and will be commented on later.

Van Hoof (1957) saw in Struve's oscillation of frequency σ_2 the radial pulsation in the fundamental mode, and in the oscillation of frequency $\sigma_1 = (\sigma_2 + \epsilon)$ the difference oscillation between this fundamental pulsation and an overtone oscillation of frequency $2\sigma_2 + \epsilon$ which at that time had not yet been detected. Despite the smallness of their difference in frequency, the two oscillations will have a fairly different run of their ξ ($= dr/r$)'s near the surface of the star. This in turn will give the two oscillations a different influence on the temperature gradient and thence on the turbulence and on the line widths.

Milone (1963a) gave support to the idea by proving that the homogeneous and the standard models will indeed find their temperature gradients increased by an adiabatic contraction and lowered by an expansion.

Following this, Van Hoof obtained photoelectric light curves of ν Eri, β Cru, θ Oph and (recently) of σ Sco in which the predicted overtone oscillation with period about $P/2$ is clearly visible around the times of minimum amplitude in the beat cycle (see Fig. 1). Further overtones were revealed by periodogram analysis of large numbers of light curves, whose period ratios reproduce the theoretical values for the polytrope $n = 3$ and $\Gamma_1 = 1.52$–1.53 which can be found by interpolation from Schwarzschild's (1941) results.

The objection has been made that several of these overtone and resonance oscillations are so weak that their reality is doubtful. But this objection ignores the fact that the corresponding amplitude ratios $\Delta m_v / \Delta m_u$ increase with increasing mode order as theory requires. Furthermore, they are detectable also in the observed velocity curves, and they yield concordant values for the stellar radius when treated by equation (3). It is moreover obvious that oscillations with small amplitudes will suffer relatively more from the smoothing out of observational errors (or what is so considered), so that the mean values found by the periodogram analysis can be expected to represent only a fraction of the real amplitudes. Table 3 shows the results of the analysis for ν Eri, the best-observed β CMa star.

TABLE 3

OSCILLATIONS OF ν ERI

Period (in days) and (in parentheses) U amplitudes (in $0^m.001$):

$P_0 = 0.173509\ (106)$	$P_{02} = 0.177386\ (060)$	$P_{03} = 0.117716\ (015)$
$P_1 = 0.116689\ (027)$	$P_{13} = 0.175381\ (040)$	$P_{14} = 0.116383\ (012)$
$P_2 = 0.087690\ (021)$	$P_{24} = P_0$	
$P_3 = 0.070140\ (006)$	$P_{35} = 0.171600\ (030)$	
$P_4 = 0.058268\ (006)$	$P_{46} = 0.169796\ (020)$	

Ratios of Y amplitude/U amplitude:

$P_0 = 0.60$	$P_{02} = 0.71$	$P_{03} = 0.75$
$P_1 = 0.69$	$P_{13} = 0.65$	$P_{14} = 0.70:$
$P_2 = 0.73$	$P_{24} = P_0$	
$P_3 = 1.:$	$P_{35} = 0.58$	
$P_4 = 1.:$	$P_{46} = 0.56:$	

Ratios P_i/P_0:

$$P_1/P_0 = 0.675 \qquad P_2/P_0 = 0.505 \qquad P_3/P_0 = 0.404 \qquad P_4/P_0 = 0.356$$

Ledoux (1958) in further advocacy of his interpretation of the β CMa phenomena in terms of non-radial oscillations, already referred to in Struve's article, has severely criticized Van Hoof's ideas. According to Ledoux, Van Hoof's forced oscillation P_1 had no physical meaning (this was written before the $P/2$-oscillation had been discovered) but should it nevertheless exist, then it would show larger turbulence variations, and hence line width changes, than does P_2. Turbulent velocities on the other hand could never attain values of over 100 km/sec, as were necessary to account for some of the observed line splittings. In the same article Ledoux also rejected Odgers' interpretation of line splittings and beats in terms of shock waves and ejected shells.

To Opolski and Ciurla (1962) the only periods that are meaningful are the short pulsation period P_2 and the long, so-called beat-period P_b. The latter is nothing but a modulation period of the former, so that the observed brightness variations obey the formula

$$\Delta m = [A_0 + A_1 \cos(2\pi\beta t)] \cos 2\pi(\alpha t + B \sin 2\pi\beta t), \tag{4}$$

where A_0 and A_1 are constants, $\alpha = 1/P_2$ and $\beta = 1/P_b$. The authors do not mention the origin of the modulation period they suggest.

An effort to check the interpretation in terms of purely radial oscillations of different modes was made by Stothers (1965) who computed model sequences for stars with masses of 15, 20, and 30 $\odot$, evolving from the main sequence to the end of H burning. He found that only the stellar envelope behaves pulsationally like a polytrope of index $n = 3$ with some characteristic Γ_1 value, and that the period ratios found by Van Hoof occur on the H-R diagram along a line well to the left of the locus of secondary contraction. The slope of the locus of constant period ratios agrees well with the observed slope of the β CMa strip but the locus itself lies rather far to the left and above this strip. According to Stothers, however, the discrepancy could be removed by an adjustment of some of the input data and by taking other opacity sources than electron scattering into account.

For the occurrence of the beats in β CMa stars, Chandrasekhar and Lebovitz (1962a) have recently proposed an explanation which deserves special attention because it is based on theoretical considerations. In a series of papers (1962b) these authors have studied the possible oscillation modes of rotating gaseous masses and distorted polytropes and have arrived at the general conclusion that among the various oscillations there exist two different pulsation modes. In the absence of rotation there is a purely radial pulsation R characterized by the Lagrangian displacements ξ in the three coordinates

$$\xi_1 = X_R \cdot x_1 \qquad \xi_2 = X_R \cdot x_2 \qquad \xi_3 = X_R \cdot x_3 \tag{5}$$

and a solenoidal oscillation S with the ξ's:

$$\xi_1 = X_S \cdot x_1 \qquad \xi_2 = X_S \cdot x_2 \qquad \xi_3 = -2X_S \cdot x_3, \tag{6}$$

where X_R and X_S are arbitrary constants. The corresponding frequencies are

$$\sigma_R = (3\Gamma_1 - 4)\frac{W}{I} \qquad \text{and} \qquad \sigma_S = \frac{4}{5} \cdot \frac{W}{I}, \tag{7}$$

W denoting the potential energy and I the moment of inertia.

In the presence of rotation around the x_3-axis these two pulsations become coupled and change their ξ's and σ's, but if the angular velocity $\Omega \to 0$ then:

(i) for $\Gamma_1 \neq 1.6$ the resulting pulsations will reduce to those defined by equations (5) and (6), and for an initial disturbance which has about spherical symmetry and which is sufficiently weak, only the radial pulsation may be excited.

(ii) For $\Gamma_1 = 1.6$ the two pulsations reduce to something entirely different, an R-oscillation with

$$\xi_1 = X_R \cdot x_1 \qquad \xi_2 = X_R \cdot x_2 \qquad \xi_3 = \frac{X_R}{A} \cdot x_3, \tag{8}$$

and an S-oscillation with

$$\xi_1 = X_S \cdot x_1 \qquad \xi_2 = X_S \cdot x_2 \qquad \xi_3 = -\frac{X_S}{B} \cdot x_3 \tag{9}$$

with $A > 2$ (increasing with increasing polytropic index n),
$\quad B < 0.2$ (decreasing with increasing polytropic index n),
and with frequencies

$$\sigma_R^2 = \sigma_0^2 - k_1 \cdot \Omega^2 \tag{10}$$

$$\sigma_S^2 = \sigma_0^2 + k_2 \cdot \Omega^2, \tag{11}$$

where σ_0 is the frequency in the absence of rotation and k_1 and k_2 are positive factors depending on n. We thus have systematically $\sigma_R < \sigma_S$ or $P_R > P_S$ and

$$\left(\frac{\Delta\sigma}{\sigma_0}\right)^2 = \frac{2\Delta P}{P_0} = k_3 \cdot \Omega^2 \tag{12}$$

with k_3 depending on n and ρ_0 (the central density). For n values from 2 to 3 and reasonable values for ρ_0, equation (12) requires rotational velocities at the equator of the order of 120–240 km/sec to yield the relative period differences $\Delta P/P = 0.02$–0.03 which are observed in β CMa stars. This now, according to Chandrasekhar and

Lebovitz, is the case we are faced with in the β CMa stars with two slightly different periods P_1 and P_2 (Struve's notation): those are simply B stars, rotating at speeds which are normal for their spectral type, but having $\Gamma_1 = 1.6$, in which a spherical symmetric disturbance has accordingly excited automatically both the R and S oscillations. The quite different deformations (8) and (9) make understandable a different effect on the line widths. At an earlier stage the authors had already advocated the identifications

$$P_2 = P_R \qquad \text{and} \qquad P_1 = P_S.$$

The interpretation must be credited for the natural way in which are explained both the occurrence of the double periodicity and the different bearing of the oscillations involved on the line widths. But a general objection which has been raised against it is that all the known members of the group show low projected rotational velocities $V_{rot} \cdot \sin i$, so that the acceptance of the interpretation implies the assumption that all these stars are viewed pole-on or nearly so. But this is an assumption for which there is no ground—though it could explain the rarity of the β CMa stars among the B stars—unless it be backed by the further assumption that the pulsations are so much weaker parallel to the equatorial plane than perpendicular to it, that they simply escape detection when the polar axis is making a sizable angle with the line of sight.

That this could really be the case is a point of view defended by Böhm-Vitense (1963) who, ascribing the longer of the two periods to the equatorial pulsation and the shorter to the polar oscillation, argued that all of the five β CMa stars with well-known double periodicity show indeed a larger amplitude for their polar pulsation $(2K_1 > 2K_2)$. Unfortunately for the latter point of view, σ Sco has since been found to violate the rule drastically, and the violation is the more meaningful that the $2K_2$ involved is the largest but one of all the velocity amplitudes known in the group. A further argument against the existence of high rotational velocities in β CMa stars was developed by Van Hoof (1964). He made the remark that for the two stars which are known members of a spectroscopic binary, namely σ Sco and 16 Lac, the mass function sets an upper limit to the orbital and to the rotational velocities if the corresponding projected velocities are known, and if the assumption is made that the orbital and equatorial planes coincide. For both stars the corresponding computations yielded for V_{rot} a maximum value which was by a factor 7–8 smaller than the values required by Chandrasekhar and Lebovitz' interpretation. Chandrasekhar (private communication) has answered this objection by announcing that a more detailed theory leads to a substantially larger period-splitting and by arguing that the tidal distortion in a member of binary system would increase the splitting effect still more.[1]

Further light is thrown upon the question by a photometric investigation of σ Sco carried out by Van Hoof (1966). This investigation, based on tens of complete photoelectric lightcurves, has settled besides other details that: (i) the ill-known secondary period P_1 is *shorter* than the main one P_2:

$$P_1 = 0^d.239671 \qquad \text{against} \qquad P_2 = 0^d.246840;$$

(ii) the secondary oscillation proceeds without any color variation:

$$(\Delta m_v)_1 = 0.022 \text{ mag.}, \qquad (\Delta m_u)_1 = 0.021 \text{ mag.}$$

The latter point proves that this oscillation cannot be a radial or nearly radial one. Equation (3), valid for a radial oscillation, can indeed be written in an approximate way:

$$1 + \frac{KP}{R} = \left(1 + 4\frac{\Delta T}{T}\right)(1 - |\Delta m_v| - |\Delta BC|) \tag{13}$$

---

[1] The critical value of $\Gamma_1 (=1.6)$ was said to be lower than 1.6 and to be a function of n.

or

$$1 + \frac{KP}{R} = (1 + k' |\Delta C|) (1 - |m_v| - k'' |\Delta C|), \tag{14}$$

where k' and k'' are positive constants, and for $\Delta C = 0$:

$$\frac{KP}{R} = -|\Delta m_v| \tag{15}$$

The opposite signs in the two members prove that the underlying hypothesis is erroneous. On the contrary, a solenoidal oscillation such as Chandrasekhar and Lebovitz' S-oscillation is not inconsistent with our observational result.

The investigation further yielded for σ Sco a radius of 13 $R_\odot$ when equation (3) was applied to the main oscillation (P_2, K_2). The value seems rather large. But an exaggerated radius value must be expected in case the pulsation has a larger amplitude in the equatorial plane than along the polar axis and if the star is observed with the equator nearly edge-on. It has already been mentioned that the use made of the mass-function makes the latter circumstance fairly certain.

Concluding we can say that (i) the sign of the inequality $P_2 > P_1$, (ii) the absence of color variation in the secondary pulsation: $(\Delta C)_1 = 0$, (iii) the large radius value derived from the main oscillation (P_2, K_2, Δm_2), all support the idea that in σ Sco

$$P_2 = P_R \qquad P_1 = P_S$$

for $\Gamma_1 = 1.6$, or whatever Chandrasekhar and Lebovitz' revised critical value for Γ_1 may be. The only reservation still to be made is that their theory has to provide a larger period-splitting factor for the moderate values of Ω that appear most probable.

But what to think then of 16 Lac where $P_2 < P_1$? No colorimetric study of this star has been made up to the present, but the sign of the inequality between the periods by itself contradicts the identifications just made for σ Sco. To claim now that $P_1 = P_R$ and $P_2 = P_S$ sounds unreasonable, and the best we can do for the moment is to postulate that in 16 Lac $\Gamma_1 \neq 1.6$. In that case the inequality between P_R and P_S depends much more on the difference ($\Gamma_1 - 1.6$) than on Ω, and the suggestion seems the more attractive in that the double inequality $P_2 \gtrless P_1$ is met also among those β CMa stars which are considered to be single stars, as Table 4 shows.

Among the latter only ν Eri (Van Hoof 1961) has had its light, velocity, and color variations analyzed. Both its P_1 and P_2 oscillations show definite color variations

TABLE 4

DATA FOR DOUBLE AND SINGLE β CMA VARIABLES

Spectroscopic binaries:	Periods	Velocity amplitudes
σ Sco	$P_2 > P_1$	$K_2 > K_1$
16 Lac	$P_2 < P_1$	$K_2 > K_1$
θ Oph*	$P_2(?) > P_1(?)$	$K_2(?) > K_1(?)$
Single Stars		
β CMa	$P_2 > P_1$	$K_2 < K_1$
12 Lac	$P_2 < P_1$	$K_2 > K_1$
ν Eri	$P_2 < P_1$	$K_2 > K_1$

* Probably a spectroscopic binary with $P_{orb} \approx 11^d$; the longer period corresponds to the larger amplitude, but it is not known for certain whether this is P_2 or P_1 (Van Hoof, unpublished)

and for both equation (3) yielded reasonable and reasonably concordant values for the stellar radius, as if the two oscillations were (at least approximately) radial (see Table 2).

These results are not inconsistent with Chandrasekhar and Lebovitz' opinion that for slow rotation and $\Gamma_1 \neq 1.6$ *only* the radial modes may be excited, and they leave room for Van Hoof's interpretation that in ν Eri, P_2 is the radial fundamental pulsation and P_1 is the difference oscillation between P_2 and the second overtone with period about equal to $P_2/2$.

The general conclusion then from the occurrence of $P_2 > P_1$ and $P_2 < P_1$ and from the detailed analysis of σ Sco and ν Eri, seems to be that Chandrasekhar and Lebovitz' theoretical investigations really help to understand the nature of the pulsations in the β CMa stars, but that contrary to their suggestion, the case $\Gamma_1 \neq \Gamma_{\text{crit}}$ occurs more often among these stars than the case $\Gamma_1 = \Gamma_{\text{crit}}$ and that the rotational velocities are actually lower than those predicted for the latter case.

As to the non-radial oscillations predicted by Cowling and Newing (1949) and by Ledoux (1951), the investigation of σ Sco disclosed in each mode a few oscillations with evenly-spaced frequencies which might perhaps represent them, but the amplitudes are so small that they play no role in the shaping of the light curve.

Chandrasekhar and Lebovitz' theoretical considerations have been worked out to a second approximation by Clement (1965a). The interesting point is that the splitting $(\Delta\sigma)^2/2$ now becomes twice as large as the older value for $n = 3$, and almost four times as large for $n = 3.5$. Moreover Γ_1 is now smaller than 1.6 and varies with n. Clement (1965b) has further established formulae by which the *real* expansion velocities in the equator and along the polar axis can be derived from the *observed* velocity amplitudes K_1 and K_2. Other formulae permit the computation of the inclination of the axis of rotation to the line of sight and of the stellar radius in case the observed amplitudes Δm_1 and Δm_2 can be freed from the contribution of the temperature changes involved. Once the radius is known, the period and the rotational speed at the equator can be computed for any assumed value of the mass of the star. Clement's formulae enabled him to find also the run of the line-width changes through the cycle, and it is worth mentioning that, although he could not explain the line splitting, he found nevertheless a double maximum in the line broadening of stars like 12 Lac and BW Vul, almost at those phases where the observations show the line splitting to be maximum. For β CMa his computations gave a single maximum in the line broadening, occurring at the phase of maximum compression in the P_2-oscillation, in full agreement with observations. It must be mentioned, however, that Clement's graphs show the line width changes to have maximum amplitude when the P_1 and P_2 oscillations are in opposition of phase, while our observations (see Fig. 2) disclosed the reverse to be true. This might mean that turbulence (or pressure?) has an even greater influence than the particular expansion and contraction patterns.

Clement's formulae can be applied to Van Hoof's photometric results on σ Sco. The Δm_R's, free from the effect of changing temperature, and hence exclusively due to the change in area of the radiating stellar disk, are easily found to be

$$(\Delta m_R)_1 = (\Delta m_1)_{\text{obs}}, \quad \text{since} \quad (\Delta C)_1 = 0 \quad \text{and hence} \quad (\Delta T)_1 = 0,$$

and

$$(\Delta m_R)_2 \cong \frac{\left[1 + \dfrac{(\Delta T)_2}{T} \right]^4}{1 + (\Delta m_2)_{\text{obs}} + \Delta BC} \quad \text{when } (\Delta T)_2 \text{ is derived from } (\Delta C)_2.$$

The two Δm_2's, combined with the semi-amplitudes of the corresponding velocity curves then yielded the results in Table 5.

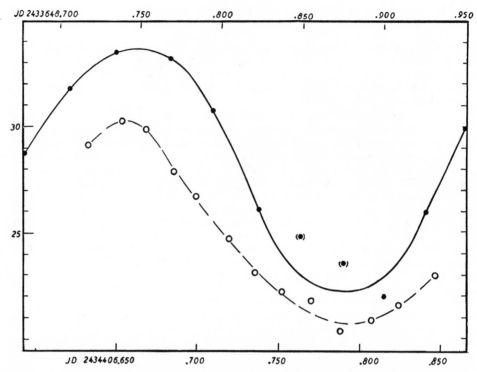

FIG. 2.—Mean half-widths of the spectral lines (O II, C II, N II, Si III, He I) of β CMa during a 6-hour cycle at maximum amplitude in the beat cycle (dots and full-drawn line), and at minimum amplitude (circles and dashed line). Abscissae are time in fractions of a day for the two curves. Ordinates are mean half-widths in units of 0.025 Å.

The radius is obviously too small and the rotational speed at the equator (ΩR) is by far too large, even for $n = 3.5$. Tolerable changes in the input data T, ΔT and ΔBC could raise the radius somewhat, but could not remove the discrepancy between the observed and the computed velocities of rotation.

In view of this failure and recalling that a much closer approximation to the probable value of R was derived on the assumption of a purely radial P_2-oscillation, the conclusion which best reconciles observations and theory is that in σ Sco also, $\Gamma_1 \neq \Gamma_{\text{crit}}$, ΩR is very low, and that the oscillations present are accordingly very close to those defined by equations (5) and (6).

TABLE 5

DATA FOR σ SCO COMPUTED WITH CLEMENT'S FORMULAE

n	$R/R_\odot$	sin i	$M/M_\odot$	ΩR (km/sec)
2	7.0	0.841	24	274
3	6.7	0.848	8.4	205
3.5	6.6	0.854	4.8	141

VI. LINE WIDENING AND SPLITTING

However nicely a proposed mechanism may account for the observed light, velocity, and color variations of the β CMa stars, it will not be acceptable if it does not provide a natural explanation for the periodic line widening or/and line splitting observed in so many of these stars. However, before it can provide a crucial test for any proposed model, the phenomenon itself must be thoroughly known observationally, and this can hardly yet be said to be the case.

What can be learned from Struve's article is:

  (i) that the widening seems related to the rotation of the star;
 (ii) that the widening or splitting follows but one oscillation, namely the (P_2, K_2);
(iii) that the widening or splitting is maximum at maximum compression in that oscillation;
 (iv) that a widening is probably an unresolved splitting; and
  (v) that the amplitude of the widening or splitting is related to K_2.

Observations made since 1952 call for the following refinements and comments on these statements:

  (i) The line width changes are *mainly* bound to the (P_2, K_2) oscillation, but (P_1, K_1) affects the lines in just the same manner, though to a much smaller extent. This can be seen from Figure 2.
 (ii) In the cases where line widening occurs, the line width changes continuously from a minimum (reached at maximum expansion in the (P_2, K_2) oscillation) to a maximum (reached at maximum compression in that oscillation), and vice-versa; in the cases where line-doubling occurs the splitting goes through two maxima which are reached respectively at about $0^P.2$ before, and about $0^P.2$ after maximum compression. The different behavior makes it hard to consider the line widenings as unresolved splittings.
(iii) Line widenings and doublings both leave the equivalent widths unaffected. This proves, as was first stated by Huang (1955) that if the changes in the line shape are due to Doppler effects, the different line-of-sight velocities must come from different areas on the stellar disk and not from elements which lie the one above the other.
 (iv) When the line width changes, the Van Hoof effect is also present.

So much for our observational knowledge of the phenomenon; these statements prove that it is based on statistical evidence rather than on an accurate description of the line profiles themselves.

As to the interpretation, Doppler effects are held responsible for the widening and the splitting-up of the lines. They result from changes either in the horizontal (or surface) motions, or in the vertical motions of elements of the stellar surface and atmosphere. As to the motions of the first kind it seems easy to realize that an increase in rotational speed of the whole or of a part of the stellar surface may *broaden* the lines, but cannot *split* them, and the same is true for surface flows such as would occur in the case of pure S-oscillations or of Cowling and Newing's g-oscillations. Changes in the vertical (or radial) motions on the stellar surface can, on the contrary, produce a splitting as well as a widening of the spectral lines. Therefore Van Hoof (1957) launched the idea that changing turbulence, intensified by the contraction and decreased by the expansion of the star, was the mechanism behind the scene. As already mentioned, Milone (1963) found some support from theory for this explanation. Van Hoof found afterwards (not yet published) from measurement on tracings of the high-dispersion spectrograms of β CMa that macroturbulence (+ rotation) changed indeed from 18 km/sec to 29 km/sec and back, in the course of the six-hour cycle, while microturbulence remained unaltered. Turbulences of this order are normal for early B-stars, but as Ledoux (1958) has pointed out, ordinary turbulence cannot be invoked to explain velocity-differences of over 100 km/sec, like those that are revealed by the split-up lines of σ Sco and BW Vul.

Van Hoof (1962) then suggested that a resonance oscillation of period $P' = P_f/2$ which is naturally excited and enhanced when the fundamental pulsation of period P_f and the second overtone of period $P_{ov} \approx P_f/2$ become coupled, and which has its amplitude decreasing towards the stellar surface in case $P' < P_{ov}$, might perhaps account for the line splittings. Just as the sound waves from a distant source touch the ear with a strength that depends on the direction and strength of the wind, so the sub-photospheric amplitude of this P' oscillation might reach the stellar surface nearly undiminished in the rising macroturbulent gas elements and largely damped out in the descending ones. Differences in their observed velocities of expansion and contraction will thus arise between those elements, and the differences will be magnified (or reduced) by increasing (or decreasing) turbulence. Such further details as the particular shape of the short branch in the discontinuous velocity curves, and its position below the mean-velocity level, are also accounted for by the proposed explanation.

A completely different interpretation of the line splitting has been given by G. J. Odgers (1955). According to him, the fast-running compression wave transforms into a shock wave which detaches a superficial shell from the stellar surface. This shell is lifted up to a certain height and then falls back upon the star. One of the two observed line components corresponds to the motion of the shell, the other to the motion of the photosphere. If the "flight-time" P_{fl} of the shell is *longer* than the pulsation period P_p, a shell that is falling back will encounter the next rising shell and hinder its ascent, so that a decrease in the height attained will result. In a series of successive collisions the effect will change because of the growing phase difference between the motion of the shell and the motion of the photosphere, and beats will make their appearance. If, on the other hand, $P_{fl} < P_p$, then there will be no beats, but a standstill will be observed on the velocity curve around the time of maximum compression. The latter case, according to Odgers, is realized in BW Vul.

In summary, Odgers' explanation implies either a standstill on the velocity curve *or* beats. But σ Sco shows beats *and* a standstill in its velocity curve, and Huang and Struve (1955) expressed as their opinion that "all observational results—constancy of the equivalent widths, absence of thermal effects and of an increase in microturbulence when the assumed superposed layers collide, etc—compel us to abandon the idea of putting one of the layers where the line components originate, above the other."

A concluding remark, which is not out of place, is that prior to any attempt to interpret the changes in line width, efforts should have been concentrated on finding the true character of the changes. Are they proportional to λ or to λ^2? In the case of β CMa the author found from his measurements on the tracings of coudé spectrograms that the changes in the widths obeyed the following law:

$$\log \Delta w = 2.13 \log \lambda + 0.14 \log Z + 0.55 \log \chi_c + 0.02 \log D + \text{const.} \quad (16)$$
$$(\pm 0.16) \quad (\pm 0.03) \quad (\pm 0.06) \quad (\pm 0.02)$$

where $\lambda =$ the wave length,

$Z =$ the Zeeman splitting factor,

$\chi_c =$ the excitation potential,

$D =$ the depth ($\simeq$ intensity) of the line.

While the influence of a magnetic field is not reflected in this result, a dependence on χ_c seems indicated, and the full factor 2 before $\log \lambda$ more than suggests a mechanism other than Doppler broadening. The author wants to stress however that the results rest on a rather short λ base and need confirmation.

VII. ROTATION

In searches for new members of the β CMa group, the B-type stars investigated were generally chosen on account of their having narrow spectral lines. The fact that all known β CMa stars do have narrow lines does therefore not prove at all that narrow lines are the requirement for this type of variability. Only a direct search among early B stars with moderate to broad lines (and suspected variability of their radial velocity) could settle this question.

Sharp spectral lines are indicative of small projected rotational velocity $V_{rot} \sin i$. Consequently the known β CMa stars either have $\sin i$ small, *i.e.*, are all viewed pole-on, a conclusion which has already been discarded as highly improbable, or have their V_{rot} themselves small. That the latter conclusion is nearer to the truth is the outcome of an investigation carried out by McNamara and Hansen (1961). These astronomers picked out all B stars down to mag 5.6, north of $\delta = -30°$, and lying within the same limits of spectral type and luminosity class as the β CMa stars, and compared the frequency distribution of their observed $V_{rot} \sin i$ with the frequency distribution of their V_{rot}. The investigators found a pronounced excess of low projected rotational speeds and noticed that it corresponded nicely to the number of β CMa stars, whence their conclusion. They also remeasured the rotational velocities of most β CMa stars observable from Mt. Wilson, making use only of plates taken during the sharp-line stage (see Sec. VI, Line Widening and Splitting) of the stars in question. Because of the latter circumstance and because of the larger dispersion of their spectrograms, McNamara and Hansen's results, shown in Table 6, have more weight than the values proposed earlier by Slettebak and Howard (1955) or by Huang (1953). McNamara and Hansen noticed that none of the β CMa stars investigated had a $V_{rot} \sin i > 40$ km/sec and that the stars with the lowest $V_{rot} \sin i$ did not show the beat phenomenon.

This induced them to suggest the following picture of the course of events: "A B-type star, evolving in the β CMa domain of the H-R diagram will experience instability, probably as a consequence of the exhaustion of the H fuel in its core and the structural reorganization following this exhaustion. If, at this stage, the star

TABLE 6

PROJECTED ROTATIONAL VELOCITIES $(V_{rot} \sin i)$
OF β CMA STARS

(After D. H. McNamara and K. Hansen [1961])

Star	$V_{rot} \sin i$ (km/sec)	Type of variation*
15 CMa	39	1
σ Sco	30	1,3
12 Lac	29	1,3
β CMa	28	1,3
θ Oph	26	1,3
BW Vul	26	1,4
β Cep	25	1
ν Eri	7	1,3
16 Lac	5	1
ξ^1 CMa	0	2
δ Cet	0	2
γ Peg	0	2

* (1) Variable velocity amplitude. (2) Constant velocity amplitude. (3) Beat period present. (4) No beats observed.

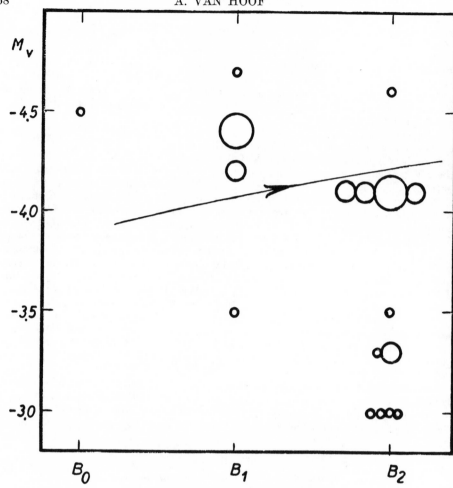

FIG. 3.—H-R diagram for known β CMa stars.

     o   for β CMa stars with $2K < 30$ km/sec
     ○   for stars with $30 < 2K < 80$ km/sec
     ◯   for stars with $2K > 80$ km/sec

The line represents the general trend of the evolutionary displacements. The M_v of σ Sco has been lowered by $0.^m2$ to account, in an arbitrary way, for the invisible component of this spectroscopic binary.

rotates very slowly, it will start pulsating in only one period and the light and velocity variations will be symmetrical with constant amplitudes. For higher rotational speeds the rotation interacts with the pulsation to produce beats in the light and radial velocity amplitudes. When the rotational velocity grows beyond a certain critical value (about 50 km/sec) the rotation completely dampens out the pulsation."

The picture certainly looks attractive but the author nevertheless wonders whether the influence of rotation has not been exaggerated and whether the type and the amplitude of the pulsations is not conditioned by something else than V_{rot}.

Be it admitted for instance that the pulsations are really related to the exhaustion, or to a critical abundance of H and hence to a critical molecular weight μ_{crit}. One would then expect the amplitude of the pulsation to be very small at the onset, to grow towards a certain maximum value and then to fade gradually out, following the growth of the actual μ towards and away from μ_{crit}. In other words, one would expect a change of the amplitude along the evolutionary tracks. In order to test this line of thought, the β CMa stars have been plotted in Figure 3 on a H-R diagram. The radii of the circles that locate their positions have been chosen in accordance with the velocity amplitude $2 K_2$. The most striking feature of the plot is not a change of the amplitudes along the evolutionary tracks, but is the monopolizing of the large amplitudes by stars of a well-defined luminosity, hence a well-defined mass. This is not so surprising (and it may provide a clue for the real cause of the pulsations) if we think of the formula

$$\frac{1 - \beta}{\beta^4} = \text{const. } M^2 \mu^4. \tag{17}$$

If the pulsations depend on Γ_1, as they certainly do, they depend on β and there will exist an optimal value for Γ_1 and for β for which the pulsation mechanism will work most efficiently. But if our premise is correct there will also exist for μ an optimal value μ_{crit}, at which the pulsations are maximized. The largest amplitudes will thus be reached when β and μ simultaneously reach their optimal values. But this, according to equation (17), can only occur in stars of a well-defined mass.

So the distribution of the amplitudes over the H-R diagram, besides supporting the vaguely sketched origin of the pulsations, seems to indicate that it is the mass much more than the rotational speed that determines the strength of the pulsations.

VIII. THE LOCATION ON THE EVOLUTIONARY TRACK

Speculations on the origin and maintenance of the pulsations and on the age of the β CMa stars force us to consider their variability as a transitory stage in the existence of a normal (?) B-type star.

The question then arises where this stage is to be located on the evolutionary track. Does it correspond to the run through the S-bend which on this track links together the stretches corresponding respectively to the H-burning phase and to the He-burning phase, or must it be placed on one of these stretches themselves? Each location has found its advocates; Struve (1955) and Stothers (1965) want it before the S bend, Kopylov (1959), Schmalberger (1960) and Bidelman (1960) at the beginning of or on the S-bend, Hitotuanagi and Takeuti (1963) beyond the S-bend in the phase of He burning.

Arguments which have been used to clarify the situation are based on:

(i) The expansion ages of open clusters which count β CMa stars among their members;
(ii) The relative frequency of β CMa stars among early B-type stars;
(iii) The secular changes of the periods of β CMa stars.

(i) The expansion age t_e of the I Lacerta association has been evaluated at 4.2×10^6 years. If this t_e can be identified with the hydrogen-burning age t_H of its member stars, 12 Lac and 16 Lac are still far from having consumed all the hydrogen in their core, which must take, according to theory, at least $1-2 \times 10^7$ years. The β CMa stage should thus take place *before* the S-bend is reached.

But Münch and Flather (1957) contest the identity $t_e = t_H$, their argument being that it took the high-latitude β CMa stars much more than 10^7 years to travel from their birthplace in the galactic plane to their present position. In that case they must lie on or perhaps even beyond the S-bend. The contradiction is perhaps more apparent than real. Stothers' investigation shows indeed that the line of constant-period ratios comes closer to the S-bend for the stars with lower masses, as the high latitude variables γ Peg and δ Cet are. The fact that these stars show no trace of beats or of overtones, and have small amplitudes, might simply mean that they have left the crossing point with the line of constant (and critical?) period-ratios far behind them, and that they are now in the fading-out stage of their fundamental pulsation, just at the end of the H-burning phase.

(ii) The relative frequency of the β CMa stars among the B-type stars must be equal to the ratio of the times spent by a B star respectively in the stages of variability and of non-variability. Consequently if the first is to be identified with the run through the S-bend, we must expect to find

$$\frac{n_\beta}{n_B} = \frac{t_S}{t_H} \approx 0.01,$$

since theoretical computations have shown that the time t_S spent in the S-bend is of the order of only 10^5 years (against 10^7 for t_H).

The observed relative frequency is however several times (6 to 11) larger, as the already-mentioned investigations by McNamara and Hansen (1953) and by the Victoria observers (Petrie and Pearce 1962) have shown. This strong inequality rules out the possibility for the S-bend to be the locus of the β CMa stars, and it indicates that the latter extends over a sizable fraction of the H-burning section.

The objection could be raised that there is a tacit assumption underlying the foregoing reasoning, namely that "all B stars have to go through the β CMa stage," and that this assumption is apparently not fulfilled since no β CMa variables are found among the B stars with broad spectral lines. But the objection has no weight, since any new β CMa variable that would be detected in the latter group could only increase the inequality and thus strengthen the conclusion already drawn.

(iii) The philosophy behind the use of secular period changes as landmarks on the evolutionary track is that, according to theory, a star expands slowly in the course of the H burning in the core, contracts during the run through the S-bend and expands again, but then at a much faster rate, when He starts burning. Thus, with the pulsations going on for a sufficiently long time, the changing densities $\bar{\rho}$ and ρ_c must affect the period because of the relation $P\sqrt{\bar{\rho}} = \text{constant}$. The observed rate of change thus characterizes the evolutionary stage of the variable star. Struve (1955) was the first to apply the idea to the β CMa stars, and collected to this end the data reproduced in Table 7.

TABLE 7

CHANGES IN THE PERIODS OF β CMa STARS
(after Struve [1955])

Star	P	ΔP/Century	Comments
β CMa	6h00m	± 0.0 sec	50 years of observation
σ Sco	5 55	$+2.3$	36 years of observation
ξ^1 CMa	5 02	$\cdots$	Interval too short
BW Vul	4 49	$+3.0$	28 years of observation
12 Lac	4 38	$+0.1$:	50 years of observation
β Cep	4 34	$+1.2$	50 years of observation
ν Eri	4 10	$+0.2$:	33 years of observation
16 Lac	4 04	$\cdots$	Complicated by binary motion
δ Cet	3 52	± 0.0	Old observations unreliable
γ Peg	3 38	$\cdots$	Interval too short

All the periods in Table 7 show a tendency to lengthen; none is shortening. This means that all β CMa stars on the list are actually expanding, and Struve, assuming that the evolution displaced the star along the narrow β CMa strip, concluded that an assumed lengthening of 1 sec/century could move the variable from the bottom to the top of the sequence in 10^6 years, a period which did not seem unreasonable. Evolution however is likely to move a star across the strip and Reddish and Sweet (1960) found from more detailed calculations than Struve's that the rate of change reported for β Cep is an order of magnitude larger than necessary to account for the star's expansion during the H burning phase. An inspection of Stothers' tabulated results (1965) leads to the same conclusion.

Could this mean that the β CMa stars are in the phase of rapid expansion which characterizes He burning? What to think then of the opposite conclusion drawn from the frequency considerations? Our answer to this apparent contradiction is that most of the rates on Struve's list—if not all—are badly in need of confirmation, and cannot *now* be said to measure cosmological expansions or contractions. In the case of σ Sco for instance, Van Hoof (1966) found that the period has been on the decrease since 1955 and that its value in 1964–65 was scarcely different from the value it had in 1916, so that little is left of the secular increase of 2.3 sec/century reported by Struve.

In α Lup we have another example of a β CMa star having changed the sign of its period variation since the time of its discovery (Van Hoof, 1966). It seems therefore sound to conclude that one should not rely too much on the figures of Table 7, and that decades of further intensive observation are needed before we shall be able to distinguish between erratic or periodic changes of the period, and secular increases or decreases having cosmological significance.

On the whole then, the various criteria seem to favor the location of the β CMa stage *before* the S-bend of the evolutionary track.

IX. GENERAL CONCLUSIONS

More than fifteen years have elapsed since Struve wrote his article for the *Annales d'Astrophysique*. The "fascinating" field of investigation, which he thought to have discovered, has since been brought to fruition, and progress in it has been gratifying. Rapid and impressive advances have been made in our knowledge of the β CMa stars, thanks to the brightness of these objects which permit one to get accurate photometric and spectroscopic data, and thanks also to the shortness of their periods which produces in a reasonably short time observable cumulative effects of phenomena so small that they could not be discerned otherwise.

But together with more information, the accumulating observations have also produced growing evidence of the extreme complexity of the variations in the β CMa stars and of the mechanisms behind these variations. Confronted with this complexity, we cannot say that the β CMa phenomena are as yet thoroughly understood, and we must for the present content ourselves with the following synthesis:

"β CMa stars are essentially pulsating B stars of low rotational speed. Pulsationally they behave like polytropes of index $n = 3$ and with $\Gamma_1 \approx 1.52$. A value of Γ_1 near this critical value, and a molecular weight μ_c near another still-unknown critical value, are the conditions for the development and the maintenance of the pulsations. The pulsations themselves are either radial or very nearly so (ν Eri) or they resemble Chandrasekhar and Lebovitz' R- and S-type pulsations (σ Sco). The difference is a matter of Γ_1-value and of rotational speed. Overtone oscillations occur in most of the stars that show beats and whose variations cannot be accounted for by two simple since waves with slightly different periods. Strong radial convection streams are the most probable cause of the periodic line widening or doubling. The β CMa stars are still in the H-burning phase of their existence."

To those who want to find out whether the above picture is somewhat more than a caricature we recommend priority being given to the following investigations:

On the theoretical side: computation of evolutionary sequences of detailed stellar models; computation of line profiles for pulsations of various types. On the observational side: intensive photometric and spectrographic study of 12 Lac and θ Oph, which are members of spectroscopic binaries; analysis of split-up line profiles on high-dispersion spectrograms; search for new β CMa stars among the B0. 5 to B 2-III-IV stars with spectral lines of medium width; repeated and accurate determinations of the periods of all known β CMa stars.

If these investigations should bring growing evidence for the truth of the above synthesis, then they would stand as tributes to the great astronomer whose memory we are honoring, and whose genius very early saw the β CMa phenomenon as the combined effect of pulsation and rotation.

Had Struve lived a few years longer and seen the discovery of completely new types of oscillations, and the opening of ways to disclose such well-hidden characteristics as the internal structure, the inclination of the axis of roation upon the line of sight, the age of the β CMa stars and perhaps the ultimate reason for the excitation and maintenance of the pulsations, then he would no doubt have felt himself well paid for the energy he spent years ago to revive the interest of the astronomical world in these puzzling objects.

REFERENCES

Bidelman, W. P. 1960, quoted by Struve, *Mem. Soc. Roy. Sci. Liège*, 5th Ser., **3**, 17.
Blaauw, A. and Savedoff, M. 1953, *Bull. Astr. Inst. Neth.*, **12**, 69.
Böhm-Vitense, E. 1963, *Pub. A.S.P.*, **75**, 154.
Chandrasekhar, S. and Lebovitz, N. R. 1962a, *Ap. J.*, **136**, 1105.
———. 1962b, *Ap. J.*, **136**, 1069, 1032.
Clement, M. J. 1965a, *Ap. J.*, **141**, 210.
———. 1965b, *Ap. J.*, **141**, 1443,
Cowling, T. G. and Newing, R. A. 1949, *Ap. J.*, **109**, 149.
De Jager, C. 1953, *Actes du Cong. Luxem. As. Fr. Av. Sci.*, 88
———. 1962, *Bull. Astr. Inst. Neth.*, **17**, 1.
Hitotuanagi, Z. and Takeuti, M. 1963, *Sci. Rep. Tohoku Univ.*, Ser. I, **47**, 169.
Hogg, A. R. 1957, *M.N.R.A.S.*, **117**, 1.
Huang, S.-S. 1953, *Ap. J.*, **118**, 285.
———. 1955, *Pub. A.S.P.*, **67**, 22.
Huang, S.-S. and Struve, O. 1955, *Ap. J.*, **122**, 101.
Kopylov, I. M. 1959, *Pub. Crimean Ap. Obs.*, **21**, 71.
Ledoux, P. 1951, *Ap. J.*, **114**, 373.
———. 1958, *Ap. J.*, **128**, 336.
Lynds, C. R. 1959, *Ap. J.*, **130**, 577.
McNamara, D. H. 1953, *Pub. A.S.P.*, **65**, 286.
———. 1956, *Pub. A.S.P.*, **68**, 263
McNamara, D. H. and Hansen, K. 1961, *Ap. J.*, **134**, 207.
McNamara, D. H., Struve, O., and Bertiau, F. C. 1955, *Ap. J.*, **121**, 326.
Miczaika, G. R. 1952, *Ap. J.*, **116**, 99.
Milone, L. A. 1963a, *Bol. Inst. Mat. Astr. Fis. Cordoba*, **1**, 3.
———. 1963b, *Bol. Inst. Mat. Astr. Fis. Cordoba*, **1**, 43.
Münch, G. and Flather, E. 1957, *Pub. A.S.P.*, **69**, 142.
Odgers, G. J. 1955, *Pub. Dom. Ap. Obs. Victoria*, **10**, 215.
Opolski, A. and Ciurla, T. 1962, *Inform. Bull. Var. Stars*, No. 3; *Kl. Veröff. Bamberg*, Nr. 34, 83.
Pagel, B. E. J. 1956, *M.N.R.A.S.*, **116**, 10.
Petrie, R. M., 1954a, *J. Roy. Astr. Soc. Can.*, **48**, 185.
———. 1954b, *Pub. Dom. Ap. Obs. Victoria*, **10**, 39.
Petrie, R. M. and Pearce, J. A. 1962, *Pub. Dom. Ap. Obs. Victoria*, **12**, 1.
Reddish, V. C. and Sweet, P. A. 1960, *Mem. Soc. Roy. Sci. Liège*, 5th Ser., **3**, 263
Rodgers, A. and Bell, M. A. 1962, *Observatory*, **82**, 26.
Schmalberger, D. C. 1960, *Ap. J.*, **132**, 591.
Schwarzschild, M. 1941, *Ap. J.*, **94**, 245.
Slettebak, A. and Howard, R. F. 1955, *Ap. J.*, **121**, 102.

Stebbins, J. and Kron, G. E. 1954, *Ap. J.*, **120**, 189.
Stothers, R. 1965, *Ap. J.*, **141**, 671.
Struve, O. 1955, *Pub. A.S.P.*, **67**, 29.
Struve, O., McNamara, D. H., and Zebergs, V. 1955. *Ap. J.*, **122**, 122.
Van Hoof, A. 1957, *Pub. A.S.P.*, **69**, 308.
———. 1961, *Zs. f. Ap.*, **53**, 124.
———. 1962a, *Kl. Veröff. Bamberg*, Nr. 34, 68.
———. 1962b, *Zs. f. Ap.*, **54**, 244.
———. 1962c, *Zs. f. Ap.*, **54**, 255.
———. 1964, *Zs. f. Ap.*, **60**, 194.
———. 1965, *Kl. Veröff, Bamberg*, **4**, Nr. 40, 149.
———. 1966, *Zs. f. Ap.*, **64**, 165.
Van Hoof, A. and Struve, O. 1953, *Pub. A.S.P.*, **65**, 158.
Van Hoof, A., De Ridder, M., and Struve, O. 1954, *Ap. J.*, **120**, 179.
Walker, M. F. 1952, *Astr. J.*, **57**, 227.
———. 1954, *Ap. J.*, **120**, 58.

10. THE ROTATION OF THE STARS

O. STRUVE: On the Axial Rotation of the Stars. *Astrophysical Journal*, **72**, 1, 1930.

ROBERT P. KRAFT: Stellar Rotation.

ON THE AXIAL ROTATION OF STARS

By OTTO STRUVE

ABSTRACT

Broad and *shallow* absorption *lines* in stellar spectra are shown to be *due to axial rotation*. The evidence is based upon the following facts: (*a*) the *broadening depends upon wave-length*, as required by the Doppler principle; (*b*) in *spectroscopic binaries* there is a distinct *correlation* between *line width* on one side and *period and amplitude* on the other; and (*c*) as was shown by Elvey, the contours of *"dish-shaped" lines* agree well with the theoretical shapes of lines of rapidly rotating stars.

A survey of the spectra of stars of various types shows that *rapid rotation is peculiar to the earliest spectral classes*. The B- and A-stars display the greatest tendency toward rapid rotation. In the F's the proportion of rapidly rotating stars is smaller, and in classes G, K, and M no cases of rapid rotation have been observed in single stars.

The effect of *rotation* upon the contours of *components of spectroscopic binaries* is discussed. In α Virginis the stronger component is found to rotate with an equatorial velocity of 200 km/sec., while for the fainter component the rotational velocity is not more than about 50 km/sec. This indicates a marked *difference in the size of the components*. The ratio of the radii is roughly 4 to 1.

The contours of lines in η Ursae Majoris suggest an equatorial velocity of about 200 km/sec. This *single star* is thus *similar to* the stronger component of α *Virginis*. Both stars are found to be *stable*, but the characteristic quantity $\omega^2/2\pi\gamma\rho$ is not far from the critical value where the Maclaurin ellipsoids of rotation lose stability. It is suggested that there is a *transition between close spectroscopic binaries and rapidly rotating single stars* in the earliest spectral classes. Observations do not establish the direction in which this transition proceeds.

I. INTRODUCTION

Many theoretical discussions of the origin of spectroscopic binaries and of other cosmogonic problems have been based upon the dynamical consequences of rapid axial rotation of single stars. The fission theory originated by Sir George Darwin and developed by Sir James Jeans[1] is built upon the assumption that stars lose energy

[1] *Astronomy and Cosmogony*, Cambridge, 1928.

by radiation and are therefore forced to contract. In doing so they must preserve their angular momentum and this can be accomplished only if the angular velocity is increased. A simple computation shows that an average star must rotate at a very high speed before instability can become dangerous. Not until the equatorial velocity of a typical B-type star has reached a value of the order of several hundred kilometers per second is there any danger of breakup. Indeed, F. R. Moulton[1] has shown that the sun must contract until its equatorial radius is 11 miles and its mean density 3×10^{13} on the water standard, before the Maclaurin ellipsoids of rotation become unstable. Moulton concludes that "the oblateness of the sun can never approach that for which the Jacobian figures of equilibrium branch." The difficulty obviously lies in the fact that for the sun with its slow rate of rotation the angular momentum is small, and that an enormous contraction would be required to increase the angular velocity to the critical value.

The problem would assume an entirely different aspect if it could be shown observationally that there are in existence stars, with normal average densities, for which the rotational velocities are large, so large indeed that the characteristic value of $\omega^2/2\pi\gamma\rho$ is not far removed from the critical point where instability sets in.

Until recently such observational evidence was completely lacking. It seems to have been the opinion of many of the leading observational astronomers that stars rotate with equatorial velocities of a few kilometers per second at the most. The relative velocity at the equatorial limbs of the sun amounts to only 3.9 km/sec., and by analogy it was inferred that similar conditions must hold for the stars in general.[2]

---

[1] *Carnegie Institution of Washington, Publication* No. 107, 150, 1909. Also *Astrophysical Journal*, **29**, 1, 1909.

[2] J. Evershed says in *Monthly Notices of the Royal Astronomical Society*, **82**, 395, 1922: ". . . . The angular speed [of Sirius] will be over three times that of the Sun, a complete rotation taking eight days or less. This high speed of rotation seems improbable considering that Sirius is in an earlier stage of evolution, and is therefore less condensed than the Sun. . . . ." Similarly J. A. Carroll expresses the opinion (*ibid.*, **88**, 555, 1928): ". . . . We are thus able to say that no stars have been found rotating with an equatorial velocity much greater than say 50 km/sec. and that . . . . the rotational speed must have been more nearly of the order of 10 km/sec. or less." The same idea is stated by H. C. Vogel (*Astronomische Nachrichten*, **90**, 75, 1877): "Ein Aequator-

However, in the light of observations made within the past few years[1] our early conceptions of this subject must be radically changed. It appears that rapidly rotating stars are by no means rare exceptions. They are quite common in the earlier spectral classes, and their equatorial velocities occasionally exceed 200 or even 250 km/sec. Such velocities are characteristic for many single stars, as well as for spectroscopic binaries of short period and large amplitude. This naturally leads to the inference that there is a continuous transition between single stars having rapid axial rotations and close spectroscopic binaries of short period.

The observations do not indicate in which direction this transition proceeds. It remains uncertain whether a rapidly rotating single star breaks up into a binary (as would follow from the fission theory), or whether the components of a binary, by falling together, give rise to a rapidly rotating single star. Nevertheless, it is of considerable cosmogonic interest that the stars of earliest spectral classes—A, B, and probably O—reveal the greatest tendency toward rapid rotation. It is in these types, then, that we should expect the binaries to originate, if we were to accept the fission theory. The rotational velocities of many single stars are of the same order of magnitude as those shown by components of the closest-known spectroscopic binaries, and are not much inferior to those predicted theoretically for the "break-up" of a single star.

This paper is not directly concerned with the fission theory. It is well known that Professors F. R. Moulton[2] and W. D. MacMillan[3] have raised a number of important objections to it, and it is not here possible to pursue the problem of the origin of double stars farther. Neither do we touch upon the question as to how the rapid rotations in single stars have originated. It is not easy to reconcile

---

punkt von α Aquilae . . . . würde die immer noch ansehnliche Geschwindigkeit von 25 geogr. Meilen haben. Es erscheinen diese Geschwindigkeiten, zumal im Vergleich mit der Rotationsgeschwindigkeit unserer Sonne (Aequatorpunkt 0.27 geogr. Meilen) als im hohen Grade unwahrscheinlich." See also the article by K. Walter in *Die Sterne*, **10**, 9, 1930.

[1] Shajn and Struve, *Monthly Notices of the Royal Astronomical Society*, **89**, 222, 1929. This paper contains references to earlier investigations on the subject of stellar rotation; C. T. Elvey, *Astrophysical Journal*, **71**, 221, 1930; *ibid.*, **70**, 152, 1929.

[2] *Op. cit.*, pp. 156, 160.   [3] *Science*, **62**, 69, 1925.

the facts with the contraction hypothesis, and we must at present be content with the purely observational result that rapid rotations can and do occur.

II. THE INTERPRETATION OF LINE CONTOURS

The evidence concerning rotation depends primarily upon the interpretation of the contours of stellar absorption lines. It is now known that there are three major factors[1] which influence the shape of these contours, viz., abundance of atoms, molecular Stark effect, and axial rotation.

a) The contour of an absorption line in a scattering atmosphere without collisions is known from theoretical considerations. Observations have shown that this "Unsöld contour" is closely obeyed by the majority of spectral lines in late-type stars. Small corrections due to the influence of collisions have been computed by Unsöld and tend to make the agreement between observation and theory even better. The abundance of atoms in the right state to absorb a given line depends mainly upon the temperature and pressure of the reversing layer. The actual percentage of occurrence of any one element in a stellar atmosphere is remarkably constant along the spectral series. For many lines the effect of absolute magnitude is very small and can be neglected, in which case the total amount of energy absorbed in any one line is a function of spectral type alone.

b) The hydrogen lines show anomalous contours in all spectral classes. The same is true of the helium lines in many B-type stars. These anomalies are due to Stark effect, produced by the electrical fields of neighboring ions. The resulting broadening of the lines is not the same in all stars; it varies with the pressure and is consequently related to absolute magnitude. However, as in the case of

---

[1] There are many physical factors which influence the contour of a line in the laboratory. Of these only the molecular Stark effect is believed to be effective in stellar atmospheres. Doppler effect due to temperature agitation is probably not appreciable (Vasnecov, *Vestnik Kral. Ces. Spol. Nauk*, **2**, 1927). Most of the other effects observed in the laboratory are due to direct interaction of the atoms. Such effects are improbable in the stars on account of the low pressures in the reversing layers. The Stark effect alone is rendered appreciable because of the high state of ionization. See also p. 5, n. 2.

abundance, the effect of absolute magnitude is comparatively small, and can be neglected if we consider only the major features of a stellar spectrum.

c) A considerable number of stars belonging to classes O, B, and A, and occasionally to class F, show broad and hazy lines for all elements, wholly different in shape from the narrow and deep contours produced by the normal absorption coefficient of Unsöld. C. T. Elvey[1] has shown that these "dish-shaped" lines agree with the contours computed by G. Shajn and the writer for rapidly rotating stars.

In section VI we shall compare the contours of certain lines of a single star (η Ursae Majoris) with those of a spectroscopic binary (a Virginis). The agreement is so satisfactory that there remains no doubt that both are produced by the same cause. Since in the binary this cause is known to be rotation[2] the single star, too, must be rapidly rotating.

III. ROTATION AND SPECTRUM

It has been customary at the Yerkes Observatory to classify all spectra according to the character of the lines. This classification was originally introduced by Professor E. B. Frost in order to provide a rough criterion as to the precision to be expected from measurements of radial velocity. Since "poor" lines are invariably dish-shaped, and since even the absence of strong lines can be interpreted as being due to excessive diffuseness, it appears that this classification is a good measure of rotation, or rather of the component of the rotational velocity in the line of sight. The distribution of all stars of types B and A observed at Yerkes[3] is shown in Tables I and II.

[1] *Astrophysical Journal*, **70**, 152, 1929.

[2] This follows from the correlation between line width and a function of period and amplitude (*Monthly Notices of the Royal Astronomical Society*, **89**, 225, 1929):

$$v = \text{const.} \frac{r \cdot K}{P^{\frac{2}{3}}}.$$

Evershed has suggested that wide and hazy lines may be due to Doppler effect caused by violent convection currents (*ibid.*, **82**, 395, 1922). This may be a contributory cause, but it does not explain the correlation just noted. The latter strongly suggests that we are dealing principally with Doppler effect due to rotation.

[3] *Astrophysical Journal*, **64**, 9, 1926; *Publications of the Yerkes Observatory*, **7**, Part I, 10, 1929.

It will be seen that the proportion of stars classified as "few, poor" is very great among the B's as well as among the A's. The number of F stars observed is too small to justify a complete tabulation, but from the existing spectrograms it appears that the proportion of dish-shaped lines is much smaller than in the B's and A's. Finally, in spectral classes G, K, and M we have never observed any stars having very wide lines. This refers mostly to the giants, since there have been only few dwarfs of late types on our program.

TABLE I

B Stars

Character of Lines	Number of Stars
Many, good...............	45
Few, good..................	59
Many, fair.................	17
Few, fair...................	101
Many, poor.................	7
Few, poor..................	139

TABLE II

A Stars

Many, good...............	118
Few, good..................	37
Many, fair.................	71
Few, fair...................	59
Many, poor.................	35
Few, poor..................	180

It seems safe to say that rotational speed is a function of spectral type, the fastest rotations occurring in the earliest types.

In the following sections we shall assume that all dish-shaped lines are caused by rotation alone and that, furthermore, broadening due to abundance and Stark effect is constant within any given spectral subdivision. That this assumption is permissible results from the fact that among the stars of early types variations due to dish-shaped character are by far the most important of any observed. Minor differences in abundance or pressure broadening depending upon the absolute magnitude can be neglected if we limit our discussion to stars having very flat and broad lines.[1]

[1] The assumption that Stark effect does not vary much within any one spectral subdivision is a modification of certain ideas put forward in my first article on Stark

IV. LINE WIDTH AND WAVE-LENGTH

Consider a rapidly rotating star. The lines are widened by amounts depending upon the projection of the equatorial velocity upon the line of sight. According to Doppler's principle the amount of widening is

$$\Delta\lambda = \lambda \frac{v \cdot \sin i}{c} . \tag{1}$$

But the wave-length in A.U. is related to the readings on the plate in millimeters by Hartmann's formula

$$\lambda - \lambda_0 = \frac{k}{S_0 - S} . \tag{2}$$

Differentiating this we get

$$\Delta\lambda = \frac{k}{(S_0 - S)^2} \cdot \Delta S = \frac{(\lambda - \lambda_0)^2}{k} \Delta S .$$

Substituting this into (1) we find

$$\Delta S = \frac{\lambda \cdot v \cdot \sin i \cdot k}{c(\lambda - \lambda_0)^2} . \tag{3}$$

Consequently the width of a line broadened by Doppler effect should show the following proportionality:

$$\Delta S \infty \frac{\lambda}{(\lambda - \lambda_0)^2} . \tag{4}$$

---

effect (*Astrophysical Journal*, **69**, 185, 1929). It was suggested there that the diffuse appearance of the hydrogen and helium lines in many stars might be due to electric fields. It now appears that this explanation is correct for the hydrogen lines only (C. T. Elvey, *ibid.*, **71**, 191, 1930; E. T. R. Williams, *Harvard College Observatory Circular*, No. 348, 1930). The variation in width of the helium lines due to Stark effect is comparatively small, though real (cf. the contours in γ Pegasi and in 67 Ophiuchi), and the dish-shaped character must be ascribed almost wholly to rotation. The existence of an effect of absolute magnitude for the hydrogen lines is largely due to changes in Stark effect, since here Stark effect is very pronounced and rotation is little effective. In the helium lines the relationship of character (*s* or *n*) to absolute magnitude must be due to a real tendency of the more luminous stars to rotate slower than the less luminous stars.

For single-prism spectrograms taken with the Bruce spectrograph of the Yerkes Observatory the constant λ_0 is approximately equal to 2300 A.

I have measured the line widths in a microphotometer tracing of a spectrogram of α Virginis. This is a binary showing two very unequal components, but the particular plate selected for measurement shows only one set of lines, the fainter component being superposed over the stronger.

The results of the measurements, given in column 3, clearly show a relationship to λ in rough agreement with formula (4). However,

TABLE III

RELATION BETWEEN LINE WIDTH AND WAVE-LENGTH

Wave-Length	Central Intensity	Width	Corrected Width
3926..........	0.11	7.0 mm	7.0 mm
4009..........	.06	5.5	6.2
4026..........	.13	7.0	5.3
4121..........	.09	5.0	5.0
4144..........	.10	5.0	5.0
4388..........	.09	5.0	5.0
4471..........	.13	7.0	5.0
4481..........	.04	3.5	4.6
4552..........	0.05	3.5	4.2

better results can be obtained if small corrections are introduced to take care of differences in central intensity. By a simple graphical procedure all line widths were reduced to a uniform central intensity of 0.10. The values thus obtained are given in the last column. The decrease in width with wave-length is very nearly that required by formula (4). The evidence is in favor of the rotation hypothesis, since it affects all lines.

V. APPLICATION TO THE STUDY OF SPECTROSCOPIC BINARIES

Figure 1 shows a microphotometer tracing of the spectrum of 67 α Virginis near maximum separation of the lines. Two components are visible for each of the two lines, λ 4472 and λ 4481. It is at once obvious that the shapes of the lines are not the same. The violet components of both lines are shallow and very broad, while the red components are much narrower.[1] Figure 3 shows the contours evaluated by means of the sensitometer intensities given at the right margin of Figure 1.

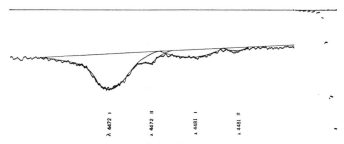

FIG. 1.—Microphotometer tracing of Plate R 1788 of α Virginis, taken on April 24, 1930, at 5^h 35^m U.T. The sensitometer intensities, of which only nine are shown near the right-hand margin, correspond to the following differences in intensity, expressed in stellar magnitudes (from the top): 0.00; 0.39; 0.70; 1.13; 1.47; 1.93; 2.26; 2.75; 3.11; 3.45; 3.86; 4.23; 4.48; 5.06.

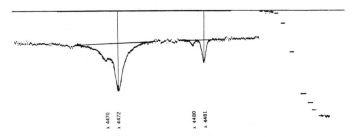

FIG. 2.—Microphotometer tracing of Plate R 1599 of γ Pegasi. Note the sharpness of the lines. The forbidden helium line at λ 4470 is visible to the right of the strong helium line λ 4472. The line at λ 4480 is due to *Al* III. The plate was taken on September 21, 1929, at 4^{h}38^m U.T.

The conclusion is obvious: One component of the binary is in rapid rotation while the other is not. This is a rather striking result and it leads to several interesting consequences.

The spectroscopic elements of α Virginis, as derived by R. H. Baker,[2] are:

---

[1] This is true of all other lines not shown in the figure. The narrow component is particularly well visible in *He* 4713. It has also been measured in *He* 4026 and *He* 4388.

[2] *Publications of the Allegheny Observatory,* **1**, 65, 1909. Measures of our spectrograms agree well with Baker's orbit.

Velocity of system.................. $+1.6$ km/sec.
Period........................ 4.01416 days
Eccentricity...................... 0.10
Time of periastron passage......... 1908 Jan. 14. 846
Longitude of periastron............ 328°
K_1............................. 126.1 km/sec.
K_2............................. 207.8 km/sec.
$a_1 \sin i$......................... 6,930,000 km
$a_2 \sin i$......................... 11,400,000 km
$m_1 \sin^3 i$........................ 9.6⊙
$m_2 \sin^3 i$........................ 5.8⊙

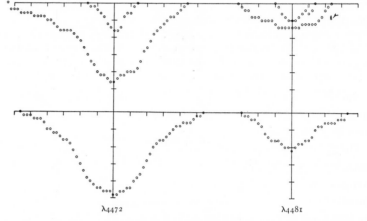

λ4472   λ4481

FIG. 3.—Observed contours of lines in α Virginis (top) and η Ursae Majoris (bottom). The contours of both components of α Virginis are shown. One unit in the abscissa corresponds to 1.28 A.U., and one unit in the ordinate to an absorption of 0.05 of the continuous spectrum.

The orbital period is so short that the rotational periods are doubtless equal to it. Consequently the two components have identical angular velocities. Since

$$v_1 = r_1 \omega_1 \sin i ,$$
$$v_2 = r_2 \omega_2 \sin i ,$$
$$\omega_1 = \omega_2 ,$$

the disparity in the v's must mean a disparity in the radii. We arrive at the conclusion that the stronger component of α Virginis (violet in Fig. 1) has the greater diameter. Information concerning stellar radii can thus be obtained by a purely spectroscopic method.

The binary is not resolved visually; consequently the light of both components enters the slit of the spectrograph simultaneously. Suppose the radial velocities are such that the two components are completely resolved. The continuous spectrum of one star will overlap the line of the other. Let the intensities of the continuous spectra outside the lines be i_1 and i_2 and let the real intensities within the two lines, in the absence of overlapping, be given by

$$j_1 = f_1(\lambda) \qquad \text{and} \qquad j_2 = f_2(\lambda) .$$

The spectral types of the two components are the same. Consequently, if there were no rotation, we should have

$$\frac{j_1}{i_1} = \frac{j_2}{i_2} .$$

It should be noted here that the contours are usually expressed in units of the intensity of the continuous spectrum. Hence the fractions in the foregoing expression.

The rotational effect may not be the same in both stars, and the foregoing equality will not, in general, be fulfilled. But rotation changes only the shape of the contour, leaving the total amount of absorbed energy unaffected. Consequently the integrals taken over the functions j_1/i_1 and j_2/i_2 should be identical:

$$A = \int_{-\infty}^{+\infty} \left(1 - \frac{j_1}{i_1} \right) d\lambda = \int_{-\infty}^{+\infty} \left(1 - \frac{j_2}{i_2} \right) d\lambda .$$

In reality we do not observe j_1 and j_2 separately. Both components are photographed simultaneously, and the plate records the combined effect of the lines and of the continuous spectra. As a result of this overlapping we observe for the two components the quantities A_1 and A_2, which are not, in general, identical:

$$A_1 = \int_{-\infty}^{+\infty} \left(1 - \frac{j_1 + i_2}{i_1 + i_2} \right) d\lambda = \frac{i_1}{i_1 + i_2} A , \tag{5}$$

$$A_2 = \frac{i_2}{i_1 + i_2} A . \tag{6}$$

From these

$$\frac{A_1}{A_2} = \frac{i_1}{i_2} .$$

The observed areas of the contours of the two components are proportional to the intensities of their continuous spectra. Since the spectral types are the same, we obtain the difference in absolute magnitude:

$$M_1 - M_2 = -2.5 \log \frac{i_1}{i_2} = -2.5 \log \frac{A_1}{A_2} . \qquad (7)$$

By means of this formula we compute the values of ΔM if A_1 and A_2 are known from the observations.

We shall now reconstruct from the observed contours the real contours j_1 and j_2, which the lines would have had if overlapping with the continuous spectrum did not occur. It follows from (5) and (6) that this is accomplished by multiplying the ordinates of the observed contours by $(1+[A_2/A_1])$ and by $(1+[A_1/A_2])$, respectively.

The numerical evaluation of A_1 and A_2 from the curves of Figure 3 gives roughly:

$$\frac{A_1}{A_2} = 8.8 .$$

Consequently[1]

$$M_2 - M_1 = 2.4 \text{ mag.}$$

Figure 4 shows the corrected contours as they would have been without overlapping. The difference in shape is very marked, confirming that the rotational effect is not identical in the two components.

We now proceed to evaluate the rotational velocity. Let us assume that the shape of the line λ 4472, not affected by rotation, is that obtained by J. Pauwen[2] for the star γ Pegasi (Fig. 2). The

---

[1] It may be noted that the lack of broadening of the fainter component, its greater "compactness", makes it possible to observe it even though $(M_1 - M_2)$ is rather large. If the second component were as broad and diffuse as the primary it could not have been seen on the background of the continuous spectrum. The actual value obtained, 2.4 mag. should be considered as a rough approximation only. It is probable that since the line is very faint even on our best plates, the measured area A_2 is slightly too small. That this is so may be seen from the fact that a line can be lost completely if it is too diffuse and broad.

[2] *Astrophysical Journal*, **70**, 263, 1929.

spectral types of α Virginis and γ Pegasi are approximately the same. Measurement of the areas shows that the line in γ Pegasi is not as strong as the line in α Virginis. In order to make the two areas identical, we multiply all ordinates of the contour for γ Pegasi by 1.8. The resulting contour is shown in Figure 5. We proceed in a manner similar to that used by Shajn and the writer, which was also

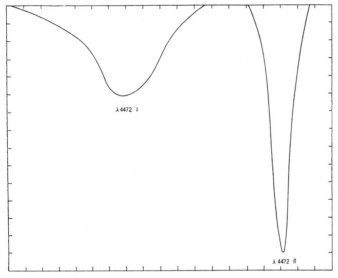

FIG. 4.—Corrected contours of the two components of α Virginis. The curves show the shapes of the lines as they would have been in the absence of overlapping with the continuous spectra. The units in the abscissa and in the ordinate are the same as in Fig. 3.

employed by Elvey. Consider the disk of the star, and imagine that the x-axis lies in the equatorial plane. If the equatorial velocity is $v \sin i$ then any point on the disk has a projected velocity

$$\frac{x}{r} \, v \cdot \sin i \; ,$$

where r is the radius. Imagine that the star is subdivided into an infinite number of sections, parallel to the axis of rotation. Each section gives an intensity $j = f(\lambda - \lambda_0)$, where

$$\lambda_0 = \lambda'_0 + \lambda'_0 \frac{v \cdot x}{r} \sin i \; .$$

The area of each section is $2\sqrt{r^2-x^2}\,dx$. Multiplying j by this area and integrating over x we obtain the intensity of the light from the whole disk. To express this, as is usual, in units of the continuous spectrum, we divide the result of the integration by the integrated intensity of the continuous spectrum, which is equal to $\pi r^2 i_1$.

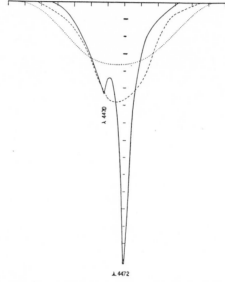

FIG. 5.—Effect of rotation upon the contour of the line λ 4472 The full line represents a non-rotating star. The two other curves correspond to equatorial velocities of 170 and 340 km/sec. The units in the abscissa and in the ordinate are the same as in Fig. 3.

Since the shape of the function f is not algebraically known we perform the integration graphically. The disk is divided into finite sections (forty in this case), each having an area of

$$a = 2\int_{x_1}^{x_2}\sqrt{r^2-x^2}\,dx = \left\{x\sqrt{r^2-x^2}+r^2\arcsin\frac{x}{r}\right\}_{x_1}^{x_2}.$$

Each section contributes a certain intensity $j=f(\lambda-\lambda_0)$ proportional to its area a. Rotation shifts each of these contributory contours by an amount equal to

$$\frac{x_1+x_2}{2r}\,v\cdot\sin i \ .$$

We thus have a number of contours,

$$j_n = f\left[\lambda - \left(\lambda_0' + \lambda_0' \frac{x_1 + x_2}{2r} \, v \cdot \sin i\right)\right] a_n \ .$$

The final curve is obtained by taking:

$$I = \frac{1}{\pi r^2 i_1} \sum j_n a_n \ .$$

This summation has been made for two arbitrary values of $v \sin i$, viz., 340 and 170 km/sec. (Fig. 5). It will be seen by superposition of the contour of the violet component of Figure 3 with the curves of Figure 5 that $v \sin i = 200$ km/sec. gives a good representation.

We have neglected here the effect of darkening at the limb. From the computations of Shajn and the writer it would seem that this is not important. It might perhaps tend to increase slightly the value of $v \sin i$.

The second component has almost no rotation. Since it is doubtful whether it would be possible to measure by this method a rotational velocity of less than 50 km/sec. we can only state that the rotation of the second component does not much exceed this value. If we tentatively assume $v_2 \sin i = 50$ km/sec., we obtain

$$\frac{v_1}{v_2} = \frac{r_1}{r_2} = 4 \ .$$

This is in good agreement with expectation. We have for the two components[1]

$$\log r_1 = \frac{5900}{T_1} - 0.2 M_1 - 0.02 \ , \tag{8}$$

$$\log r_2 = \frac{5900}{T_2} - 0.2 M_2 - 0.02 \ , \tag{9}$$

where T_1 and T_2 are the temperatures and M_1 and M_2 the visual absolute magnitudes. In this case $T_1 = T_2$ and $M_1 - M_2 = -2.4$. By subtraction:

$$\log \frac{r_1}{r_2} = 0.2 (M_2 - M_1) \ .$$

[1] Russell, Dugan, and Stewart, *Astronomy*, **2**, 738, 1927.

Consequently $r_1/r_2 = 3$, in fair agreement with the value obtained before.

While too much stress should not be placed upon the numerical values of $v \sin i$, it is clear that interesting information is contained in the contours of the components of spectroscopic binaries. For example, if the two corrected contours, j_1 and j_2, are similar, we can safely say that the diameters of the two stars are also about the same. Of course, this applies only to double stars with short periods, since it is essential that the periods of revolution and of rotation are the same. We do not know exactly at which stage this equality begins to break down. However, there is very little doubt that it is fulfilled in all binaries with periods of the order of a few days.

VI. ROTATION IN η URSAE MAJORIS

Figures 6 and 3 contain the original tracing and the contours of λ 4472 and λ 4481 for the star η Ursae Majoris. This is not a spectroscopic binary of short period and large amplitude[1] like α Virginis. The value of $v \sin i$ is again close to 200 km/sec. We are thus dealing here with a single star of the same type and presumably the same luminosity as the stronger component of α Virginis. Now this latter star is what Jeans calls a very young binary. Using equations (8) and (9) and substituting $T_1 = T_2 = 20,000°$ and $M_1 = -2.0$, we find[2]

$$r_1 = 3 \times 10^6 \text{ km },$$

$$r_2 = 1 \times 10^6 \text{ km }.$$

From the orbit we have

$$a_1 \sin i = 7 \times 10^6 \text{ km },$$

$$a_2 \sin i = 11 \times 10^6 \text{ km }.$$

[1] It was announced as a spectroscopic binary by L. L. Mellor (*Publications of the Observatory of the University of Michigan*, **3**, 72, 1923) but the range is small and the period is probably long. None of our plates shows any indication of duplicity. Consequently the observed spectrum refers to a single star.

[2] Since $r_1 = 3 \times 10^6$ km and $P = 4.0$ days, the theoretical equatorial velocity of rotation should be $v_1 = 2\pi r_1/P = 55$ km/sec. This seems to indicate that we have used too faint a magnitude for M_1 in (8).

The inclination[1] is probably close to 90°, so that

$$a_1 + a_2 = 18 \times 10^6 \text{ km} .$$

The distance between the surfaces of the two components is only about 3.5 times the sum of their radii. Consequently it is probable that the angular rotational velocity of the stronger component of α Virginis is not far from the critical value at which a double star merges into a rapidly rotating single body. For η Ursae Majoris the

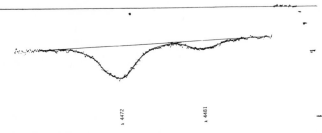

λ 4472   λ 4481

FIG. 6.—Microphotometer tracing of Plate R 1774 of η Ursae Majoris, taken on March 3, 1930, at $6^h 35^m$ U.T.

angular velocity is of the same order of magnitude. We conclude therefore that this star is also rather close to the critical stage where break-up may occur.[2]

VII. STABILITY

It may be of interest to compute the value of the quantity $\omega^2/2\pi\gamma\rho$, which is characteristic for the state of stability of a rotating star. Following Tissérand's method,[3] we have for the star

$$\frac{\omega_1^2}{2\pi\gamma\rho_1} = \frac{2\pi}{\gamma\rho_1 P_1^2} ,$$

[1] α Virginis was tentatively announced as an eclipsing variable by J. Stebbins (*Astrophysical Journal*, **39**, 475, 1914). However, in a later publication Stebbins did not include this star among those which are definitely known to be eclipsing variables (*Publications of the Washburn Observatory*, **15**, 56, 1928). Professor Stebbins has informed me that the question is not definitely settled, since there are few suitable comparison stars of similar spectral type in the vicinity of α Virginis.

[2] The angular velocity $\omega = 2\pi/P$ may be considered as a measure of the distance between the two components of a double star. Consequently, whatever the origin of the binary, there is a critical value of ω such that the components are just in contact. At this stage the double star ceases to exist as such and should be considered as a single body.

[3] *Traité de Mécanique Céleste*, **2**, 92, 1891.

and for the earth,

$$\frac{\omega_2^2}{2\pi\gamma\rho_2} = \frac{2\pi}{\gamma\rho_2 P_2^2} = 0.00230 \ .$$

Consequently

$$\frac{\omega_1^2}{2\pi\gamma\rho_1} = 0.00230 \left(\frac{P_2}{P_1}\right)^2 \left(\frac{\rho_2}{\rho_1}\right) \ .$$

For the earth $P_2 =$ day, while for α Virginis $P_2 = 4$ days. The density of the earth is 5.5 gr/cm^3. Consequently

$$\frac{\omega_1^2}{2\pi\gamma\rho_1} = \frac{0.00079}{\rho_1} \ .$$

The mean density of the B-type stars ranges from about 0.01 to 0.1 gr/cm^3. Using the two limits, we find

$$0.0079 \leq \frac{\omega_1^2}{2\pi\gamma\rho_1} \leq 0.079 \ .$$

If the laws deduced by Jeans and others for rotating homogeneous liquid bodies are applicable to the stars, the foregoing result would indicate that the star is stable and that the figure of equilibrium is probably a Maclaurin spheroid, the meridional cross-section of which has an eccentricity of not more than about 0.5. Since the density of α Virginis is not known, the quantity $\omega^2/2\pi\gamma\rho$ cannot be determined any closer. However, a shortening of the period to about 2.8 days, with constant dimensions and density, would bring the angular velocity rather dangerously close to the point where the Maclaurin ellipsoids lose stability. It seems quite possible that such stars exist. In fact, α Virginis is only one of many examples of rapid rotations, and it is not at all improbable that such rapidly rotating binaries as V Puppis and μ^1 Scorpii approach the critical point even closer than α Virginis.

YERKES OBSERVATORY
May 22, 1930

STELLAR ROTATION*

Robert P. Kraft

Lick Observatory
University of California, Santa Curz

I. INTRODUCTION

Brief Survey of the Problems

Theoretical studies of rotating stellar configurations, beginning with the pioneering work of von Zeipel (1924), have given ample reason to believe that rotation not only induces an aspect dependence in the emergent flux, but also that it modifies the celebrated theorem of Russell and Vogt, according to which the mass and chemical composition determine uniquely the luminosity and radius of a chemically homogeneous star. Current interpretations of spectroscopic and photoelectric observations support these results and show that the position of a main sequence star in the H-R diagram is affected by rotation. From one point of view, this is perhaps somewhat surprising if we consider the kinetic energy E_{rot} tied up in stellar rotation; viz.,

$$E_{rot} = \frac{4\pi}{3} \int_0^R r^4 \rho(r) \omega^2(r) \, dr, \tag{1}$$

where ρ is the density and ω is the angular velocity at a distance r from the center of a star of radius R. For a typical A-type star with an equatorial rotational velocity of 200 km/sec, $E_{rot} \approx 10^{46}$ erg, if we assume that ω is not a function of r. Though this number is of the same order as the total ionization energy, it is at least three orders of magnitude smaller than the potential and thermal energies, and five orders smaller than the total available nuclear energy. If the rotational kinetic energy could be made available in some other form, the star obviously would not be much affected. However, since the energy generation rate is highly temperature sensitive, the levitational effect of rotation, even in the core, plays a leading role in determining the luminosity of a rotating star, and this effect could be especially well marked if the angular velocity were not constant, but rather increased with decreasing distance from the center.

A different manifestation of the role played by rotation in stellar atmospheric and possibly in stellar structure problems is found in the peculiar and metallic-line A-type stars. Together with otherwise normal stars, these objects populate that portion of the main sequence to which stars of rapid rotation are confined, yet their spectra are characterized, among other things, by narrower than average lines. Recent results show rather conclusively that most of the truly slow rotators within the domain of rapid rotators are Am's and Ap's, and it is of interest to discuss the relation of this fact to their abundance anomalies, magnetic field strengths, and binary characteristics.

Of equal or greater significance is the problem of angular momentum and star formation. It has sometimes been stated that a star collapsing from a typical gas cloud will develop a rotational velocity of the order of the speed of light if angular

* This article was started while the author was a visiting professor at the Astronomy Department, Columbia University, in November 1965. I wish to record here my appreciation to Dr. L. Woltjer for his kind hospitality and to Dr. K. Prendergast for some valuable discussions.

momentum is conserved and if the gas cloud reflects the rotation of the galaxy. Questions arise, however, regarding the validity of this picture, especially if stars are actually formed in groups in which the distribution of angular momentum among various modes of motion must be accounted for. Once formed, however, a star might be thought to have—and to maintain thereafter—a certain definite angular momentum. Theories of star formation customarily must predict the mass-frequency function $N(\mathfrak{M})d(\mathfrak{M})$, but they can also be required to give the correct form of $\langle J(\mathfrak{M})\rangle$, the mean angular momentum for stars of mass $\mathfrak{M}$. A leading task of spectroscopic observation is to provide this function, and though it is obviously convenient to study its form on the main sequence, the relationship between this and the desired initial form may be obscured by stellar angular momentum losses.

Unfortunately, even the main sequence form of $\langle J(\mathfrak{M})\rangle$ cannot be directly observed. For one thing, only the absolute magnitude is observable, and one must pass through some form of the mass-luminosity relation to determine $\mathfrak{M}$. Further, for any given star, spectroscopic observation gives us $v \sin i$, the equatorial rotational velocity projected on the line of sight, rather than the true rotational velocity v. For a number of stars of given luminosity L, we can obtain $\langle v \sin i\rangle$, and this in turn determines $\langle v\rangle = 4/\pi \; \langle v \sin i\rangle$ (Chandrasekhar and Münch 1950), provided the orientation of rotational axes is distributed at random. Even so, the step from $\langle v\rangle$ to $\langle J\rangle$ requires one further assumption. The angular momentum J of a star of radius R is

$$J = \frac{8\pi}{3} \int_0^R r^4 \rho(r)\omega(r) \, dr. \tag{2}$$

It is convenient to assume the star is a rigid rotator, in which case

$$J = \frac{8\pi v}{3R} \int_0^R r^4 \rho(r) \, dr, \tag{3}$$

but the physical validity of this assumption remains unknown. Integration over the density distribution of main-sequence stellar models then gives J and, therefore, $\langle J(\mathfrak{M})\rangle$ for a given $\langle v(\mathfrak{M})\rangle$.

Closely related to this problem is obviously the change in v induced by stellar evolution if we assume that J is constant with time. Indeed, for sufficiently large groups of stars we may test for evolutionary compatibility between the development of the luminosity and radius on the one hand, and the mean rotational velocity on the other, under the assumption of a fixed $\mathfrak{M}$ and $\langle J\rangle$; we can pursue the evolutionary tracks both forward and backward in time from the main sequence. As we follow a track, however, we have to make an assumption about the manner in which J is conserved. We ask: does a star always rotate as a rigid body with concomitant radial exchange of angular momentum, or does each infinitesimal shell conserve its own angular momentum as the radius changes? Or does a star follow a course characterized by a more complicated radial redistribution of its angular momentum?

For any spectroscopically isolatable group, we can consider both the total angular momentum and total energy tied up in rotational and translational motion. Struve (1950) pointed out that E_{rot} and E_{trans} for early-type stars were roughly the same order of magnitude, suggesting the possibility of equipartition between rotational and translational energy. On the other hand, $E_{\text{trans}} \gg E_{\text{rot}}$ for stars of spectral type dG and later. For all kinds of stars in binary systems, it is generally true that $J_{\text{orb}} > J_{\text{rot}}$ by at least one order of magnitude. Since binaries constitute something like one-half of all stars of types dK and earlier, it is of interest to ask whether the rotations of stars in binary systems are the same as single stars, and also whether, at any point in the evolution of a binary, exchange can take place between J_{orb} and J_{rot}. For typical binaries of early type, $|E_{\text{orb}}| \approx E_{\text{rot}}$, but $|E_{\text{orb}}| \gg E_{\text{rot}}$ for binaries of type dG and later, unless they are W UMa stars.

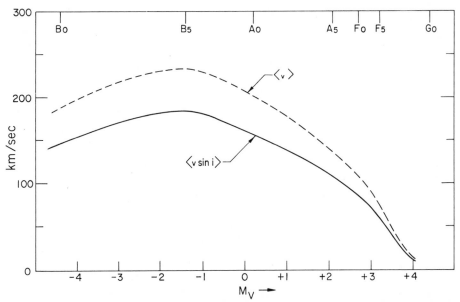

Fig. 1.—$\langle v \sin i \rangle$ as a function of M_V for main-sequence stars, after Abt and Hunter (1962). The curve for $\langle v \rangle$ is derived from $\langle v \sin i \rangle$ on the assumption of a random orientation of rotational axes.

In the thirty-five years since Otto Struve and his associates demonstrated conclusively the existence of stellar rotation, studies of line profiles have been systematically used to derive rotational velocities, especially for early-type stars of the main sequence. Basic techniques for the determination of $v \sin i$ from line profiles have been reviewed *in extenso* by Huang and Struve (1960); the reader is referred to this article for details. Here we focus attention on the problems delineated in the preceding paragraphs as they might be treated from an observer's point of view. With the difficult theoretical questions involved in providing a satisfactory interior model for a rotating star, particularly the proper description of $\omega(r)$, the writer is not competent to deal; some of the more recent theoretical results will, however, be quoted uncritically.

II. $\langle v \sin i \rangle$ AND $\langle J(\mathfrak{M}) \rangle$ FOR MAIN-SEQUENCE STARS

(a) Derivation of $\langle J(\mathfrak{M}) \rangle$ for Field Stars

Roughly 1500 estimates of $v \sin i$ are available from the fundamental lists by Slettebak (1949, 1954, 1955), Slettebak and Howard (1955), and Herbig and Spalding (1955); of these, approximately 375 correspond to normal main-sequence or near main-sequence stars of type O to G0. [A comprehensive summary of all known estimates of $v \sin i$ has been given by Boyarchuk and Kopylov (1964).] Dividing this material into appropriate small intervals of spectral type, Abt and Hunter (1962) derived the function $\langle v \sin i \, (M_V) \rangle$, which is illustrated in Figure 1. Considered in the direction of decreasing brightness, the leading features of the function are the increase to a broad maximum of 180 km/sec near B5V, followed by a decline, rather sharp below F0V, to the limit of resolution near F5V. The latter is set essentially by the dispersion, and has a value close to 25 km/sec and 15 km/sec for the investigations of Slettebak and of Herbig, respectively. Little is known in detail about the rotations of main-sequence stars of spectral type G and later except that they are small, *i.e.*, less than 20 to 25 km/sec (Herbig and Spalding 1955; Wilson, private communication).

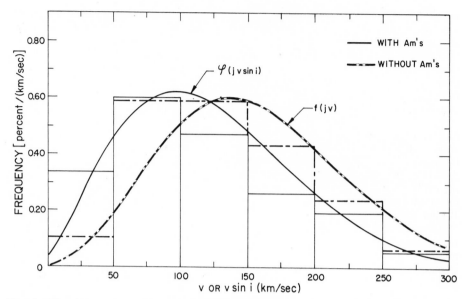

Fɪɢ. 2.—The frequency function of rotational velocities for 72 main-sequence stars of types A3 to F2, after Deutsch (1965). The histogram of solid lines refers to all stars; the histogram of dot-dash lines refers to the same sample with Am stars omitted. The curves refer to the Maxwell-Boltzmann law in v and in $v \sin i$.

Since the observed $v \sin i$'s show no apparent correlation with galactic latitude or longitude (Struve 1950), it seems reasonable to assume that the rotational axes are distributed at random. Supporting evidence for this was found in Slettebak's (1949) study of Be stars, objects which are rotating so fast that matter is ejected in the equatorial plane (cf. also Struve 1951). In such a sample, it is certain that all objects are really rotating rapidly, and one can test for a random distribution in i freed from the influence of stars that are, in fact, slow rotators. The function $\langle V(M_V) \rangle$ for normal field dwarfs, on the assumption of a random orientation of axes, is also illustrated in Figure 1.

One may now inquire how the rotational velocities are distributed about the mean, when the stars are grouped in small intervals of M_V. Deutsch (1967) has emphasized that, on statistical mechanical principles, one would expect a Maxwell-Boltzmann distribution of v, but most discussions (Huang 1953; Huang and Struve 1960) have indicated that the observations do not conform to this view. These studies showed that when B-, A-, and F-type stars are grouped independently, the distribution function tends to a maximum in each group as $v \to 0$, whereas a Maxwell-Boltzmann law would require it to vanish (cf. Huang and Struve 1960, p. 328). Deutsch (1967) has criticized these conclusions, and finds that for a group of seventy-two main-sequence stars of types A3 to F2 the distribution of v, when based on Slettebak's measures, is essentially Maxwellian. This is illustrated in Fig. 2. We note that if the true distribution is $f(jv) = 4\pi^{-1/2}(jv)^2 \exp[-(jv)^2]$, where $j = [\langle (v \sin i)^2 \rangle]^{-1/2}$, then the observed distribution, for a random orientation of axes, is given by

$$\varphi(jv \sin i) = 2(jv \sin i) \exp[-(jv \sin i)^2];$$

it is convenient to study the observations using this latter function.

It will be seen from Figure 2 that, as $v \sin i$ goes to zero, φ reaches a maximum and then tends toward zero. Deutsch included in his sample fourteen Am stars, all of which have $v \sin i < 100$ km/sec. If one should choose to omit these stars on the grounds that most, or all, are binaries (Abt 1961), one still finds a satisfactory fit to a Maxwell-Boltzmann distribution (cf. Fig. 2). Although Brown (1950) has pointed out that the inversion of the integral equation to recover f from φ is not necessarily unique, these results certainly indicate that there is nothing about the rotations of stars later than A2V that is incompatible with a Maxwellian distribution. Support for this is found in Kraft's (1965a) study of the rotations of stars in the Hyades, wherein the distribution of φ appeared to be compatible with a Maxwell-Boltzmann law.

Earlier than A2V, Deutsch again suggests a velocity distribution that is Maxwellian, but which is interrupted by two very sharp peaks of overpopulation of sharp-lined stars at B2V and at A0V; the latter peak was first pointed out by Conti (1965). An interpretation of this effect has been given by Deutsch (1967), and will be discussed in Sec. IVc. The explanation of the apparent discrepancy between Deutsch and Huang regarding the shape of the velocity distribution function probably results from the latter's inclusion of these two sharp peaks into the spread-out statistics of *all* A- and B-type stars.

From the run of $\langle v \rangle$ shown in Figure 1 we can now derive $\langle J(\mathfrak{M}) \rangle$. We take the mass-luminosity law given by Harris, Strand, and Worley (1963), and the stellar models of main-sequence stars tabulated by Schwarzschild (1958); the assumption is made that the stars are rigid rotators. A plot of $\log \langle \mathscr{J}(\mathfrak{M}) \rangle = \log \langle J(\mathfrak{M})/\mathfrak{M} \rangle$ is given as a function of $\log \mathfrak{M}$ in Figure 3. The quantity shows a sharp break at mass

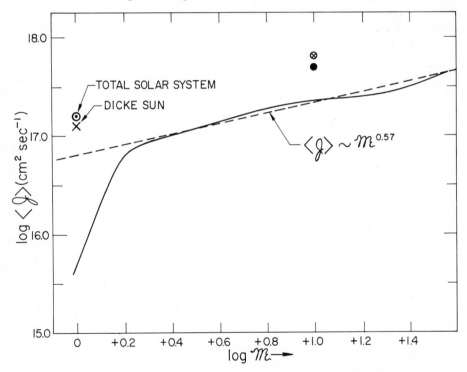

Fig. 3.—$\langle \mathscr{J}(\mathfrak{M}) \rangle = \langle J(\mathfrak{M})/\mathfrak{M} \rangle$ as a function of $\log \mathfrak{M}$ for main-sequence stars. The angular momentum per gram of the total solar system and of the Sun with the $\omega(r)$ distribution proposed by Dicke are shown. The crossed circle represents the angular momentum per gram of the Härm-Rogerson B-type star described in the text if rigid-body rotation is assumed. The filled circle is the same star with the extreme $\omega(r)$ distribution given by Roxburgh and Strittmatter (1966a).

1.5 $\mathfrak{M}\odot$; its behavior below mass 1.0 $\mathfrak{M}\odot$ is not known with any precision. Above mass 2.0 $\mathfrak{M}\odot$, the run of $\langle \mathscr{J} \rangle$ is quite flat and goes approximately as $\mathfrak{M}^{0.57}$. Here it is nearly true that the stars are given, on the average, an amount of angular momentum simply in proportion to their masses.

(b) Possible Causes of the Break in $\langle \mathscr{J}(\mathfrak{M}) \rangle$

Throughout the preceding discussion, we have assumed that stars are rigid rotators. We may now inquire whether a relaxation of that condition might remove the break in $\langle \mathscr{J}(\mathfrak{M}) \rangle$ either by increasing $\mathscr{J}$ among stars of the lower main sequence, decreasing $\mathscr{J}$ among stars of the upper main sequence, or both. The first possibility would manifest itself in solar-type stars if $\omega(r)$ increased as r decreased. Roxburgh (1964a, b) has considered "steady state" solutions of the structure equations for rotating stars in which the critical minimal rotational velocity is set by the condition that the meridionally circulated material moves a distance equal to the stellar radius in a time less than the star's lifetime on the main sequence. This condition is not satisfied for stars with $\mathfrak{M} < 1.5$ $\mathfrak{M}\odot$ as long as one assumes rigid-body rotation, and Roxburgh (1964b) asks whether some property of the internal structure or energy generation might lead to a rapid increase in $\omega(r)$ as $r \to 0$, thus accounting for the apparent slow rotation. The steady-state solutions for stars with $\mathfrak{M} < 1.5$ $\mathfrak{M}\odot$ lead, in fact, to an increase of ω from surface to center by a factor of about 2.5, but this is insufficient to change $\mathscr{J}$ significantly.

In the case of the Sun, Dicke (1964) advanced an interesting hypothesis which requires a sharp increase of $\omega(r)$ with depth (cf. also Temesvary 1952). He noted that one of the three tests of general relativity, viz., the gravitational redshift, actually can be predicted by other gravitational theories; a second, the deflection of starlight by massive bodies, is too inaccurately known to provide a good test, and thus only the precession of the perihelion of Mercury's orbit remains as apparently incontrovertible support for general relativity. He suggested, however, that, if the Sun were actually very slightly oblate, e.g., by $0\rlap{.}''05$, 10 percent of the observed precession would be due to the distortion of the solar gravitational field. Certain scalar-tensor theories of gravitation require just such a reduction, amounting at most to about 90 percent of the Einstein value. An oblateness of $0\rlap{.}''05$ corresponds to a solar rotation period of 25.5 hours. Dicke required that the outer convection zone, which has lost angular momentum by way of the solar wind, has become decoupled from the inner radiative region. In connection with this proposal, we computed the angular momentum of a solar-type star on the assumption that the surface rotates at $v = 2$ km/sec, the convection zone (outer 13 percent) rotates as a whole with this value, and the inner 87 percent (containing 99 percent of $\mathscr{J}$) rotates as a rigid body with $v = 42$ km/sec (at the surface of the zone, corresponding to $P = 25.5$ hours). The model for one solar mass tabulated by Schwarzschild (1958) was used. The angular momentum is increased by a factor of about 20 (cf. Fig. 3), and becomes nearly the same as that contained in the entire solar system.

In a later communication, Dicke and Goldenberg (1967) reported that the isophotes of solar limb brightness were in fact nonspherical by an amount compatible with a polar flattening of

$$\Delta R = R_{\text{equatorial}} - R_{\text{polar}} = 34 \text{ km};$$

the corresponding oblateness of $0\rlap{.}''042$ is in agreement with earlier expectation. However, Dicke's interpretation of the phenomenon as indicating the existence of a quadrupole moment produced by a rapidly rotating solar interior was criticized most notably by Goldreich and Schubert (1967) and by Howard, Moore, and Spiegel (1967), who argued that the implied differential rotation would be damped out by a "spin-down" process on a time-scale short compared with the nuclear, and even the contraction, age of the Sun. These arguments were countered by Dicke (1967), but until the matter can be clarified to the satisfaction of all concerned, it seems best to regard the solar angular momentum as an essentially unknown quantity.

The contrary alternative to the considerations of the preceding paragraphs is that early-type stars may have smaller angular momenta than those implied by Figure 3 owing to a decline in $\omega(r)$ as $r \to 0$. This can be explored using a model advanced by Roxburgh and Strittmatter (1965a). In contrast to the aforementioned solution (Roxburgh 1964b), the introduction of a toroidal magnetic field leads to an outward increase of angular velocity for a steady-state solution. Using the Härm and Rogerson (1955) stellar model with electron-scattering opacity (the same as that used by Roxburgh and Strittmatter), and the physical conditions appropriate to a B3V star ($\mathfrak{M} = 10\ \mathfrak{M}\odot$, $R = 2.5 \times 10^{11}$ cm, $v = 230$ km/sec, $\omega_{\mathrm{surf}} = 1.04 \times 10^{-4}$ sec^{-1}), one finds that, even for the extreme case of a very large core magnetic field, J is decreased only to 77 percent of its value for the rigid-body case (cf. Fig. 3). It is clear, therefore, that no presently conceived rotating model with a nonconstant ω can significantly lower $\langle \mathscr{J}(\mathfrak{M}) \rangle$ for the stars of the upper main sequence.

If $\langle \mathscr{J}(\mathfrak{M}) \rangle$ for stars of large mass correctly reflects the primordial, or at least initial main sequence, state of affairs, one may inquire whether the stars below the break share this reflection, or whether some additional loss of angular momentum might have occurred after the objects had in fact reached the main sequence. Spitzer (1956), for example, proposed that a stellar magnetic field, extending into the surrounding interstellar medium, would tend to drag that medium as the star rotated, and cause a net braking action on the star itself. However, the mechanism would be more effective in H II than H I regions, and would therefore work the better the earlier the spectral type of the star (cf. also Section VI).

Schatzman (1959, 1962), on the other hand, has advanced a very efficient mechanism for loss of angular momentum that is appropriate for late-type stars. It is suggested that, during solar-like flares, jets of material carry the magnetic field to distances several stellar radii above the surface, where the magnetic tension is no longer large enough to confine the material; from that point, the matter essentially leaves the star and carries away angular momentum. The leading point is that, as long as the magnetic field controls the jet, the matter is forced to turn with the star rather than to pursue a Keplerian orbit. For typical values of magnetic field in solar spots and typical solar ejection velocities, Schatzman obtains the expression

$$\left(\frac{\omega}{\omega_0}\right)^{2/5} = 1 - 1.08 \times 10^3 \frac{|\Delta\mathfrak{M}|}{\mathfrak{M}_0} \tag{4}$$

where $|\Delta\mathfrak{M}|$ is the mass loss, and the subscripted quantities are the original values of mass and angular velocity. It is seen that a quite small $|\Delta\mathfrak{M}|$ can lead to a large loss of angular momentum.

The theory then connects jet or flare activity with the presence of chromospheres; these, in turn, are postulated to exist only in stars having well-developed subsurface convection zones. The transition between stars with and without such convection zones occurs in spectral-type F; thus stars later than type F have, in Schatzman's view, undergone considerable magnetic braking. It is possible, in fact, to divide the H-R diagram with a nearly vertical line into two regions; to the right of the line are stars with well-developed convection zones, chromospheres, "activity," and concomitant loss of angular momentum. If the presence of Ca II emission in the spectrum of a star is a sign of chromospheric development, then Schatzman's theory receives support from the discovery (Wilson and Bappu 1957) that Ca II emission is present in the spectra of stars of all luminosities, but only of types later than F5 or G0. It should be noted that if a star condenses from an interstellar gas cloud as outlined in Sec. I of this article, its angular momentum is of the order of 100 times that of the most rapidly rotating main-sequence star. But all stars, whether of high or low mass, pass through the right-hand region of the H-R diagram during the stage of gravitational contraction to the main sequence. The Schatzman mechanism may provide the means therefore by which stars can dispose of a considerable fraction of their initial angular momentum.

An alternative mechanism for disposal of angular momentum is provided by stellar winds, a concept generalized from the known existence of the solar wind. Brandt (1966) has given evidence, from a comparison of the directions of the tails of direct and retrograde comets, that the solar wind carries a net flux of angular momentum and produces a torque on the sun sufficient to halve its angular momentum in about 5×10^9 years, provided the sun is in uniform rotation. If the Dicke–Goldenberg (1967) interpretation of their solar oblateness observations is correct, then only the outer convection zone is slowed, and the deceleration time drops by two orders of magnitude.

Generalization of these concepts (cf. Wilson 1966 a, b) suggests that stars undergo deceleration, at least of their observable surface layers, as a result of the torque exerted by stellar winds in the presence of a magnetic field. The instantaneous rate of deceleration depends, in accordance with the solar wind model advanced by Weber and Davis (1967), on the initial rotational velocity, the velocity and density of the wind, and the strength of the magnetic field. The length of time over which a torque is exerted can be identified with the time a star spends in regions of the H-R diagram where sub-surface hydrogen convection zones give rise to concomitant chromospheres and winds. Thus stars destined to become main sequence objects earlier than F5V probably lose angular momentum in their convective contraction stages (Hayashi 1966), but, upon arriving in a stage of radiative equilibrium in the outer layers, support insignificant winds after main sequence occupation. These stars maintain a high proportion of their initial main sequence angular momentum. On the other hand, stars destined to arrive on the main sequence at positions later than F5V lose angular momentum at all evolutionary stages, even after taking up main-sequence residence: Theoretical support for this argument is derived from the work of Baker (1963) and of Demarque and Roeder (1967) who independently showed that the hydrogen convection zone in main sequence stars diminished rapidly in thickness with increasing luminosity, and essentially cut off among the middle to early F's.

Another kind of evidence in support of this general picture comes from the observational studies by Wilson (1966a), who suggests that stars with active chromospheres, and presumably torque-exerting winds, may be usefully defined as those showing Ca II emission (K2 and H2) on spectrograms of dispersion 10 Å/mm. (By this criterion, the Sun's chromosphere is relatively "inactive" since dispersions of 2 Å/mm or higher are required to detect K2 emission.) At 10 Å/mm dispersion, corresponding to a limiting resolution of around 12 km/sec in $v \sin i$, Wilson found that, as one passes down the main sequence, slow rotation sets in rather abruptly (near F5V) at a value of (B — V) less than 0.02 mag away from the place where K2 first puts in an appearance. Among the stars later than F5V less than 10 percent show K2 at 10 Å/mm dispersion; the presumption is strong that these are the younger-than-average field stars, a surmise confirmed by their close delineation of the zero-age main sequence in the Strömgren c_1 vs b-y photometry and by the fact that stars F5V and later in young galactic clusters almost invariably show K2. Indeed Wilson found (1964, 1966b) that the strength of K2 declined with advancing nuclear age in the galactic clusters Pleiades (age 4×10^7 years), Hyades, Coma, and Praesepe (ages near 4×10^8 years).

Wilson's work was refined and extended by Kraft (1967),b who worked at higher dispersion and therefore higher resolution in $v \sin i$. He found that, among stars later than F5V, those with K2 rotated on the average faster than those without, and that

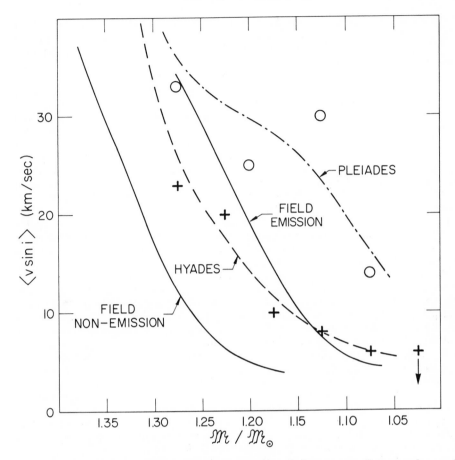

FIG. 4.—⟨$v \sin i$⟩ as a function of $\mathfrak{M}/\mathfrak{M}\odot$ for field and cluster stars. Crosses and open circles correspond, respectively, to mean values for the Hyades and Pleiades. ⟨v⟩ at a fixed value of $\mathfrak{M}/\mathfrak{M}\odot$ is equal to $4/\pi$ ⟨$v \sin i$⟩ if we assume a random orientation of rotational axes.

stars in galactic clusters also rotated faster than field stars without K2. The most rapid rotations were found in the Pleiades, the youngest cluster investigated. The run of ⟨$v \sin i$⟩ as a function of $\mathfrak{M}/\mathfrak{M}\odot$ for four samples of stars, grouped by age, is shown in Figure 4, and ⟨v⟩ at $\mathfrak{M}/\mathfrak{M}\odot = 1.20$ as a function of age is given in Table 1 This result suggests strongly that stars of mass $\mathfrak{M}/\mathfrak{M}\odot = 1.20$ with ages near 4×10^7 years (that of the Pleiades) have a mean rotation of 40 km/sec, and that this rotation decays by a factor of 2 in a time equal to the age of the Hyades, about 4×10^8 years, presumably in response to torques exerted by stellar winds.

An "observed" decay time of the order of 10^8 years for the rotations of solar-type Pleiades stars cannot, however, be construed as supporting the Dicke–Goldenberg (1967) model·for the distribution of angular velocity in the solar interior. We have already noted that the torque of the solar wind will reduce the solar rotation by a factor of 2 in a time of the order of 10^8 years if only the outer convection zone need be decelerated. However the agreement of these two time-scales is fortuitous, since we have no way of knowing the velocity and density of the solar wind or the strength of the solar magnetic field when the sun was as young as Pleiades stars are now.

It has also been argued that the formation of planetary systems is responsible for the slow rotation of late-type stars; one finds that if all the angular momentum tied up in the planets were returned to the sun, $\mathscr{J} \odot = 1.6 \times 10^{17}$ cm²/sec. In Figure 3 we plot this value and find that it is only a factor 2.5 above the extension of $\langle \mathscr{J} \rangle \approx \mathfrak{M}^{0.57}$. Huang (1965), for example, treated the problem of planetary formation by suggesting that all stars undergo some kind of (unspecified) braking, the degree of which depends on the strength of the braking mechanism and the time over which it acts. He showed that, starting from a Maxwellian velocity distribution, braking action will lead to velocity distributions which depart from the Maxwellian law in the sense of having too many stars with low values of $v \sin i$, a result in accordance with Huang's (1953) earlier conclusions about the observations. This departure becomes more and more severe as one advances to later spectral type, and Huang therefore suggested that some braking mechanism, such as that advanced by Schatzman, is more efficient in stars of types F and G than in types B and A. He concluded that sufficient material might be ejected from the progenitors of late-type stars to form a primordial planetesimal nebula, but the force of the argument is weakened by the relatively poor resolution of the observed rotational velocities (Huang 1953) on which it is based. A related picture of the formation of the solar system was discussed earlier by Hoyle (1960), who required that the contracting sun become rotationally unstable. The equatorial disc so formed recedes from the sun to the present characteristic distance of the planets by the action of a connecting magnetic field that transports angular momentum from protosun to disc. Poveda (1965) noted also that, if such discs actually exist, they could be opaque for a short period of time if viewed edge-on. Thus at a given mass, a contracting T Tau star may have a wide range of apparent magnitudes depending on aspect, and this is, in fact, observed in young clusters such as NGC 2264. On the other hand, a spread in the time of formation of stars could also induce such a scatter in M_V.

We summarize this section by noting that the formation of a planetary system is by no means the only mechanism by which the slow rotation of late, compared with early, type stars can be explained. Even without planets, main sequence stars later than F5V would achieve slow rotation on the nuclear time-scale because of angular momentum losses induced by stellar winds. Indeed we could justifiably argue that the break in $\langle v \sin i \rangle$ near F5V has nothing whatever to do with planetary formation, and that a main sequence star earlier than F5V is as likely to have planets as one later.

(c) $\langle v \sin i \rangle$ for Stars in Galactic Clusters

The preceding two sections summarize the literature establishing, among other things, (1) the mean dependence of $v \sin i$ on M_V for the main-sequence field stars, (2) the likelihood that $\langle J(\mathfrak{M}) \rangle$ for early-type stars is relatively unaffected by physically plausible departures from $\omega(r) = $ constant, and (3) the proposition that the orientation of rotational axes in space is entirely at random. The last conclusion indicates that individual stars do not remember the orientation of the angular momentum vector of the galaxy as a whole. In all probability the gas clouds out of which they were formed had random angular momenta as a result of turbulent motions or collisions (cf. Huang and Struve 1954). We may now inquire whether the stars of galactic clusters share these properties.

The various determinations of $\langle v \sin i (M_V) \rangle$ for cluster main-sequence stars are illustrated in Figure 5; the sources of the rotational velocities, all on Slettebak's system, are given in the caption. Abt and Hunter (1962) have given the most extensive list of velocities for cluster stars, and their mean field rotation function, repeated from Figure 1, is also shown in Figure 5. Though the amount of observational material is limited, several fairly definite conclusions can be made.

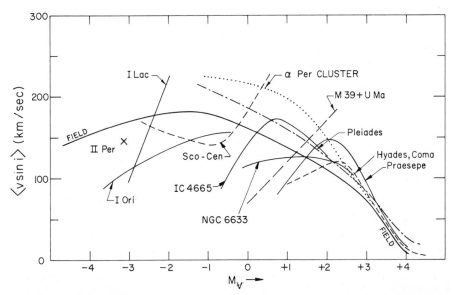

FIG. 5.—⟨$v \sin i$⟩ as a function of M_V for stars in galactic clusters. The mean field line from Fig. 1 is also shown. The sources of the rotational velocities are as follows. Field: Abt and Hunter (1962); I Ori: McNamara (1963); II Per: Abt and Hunter (1962); I Lac: Abt and Hunter (1962); α Per: Kraft (1967a); Pleiades: Anderson, Stoeckly, and Kraft (1966), Kraft (1967b); IC 4665: Abt and Chaffee (1967); M39 + UMa: Meadows (1961), Abt and Hunter (1962); Hyades: Kraft (1965a); Coma: Kraft (1965a); Praesepe: Dickens, Kraft, and Krzeminski (1968); Scorpio-Centaurus Association (Slettebak 1968); NGC 6633: Kraft (unpublished).

First, with the exception of the Pleiades, the α Per cluster, and possibly the Sco-Cen association, the curves show a marked turn-down in the mean rotation as M_V decreases toward the brightest stars in each cluster. This can probably be interpreted as resulting from slight evolution away from the main sequence with angular momentum conserved, at least in the cases of Hyades, Coma, Praesepe, NGC 6633, IC 4665, and possibly I Ori. The very large effect makes this explanation improbable for M39 and I Lac, however. Second, for all clusters having extensive observations to faint M_V's, the mean rotation of the faintest stars is not less than that of the field. The only possible exception to this is I Ori, but ⟨$v \sin i$⟩ for this association has not been carried below B9V, where it very nearly comes up to the mean field value. Extension below B9V would be difficult because of the well-known paucity of members near the critical lower "contraction" turnoff (Parenago 1954). Third, the ⟨$v \sin i$⟩ curves for Hyades, Praesepe, and Coma clusters having about the same age (≈ 1 to 5×10^8 years), are quite similar. Fourth, ⟨$v \sin i$⟩ for the Pleiades is everywhere larger than the field, and is larger than ⟨$v \sin i$⟩ for the Hyades even for stars as late as F5V and G0V, as noted in the preceding section. Finally, ⟨$v \sin i$⟩ for the α Per cluster, which has a main-sequence color-magnitude diagram (Mitchell 1960) identical with that of the Pleiades (Johnson and Mitchell 1958), agrees fairly closely with that of the Pleiades to spectral types as late as the investigation has been carried (F5V). The curves of Figure 5 extend considerably the material available to Abt and Hunter, but do not lend much support to their tentative proposal that the deficiency (compared with the field) of angular momentum among the bright stars of a cluster is balanced by a corresponding excess among the faint.

Smith and Struve (1944) first discovered that the rotational velocities of the

brightest Pleiades stars were unusually large, and Struve (1950) suggested that the
rotational axes might be preferentially oriented in space with $i \approx 90°$. Miss Van Dien
(1948), however, concluded that the distribution of rotational velocities in the Pleiades
was more nearly compatible with a Maxwell-Boltzmann distribution (implying a
random orientation of axes) than with a distribution based on a Boltzmann energy
formula together with a unidirectional orientation of axes. The latter is Gaussian,
i.e., corresponds to a one-dimensional Boltzmann law, about some (non-zero) mean
value, and has the form

$$F \approx \exp\left[-h^2(\langle v \rangle - v)^2\right]. \tag{5}$$

The observed distribution then becomes

$$\Phi \approx \exp\left[-k^2(\langle v \sin i \rangle - v \sin i)^2\right] + \exp\left[-k^2(\langle v \sin i \rangle + v \sin i)^2\right] \tag{6}$$

where the two exponential terms correspond to stars with angular momentum
vectors of opposite sign. Miss Van Dien preferred the Maxwell-Boltzmann distribution
to the expression given by equation (6) as a fit to the Pleiades data, but the discrimin-
ation within the errors is not very convincing. However, because of the second term,
(6) always leads to an excess of stars of slow apparent rotation over that given by the
Maxwell–Boltzmann formula. On the other hand, if we should insist on the stronger
condition that all angular momentum vectors not only be aligned but show the same
sign, then we would require that, as $v \sin i \to 0$, $\Phi \to 0$, and the second term of (6)
disappears. In that case, Φ is strictly Gaussian. Both of these cases leads to a fit of
the distribution of Hyades-Coma (Kraft 1965a) rotational velocities that is distinctly
poorer than that provided by a Maxwell-Boltzmann law. Further support for the
view that the rotational axes in clusters are distributed at random comes from an
inspection of the inclinations of the orbits of visual and spectroscopic binaries in
Coma and the Hyades; these too appear to be distributed at random (Kraft 1965a).

The somewhat more special condition required to maintain that the stars of clusters
remember the orientation of the angular momentum vector of the galaxy is more
easily dismissed by comparing the Hyades and Praesepe with Coma, the only galactic
cluster seen at high galactic latitude ($b = 85°$). We would expect to see the Coma stars
pole-on under this hypothesis. Yet the $\langle v \sin i \rangle$ curve of Coma is indistinguishable from
that of the Hyades (Kraft 1965a), and it would be most remarkable if $\langle v \rangle$ and $\sin i$
accidentally arranged themselves to produce this coincidence. To summarize, we find
that the assumption of a random orientation of rotational axes seems to fit best our
present knowledge of stellar rotations in galactic clusters, but the evidence is not
strictly compelling.

In the preceding section, evidence was offered in support of the proposal that, for
stars later than F5V, the decline in $\langle v \sin i \rangle$ with time was a direct result of angular
momentum losses generated by stellar winds. This would not explain, however, why
early-type stars in α Per and the Pleiades rotate on the average 20 percent faster
than the presumably somewhat older field stars of similar spectral type. Abt and
Hunter (1962) suggested, however, that whenever there was a paucity of spectroscopic binaries in a cluster, the cluster stars would have larger than average rotations;
the Pleiades is believed to have few spectroscopic binaries (Smith and Struve 1944;
Abt, Barnes, Biggs, and Osmer 1965), at least of large amplitude, in comparison with the
field. A paucity of spectroscopic binaries is also characteristic of the α Per group
(Heard and Petrie 1967). We note also that the stars of IC 4665, a cluster with age
similar to the Pleiades, are unusually sharp-lined, and a slightly larger than average
number are spectroscopic binaries (Abt and Chaffee 1967); its $\langle v \sin i \rangle$ function runs
rather low in Figure 5. One might suppose that the detection of small-amplitude spectroscopic binaries is the easier the sharper the lines, and a careful intercomparison of
velocity amplitudes and line widths for all three clusters should be made, preferably

with the same spectrographic equipment. However, Abt (1965), in his study of the frequency of binaries among normal A-type stars noted that the scatter in velocities for nineteen constant-velocity stars with $0 < v \sin i < 100$ km/sec was 2.04 km/sec per plate, whereas for ten constant-velocity stars with $150 < v \sin i < 240$ km/sec the scatter was only slightly larger, $viz.$, 2.74 km/sec. We conclude, therefore, that the Abt-Hunter hypothesis, though not definitely proved, cannot be easily dismissed since much of the radial-velocity data in clusters is based in fact on the highly uniform treatment of Abt and his associates.

An interesting cosmogonic implication possibly follows from the similarity in $\langle v \sin i \rangle$ for α Per and the Pleiades. We have already noted that both clusters have a paucity of (detectable) binaries. Eggen (1965) has remarked that the color-magnitude diagrams and the galactic velocity components U, V, W are identical. IC 2391, IC 2602, and possibly NGC 2516 also have these U, V, W components and similar color-magnitude diagrams, but nothing is known of the rotational velocities in these clusters. Considered together, these facts suggest that all five clusters may have fragmented from a super-cluster some 10^7 years ago, as suggested by Eggen (1965). Since the age of IC 4665 is about the same as these, but its $\langle v \sin i \rangle$ function is very different, it would obviously be of interest to determine its space motion.

III. STELLAR ROTATION AND STELLAR EVOLUTION

(a) Post-Main-Sequence Evolution

The change in v to be expected for a given star as a result of evolution depends on how J is conserved along an evolutionary track. Oke and Greenstein (1954) considered two cases that might be thought to bracket the possibilities: (1) no radial exchange of angular momentum ($i.e.$, each shell conserves its own angular momentum); (2) complete radial exchange of angular momentum ($i.e.$, rigid-body rotation). In case (1) we have $v_f/v_i = R_i/R_f$, where the subscripts i and f refer to the initial and final states, but in case (2) we must compute the full integral through the stellar model [eq. (3)]. Oke and Greenstein adopted the Sandage-Schwarzschild (1952) models and tracks. They concluded that, for a given initial rotational velocity, the final velocity for case (2) always exceeds that for case (1) by about a factor of 2. The recent papers by Iben (1965b, 1966a, b), giving improved evolutionary tracks, do not contain enough details of the $\rho(r)$ distribution to compute the angular momentum integral, but a calculation of the writer for one of the giant models of Kippenhahn and his associates (Hofmeister, Kippenhahn, and Weigert 1964) confirms the Oke-Greenstein conclusion.

Both Oke and Greenstein and Sandage (1955) agree that the rotations of stars of classes III and IV, earlier than G5, are compatible with the view that they have evolved from the main sequence along roughly horizontal tracks with angular momentum conserved. No decision was reached whether case (1) or case (2) was a better fit to the data. Abt (1957, 1958) considered the A- and F-type stars of classes II and Ib, some of which are characterized by relatively narrow lines. Again, Sandage-Schwarzschild-type tracks were assumed. Because of the narrowness of the lines in Ib's, no firm decision could be reached regarding the source of line broadening. Abt concluded that radial macroturbulence with a formal rms. velocity $\sigma \approx 2/3\ v \sin i$ gave as good a fit to the observed profiles as did rotation; this is illustrated in the case of α Per (F5 Ib) in Figure 6. For class II stars, the large values of line-width left little doubt that rotation is the source of broadening. For both kinds of stars, Abt concluded that, if they descended from main-sequence B-type stars, angular momentum conserved in shells (case 1) was more nearly compatible with the observations than rigid-body rotation. Kraft (1966), however, pointed out that the dispersion in the rotational velocities of A- and F-type Ib stars is not appropriate to a Maxwellian distribution, but is too narrow by a factor of at least 3; the velocities of A- and F-class II stars are, however, distributed by a Maxwell-Boltzmann law. This suggests that macroturbulence alone is responsible for the line profiles of the more luminous group. How this difference between class II and class Ib stars comes about is explained below.

Among K-type giants we experience the first difficulties with the hypothesis that angular momentum is conserved during post-main-sequence evolution. In Figure 7 we plot Iben's (1965b, 1966a, b) evolutionary tracks transformed to the observed M_V

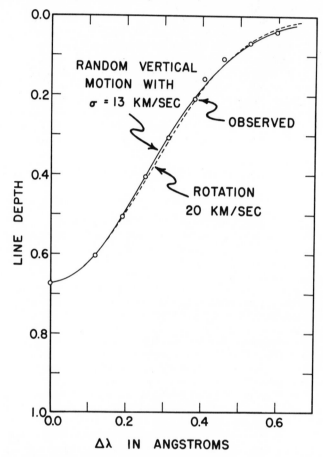

FIG. 6.—The line profile of $\lambda4508$ of Fe II in α Per (F5 Ib). The computed profiles for rotation and for radial macroturbulence are not distinguishable (after Abt 1957).

versus $(B - V)$ plane (Kraft 1966); also shown are the mean isorotational contours in $v \sin i$ expected for conservation of angular momentum in shells. Seven out of every ten K giants should have descended from F-type stars along M67- and NGC 752-like tracks (Sandage 1957), and these should have very slow rotations, partly because their progenitors have slow rotation, and partly because the change in radius is very large. The other 30 percent of K giants should have come, however, from late B's or early A's and should, therefore, have projected rotations of 15 to 50 km/sec. These numbers would be twice as large if the stars evolved as rigid rotators. Yet no K-type giants are known with $v \sin i > 10$ km/sec. This is particularly well shown by the K-type

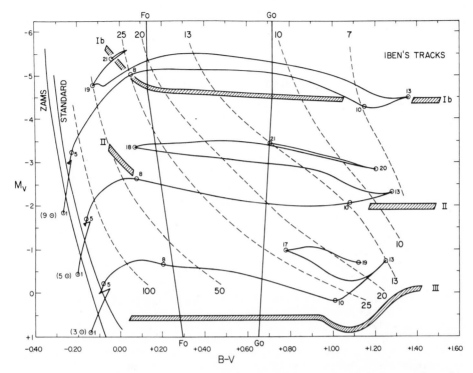

FIG. 7.—The evolutionary tracks of Iben (1965*b*, 1966*a*, *b*) for masses 3, 5, 9 𝔐⊙ transformed to the M_V vs. $B - V$ plane. Numbers on the tracks refer to the model numbers given by Iben. Dashed lines give the predicted rotational velocities for angular momentum conserved in shells, given the $\langle v \sin i \rangle$ curve of Fig. 1 for main-sequence stars.

giants of galactic clusters in which the breakoff is near the spectral type A0. In the Hyades, for example, $\langle v \sin i \rangle$ (predicted) ≈ 30 km/sec, but the observed value of $v \sin i$ for each of the four K-type giants is ≤ 6 km/sec (Kraft 1965*a*). In NGC 6633 the five yellow giants each have $v \sin i \leq 10$ km/sec (Kraft, unpublished), yet we predict from Figure 7 that $\langle v \sin i \rangle \approx 30$ km/sec. Similar conclusions can be reached for the three yellow giants of NGC 2281 (Kraft, unpublished). These results are summarized in Table 1. We conclude that either angular momentum is lost to stars somewhere in the Hertzsprung gap, or else it is transported inward from the surface regions.

Again it is of interest that loss of angular momentum from the layers accessible to observation sets in among giant stars that are late enough to have chromospheres, as indicated by the presence of Ca II emission (Wilson and Bappu 1957). We therefore suspect that the anomalously slow rotation of K-type giants is a result of mass loss and operation of the Schatzman or the stellar wind-type mechanism. Mass loss among stars in this part of the H-R diagram has been demonstrated observationally by Deutsch (1956, 1960), but it is of a secular variety. Whether the kind of "jet" ejection required by the Schatzman theory actually exists in K-type giants is not known.

If the mass loss is not sufficiently large to affect the evolutionary tracks, we can conclude from the time scales associated with Iben's calculations that the A and F stars of class I*b* are almost all descendants of the K supergiants, but the A and F stars of class II are, for the large part, passing for the first time from left to right in the

TABLE 1

ROTATIONAL VELOCITIES OF GIANTS IN GALACTIC CLUSTERS WITH "HORIZONTAL" EVOLUTION

Cluster	Star	$(B - V)°$	$M_V°$	$v \sin i$ (km/sec) Observed	Predicted (Shells)	Predicted (Rigid)
Hyades	γ Tau	0.99	+0.32	$\leq$ 5	30	60
	δ Tau	0.98	+0.33	$\leq$ 5	30	60
	ε Tau	1.01	+0.22	$\leq$ 5	25	50
	θ^1 Tau	0.96	+0.50	$\leq$ 5	35	70
NGC 6633	No. 67	1.26	−0.95	$\leq$ 6	13	26
	116	0.96	+0.05	$\leq$ 10	25	50
	122	0.93	+0.41	$\leq$ 10	35	70
	134	0.87	+0.70	$\leq$ 10	40	80
	140	0.89	+0.53	$\leq$ 10	35	70
NGC 2281	No. 18	1.05	+0.50	$\leq$ 10	20	40
	55	0.89	+0.12	$\leq$ 10	30	60
	63	1.25	−1.52	$\leq$ 10	10	20

H-R diagram. This seems confirmed (Kraft 1966) by a comparison of observed star counts with counts predicted from the evolutionary lifetimes based on Iben's theory and Schmidt's (1963) luminosity function for the main sequence. If the A and F Ib stars descend, therefore, from stars which have lost a great deal of angular momentum, we can then understand why their line profiles do not show a Maxwellian distribution: the rotations are, in fact, so slow that the profiles are dominated entirely by macroturbulence.

Support for the view that the line profiles of supergiants of classes F, G, and K are dominated by macroturbulence and not rotation comes from studies of the line profiles as a function of line strength, *i.e.*, as a function of depth in the atmosphere (Rodgers and Bell 1964, 1965; Bonsack and Culver 1966). These investigators found a marked dependence of line width on line strength in the sense that weak lines were systematically narrower than strong by an amount clearly incompatible with a rotational interpretation of the profiles.

Most of the regular intrinsic variables such as cepheids, δ Sct stars, and β CMa variables are encountered in post-main-sequence evolutionary stages and it is appropriate to discuss their rotational velocities here. Preston (1965) advanced the view that cepheids had very sharp lines because they were descended only from main-sequence stars with small values of v; we have shown above, however, that the profiles are explicable as a result of mass-loss in the K-supergiant stage, together with the acceptance of the validity of Iben's evolutionary tracks. The general idea of incompatibility between rotation and pulsation was advanced also by McNamara and Hansen (1961) for β CMa stars and by McNamara (1961) for δ Sct variables. These authors noted that, in both cases, the variables had lines unusually sharp in comparison with non-variables in the same part of the H-R diagram. Later results reduced the force of these arguments, however. Danziger and Dickens (1967), on the one hand, discovered a few δ Sct stars with $v \sin i > 100$ km/sec, and Hill (1967), on the other hand, found some new β CMa stars with very wide lines, two in fact with $v \sin i > 300$ km/sec. Christy (1967) explored the possibility of abandoning pulsation altogether as a mechanism for light and velocity variability in β CMa stars. Following Ledoux (1951), he proposed that the variations might be understood in terms of a wave travelling around the axis of a rotating star, and suggested that the appropriate instability might be generated among B-type stars in that region of the H-R diagram where β CMa variables are located. A test of this hypothesis requires a determination of line profiles around the cycle of some β CMa variable, a problem rendered difficult by the need for very fine time resolution.

(b) Pre-Main-Sequence Evolution

Little is known about the rotational velocities of stars in pre-main-sequence contraction; they are all quite faint for study at coudé dispersions. Walker (1956) noted that some of the late-type stars lying above the main sequence in the very young cluster NGC 2264 appeared to have fuzzy lines at dispersions near 85 Å/mm. The star with the broadest lines turned out later, however, not to be a member of the cluster (Herbig 1962; Vasilevskis, Sanders, and Balz 1965). Herbig (1957, 1962) reported on the rotational velocities of three T Tau stars in the Taurus dark clouds and on one other star of related type; these are distinctly moderate, running in the 20–50 km/sec range. Herbig concluded, on the basis of the evolutionary tracks computed by Henyey, LeLevier, and Levée (1955), that these variables would arrive on the main sequence among the F-type stars and would rotate with velocities similar to those presently observed if angular momentum were conserved in shells; the conclusion would not be altered significantly if the stars conserved their angular momenta as rigid bodies.

Subsequent observational and theoretical work which had transpired prior to 1965 indicated however that Herbig's conclusion could be maintained only with the greatest difficulty. First, it seemed likely that the stars were later in spectral type than Herbig supposed: the classifications had been taken from the work of Joy (1945), based on small-scale spectrograms. New classifications by Herbig, reported by Kuhi (1964) showed that the spectral types of RY Tau and T Tau were later than G—in fact, are near K0-K2. The emission-line- and reddening-freed colors of Smak (1964) also indicated spectral types in the range of early K. Second, the improved evolutionary tracks of Hayashi (1961), which take convection into account, indicated that for stars as late as K0 the evolution to the main sequence in the H-R diagram was nearly vertically downward. The main sequence descendants of these T Tau variables would then have been stars of type K2V rather than type F, and even for the case of angular momentum conserved as in rigid-body rotation, the predicted rotations of these K2V stars would have been 5 to 10 times too fast.

Results obtained still more recently, however, throw the subject into renewed confusion. A more thorough treatment of the evolutionary tracks by Iben (1965a) shows that T Tau stars as late as K0 are not only more massive than those computed by Hayashi, but subsequently develop radiative cores, and approach the main sequence horizontally. Herbig's three T Tau stars in this case lie almost exactly along Iben's track for $\mathfrak{M} = 1.25\,\mathfrak{M}\odot$, and though the predicted rotations for the main sequence descendants are still a factor of 2 to 4 times too large for the case of angular momentum conserved in shells, the assumption of rigid body rotation brings them comfortably into the expected range.

There are, nevertheless, the following factors which further cloud the issue:

(1) According to Iben (1965a), the exact location in $\log T_e$ of the vertical convective track is quite sensitive to the choice of metal abundance;

(2) The variability of T Tau stars leads to uncertainty in their location in the H-R diagram;

(3) The Iben tracks map the T Tau stars of Herbig into a position near F5V on the main sequence where $\langle v \sin i \rangle$ is changing rapidly; thus the uncertainties induced by (1) and (2), among others, may render the above conclusion meaningless;

(4) T Tau stars probably lose a considerable fraction of their mass (Kuhi 1964, 1966) during the contraction stage, and thus both the evolutionary tracks may be inapplicable and appreciable angular momentum may be lost in any case;

(5) The discovery of large infrared excesses (Mendoza 1966, 1968) in T Tau stars, the interpretation of which is quite uncertain, coupled with the previously well-known violet excesses (Smak 1964), may invalidate the transformations from M_V to M_{bol} and from $(B - V)°$ to log T_e.

At present it is attractive to view the large infrared excesses as resulting from a nebula surrounding the T Tau star (Mendoza 1966). If this should be identifiable with a pre-planetesimal disc, the discussions by Schatzman, Poveda, and Hoyle, already mentioned, would clearly be relevant.

IV. INFLUENCE OF ROTATION ON THE LUMINOSITIES AND TEMPERATURES OF MAIN-SEQUENCE STARS

(a) Theoretical Results

A rotating star, distorted from spherical shape, has a surface gravity smaller at the equator than at the poles. By von Zeipel's (1924) theorem, both the local effective temperature and surface brightness are therefore lower at the equator than at the poles. It follows that the mean luminosity and temperature will be a function of the angle between the direction of observation and the axis of rotation (the aspect effect). Moreover, in comparing a rotator with a non-rotator of the same mass, we would expect to find that, in principle, the former suffers a slight change in total luminosity, the gravitational potential and therefore the central pressure and temperature being altered slightly by the rotation. Consequently, the theoretical problem divides naturally into two parts: (1) what is the internal structure and energy generation of a rotating stellar configuration, and (2) what is the quality of the integrated radiation as a function of aspect when that structure is surmounted by a suitably realistic model atmosphere?

The problem of the internal structure of a rotating star has been extensively investigated, and the reader is referred to the paper by Roxburgh, Griffith, and Sweet (1965) for a bibliography. Of the papers prior to 1965, only that by Sweet and Roy (1953) concerned itself with the observable effects of rotation, but these authors limited themselves to the case of slow rotation using a first-order perturbation analysis. They derived the following expressions for mean luminosity, effective temperature, and gravity over the surface of a rotating star:

$$M_{bol} = M_{bol}(0) \begin{cases} +0.569 \ \alpha_1 & \text{(edge-on)} \\ -0.427 \ \alpha_1 & \text{(pole-on)} \end{cases}$$

$$\log T_e = \log T_e(0) \begin{cases} -0.065 \ \alpha_1 & \text{(edge-on)} \\ -0.040 \ \alpha_1 & \text{(pole-on)} \end{cases} \qquad (7)$$

$$g = g(0) \times \begin{cases} (1 - 1.91 \ \alpha_1) & \text{(edge-on)} \\ (1 + 0.22 \ \alpha_1) & \text{(pole-on)} \end{cases}$$

where the quantities evaluated at (0) refer to a non-rotating star of the same mass, and the first and second forms of the equation refer, respectively, to a star viewed perpendicular to the axis of rotation and viewed pole-on. The quantity α_1 is given by

$$\alpha_1 = \frac{v^2}{Rg(0)} \left[1 + 0.772 \ \frac{v^2}{Rg(0)} \right]. \qquad (8)$$

These equations indicate that a rotating star is not only cooler than a non-rotator, regardless of aspect, but also that the rotator is brighter or fainter than the non-rotator, according as it is seen pole-on or equator-on, respectively. The changes involved are quite small, amounting at most to a few hundred degrees in T_e and a few tenths of a magnitude in M_{bol}.

The approximations upon which the Sweet and Roy theory is based break down, however, for rapid rotation, and a recent paper by Roxburgh, Griffith, and Sweet (1965) considers a more realistic model. The star is divided into two regions. In the inner part, which contains most of the mass, the ratio of centrifugal force to gravity is everywhere < 1. A first-order perturbation analysis is valid here, similar to that employed by Sweet and Roy. In the outer part, the density is, however, so low that the gravitational potential ψ is due entirely to the matter of the inner region, and is therefore derivable as a solution of Laplace's equation. Once ψ is known, the structure equations are solved, and the outer solution is fitted to the inner solution at an appropriately chosen interface. Models with electron-scattering opacity were taken, and it was found that, at a given mass, the integrated luminosity L of a rapid rotator is about 25 percent less than that of a non-rotator. Presumably the rotational term in ψ produces a net levitational effect in the core and slightly reduces thereby the rate of energy generation. In a subsequent paper, Roxburgh and Strittmatter (1966)

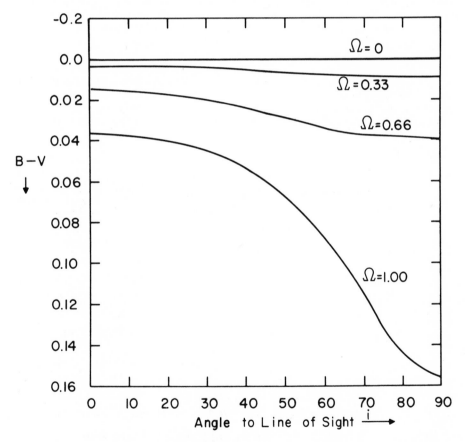

FIG. 8.—The effect of rotation on $B - V$ as a function of aspect, after Roxburgh and Strittmatter (1965b). The quantity Ω^2 gives the ratio of centrifugal force to gravity at the equator. All calculations are based on rigidly rotating models.

studied the changes in M_{bol} and log T_e for models with Schwarzschild, Kramers, and electron scattering opacities, and for models having a varying magnetic field strength.

The more recent treatments confirm and extend the Sweet and Roy (1953) results qualitatively, by showing that the main sequence of rotating stars would be expected to lie to the right in the H-R diagram relative to the main sequence of non-rotators. This result was projected into the observational color-magnitude diagram by Roxburgh and Strittmatter (1965b), who studied gray atmospheres surmounting the uniformly rotating models of Roxburgh, Griffith, and Sweet (1965). In Figures 8, 9, and 10, the observational effects on M_V and $B - V$ are represented; it should be noted that in Figure 10 the displacement vectors are for stars at the limit of rotational stability and are therefore appropriate only to the extreme situation. The quantity Ω^2 is the ratio of centrifugal force to gravity at the equator, $i.e.$,

$$\Omega^2 = \left[\frac{v^2}{R_e}\right] \bigg/ \left[\frac{G\mathfrak{M}}{R_e{}^2}\right], \tag{9}$$

and is physically the same as the α_1 of Sweet and Roy (1953) [Eq. (8)]. If we consider instead of the extreme case the main-sequence stars of average rotation ($\langle v \rangle$ of Figure 1), then $\Omega \approx \frac{1}{3}$ down to the critical break at $\mathfrak{M} \approx 1.5\ \mathfrak{M}\odot$. The vectors of Figure 10 are scarcely discernible for an Ω of this value, and really significant

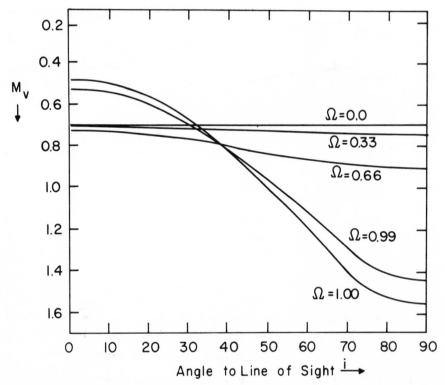

Fig. 9.—The effect of rotation on M_V as a function of aspect, after Roxburgh and Strittmatter (1965b).

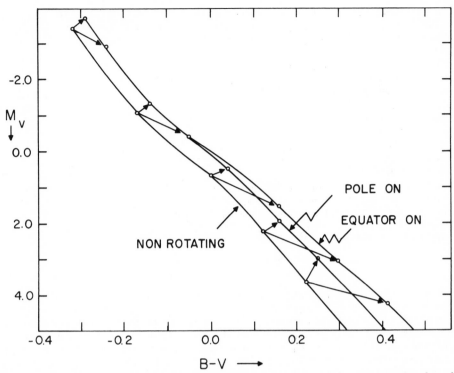

FIG. 10.—The effect of rotation on the position of the main sequence, after Roxburgh and Strittmatter (1965*b*).

displacements take place only for stars with $v \approx 1.5 \langle v \rangle$ or more. Considerably larger effects were found, however, if one studied instead non-uniformly rotating, necessarily magnetic stars (Roxburgh and Strittmatter 1966). These results will be considered in more detail in the next Section.

Peters, Poland, and Wrubel (1966) considered approximate departures from grayness by asuming κ_v, the monochromatic absorption coefficient, varies with frequency but that $\kappa_v/\bar{\kappa}$ is constant with optical depth. For application to A- and F-type stars, they included opacities of H^-, H, and the metals. In calculations of this type, which do not concern themselves with the integrated luminosity, one can study the effect of rotation in diagrams where the observable quantities are sensitive to temperature and surface gravity, as, for example, in the $U - B$ versus $B - V$ diagram or in the Strömgren-Perry $b - y$ versus c_1 diagram. Peters *et al.* found that moderate rotations ($\Omega \approx 0.6$) produced displacements of a few hundredths of a magnitude in the $U - B$ versus $B - V$ diagram among late A- and early F-type stars, in the sense that large $v \sin i$'s were associated with an ultraviolet deficiency, and small $v \sin i$'s with an ultraviolet excess.

Among early A-type and B-type stars, departures from grayness are more serious and have been considered by Collins (1963, 1965) and more recently by Hardorp and Strittmatter (1968*a*). The former fitted the local g and T_e to each surface element of a rotating star and integrated the emergent intensity over the visible surface to obtain the monochromatic continuum radiation as a function of aspect. Collins found that the gray atmosphere was a good approximation in the visible part of the spectrum, but a poor one in the ultraviolet. Furthermore, he predicted that the small flux changes induced by rapid rotation in the visible spectral region would be accompanied by flux changes of the order of a magnitude in the ultraviolet below both the Balmer and the Lyman jumps.

Hardorp and Strittmatter (1968a) used the non-gray model atmospheres of Mihalas (1964, 1965, 1966) and the specification of the local gravity and the distortion of shape from the interior models of Roxburgh, Griffith, and Sweet (1965). In the M_V vs $B - V$ plot the results of the gray atmosphere calculation (Roxburgh and Strittmatter 1965b) were essentially confirmed among B-, A-, and F-type stars, viz., that rapid rotators viewed at different aspects lie on a sequence parallel to, but displaced to the right of, the zero-rotation main sequence, and thus that the displacement depends mostly on v^2 and not much on i. On the other hand, contrary to the situation among late A's and F's, it was found that, in the $U - B$ vs $B - V$ diagram, rotation projects stars essentially along, rather than at a considerable inclination to the main sequence. Similar results were found in the Strömgren c_1 vs $(b - y)$ diagram, indicating that rotation generates no discernible effects in the conventional color-color diagrams for early-type stars. Similar results had been noted earlier by Collins and Harrington (1966) whose work will be discussed more extensively in the following section.

(b) Observational Results (Normal Stars)

The theoretical papers predict effects so small that only highly accurate measurements would be presumed capable of revealing the influence of rotation on colors and magnitudes in the ordinary visible part of the spectrum. The main-sequence stars in the general field have the complication, however, that their colors are affected by differences in the hydrogen to metal ratio (Eggen and Sandage 1962); in only a very small proportion of cases, the metal line strengths and therefore the colors may also be affected by differences in microturbulence (Conti and Deutsch 1966; Barry 1967; Kraft, Kuhi and Kuhi 1968). Unique values of ultraviolet excess, the commonly accepted measure of line weakening, seem however to characterize stars in clusters, which therefore provide the ideal testing ground for the theory provided differential reddening is negligible or removable.

McNamara and Larsson (1962) noted that, among the B-type stars of the Orion association, stars with large values of $v \sin i$ had weaker $H\beta$ lines than stars with small $v \sin i$ at fixed $U - B$. The $H\beta$ line strengths had been taken from the interference-filter photometry of Crawford (1958), and Guthrie (1963) showed that the effect was not due to the use of an $H\beta$ filter with too narrow a half-width, a conclusion confirmed by Abt and Osmer (1965). Abt (quoted by McNamara and Larsson) suggested that $H\beta$ was weakened because a large rotation lowers the effective gravity; it is known that, for a given $U - B$, giants have weaker $H\beta$'s than dwarfs—an effect which is attributable to reduced gravity (Crawford 1958). Guthrie (1963) explained the I Ori observations using the theory of Sweet and Roy (1953). We note from equation (7) that the coefficient of α_1 for a pole-on rotator is smaller (in absolute value) than that of an edge-on rotator; thus the effective gravity of the former does not much differ from a non-rotator. Guthrie estimated the change in $U - B$ and β that might be anticipated from equation (7). The displacement vectors show that a pole-on rotator lies almost on the same main sequence as a nonrotator in a β versus $U - B$ diagram; only the edge-on (equator-on) rotator shows a significant deviation, thus explaining why the displacements $\delta\beta$ should be correlated with the displacements of $\delta(v \sin i)$.

Collins and Harrington (1966) studied this problem in much greater detail with a series of models covering the range of B-type stars. They calculated $H\beta$ line strengths and profiles as a function of $v \sin i$, as well as UBV colors. The macroscopic effects of shape distortion, aspect, gravity darkening, and limb darkening were considered, and in the hydrogen line formation, such microscopic effects as pure absorption, noncoherent resonance scattering, and coherent electron scattering. Effects in the color-magnitude and color-color diagrams similar to those found by Hardorp and Strittmatter (1968a) were found, as already mentioned, but in addition the strength of Hβ at a fixed $(U - B)$ was found to decrease by as much as 50 percent for stars in rapid rotation compared with non-rotators.

The possibility that the line spectrum of a B-type star might be affected by rapid rotation was first suggested by Huang and Struve (1956) in the case of 20 Tau (Maia), a Pleiades B-type star with $v \sin i = 35$ km/sec (Abt and Hunter 1962). They noted that the unusually large range of excitation exhibited by the line spectrum might be due to the range in temperature associated with a rapid rotator seen pole-on; in this picture the He I lines are produced only in the center of the disk and are weakened by the contribution of purely continuous radiation at the limb. Guthrie (1965) extended these considerations and suggested that, among B-type stars, rapid rotators seen pole-on could be recognized by the presence both of weak He I and low excitation-low surface gravity lines such as Ti II, Sr II, and Fe II. Examples are HD 37058 and α Scl, as well as 20 Tau.

This suggestion was criticized on a number of grounds. First, according to Searle and Sargent (1964), the lines of C II, which have ionization and excitation potentials similar to those of He I, are not weakened in Maia or α Scl. Second, Sargent and Strittmatter (1966) noted that the sharp-lined Be stars, which must surely be rapid rotators seen pole-on, do not exhibit weakened He I for their colors. And finally according to these same authors, the four known weak He stars in Orion have small $v \sin i$ and lie essentially below the main sequence as defined by the normal stars. In view of the arguments put forward in the preceding section, one can therefore interpret the weak He stars as a group of truly slow rotators, rather than as rapid rotators seen pole-on. This general picture received support from the work of Hardorp and Strittmatter (1968b), who computed the change in strength of He I $\lambda5876$ and $\lambda4121$ at a fixed $b - y$ as a function of v for pole-on rotators. The calculation was based on the rather detailed model atmospheres for rotating stars already discussed (Hardorp and Strittmatter 1968a). At v corresponding to ejection of matter in the equatorial region (i.e., $\Omega \approx 1$), the decline in He I line strength turned out to be only about 20 percent, much too small to account for the anomalous He I line intensities in Maia.

Searle and Sargent (1964) and later Sargent and Strittmatter (1966) preferred the interpretation that the weak He stars were an extension of the Ap stars into bluer colors; the latter share with the former small values of $v \sin i$. The anomalous decline in He abundance in this view must therefore be confined to the atmosphere; otherwise the stars would appear to the right, rather than the left, of the normal main sequence.

Returning to a consideration of A- and F-type stars, we recall that the predictions of changes in color and magnitude based on gray atmospheric models are a good approximation (Roxburgh and Strittmatter 1965b), and that these are in rough qualitative agreement with the even more elementary treatment by Sweet and Roy [cf. eq. (7)]. In particular, Kraft and Wrubel (1965) considered the Hyades, which has no main-sequence stars earlier than A3V. They noted that the change in absolute magnitude implied by equation (7) was too small to be detected within the uncertainties introduced by the extension of the Hyades along the line of sight. The remaining gravity and temperature equations were tested in both the $U - B$ versus $B - V$ (Johnson, Mitchell, and Iriarte 1962), and in the $b - y$ versus c_1 (Crawford and Perry 1966) diagrams. The gravitational effect in the U, B, V case can be seen in the work of Eggen and Sandage (1964), where the trajectories in $U - B$ versus $B - V$ for class V and class III stars are illustrated. For F-type stars with $B - V > +0.275$, class III stars show a marked ultraviolet deficiency relative to class V stars; bluer than $B - V \approx +0.275$, the two trajectories cross over and $\delta(U - B)$ is no longer a sensitive measure of surface gravity. For the redder group of Hyades stars, Kraft and Wrubel found a marked correlation between rotation and departures from the main-sequence relation in the sense that stars with $v \sin i$ large showed ultraviolet deficiencies (i.e., were more like giants) and stars with $v \sin i$ small showed ultraviolet excesses. The degree of departure was confirmed by the non-gray rotating atmospheres of Peters, Poland and Wrubel (1966). The observed effects are better illustrated however in the Strömgren-Crawford system because c_1 is sensitive to gravity everywhere throughout the range of A- and F-type stars. In Figure 11 we illustrate the departures δc_1 from the mean Strömgren–Crawford relation for the Hyades main sequence as a function of $Y = v \sin i / \langle v \sin i \rangle$ for each star not known to be a spectroscopic binary. In Figure 12 we find the vectors δc_1 predicted from the Sweet and Roy equations for an A-type star, and these are seen to be of the right order to explain Figure 11. We note again that, as in the case of Guthrie's study of the B-type stars, pole-on rapid rotators are projected onto the sequence of non-rotators, and this explains why δc_1 correlates with $v \sin i$.

Strittmatter (1966) considered rotation in the M_V versus $B - V$ diagram on the basis of the paper by Roxburgh and Strittmatter (1965b) already cited. Fixing attention on a given $B - V$, Strittmatter defined $\Delta_0 M_V$ as the deviation of a rotator above the main sequence defined by non-rotators. The Roxburgh-Strittmatter

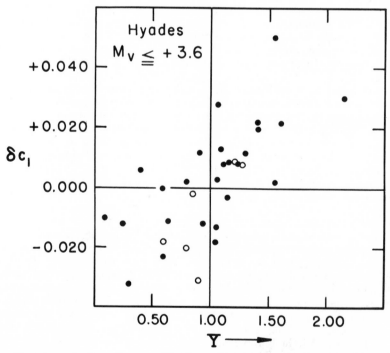

Fig. 11.—Departures δc_1 from the mean Hyades main sequence as a function of $Y = v \sin i / \langle v \sin i \rangle$, after Kraft and Wrubel (1965). Stars of large projected rotation have $Y > 1$ and stars of small projected rotation have $Y < 1$.

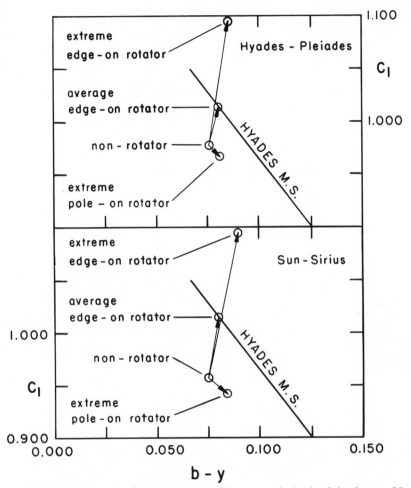

FIG. 12 —Theoretical departures δc_1 as a function of $b - y$, on the basis of the theory of Sweet and Roy (1953), after Kraft and Wrubel (1965). The "average" rotator has $Y = 1$, and the "extreme" rotator is defined as a star with $Y = 1.5$.

theory indicated that

$$\Delta_0 M_V \approx \Omega^2 = v^2 \left(\frac{R}{G\mathfrak{M}}\right) \quad [\text{fixed } (B - V)], \tag{10}$$

i.e., the departures $\Delta_0 M_V$ depend almost not at all on the aspect, and go nearly in proportion to v^2, the factor $(R/G\mathfrak{M})$ being a very slowly varying function on the main sequence. For any particular star we have

$$\Delta_0 M_V = bv^2, \tag{11}$$

where b is a constant given by theory. If we consider a small group of stars with nearly the same $B - V$, and if there is a random orientation of axes, we can write $\langle \Delta_0 M_V \rangle = \frac{3}{2} b \langle (v \sin i)^2 \rangle$; whence

$$\Delta_0 M_V - \langle \Delta_0 M_V \rangle = \frac{3}{2} b [(v \sin i)^2 - \langle (v \sin i)^2 \rangle]. \tag{12}$$

In applying equation (12), Strittmatter considered the stars of Praesepe, a cluster sufficiently far away that back-to-front magnitude differences are negligible. Making use of Johnson's (1952) photometry and Treanor's (1960) values of $v \sin i$, he found a good correlation between $Q = (v \sin i)^2 - \langle (v \sin i)^2 \rangle$ and $\Delta_0 M_V - \langle \Delta_0 M_V \rangle$. It is clear that a projection "backward" through equation (12) will locate approximately the "zero-rotation" main sequence. The constant b depends on the model adopted for the rotating star, and we give in Table 2 values taken from the work of Strittmatter and

TABLE 2

DETERMINATION OF ROTATIONAL CONSTANT b

Source	$b \times 10^5$
(1) Models of rigid rotators	0.2–0.3
(2) Models of non-uniform rotation	
(a) slow rotators	1.65
(b) rapid rotators	1.35
(3) Normal Praesepe stars	1.1

Sargent (1966) and of Roxburgh and Strittmatter (1966a). For the two types of non-uniformly rotating stars, the first line refers to slow rotators with negligible magnetic fields, and the second line to stars in which the centrifugal and magnetic forces are comparable. It will be seen that the Praesepe stars fit best to the theoretical models that do not rotate as rigid bodies, and the more detailed and more accurate treatment of rotating stellar models by Faulkner, Roxburgh, and Strittmatter (1968) confirms this view.

The results obtained by Strittmatter (1966) in Praesepe were considered somewhat problematical by Dickens, Kraft, and Krzeminski (1968). They noted that the expected corresponding correlation between $\delta(U - B)$ and $Y = v \sin i / \langle v \sin i \rangle$ was either weak or non-existent in Praesepe, that the rotational velocities used by Strittmatter were insufficiently accurate for the purpose (cf. McGee, Khogali, Baum, and Kraft 1967), and that some cluster stars with $v \sin i$ departing most sharply from the mean had not been observed photoelectrically by Johnson. The more extensive photometry and more accurate rotational velocities of these authors led, however, to a poorer correlation of Q with $(v \sin i)^2 - \langle (v \sin i)^2 \rangle$ than that found by Strittmatter, and to the suggestion, as yet unconfirmed, that Praesepe might suffer a very slight differential reddening.

Kraft and Wrubel (1965) pointed out that color-color and color-magnitude diagrams of stars in clusters could be used to disentangle aspect from true rotation without recourse to the observed values of $v \sin i$, provided differential reddening and line-of-sight extension were negligible. In a c_1 versus $b - y$ plot, slow rotators and rapid rotators seen pole-on are indistinguishable, but lie low in the diagram (i.e., c_1 is large for a given $b - y$). On the other hand, in an M_V versus $b - y$ (or $B - V$) diagram, rapid rotators always lie to the right, i.e., are too bright at a given $b - y$ compared with the average rotator. Thus from the sample of stars in the first diagram that have $v \sin i$ small, one can select from the second diagram those with i near zero and v large. Roxburgh, Sargent, and Strittmatter (1966) advanced essentially the same idea, but substituted use of the observed $v \sin i$ for the c_1 versus $b - y$ diagram. The use of these methods must, however, be restricted to stars of type A and F since as already noted (Collins and Harrington 1966, Hardorp and Strittmatter 1968a), rotation does not displace B-type stars in a color-color plot.

(c) Rotation of Be Stars

It has been known for some time that the stars of most rapid rotation have values of v at least 20 percent smaller than expected for equatorial breakup; yet the interpretation of the emission-line spectra of Be stars seems to demand an equatorial ejection of matter. The problem was considered by Slettebak (1949, 1966) who suggested that the "observed" values of $v \sin i$ for very rapid rotators might be too small because of the neglect of possible non-uniform rotation and of gravity darkening; at the same time the rotational velocity required for ejection might, because of atmospheric turbulence, be less than the critical velocity determined from equating the surface gravity to the centrifugal force at the equator.

Hardorp and Strittmatter (1968b) studied this point in more detail on the basis of model atmosphere considerations previously described (cf. Section IVa), and concluded that the neglect of gravity darkening did indeed lead to an underestimate of the rotational broadening and further the earlier assumption that the equivalent width of a line remained constant with increasing v was not valid, especially near the limiting equatorial breakup velocity. Variations in the "apparent" abundance of an element because of changes in the temperature and gravity over the surface were found to introduce an additional complication into the determination of $v \sin i$. Hardorp and Strittmatter concluded that, over the B, A, and F-type stars, the most rapid rotators were in fact at the limit of equatorial rotational breakup, and recourse to arguments about turbulent motions was therefore unnecessary. It should be noted that the corresponding increases in $v \sin i$ apply only to the relatively few stars near the breakup limit; moderate rotators remain unaffected. Thus since, for example, Be stars constitute only about 10 percent of the total population of B-type stars (Slettebak 1966), the mean curves exhibited in Figures 1, 3, and 5, which are based on the original Slettebak data, are only slightly affected.

(d) Rotation of Am and Ap Stars

The discovery by Abt (1961) that all, or most, metallic-line A-type stars (Am) are spectroscopic binaries leaves little doubt that their unusually narrow absorption lines imply truly slow rotation. Since the distribution of their orbital angular momentum vectors is apparently at random, it is difficult to believe that they could be characterized as pole-on rapid rotators.

The case of the peculiar A-type stars (Ap) is not so easily dismissed. The frequency of binaries among them (Deutsch 1958) does not exceed that of ordinary stars of similar luminosity, but it should be admitted that a thorough observational study has not been carried out. The most powerful argument in favor of the view that the Ap stars are a group of relatively slow rotators is the period-line-width relation (Deutsch 1958; Steinitz 1964) found among the subgroup of Ap's that are periodic spectrum variables. This has the form $v \sin i < 125/P$, where P is in days, the other numbers are in km/sec, and the equality occurs when $i = 90°$. The relationship assumes that the variable spectrum is modulated by the stellar rotation, and the constant is determined from the average properties of early A-type stars. Steinitz (1964) found no spectrum variable that failed to satisfy the inequality, even among those formerly "irregular" magnetic variables (Babcock 1958) discovered by him to have definite magnetic (and spectrum) periods. Since the periods rarely are less than one day, the spectrum variables are evidently a group of stars of rather slow, but not very slow rotation; in other words, the value 125 km/sec in the above inequality is small compared with the mean v of about 225 km/sec for normal stars of approximately the same luminosity.

Deutsch (1967) considered the problem of the rotation of Ap stars from the point of view of the statistics of stellar rotation near A0V. From a sample of 84 main-sequence stars with known rotations and colors in the range $-0.05 < (B - V) < +0.10$, he found that the rotational velocity distribution has three times too many stars with $v \sin i < 50$ km/sec (cf. Section IIa) to be representable by a Maxwell-Boltzmann law. He suggested that near A0 on the main sequence there are, in fact, two populations, P_Y and P_O, with very different Maxwellian velocity distributions. The former has $1/j_Y = 170$ km/sec, and the latter is characterized by $1/j_O = 25$ km/sec. Of the 24 stars with $v \sin i < 50$ km/sec, Deutsch assigns 6 to P_Y and 18 to P_O on the basis of the distribution functions. The questions then arise: (1) what is the significance of the "extra" group of slow rotators; and (2) is there any spectroscopic means of discriminating between the stars of P_O and the sharp-lined stars of P_Y?

In the sample in question, there are in fact exactly six peculiar stars—five Ap's and one Am—and Deutsch identified these with the sharp-lined stars of P_Y. The peculiar A-type stars therefore represent, on this view, the entire group of relatively slow rotators that belong to the population of conventional main-sequence stars. The P_O population is identified with a group of post-red giants, mapped into this near-main-sequence position along evolutionary tracks perhaps similar to the horizontal branch of globular clusters. The slow rotation in P_O would then result from loss of angular momentum in the red-giant stage, an argument somewhat analogous to that advanced by Kraft (1966) to explain the line profiles of A- and F-type supergiants. It should be noted that, if Deutsch's identifications are correct, then the assignment of the Ap's to the group of primordial main-sequence stars is exactly opposite to that required by the most recent nucleogenesis theory (cf. Fowler, Burbidge, Burbidge, and Hoyle 1965) that attempts to explain the peculiar abundances.

More direct evidence that the Ap stars are truly slow rotators was offered by Abt, Chaffee, and Suffolk (1967) who obtained estimates of $v \sin i$ for the 63 brightest objects. These velocities were found to average only about 22 percent that of normal A-stars in the same temperature range. They found that the idea that Ap stars are normal stars seen pole-on is incompatible with the frequency distribution of $v \sin i$, and thus if both groups have a random orientation of rotational axes, there are no normal stars between B7V and A0V with $v < 100$ km/sec and no Ap stars with $v > 150$ km/sec.

If then the Am's and Ap's represent the main-sequence stars of truly slow rotation, and if we remove from their colors and magnitudes the excess blanketing of metallic lines in comparison with "normal" stars of the same temperature, the rectified color-magnitude diagram of these stars should then represent closely the zero-rotation main sequence. Kraft (1965b) pointed out that if Ap stars were, in fact, slow rotators, they would require a slight correction ($+0.01$ to $+0.02$ mag.) in $B - V$ to bring them into proper comparison with normal stars of moderate rotation, on the basis of the theory of Sweet and Roy. Strittmatter and Sargent (1966) carried these corrections out in detail for Hyades, Praesepe, and Coma Am and Ap stars using the models of Roxburgh and Strittmatter (1965a); the de-blanketing corrections were those of Baschek and Oke (1965). The latter corrections move the Am's and Ap's to the left of the main sequence defined by the normal rotating cluster stars, in agreement with the expectation from theory. The observed shifts agree more satisfactorily with the theoretical calculations of nonuniformly rotating—as opposed to rigidly rotating—models, and movements of several hundredths of a magnitude in $B - V$ seem to be predictable on the basis of the former kind of model. If these considerations are valid, the interpretation of the shapes of cluster breakoffs on the basis of stellar hydrogen exhaustion needs adjustment. In addition, one cannot argue that the failure of the blanketing-corrected positions of Ap stars to satisfy the main sequence of average rotators is evidence that Ap stars are post-red giants (Fowler et al. 1965).

V. ENERGY AND ANGULAR MOMENTUM IN GROUPS OF STARS AND IN BINARY SYSTEMS

(a) Rotational and Translational Kinetic Energy of Stars

As mentioned in Section I, Struve (1950) pointed out that $E_{\rm rot}$ and $E_{\rm trans}$ were approximately the same for early-type stars. In Figure 13 we plot $\langle \mathscr{E}_{\rm rot} \rangle = \langle E_{\rm rot}/\mathfrak{M} \rangle$ and $\mathscr{E}_{\rm trans} = E_{\rm trans}/\mathfrak{M}$ as functions of $\mathfrak{M}$ for main-sequence stars, where $\langle \mathscr{E}_{\rm rot} \rangle = \frac{1}{2}(v/R)\,\langle \mathscr{J}_{\rm rot} \rangle$ with $\langle \mathscr{J}_{\rm rot} \rangle$ taken from Figure 3, and $\mathscr{E}_{\rm trans}$ is derived from the characteristic translational velocities tabulated by Delhaye (1965). As before, we assume that

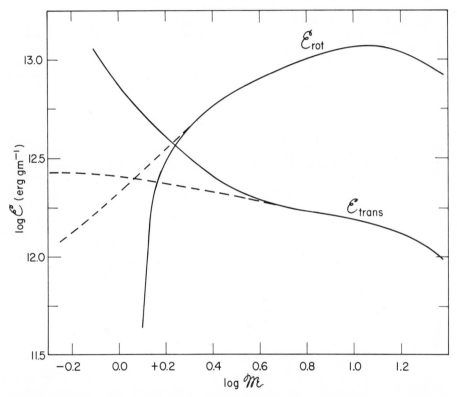

Fig. 13.—$\langle \mathscr{E}_{\rm rot} \rangle$ and $\mathscr{E}_{\rm trans}$ as functions of $\mathfrak{M}$ for main-sequence stars. The solid curves are based on Fig. 1 and the raw tabulated data of Delhaye (1965). The dashed curves are proposed reconstructions representing the primordial distributions of $\langle \mathscr{E}_{\rm rot} \rangle$ and $\mathscr{E}_{\rm trans}$.

stars rotate as rigid bodies. The plot confirms the conclusion reached by Struve, with $\langle \mathscr{E}_{\rm rot} \rangle$ exceeding $\mathscr{E}_{\rm trans}$ at most by a factor of about 5 for stars with $\mathfrak{M} > 3\mathfrak{M}\odot$; whether the factor of 5 is significant mainly depends on the validity of the assumption ω is constant with r. Below mass $\mathfrak{M} = 1.5\,\mathfrak{M}\odot$, $\mathscr{E}_{\rm rot}$ drops off sharply, and $\mathscr{E}_{\rm trans}$ increases.

It is doubtful, however, if the relations shown in Figure 13 represent satisfactorily the initial main sequence ratio $\langle \mathscr{E}_{\rm rot} \rangle/\mathscr{E}_{\rm trans}$ for at least two reasons: (1) $\langle \mathscr{E}_{\rm rot} \rangle$ probably declines along with $\langle \mathscr{J}_{\rm rot} \rangle$ owing to postprimordial loss of angular momentum below $\mathfrak{M} = 1.5\,\mathfrak{M}\odot$, as explained earlier; (2) small-mass stars may have experienced some translational acceleration as a result of forces acting over a long period of time, as, for example, by encounters with interstellar gas clouds. It seems best to ask the question: how do $\langle \mathscr{E}_{\text{rot}} \rangle$ and $\mathscr{E}_{\text{trans}}$ compare in a sample of main-sequence stars of sufficient youth that no loss of angular momentum has yet occurred, and wherein $\mathscr{E}_{\text{trans}}$ reflects the kinetic energy associated with the epoch of formation? In Figure 13 we plot a curve for $\langle \mathscr{E}_{\text{rot}} \rangle$ which corresponds to $\langle \mathscr{J}_{\text{rot}} \rangle \approx \mathfrak{M}^{0.57}$, extended to small masses, as in Figure 3, and a corrected curve for $\mathscr{E}_{\text{trans}}$ which passes from the value previously given at $\mathfrak{M} = 3\mathfrak{M}\odot$ to a value corresponding to the characteristic motion of dMe stars (Vyssotsky 1957). With these adjustments, it will be seen that $\langle \mathscr{E}_{\text{rot}} \rangle$ and $\mathscr{E}_{\text{trans}}$ are roughly of the same order of magnitude at all values of $\mathfrak{M}$, but it should be noted that E_{trans} itself, when so adjusted, no longer satisfies the expectation that it is independent of $\mathfrak{M}$. It is difficult to attach physical significance to this conclusion about E_{trans}, however, since the samples of B-type and dMe-type stars are drawn from widely differing volumes of space. Regardless of the exact behavior of E_{trans} of as function of $\mathfrak{M}$ in a fixed volume of space, it remains true, nevertheless, that the major discrepancy between $\mathscr{E}_{\text{trans}}$ and $\langle \mathscr{E}_{\text{rot}} \rangle$ can be eliminated by a restoration of angular momentum to the stars of the late main sequence.

(b) Angular Momentum and Energy of Binary Stars

We turn in conclusion to a comparison of orbital with rotational angular momentum and energy in binary stars, a matter of some interest in connection with the following picture. Consider the formation of stars in a cluster from the condensation

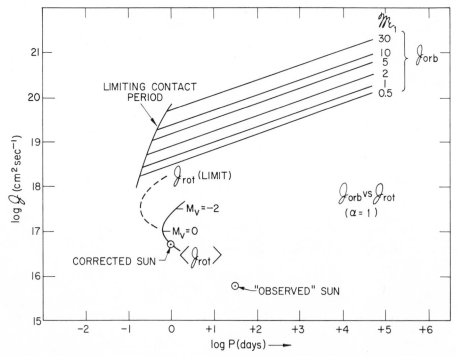

FIG. 14.—The angular momentum (per gram) of single stars and binaries as a function of period. $\mathfrak{M}_1$ is the mass of the primary. The "corrected Sun" is the position the Sun would take if its angular momentum were that of the line $\langle \mathscr{J}(\mathfrak{M}) \rangle \sim \mathfrak{M}^{0.57}$ in Fig. 3. $\mathscr{J}_{\text{rot}}$ (limit) refers to stars rotating with equatorial velocity equal to the escape velocity. In this figure, the mass ratio α for the binaries is unity.

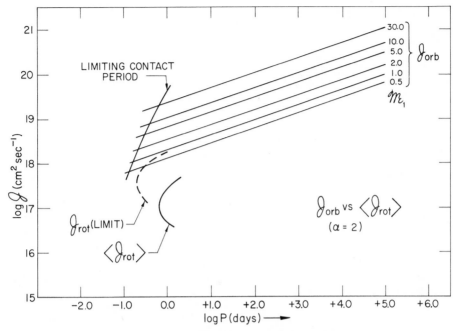

Fɪɢ. 15.—Same as Fig. 14, except that $\alpha = 2$.

of globules having a certain mass spectrum. We imagine that some globules have an angular momentum J so large that, on contraction, fission takes place; Roxburgh (1966), in considering completely convective contracting stars, has concluded that fission is physically possible and leads probably to the formation of W UMa stars. (We will return to the details of this paper later.) Globules having J below a certain critical value form single stars of rapid rotation.

At the present moment, studies of $\langle \mathscr{J}(\mathfrak{M}) \rangle$ permit us to determine for the cluster the function $N(\mathscr{J})\, d\mathscr{J}$, the number of "objects"—single, double, or multiple—with total angular momentum per gram lying between $\mathscr{J}$ and $\mathscr{J} + d\mathscr{J}$. We might have expected that, when the individual stars of low mass were corrected for possible subsequent loss of angular momentum through the mechanisms suggested in preceding sections, $N(\mathscr{J})$ would pass smoothly across $\mathscr{J}_{\text{crit}}$. In other words, the rotational angular momentum of single stars near rotational instability would pass over smoothly to the contact binaries of smallest $\mathscr{J}_{\text{orb}}$. Observations show, however, that this is most emphatically not the case for the majority of binaries, as we may see from inspection of Figures 14 and 15. When the mass ratios $\alpha = \mathfrak{M}_1/\mathfrak{M}_2$ are near 1 or 2, one finds that even the smallest values of $\mathscr{J}_{\text{orb}}$ exceed the limiting value of $\mathscr{J}_{\text{rot}}$ by an order of magnitude at a fixed mass. It is difficult to escape the conclusions that $N(\mathscr{J})$ is definitely bimodal, with two nearly equal lobes, if binaries in fact constitute approximately one-half of all stars.

One way in which we might restore a smooth transition between $\mathscr{J}$ of binaries and $\mathscr{J}$ of single stars is illustrated by Figure 16, in which $\alpha = 10$. If all "single" stars were really binaries of large mass ratio, we would find $\mathscr{J}_{\text{orb}}$ of such systems comparable with $\mathscr{J}_{\text{rot}}$ of single stars. The function $N(\mathscr{J})$ so restored could then increase smoothly from zero at $\mathscr{J} = 0$ to some maximum, and fall asymptotically to zero as $\mathscr{J} \to \infty$. Whether this is true can scarcely be established by spectrographic observations since $\alpha = 10$

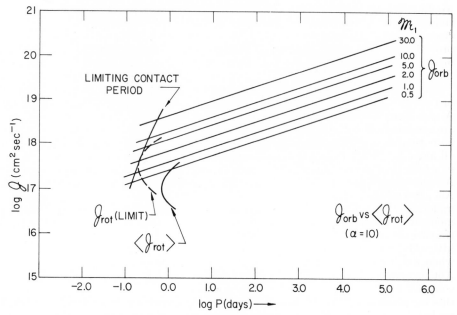

FIG. 16.—Same as Fig. 14, except that $\alpha = 10$.

or larger would lead to single-lined binaries with extremely small amplitudes. The writer nevertheless is impressed by the fact that when he recently made a coudé investigation of the radial velocities of F-type dwarfs in the Hyades, at least three new velocity variables were found among those with sharp lines (Kraft 1965a). The sample of spectra was obtained for a purpose other than radial-velocity study, and it is not known how many cases could actually be uncovered in a really systematic survey.

One can easily show that the absolute value of the orbital energy is given by

$$|E_{\rm orb}| = \mathfrak{M}_1 \mathscr{J}_{\rm orb}\, \pi/P. \tag{13}$$

For a fixed $\mathfrak{M}_1$ and α, $\mathscr{J}_{\rm orb} \approx P^{1/3}$, so that

$$|E_{\rm orb}| \approx P^{-2/3} \tag{14}$$

and thus $|E_{\rm orb}|$ does not change its order of magnitude for changes of period by factors up to 10. Consider a typical binary system in which $\mathfrak{M}_1 = 1.5\ \mathfrak{M}\odot$, $\alpha = 2$, and $P = 10$ days. Then $\mathscr{J}_{\rm orb} = 6 \times 10^{18}\ {\rm cm}^2\ {\rm sec}^{-1}$, and $|E_{\rm orb}| = 6 \times 10^{46}$ erg. On the other hand, if the primary has the average rotational velocity, $\langle \mathscr{J}_{\rm rot} \rangle = 1.0 \times 10^{17}\ {\rm cm}^2\ {\rm sec}^{-1}$, but $E_{\rm rot} = 1.2 \times 10^{48}$ erg. Thus $|E_{\rm orb}|/E_{\rm rot} \approx 5$, but $\mathscr{J}_{\rm orb}/\langle \mathscr{J}_{\rm rot} \rangle \approx 60$, i.e., the (negative) orbital and rotational energies are of the same order of magnitude, but $\mathscr{J}_{\rm orb}$ greatly exceeds $\langle \mathscr{J}_{\rm rot} \rangle$. If P increases, $|E_{\rm orb}|$ declines, but $\mathscr{J}_{\rm orb}$ increases. One can say that, through the entire range of periods of binaries, the (negative) orbital energy is of the same order as the rotational energies of single stars. Thus, if the energy of rotation of the components of binaries could somehow be transformed into orbital energy, an increase in orbital period or even dissolution of the binary might be energetically possible, provided the mechanism could also increase the angular momentum by a large amount. The situation would be relieved somewhat if the rotational velocities were initially near the limit of gravitational instability (cf. Figures 14 and 15), but $\mathscr{J}$

would still need to increase by nearly an order of magnitude. One could imagine here that $\mathscr{J}$ might be increased as a result of unidirectional mass loss, but in the case of close binaries, attempts to calculate the sign and amount of period change have proved somewhat disappointing (cf. Huang 1956, 1963). Further consideration of mass loss and exchange in close binaries takes us too far afield, however, and will not further be considered.

The problem of exchange between orbital and rotational J is, as Abt (1961, 1965) has shown, of great importance in understanding the rotations of Am versus normal A-type stars. Abt's studies indicate that (1) most or all of the 25 Am stars investigated are spectroscopic binaries, the great majority of which have $P < 100^d$; and (2) of the 55 normal A-type stars investigated, only 30 percent are spectroscopic binaries, and all have $P > 100^d$. It may be said then that the Am's represent a group of binaries in which the majority have separations of several stellar radii, whereas the normal A's either are single or are widely separated pairs. We have already mentioned that Am stars have sharp lines and rotate systematically slower than normal A's. We can understand qualitatively this difference if we postulate that rotational angular momentum is transferred to orbital angular momentum in the Am's, but no such transfer occurs for normal A's. It seems likely that, at the present separation of Am's, little tidal interaction could be expected, but this was presumably greatly enhanced when the stars had much larger radii before they reached the main sequence. Thus Abt proposes that Am binaries underwent transformation of J_{rot} into J_{orb} during the contraction stage, but that binaries among normal A's were always too far apart for this interaction to have been important. We note that this transfer of J_{rot} into J_{orb} probably did not greatly affect the *original* J_{orb} for reasons cited in the preceding paragraph. This attractive hypothesis, nevertheless, leaves unexplained the observed slow rotations of the 24 percent of Am binaries that have $P > 100^d$. Consistent with the view that binaries with $P < 100^d$ might have experienced tidal interaction at some point in their evolution are the observed $v \sin i$'s for the components of visual binaries (Slettebak 1963); these are entirely normal.

Supporting evidence that slow rotations generally characterize stars in close, but not contact binaries is found in a study of main-sequence eclipsing binaries by Koch, Olson, and Yoss (1965). In a sample of 19 eclipsing stars with periods between 1 and 12 days, they found that the rotational velocities were ordinarily somewhat larger than expected for synchronism with the orbital rotation, but still were a factor of about 2 below the field rotation curve given in our Figure 3. A certain amount of evidence suggests that contact binaries, such as W UMa stars, have orbital and rotational periods nearly the same (Struve 1950), but exceptions have been noted (Struve 1950; Mitchell 1954; Limber 1963).

The fission theory of the origin of close binary stars has recently been revived by Roxburgh (1966). He considers a rotating star which remains fully convective along an evolutionary track on the H-R diagram that descends vertically until a radiative core develops shortly before the star reaches the main sequence (Hayashi 1961; Iben 1965a). It is expected that fully convective stars will rotate as rigid bodies during contraction because the convective motions transport angular momentum as well as heat. Roxburgh shows that such stars will, on reaching rotational instability, shed matter at the equator, and will continue to do so as evolution proceeds. The radiative core, however, on uncoupling itself from the convective superincumbent layers, rotates with a much faster angular velocity than would have been the case if convection could have transported out the angular momentum. He then shows that the core reaches the limit of dynamical instability for breakup, and the central region splits into two components, provided the mass of the star is greater than $0.8\mathfrak{M}_\odot$. A critical point of the theory is that, according to Jeans (1929), a compressible fluid with effective polytropic index >0.8 and which is rotating uniformly does not break up as ω increases above some critical value, but rather sheds matter in the equatorial plane; only a rotating liquid actually breaks up for sufficiently large ω. Roxburgh argues that in the central radiative core, ρ is only a very slow function of r, and therefore the core behaves as a liquid. In support of Roxburgh's theory is the fact that the lower mass limit predicted for breakup, $viz.$, $\mathfrak{M} = 0.8\,\mathfrak{M}\odot$, is very close to the lower limit observed for total masses of W UMa systems (Kopal and Shapley 1956).

VI. SUMMARY AND PROSPECTUS

The work of Struve and his associates established that the slow rotation of the Sun was uncharacteristic of certain groups of stars. It also sketched in broad outlines the domain of the H-R diagram in which rotational velocities were large, and discussed the fission theory of the origin of binaries from a comparison of orbital and rotational angular momenta. In this paper we have traced the subsequent development of the subject from the point of view of an optical observer, stressing especially the relevance of rotation to star formation, stellar energy generation, stellar evolution, cosmogony, and certain types of peculiar stars. Many problems remain unsolved, but one notes with satisfaction a renaissance of activity, both theoretical and observational, with the appearance of more than 50 papers on stellar rotation during the past five years.

We suggest now several problems, principally observational, that are likely to receive attention during the next few years. The first is tractable only if a large telescope is available. We have seen that loss of angular momentum in stars contracting to the main sequence can neither be established nor disproved on the basis of existing observations: the evolutionary tracks are not sufficiently well known and the stars in question are too close to the horizontal break resulting from the development of a radiative core. But rotational velocities could be obtained from plates of rather modest dispersion (25 to 50 Å/mm) for contracting stars of still later spectral type. These will presumably reach the main sequence among stars of very slow rotation. Observational results are possible in the Taurus cloud and in NGC 2264. Considerable mass loss among these stars would, of course, affect the evolutionary tracks, but it is not plausible to expect such losses to bring the stars onto the main sequence among the rapid rotators. The question of whether very late-type contracting stars have appreciable angular momentum is obviously of cosmogonical, as well as evolutionary, interest. The advisability of such observations has previously been noted by Herbig (1962).

A second problem related to this concerns the rotation of late main sequence stars of types K and M. Among field stars of these kinds, $v \sin i$ does not exceed 20 or 25 km/sec, and for many, $v \sin i < 10$ km/sec. In these cases, the maximum rotational velocity is set simply by the limiting resolution of the spectrograph. It seems likely, however, that the field stars that can be observed, which are all within 50–100 parsecs, are relatively old; none are found with ages less than or equal to that of the Pleiades simply because these could not have escaped the star formation regions in so short a time. We can ask, for example, what is the rotational velocity of a K or M dwarf with an age of only 4×10^7 years? Such a star would be found in the Pleiades at m_V between 14 and 17, much too faint for a high dispersion study of its rotational velocity.

Of the field stars, the ones with hydrogen and Ca II emission, (the dKe and dMe stars) are known to be younger than the ones without emission lines. For a few of the former, Krzeminski (1968) has found small, periodic light variations with a time-scale of a few days. The most reasonable explanation of the phenomenon consists of the rotational modulation of a spotty surface; in that case, the rotational velocities run as high as 20 km/sec. It is conceivable that similar observations of late-type Pleiades members would yield even higher rotational velocities.

Turning to the other end of the evolutionary sequence, we inquire about the rotational velocities of white dwarfs. If angular momentum were conserved, the white dwarf descendant of a typical A-type star would have to rotate with a velocity near 10,000 km/sec. A rotation of this order would produce line broadening of the same order as the Stark broadening of the hydrogen lines. In their study of rotating white dwarf configurations, Ostriker and Bodenheimer (1968) suggested that DC white dwarfs might in fact be identified with rapidly rotating degenerate configurations; it does not seem likely however that these could be identified as DA white dwarfs seen "edge-on." In this picture DC stars must either be hydrogen-poor or have temperatures sufficiently low that the hydrogen lines are not prominent. This view becomes even more compelling if the red giant progenitors of white dwarfs lose mass and therefore a large fraction of their initial angular momentum, as suggested in Section IIIa.

Ostriker and Bodenheimer concluded that rapidly rotating white dwarfs could have masses considerably above the electron-degenerate mass limit of $1.2 \, \mathfrak{M}\odot$, but no estimate was made of total luminosity or of changes in luminosity and color with aspect. Thus at the present moment it is not possible to devise satisfactory observational tests for the existence of these objects.

We return finally to a consideration of the peculiar A-type stars. There are really three basic problems presented by these objects. The first centers around the nuclear physics of the peculiar abundances; this has received perhaps the most attention. And it has been possible to see, to some extent, how this problem might be related to the presence of large coherent magnetic fields, at least for those abundance anomalies that result from spallation processes. On the other hand, the abundance peculiarities associated with r-processes are difficult to understand unless the Ap's have gone through post-main-sequence evolutionary processes, and this seems to be ruled out both by their presence in certain very young clusters (cf. Kraft 1967c), their rotation- and blanketing-rectified position in the H-R diagram (cf. Strittmatter and Sargent 1966, Kraft 1965b), and by the apparently normal main-sequence mass of HD 98088, an Ap star that is a component of a double-line spectroscopic binary (Abt, Conti, Deutsch, and Wallerstein 1966). A third property must, however, be taken into account, and that is the fact that, in their main sequence temperature range, the peculiar A-stars are the entire group of slow rotators in a part of the H-R diagram characterized by rapid rotation. One could suppose, for example, that the slow rotation results from a torque exerted by ionized interstellar matter near the star as it drags on the rotating magnetic field; this would be analogous to the picture by Spitzer (1956) already alluded to. This would not, however, explain the anomalous abundances resulting from r-processes. On the other hand, van den Heuvel (1968) suggested that all Ap's are binaries with white dwarf companions of very small mass. The latter were assumed once to be the primaries which passed through the r-processing stage, swelled up to fill their critical inner Lagrangian surface, and spilled the r-processed material over to the secondary; the latter gained mass and moved up the main sequence to become the presently observed peculiar A-type star. The present slow rotation follows as a result of the mass-exchange and variation of the orbital elements. While this picture explains some of the features of Ap-stars, it is not clear in what way the magnetic field is generically relevant. But whatever the deficiencies of these various pictures, it seems clear that any successful comprehensive theory of peculiar A-type stars must explain all three of the leading features: anomalous abundances, magnetic fields, and slow rotation.

I am indebted to Drs. John Heard, Peter Strittmatter, W. L. W. Sargent, John Faulkner, Helmut Abt, Armin Deutsch, Olin Wilson, and George Collins for communicating results in advance of publication. To Armin Deutsch, Helmut Abt, Olin Wilson, and George Preston, in particular, I express my deepest appreciation for their kind advice and apt criticism at various phases of the investigation.

REFERENCES

Abt, H. 1957, *Ap. J.*, **126**, 503.
———. 1958, *Ap. J.*, **127**, 658.
———. 1961, *Ap. J. Suppl.*, **6**, 37.
———. 1965, *Ap. J. Suppl.*, **11**, 429.
Abt, H., Barnes, R., Biggs, E., and Osmer, P. 1965, *Ap. J.*, **142**, 1604.
Abt, H. and Chaffee, F. 1967, *Ap. J.*, **148**, 459.
Abt, H., Chaffee, F., and Suffolk, G. 1967, *Astr. J.*, **72**, 783.
Abt, H., Conti, P., Deutsch, A., and Wallerstein, G. 1966, *Astr. J.*, **71**, 843.
Abt, H. and Hunter, J. 1962, *Ap. J.*, **136**, 381.
Abt, H. and Osmer, P. 1965, *Ap. J.*, **141**, 949.
Anderson, C., Stoeckly, R., and Kraft, R. 1966, *Ap. J.*, **143**, 299.
Babcock, H. 1958, *Ap. J.*, **128**, 228.
Baker, N. 1963, "The Depth of the Outer Convection Zone in Main Sequence Stars," NASA preprint (unpublished).
Barry, D. 1967, *Ap. J.*, **148**, L87.
Baschek, B. and Oke, J. 1965, *Ap. J.*, **141**, 1404.
Bonsack, W. and Culver, R. 1966, *Ap. J.*, **145**, 767.
Boyarchuk, A. and Kopylov, I. 1964, *Bull. Crimean Ap. Obs.*, **31**, 44.
Brandt, J. 1966, *Ap. J.*, **144**, 1221.
Brown, A. 1950, *Ap. J.*, **111**, 366.
Chandrasekhar, S. and Münch, G. 1950, *Ap. J.*, **111**, 142.
Christy, R. 1967, *Astr. J.*, **72**, 293.
Collins, G. 1963, *Ap. J.*, **138**, 1134.
———. 1965, *Ap. J.*, **142**, 265.
Collins, G. and Harrington, J. 1966, *Ap. J.*, **146**, 152.
Conti, P. 1965, *Ap. J.*, **142**, 1594.
Conti, P. and Deutsch, A. 1966, *Ap. J.*, **145**, 742.
Crawford, D. 1958, *Ap. J.*, **128**, 185.
Crawford, D. and Perry, C. 1966, *Astr. J.*, **71**, 206.
Danziger, I. and Dickens, R. 1967, *Ap. J.*, **149**, 55.
Delhaye, J. 1965, in *Stars and Stellar Systems*, Vol. **V**, *Galactic Structure*, ed. A. Blaauw and M. Schmidt (Chicago: Univ. of Chicago Press), p. 61.
Demarque, P. and Roeder, R. 1967, *Ap. J.*, **147**, 1188.
Deutsch, A. 1956, *Ap. J.*, **123**, 210.
———. 1958, in *Handbuch der Physik*, **51**, ed. S. Flügge (Berlin: Springer-Verlag), p. 689.
———. 1960, in *Stars and Stellar Systems*, Vol. **VI**, *Stellar Atmospheres*. ed, J. Greenstein (Chicago: Univ. of Chicago Press), p. 543.
———. 1967, *The Magnetic and Related Stars*, ed. R. Cameron (Baltimore: Mono Book Corp.), p. 181.
Dicke, R. 1964, *Nature*, **202**, 432.
———. 1967, *Ap. J.*, **149**, L121.
Dicke, R. and Goldenberg, H. 1967, *Nature*, **214**, 1294.
Dickens, R., Kraft, R., and Krzeminski, W. 1968, *Astr. J.*, **73**, 6.
Eggen, O. 1965, in *Ann. Rev. Astr. Ap.*, **3**, (Palo Alto: Annual Reviews), p. 235.
Eggen, O. and Sandage, A. 1962, *Ap. J.*, **136**, 735.
———. 1964, *Ap. J.*, **140**, 130.
Faulkner, J., Roxburgh, I., and Strittmatter, P. 1968, *Ap. J.*, **151**, 203.
Fowler, W., Burbidge, E., Burbidge, G., and Hoyle, F. 1965, *Ap. J.*, **142**, 423.
Goldreich, P. and Schubert, G. 1967, *Ap. J.*, **150**, 571.
Guthrie, B. 1963, *Pub. Roy. Obs. Edinburgh*, **3**, 84.
———. 1965, *Pub. Roy. Obs. Edinburgh*, **3**, 263.
Hardorp, J. and Strittmatter, P. 1968a, *Ap. J.*, **151**, 1057.
———. 1968b, *Ap. J.*, **153**, 465.
Härm, R. and Rogerson, J. 1955, *Ap. J.*, **121**, 439.
Harris, D., Strand, K. Aa., and Worley, C. 1963, in *Stars and Stellar Systems*, Vol. **III**, *Basic Astronomical Data*, ed. K. Aa. Strand (Chicago: Univ. of Chicago Press), p. 273.
Hayashi, C. 1961, *Pub. Astr. Soc. Japan*, **13**, 450.
———. 1966, in *Ann. Rev. Astr. Ap.*, **4** (Palo Alto: Annual Reviews), p. 171.
Heard, J. and Petrie, R. 1967, *I.A.U. Symp. No. 30*, p. 179.
Henyey, L., LeLevier, R., and Levée, R. 1955, *Pub. A.S.P.*, **67**, 154.
Herbig, G. 1957, *Ap. J.*, **125**, 612.
———. 1962, in *Advances in Astronomy and Astrophysics*, **1** (New York and London: Academic Press), p. 47.
Herbig, G. and Spalding, J. 1955, *Ap. J.*, **121**, 118.

Hill, G. 1967, *Ap. J. Suppl.*, **14**, 263.
Hofmeister, E., Kippenhahn, R., and Weigert, A. 1964, *Zs. f. Ap.*, **60**, 57.
Howard, L., Moore, D., and Spiegel, E. 1967, *Nature*, **214**, 1297.
Hoyle, F. 1960, *Quart. J.R.A.S.*, **1**, 28.
Huang, S.-S. 1953, *Ap. J.*, **118**, 285.
——. 1956, *Astr. J.*, **61**, 49.
——. 1963, *Ap. J.*, **138**, 471.
——. 1965, *Ap. J.*, **141**, 985.
Huang, S.-S. and Struve, O. 1954, *Ann. d'Ap.*, **17**, 85.
——. 1956, *Ap. J.*, **123**, 231.
——. 1960, in *Stars and Stellar Systems*, Vol. **VI**, *Stellar Atmospheres*, ed. J. L. Greenstein (Chicago: University of Chicago Press), p. 321.
Iben, I. 1965a, *Ap. J.*, **141**, 993.
——. 1965b, *Ap. J.*, **142**, 1477.
——. 1966a, *Ap. J.*, **143**, 483.
——. 1966b, *Ap. J.*, **143**, 505.
Jeans, Sir James. 1929, *Astronomy and Cosmogony* (Cambridge: Cambridge Univ. Press).
Johnson, H. 1952, *Ap. J.*, **116**, 640.
Johnson, H. and Mitchell, R. 1958, *Ap. J.*, **128**, 31.
Johnson, H., Mitchell, R., and Iriarte, B. 1962, *Ap. J.*, **136**, 75.
Joy, A. 1945, *Ap. J.*, **102**, 168.
Koch, R., Olson, E., and Yoss, K. 1965, *Ap. J.*, **141**, 955.
Kopal, Z. and Shapley, L. 1956, *Elements of Binary Stars* (Jodrell Bank Annals, Vol. I).
Kraft, R. 1965a, *Ap. J.*, **142**, 681.
——. 1965b, Remarks presented at Conference on Stellar Evolution, La Jolla, California.
——. 1966, *Ap. J.*, **144**, 1008.
——. 1967a, *Ap. J.*, **148**, 129.
——. 1967b, *Ap. J.*, **150**, 551.
——. 1967c, in *The Magnetic and Related Stars*, ed. R. Cameron (Baltimore: Mono Book Corp.), p. 303.
Kraft, R., Kuhi, L., and Kuhi, P. 1968, *Astr. J.*, **73**, 221.
Kraft, R. and Wrubel, M. 1965, *Ap. J.*, **142**, 703.
Krzeminski, W. 1968. Paper read at *AAS Faint Star Symposium*, Charlottesville, Va.
Kuhi, L. 1964, *Ap. J.*, **140**, 1409.
——. 1966, *Ap. J.*, **143**, 991.
Ledoux, P. 1951, *Ap. J.*, **114**, 373.
Limber, D. 1963, *Ap. J.*, **138**, 1112.
McGee, J., Khogali, A., Baum, W., and Kraft, R. 1967, *M.N.R.A.S.*, **137**, 303.
McNamara, D. 1961, *Pub. A.S.P.*, **73**, 269.
——. 1963, *Ap. J.*, **137**, 316.
McNamara, D. and Hansen, K. 1961, *Ap. J.*, **134**, 207.
McNamara, D. and Larsson, H. 1962, *Ap. J.*, **135**, 748.
Meadows, A. 1961, *Ap. J.*, **133**, 907.
Mendoza V., E. 1966, *Ap. J.*, **143**, 1010.
——. 1968, *Ap. J.*, **151**, 977.
Mihalas, D. 1964, *Ap. J. Suppl.*, **9**, 321.
——. 1965, *Ap. J.*, **141**, 564.
——. 1966, *Ap. J. Suppl.*, **13**, 1.
Mitchell, R. 1954, *Ap. J.*, **120**, 274.
——. 1960, *Ap. J.*, **132**, 68.
Oke, J. and Greenstein, J. 1954, *Ap. J.*, **120**, 384.
Ostriker, J. and Bodenheimer, P. 1968, *Ap. J.*, **151**, 1089.
Parenago, P. 1954, *Trudy Sternberg Astr. Inst.*, Vol. **25**.
Peters, J., Poland, A., and Wrubel, M. 1966, *Astr. J.*, **71**, 174.
Poveda, A. 1965, *Bols. Obs. Tonantzintla y Tacubaya*, **4**, 15.
Preston, G. 1965, *Kl. Veröff. Bamberg*, **4**, Nr. 40, p. 155.
Rodgers, A. and Bell, R. 1964, *M.N.R.A.S.*, **128**, 365.
——. 1965, *M.N.R.A.S.*, **129**, 157.
Roxburgh, I. 1964a, *M.N.R.A.S.*, **128**, 157.
——. 1964b, *M.N.R.A.S.*, **129**, 237.
——. 1966, *Ap. J.*, **143**, 111.
Roxburgh, I., Griffith, J., and Sweet, P. 1965, *Zs. f. Ap.*, **61**, 203.
Roxburgh, I., Sargent, W., and Strittmatter P. 1966, *Observatory*, **86**, 118.
Roxburgh, I. and Strittmatter, P. 1965a, *M.N.R.A.S.*, **133**, 1.
——. 1965b, *Zs. f. Ap.*, **63**, 15.
——. 1966, *M.N.R.A.S.*, **133**, 345.

Sandage, A. 1955, *Ap. J.*, **122**, 263.
———. 1957, *Ap. J.*, **125**, 422.
Sandage, A. and Schwarzschild, M. 1952, *Ap. J.*, **116**, 463.
Sargent, W. and Strittmatter, P. 1966, *Ap. J.*, **145**, 938.
Schatzman, E. 1959, *I.A.U. Symposium No.* 10, ed. J. L. Greenstein, p. 129.
———. 1962, *Ann. d'Ap.*, **25**, 18.
Schmidt, M. 1963, *Ap. J.*, **137**, 758.
Schwarzschild, M. 1958, *Structure and Evolution of the Stars* (Princeton: Princeton Univ. Press).
Searle, L. and Sargent, W. 1964, *Ap. J.*, **139**, 793.
Slettebak, A. 1949, *Ap. J.*, **110**, 498.
———. 1954, *Ap. J.*, **119**, 146.
———. 1955, *Ap. J.*, **121**, 653.
———. 1963, *Ap. J.*, **138**, 118.
———. 1966, *Ap. J.*, **145**, 121, 126.
———. 1968, *Ap. J.*, **151**, 1043.
Slettebak, A. and Howard, R. 1955, *Ap. J.*, **121**, 102.
Smak, J. 1964, *Ap. J.*, **139**, 1095.
Smith, B. and Struve, O. 1944, *Ap. J.*, **100**, 360.
Spitzer, L. 1956, private communication to E. Schatzman (1962).
Steinitz, R. 1964, *Bull. Astr. Inst. Neth.*, **17**, 504.
Strittmatter, P. 1966, *Ap. J.*, **144**, 430.
Strittmatter, P. and Sargent, W. 1966, *Ap. J.*, **145**, 130.
Struve, O. 1950, *Stellar Evolution* (Princeton: Princeton Univ. Press).
———. 1951, in *Astrophysics*, ed. J. Hynek (New York: McGraw-Hill Book Co.), p. 85.
Sweet, P. and Roy, A. 1953, *M.N.R.A.S.*, **113**, 701.
Temesvary, S. 1952, *Zs. für Naturforschung*, **7a**, 103.
Treanor, Fr. P. 1960, *M.N.R.A.S.*, **121**, 503.
van den Heuvel, E. 1968, *Bull. Astr. Inst. Neth.*, **19**, 326.
van Dien, E. 1948, *J. Roy. Astr. Soc. Can.*, **42**, 249.
Vasilevskis, S., Sanders, W., and Balz, A. 1965, *Astr. J.*, **70**, 797.
von Zeipel, H. 1924, *Festschrift für H. von Seeliger* (Berlin: Springer-Verlag).
Vyssotsky, A. 1957, *Pub. A.S.P.*, **63**, 109.
Weber, E. and Davis, L. 1967, *Ap. J.*, **148**, 217.
Walker, M. 1956, *Ap. J. Suppl.*, **2**, 365.
Wilson, O. 1964, *Pub. A.S.P.*, **76**, 28.
———. 1966a, *Ap. J.*, **144**, 695.
———. 1966b, *Science*, **151**, 1487.
Wilson, O. and Bappu, M. 1957, *Ap. J.*, **125**, 661.

11. SPECTROSCOPIC BINARIES

O. STRUVE: Spectroscopic Observations of Thirteen Eclipsing Variables. *Astrophysical Journal*, **102**, 75, 1945.

D. M. POPPER: Spectroscopic Binaries.

SPECTROGRAPHIC OBSERVATIONS OF
THIRTEEN ECLIPSING VARIABLES

OTTO STRUVE

McDonald and Yerkes Observatories

Received May 14, 1945

INTRODUCTION

The fourth catalogue of spectroscopic binaries by J. H. Moore[1] contained 365 orbits of stars regarded as true binaries. Fifty-five of these were known as eclipsing variables. In the nine years which have elapsed since the publication of Moore's catalogue, several additional orbits of eclipsing variables have been published, so that the number now available from observatories other than McDonald is between 60 and 70. The total number of eclipsing variables listed in Schneller's catalogue for 1941[2] exceeds 1100. The percentage of these stars for which spectrographic orbits have been determined is, therefore, about 6. The spectrographic material, valuable as it has been, is not sufficient for detailed statistical and individual studies of the mechanical properties of close binary systems. There is an even greater lack of information when we consider the astrophysical properties of these stars. Quite naturally, the stellar spectroscopists were primarily interested in deriving accurate velocity-curves. With the exception of a few spectacular systems, like β Lyrae, the spectra as such aroused slight interest; and the great majority of the original papers listed by Moore contain little more than the briefest descriptions of the principal features of the spectra upon which the measurements were based. Twenty years ago stellar spectroscopists were mystified by the existence of a few very unusual spectroscopic binaries among a large number of systems apparently having normal spectra. There was only *one* β Lyr, with its variable emission lines and its normal B9 and abnormal B5 sets of absorption lines; only *one* ϵ Aurigae, with its long period of twenty-seven years and its unsymmetrically broadened lines during eclipse; and only *one* W Ursae Majoris, with a period of a few hours. The unusual binaries were not only rare; they were apparently quite unrelated to one another. The past years have increased the variety of unusual spectroscopic phenomena found in close binaries of all kinds and have led to the discovery of additional systems showing peculiarities previously known in only one or two stars. Thus, we now know from the spectrographic work of Popper and the photometric observations of S. Gaposchkin that RY Scuti shares some of its features with β Lyr. We know that ζ Aurigae and VV Cephei are in many respects similar to ϵ Aur. There has also been some progress in finding the correct physical explanation of these phenomena and in linking together problems which at first had seemed to be unrelated. For example, we attribute the peculiar B5 spectrum of β Lyr to an extended atmosphere or shell, and we make use of the dilution effect to connect this problem with that of the peculiar relative intensities of He I triplets and singlets in the spectrum of ϕ Persei. Perhaps even more important is our growing realization that the abnormal spectroscopic features of β Lyr, VV Cep, ϕ Per, etc., must be regarded as extreme examples of phenomena which in a less conspicuous manner exist in many systems previously regarded as normal.

There are, however, several important astrophysical questions which cannot be answered without additional material. What are these new questions? It is, of course, not possible to segregate them entirely from questions of a dynamical nature. Although we shall be able to answer only a few of them, I shall discuss those problems and results which are serving me as a guide in the preparation of my observing programs.

ASTROPHYSICAL QUESTIONS ENCOUNTERED IN THE STUDY
OF SPECTROSCOPIC BINARIES

I. PROBLEMS CONNECTED WITH THE AXIAL ROTATION OF BINARY COMPONENTS

a) Observations of numerous systems have shown that in all stars heretofore investigated the direction of rotation is the same as the direction of orbital revolution. Is this *always* the case? If it is, the result must have an important cosmological significance.

b) A few years ago it was considered probable that in all close systems the periods of

---

[1] *Lick Obs. Bull.*, No. 483, 1936.   [2] *Kl. Veröff. Berlin-Babelsberg*, No. 22, 1940.

rotation and revolution are the same. In systems of fairly long period, excessive rotational velocities are frequent. There is one striking case (U Cephei) in which the rotational velocity of one component is much greater than is required by the condition of synchronism, although the orbital period is only 2.49 days.[3] What is the significance of these excessive rotational velocities in close systems?

c) In one system (RZ Scuti) the rotational disturbance determined from *He* I has a range of at least 250 km/sec, while that determined from *H* has a range of about 100 km/sec.[4] This suggests a stratified atmosphere with an extended and slowly rotating *H* shell. Are there other stars with a similar phenomenon?

d) The meager material now available on the spectra of close systems in which the secondary component is a relatively cool subgiant, while the primary is a normal A or B star, suggests that the rotational velocities of the cool components are small (U Cep).[3] But is this always true; and, if it is, has the phenomenon any relation to the well-known avoidance of single stars of the later spectral classes to show rapid rotations?[5]

e) It is probable that the largest rotational velocities on record occur in single stars and not in close binaries. Among different groups of stars the Be's undoubtedly have the largest rotations.[6] As a rule, they are not binaries. The explanation of this phenomenon lies probably in the fact that the majority of binary components are normal stars in size and spectrum. Thus, a B-type binary consisting of two equal components cannot have a period shorter than that which would correspond to $a = r_1 + r_2 = 2r_1$; and this places an upper limit upon the period of rotation of each component. More rapid rotational velocities can occur only in underluminous systems. We do not know enough of them to draw any conclusions from the observations.

II. PERIODIC CHANGES IN THE INTENSITIES OF THE
ABSORPTION LINES OF BINARY COMPONENTS

a) One of the most puzzling questions is that of the reported strengthening of the lines of the approaching components. Originally discovered by Bailey[7] in μ^1 Scorpii, this effect was independently described by Miss Cannon[8,9] and was found by Miss Maury in V Puppis[10] and ζ Centauri.[11] It was independently found by the writer in a Virginis[12] and later in σ Aquilae[13] and β Scorpii.[13] R. M. Petrie[14] made accurate measurements in several spectroscopic binaries and found a small effect of the character found by me, in σ Aql and 2 Lacertae. However, he did not consider the phenomenon as definitely established. On the other hand, fairly reliable measures of equivalent widths in μ^1 Scorpii by Struve and Elvey revealed no change.[15] Popper failed to find a change in V Pup[16] and ζ Cen.[17] My own observations in a Vir, σ Aql, and β Sco were so convincing that I cannot doubt the reality of the phenomenon. Nor is it reasonable to question the extensive Harvard results. Yet my own observations of μ^1 Sco were completely negative. The existence of the effect should be checked, and an attempt should be made to explain it. It is possible that the visual impression of a spectral line and the equivalent width, measured from a tracing, record different things. It is known that such differences exist. For ex-

[3] *Ap. J.*, **99**, 222, 1944. References to specific stars will usually be limited to the last paper in which a complete bibliography may be found.

[4] *Ap. J.*, **101**, 240, 1945.

[5] *Pop. Astr.*, **53**, 214, 1945.

[6] W. W. Morgan, *A.J.*, **51**, 21, 84, 1944.

[7] *Harvard Circ.*, No. 11, 1896.

[8] *Harvard Ann.*, Vol. **28**, Part II, remark 73, 1897.

[9] *Harvard Ann.*, **84**, 169, 1920.

[10] *Ibid.*, p. 178.

[11] *Harvard Circ.*, No. 233, 1922.

[12] *Ap. J.*, **80**, 369, 1934.

[13] *Ap. J.*, **85**, 41, 1937.

[14] Unpublished.

[15] *Pub. A.S.P.*, **54**, 151, 1942.

[16] *Ap. J.*, **97**, 402, 1943.

[17] *Ibid.*, p. 405.

ample, a very broad and shallow line may be completely invisible with high dispersion and may have a considerable equivalent width. Such a line can sometimes be easily seen on spectrograms of small dispersion. It is very strange that there are so few cases of this kind of variation recorded in the literature. If we could assume that the observers were focusing their attention upon the relative intensities of the components, and not simply ignoring them, we should certainly have a strong argument against the reality of the phenomenon. But previous experience would hardly justify this assumption. In stars of advanced spectral class, blending causes some very remarkable inconsistencies in the relative intensities of the components, and lack of information is not surprising. A remarkable type of variation is discussed by Sahade and Cesco elsewhere in this issue of the *Journal*.[18] The suspicion was once expressed by Van Arnam[19] that in BD−18°3295 the faint component is observed more often when its velocity is near maximum, and not near minimum, as in α Vir, σ Aql, etc.

b) Some rather strange changes in line intensities, which are strictly periodic but which cannot be attributed to the blending of two component lines, have been found in several close systems (AU Monocerotis[20] and AB Persei[21]). The suspicion exists that absorption lines formed in extended gaseous currents far above the normal reversing layers of the stars give rise to the strengthening of certain lines (*H* in AU Mon) in certain phases (just before principal minimum in AU Mon). The nature of these currents is quite obscure, but they are probably related to the currents in the system of β Lyr, which produce enormous changes in the intensities of the B5 lines.[22] It is important that more stars be found in which this phenomenon can be studied|.It is probable that these changes in line intensities are related to the distortions in the velocity-curves of several spectroscopic binaries, to be discussed later.

III. SPECTROSCOPIC PHENOMENA ATTRIBUTED TO THE REFLECTION EFFECT

a) The lines of the second component of α Vir not only vary in equivalent width, as described in Section II (a), but they are unusually narrow,[23] especially when observed on the red side. Additional spectrograms have led to the suspicion that these narrow lines are superposed unsymmetrically over very faint, diffuse lines whose rotational broadening may be comparable to that of the broad lines of the principal component.[24] It is theoretically possible that we are here concerned with the reflection effect of the primary on the secondary. There is, as yet, no other observational evidence to confirm this conclusion; nor is there an adequate theory of the formation of a line under the influence of the reflection effect. But it is probable that the influence of reflection upon the absorption lines of a binary component is much greater than upon the total light of the system.[25] The principal function of the diluted radiation of the brighter component is to ionize the outermost layers of the fainter component. This process will be regulated by the temperature of the primary and by the dilution factor. In a close system the latter may be of the order of $\beta = 0.1$ or 0.01. Consider the ionization of *He* in a system whose components have temperatures of $T_1 = 25,000°$ and $T_2 = 15,000°$ and whose atmospheres are, at their nearest point, $2r_1$ apart, where r_1 is the radius of the brighter component. Then $\beta \sim 0.06$. The ionization produced by the fainter component alone is

$$\log \frac{n_2}{n_1} = -24 \frac{5,040}{15,000} + \tfrac{5}{2} \log 15,000 - 0.48 + \log \frac{2u_2}{u_1} - \log p_e .$$

[18] *Ap. J.*, **102**, 128, 1945.

[19] *Ap. J.*, **75**, 351, 1932.

[20] Sahade and Cesco, *Ap. J.*, **101**, 235, 1945.

[21] *Ibid.*, p. 232.　　　　　　　　[23] *Ap. J.*, **72**, 15, 1930.

[22] *Ap. J.*, **93**, 107, 1941.　　　　　[24] *Ap. J.*, **86**, 198, 1937.

[25] Eddington, *Internal Constitution of the Stars*, p. 214, Cambridge, 1926.

The ionization produced by the primary alone, at the nearest point of the atmosphere of the secondary, is

$$\log \frac{n_2'}{n_1'} = -24 \frac{5{,}040}{25{,}000} + \tfrac{5}{2} \log 25{,}000 - 0.48 + \log \frac{2u_2}{u_1} - \log p_e + \log 0.06 \,.$$

The ratio of the two quantities is given by

$$\log \frac{n_2}{n_1} - \log \frac{n_2'}{n_1'} = -2.6 \,.$$

The ionization produced by the primary is very much greater than that produced by the secondary. Since all of the continuous radiation of the secondary passes through this ionized region, the corresponding lines must be very strong.

b) There has been much discussion of what appears to be a sound observational result: in many double-lined spectroscopic binaries the types of the two components are more nearly alike than is consistent with the relative surface brightnesses of the stars derived from the two light-minima. This phenomenon has been commented upon by J. S. Plaskett,[26] by Eddington,[27] and, most recently, by S. Gaposchkin.[28] For example, in TV Cassiopeiae[26] the types were found by Plaskett to be A0 and A0, but the depths of the two eclipses are 1.05 and 0.09 mag. There are other striking discrepancies, and some of them were specifically commented upon by Plaskett. A. B. Wyse[29] found that, when the surface brightnesses inferred from the observed spectra at the two minima are compared with the results from the light-curve, satisfactory agreement is obtained for three-fourths of the stars; but "the exceptions are numerous and noteworthy" and cannot be accounted for by reflection, since at principal minimum the spectrum of the secondary is not affected by reflection. It is probably necessary to distinguish between observations made at principal minimum, as were those of Wyse, and observations of double lines at maximum light, as were those of Plaskett. Among the latter type of observations, striking discordances are present for TV Cas, already mentioned, and for RS Vulpeculae, Z Herculis, GO Cygni and U Coronae Borealis. The data pertaining to these stars are as shown in the accompanying table. The spectrographic data for all five stars are based

Star	Spectra	Δm_1	Δm_2	Reference
TV Cas..............	A0+A0	1.05	0.09	Plaskett, *Dom. Ap. Obs.*, 2, 141, 1922
RS Vul..............	B8+B9	0.73	.08	Plaskett, *Dom. Ap. Obs.*, 1, 141, 1919
Z Her..............	F2+F2	0.80	.12	Adams and Joy, *Ap. J.*, 49, 192, 1919
GO Cyg..............	B9n+A0n	0.54	.23	Pearce, *J.R.A.S. Canada*, 27, 62, 1933
U CrB..............	B3+B3	1.02	0.00	Plaskett, *Dom. Ap. Obs.*, 1, 187, 1920
U CrB..............	B5+B9	1.02	0.00	Pearce, *Pub. A.A.S.*, 8, 220, 1936

upon numerous measurements by experienced observers. Pearce comments upon the appreciable difference in spectrum of the two components of GO Cyg. This star is the least abnormal in the list. Plaskett remarks in connection with TV Cas: ".... it seems probable that the lines measured as due to the second spectrum, even though they are often only on the border line of visibility, are really present. ...." In the case of RS Vul he states: ".... the second spectrum .... can only be seen and measured with great

[26] *Pub. Dom. Ap. Obs. Victoria*, 1, 141, 1919; 1, 187, 1920; 2, 141, 1922.

[27] *Internal Constitution of the Stars*, p. 214.

[28] *Variable Stars*, p. 36, Cambridge, 1938.   [29] *Lick Obs. Bull.*, No. 464, 1934.

difficulty, and it is estimated to be only one-fourth the intensity of the brighter spectrum. . . . . There can be no doubt of the reality of the second spectrum. . . . ." No doubt of any kind was recorded by Plaskett with regard to U CrB or by Adams and Joy with regard to Z Her.

Among the stars that Wyse observed at principal minimum, the greatest discrepancy occurs in RX Geminorum,[29] with types A2 and A3p and minima of 2.1 and 0.0 mag. Perhaps the peculiarity of the A3 spectrum with weak $H\beta$, sharp $H\gamma$, and strong and broad $H\delta$ and $H\epsilon$ may provide an explanation in terms of shell absorption—a phenomenon which I have observed in SX Cassiopeiae[30] and which probably accounts for the effect described under Section II (b).

It is important to observe these stars again in order to ascertain, if possible, whether the spectra of the 5 stars listed above show the same types at principal minimum as were observed for the fainter components at maximum light and whether the spectra are normal. This should permit us to distinguish between the reflection effect and shell absorption.

IV. CHANGES IN SPECTRUM DUE TO DIFFERENCES IN GRAVITY

a) There is no reliable information concerning possible differences in the spectral types of binary components dependent upon their phase angles. The reflection effect may cause differences in the spectra of the two hemispheres directed toward or away from the other component. Differences in gravity may cause the spectrum of the tidal bulges to differ from that of the preceding and following hemispheres. Observation is complicated by the overlapping of lines of two components at phase angles 0° and 180°.

b) It is conceivable that there may also be an appreciable periastron effect in close binaries of high eccentricity.

V. PROBLEMS OF EMISSION LINES

a) There seems to be a growing realization that the formation of emission lines is facilitated in binary systems. Wyse stated in 1934:[29] "The eclipsing binaries on the whole appear to have perfectly normal spectra. This applies to the faint companions as well as to the primaries. Exception is noted in an unexpected tendency of the secondaries to display bright H lines." Out of 25 secondaries observed by him, 5 had the characteristic of H emission. The types of these secondaries range from A to K. In single stars emission lines are very rare for these types. The number of binaries with H emission lines can be considerably augmented by the addition of SX Cas,[30] RX Cas,[31] etc., all of class A or later. There are a few bright-line binaries of types O and B—for example, HD 163181 = V 453 Sco,[32] 29 Canis Majoris = UW CMa,[33] and, of course, β Lyr and RY Sct. But the great majority of B-type binaries show no bright lines. Conversely, we may say that the great majority of Be stars are not binaries. The material indicates that the frequency distribution of bright H lines among binaries is entirely different from that observed in single stars.

Binaries of types G and K very often exhibit bright lines of Ca II, and the number of these cases is also considerably in excess of what might have been expected from the frequency of occurrence of these lines in single stars.

We know so little about the behavior of the bright lines in binaries that our first task should be the discovery of a greater number of systems having this peculiarity.

b) There is a marked difference in the structure and behavior of the emission lines on the one hand of H and of other lines associated with it—namely, Fe II, Ca II, and occasionally a few others—and on the other hand of Ca II when the latter appears alone in systems both components of which have types as late as G or K. In the former group

[30] *Ap. J.*, **99**, 89, 1944.
[31] *Ap. J.*, **99**, 295, 1944.
[32] *Ap. J.*, **99**, 210, 1944.
[33] *Ap. J*, **93**, 84, 1941; **82**, 95, 1935.

the lines are double and are divided by a deep absorption core. In the latter they are single (if observed with moderate dispersion), though sometimes slightly broadened. This difference is probably important, but we need more material to substantiate it.

c) The double components of the *H* emission lines and of other emission lines associated with them undergo eclipses in the stars RW Tauri,[34] SX Cas,[30] and RX Cas[31] but fail to show this effect in W Serpentis.[35] In the three systems with eclipses the emission lines originate in a ring or stream surrounding the early-type component of the binary irrespective of whether that component is the brighter (RW Tau and SX Cas) or the fainter (RX Cas). The direction of the stream motion is the same in all three systems and agrees with the rotation of the component and with the orbital revolution of system. In velocity the stream motion exceeds very greatly that determined from the absorption lines. In RW Tau the displacements of the bright lines are ± 350 km/sec, while the absorption lines are described by Joy as well defined and without any abnormal rotation effect. The periods range from 2.77 days for RW Tau to 36.6 days for SX Cas. The absence of eclipse in the bright lines of W Ser is related to the appearance in this spectrum of forbidden [*Fe* II], while only permitted *Fe* II is seen in SX Cas, RX Cas, and RW Tau. It is possible that the essential difference between W Ser and the other systems is one of size of the emitting nebulosity, but more material is required.

No eclipsing effects have been observed in the *Ca* II emission group. In one star, WW Draconis,[36] the velocity-curve of the *Ca* II emission lines agrees in phase with that of the fainter and less massive K0 component of this system, whose brighter component is of type G2. However, the range of the velocity-curve is smaller than that of the K0 star, and Joy inferred that "the secondary star is surrounded with an envelope of calcium gas which is greatly extended in the direction of the primary by tidal attraction." In AR Monocerotis,[37] where only one component of type K0 is observed, the emission lines follow the velocity-curve of that component with undiminished amplitude. But in both stars the emission lines follow the curve of that component which stands in front at principal eclipse. We need more material to confirm this feature and to explain it.

d) P Cygni type emission and absorption lines are present in several binaries, such as in β Lyr,[22] where the B5 absorption component shows a velocity of expansion; 29 CMa,[33] where such lines appear in certain phases, where there are large periodic changes in the structure and intensity of the emission lines, and where the emission lines exhibit a large "red shift"; and in the Wolf-Rayet spectroscopic binaries, where violet absorption borders are commonly present. The variety of changes in the emission lines is great; and we have, as yet, little material to go by.

e) The frequent occurrence of *Ca* II emission in close binaries is of interest in connection with the suggestion that these lines are also often bright in visual binaries.[38] But it is not quite certain whether this tendency is real.

f) Another phenomenon usually associated with wider pairs is the occurrence of intense [*Fe* II] with relatively weak permitted *Fe* II. Typical examples are Boss 1985 and WY Geminorum, both having composite spectra,[39] and the fainter, B-type component of α Scorpii.[40] It is probable that several other emission-line phenomena are connected with the binary nature of the stars. For example, the complex and variable emission lines of high ionization potential in Z Andromedae, AX Persei, CI Cygni,[41] and stars whose

[34] Joy, *Pub. A.S.P.*, **54**, 35, 1942.

[35] Bauer, *Ap. J.*, **101**, 208, 1945. It should be noted that the light-curve and the velocity-curve of this star are peculiar; and while I am inclined to agree with S. Gaposchkin and with C. A. Bauer that it is a double star, the evidence on this point is not conclusive.

[36] Joy, *Ap. J.*, **94**, 407, 1941.

[37] Sahade and Cesco, *Ap. J.*, **100**, 374, 1944.

[38] Swings and Struve, *Pub. A.S.P.*, **53**, 244, 1941.   [40] *Ap. J.*, **92**, 316, 1940.

[39] *Ap. J.*, **91**, 596, 1940; **93**, 455, 1941.   [41] Merrill, *Ap. J.*, **99**, 15, 1944.

absorption features contain TiO, and other low-temperature lines and bands, are best explained by assuming that the star is a binary, whose bright component is cool and whose hot component is so small that it must be regarded as underluminous. There is, however, no direct evidence of the duplicity of these stars (except in α Sco and Mira Ceti).

VI. DEFORMATIONS OF VELOCITY-CURVES

a) This problem has come into the foreground since it was shown, a year ago, that the unsymmetrical velocity-curve of U Cep,[3] first observed by Carpenter in 1930, agrees with the new spectrographic observations. The spectrographic values of *e* and ω are inconsistent with the photometric value of *e* cos ω determined from the spacings of the minima in the light-curve. A similar asymmetry was found in the velocity-curve of SX Cas[30] and perhaps in that of RX Cas.[31] In SX Cas there is a strong suspicion that the asymmetry arises from absorption in the same stream of gas which gives rise to the disappearing emission components described under Section V (*b*). It is noteworthy that in AU Mon,[20] where the *H* lines show periodic changes in the absorption-line intensities (see Sec. II [*b*]), the velocities derived from these lines give a curve that is unsymmetrical in the same manner as in U Cep. In the latter star, Joy and the writer have found unsymmetrica line contours at and near maximum radial velocity. This is probably the same effect as that observed in AU Mon by Cesco and Sahade. There are some other stars in which discrepancies between photometric and spectrographic values of *e* have been recorded. One such case, TX Ursae Majoris, is probably not real, according to Hiltner,[42] who finds a small spectrographic value of *e* which is in good accord with the photometric results. But in u Herculis a new spectrographic investigation by Smith[43] confirms the old value $e = 0.05$ while the light-curve gives $e \leq 0.01$. Another case is MR Cygni, where Pearce[44] found $e = 0.12$, while Miss Swope[45] found, from the light-curve, $e \leq 0.01$. Kopal[46] has suggested that these discrepancies "may quite possibly be a common, rather than an exceptional, feature of close binary systems"; the present evidence favors an explanation in terms of superposed absorption produced in a gaseous stream, as in SX Cas.

b) There is a well-known tendency among spectroscopic binaries to have values of ω in the neighborhood of 90°. The latest information concerning this phenomenon is contained in a paper by S. V. Nekrasova.[47] Her data are reproduced in the accompanying table. The phenomenon is especially pronounced for stars with small eccentricities.

ω	NUMBER OF SYSTEMS		ω	NUMBER OF SYSTEMS	
	$e > 0.10$	$e < 0.10$		$e > 0.10$	$e < 0.10$
0°– 30°.............	21	17	180°–210°.............	11	15
30 – 60.............	24	19	210 –240.............	15	7
60 – 90.............	20	12	240 –270.............	8	6
90 –120.............	30	16	270 –300.............	21	5
120 –1`0.............	19	11	300 –330.............	15	7
150 –180.............	13	7	330 –360.............	22	15

Various attempts have been made to explain it. Struve and Pogo[48] suggested the possibility of a systematic tendency in the orientations of the lines of apsides, favoring an

[42] *Ap. J.*, **101**, 108, 1945.

[43] Unpublished.

[44] *J.R.A.S. Canada*, **29**, 413, 1935.

[45] *Harvard Bull.*, No. 914, 1940.

[46] *Ap. J.*, **99**, 239, 1944.

[47] *Bull. Engelhardt Obs. Kazan*, No. 13, p. 10, 1937.

[48] *A.N.*, **234**, 300, 1928.

angle of 90° with the direction toward the galactic center. But the discovery of physically distorted velocity-curves described in Section VI (*a*) indicates that at least a part of the observed frequency distribution may be caused by these distortions.

c) We may tentatively mention here a phenomenon which played an important role in stellar spectroscopy some twenty years ago. In 1908 and 1909 Schlesinger and Curtiss noticed that in Algol and δ Librae the times of spectroscopic conjunction precede mid-eclipse by about 73 minutes and 112 minutes, respectively.[49] These discrepancies were regarded as exceeding the possible errors of the photometric and spectroscopic elements. Later, J. S. Plaskett[50] added new material to the discussion and concluded that in several other eclipsing stars the departures are either negligible or of opposite sense. In 1925 Hellerich[51] summarized the data for 14 stars. He concluded that in most systems the departures are reduced if, in place of elliptical elements, circular elements are computed from the velocity-curves and if all radial velocities obtained during the principal eclipse are omitted. In almost all systems the remaining discrepancies are then within the limits of their probable errors. Martinoff[52] has recently pointed out that the Tikhoff-Nordmann effect undoubtedly contributes to the observed discrepancies. This effect is attributed by Mustel[53] to gravity darkening at the tidal bulge, which is assumed to lag behind, as was suggested by Dugan in the case of U Cep.

VII. SHELL ABSORPTION IN ECLIPSING BINARIES

a) Several eclipsing variables show conspicuous dilution effects in their absorption spectra. The weakness of Mg II 4481 in the spectrum of ε Aur during principal eclipse[54] and the great strength of the ultimate lines of Ti II in VV Cep[55] are good examples. The dilution effect probably explains the weakness of Mg II 4481 in SX Cas,[30] and this suggests that the entire observed A-type spectrum may come from the streams rather than from the normal reversing layer of the early-type components. Dilution is prominent in the B5 spectrum of β Lyr.[22]

b) The question arises whether shell absorption may not in some cases vitiate the spectra of the fainter components observed at principal minimum. In SX Cas,[30] whose minima are $A_1 = 0.99$ mag. and $A_2 = 0.32$ mag., the spectrum at principal eclipse is G6, in agreement with prediction. But this G6 spectrum is certainly not normal, because some absorption lines continue to follow the velocity-curve of the A-type component, while those which are strengthened at mid-eclipse show the trend of the G star. I concluded, a year ago, that "the observed spectrum is a blend of a true star of a fairly late spectrum—probably G—and lines of another origin which resemble in character the shell lines of the A star and share with it the descent of the velocity-curve." Yet, the photometric evidence of a total eclipse is complete. In RX Cas[31] the behavior of the H absorption lines was even more revealing: they gave no perceptible change in velocity with phase. The K-type component predominates in this star at all phases, but the H absorption cores are clearly of shell origin. These cores, incidentally, show remarkable changes in RX Cas, as well as in SX Cas. It is exceedingly tempting to attribute to shell absorption such glaring discrepancies between spectroscopic and photometric classes as that found by Wyse[29] for RX Gem (see Sec. III [*b*]).

c) The importance of the dilution effect in estimating the size of the nebulous shells or streams in binary systems was illustrated in the case of β Lyr.[22] We require informa-

[49] *Pub. Allegheny Obs.*, **1**, 25, 1908; **1**, 129, 1909.

[50] *Pub. Dom. Ap. Obs. Victoria*, **3**, 248, 1926.

[51] *A.N.* **216**, 277, 1922; **223**, 367, 1925.

[52] *Variable Stars*, p. 62, Moscow, 1939.

[53] *Astr. J. Soviet Union*, **11**, 415, 1934.

[54] *Proc. Amer. Phil. Soc.*, **81**, 212, 1939.   [55] *Ap. J.*, **99**, 70, 1944.

tion concerning the behavior of those few lines which are sensitive to dilution: the singlets of *He* I in B stars, the lines of *Mg* II and *Si* II in A stars, and a few selected lines of *Fe* I (λ 4260, etc.)[56] in later-type systems.

VIII. THE ζ AURIGAE EFFECT

a) The line absorption caused by the extended atmosphere of an eclipsing star in the continuous spectrum of the partly eclipsed star has enabled us to study the structure of the atmosphere of the K-type component of ζ Aur.[57] Similar phenomena have been observed in VV Cep[58] and ϵ Aurigae.[59] All three are peculiar supergiant systems of relatively long period (twenty-seven years in the case of ϵ Aur). It is exceedingly important to search for other, similar cases. There is a strong indication that the phenomenon may occur even in ordinary systems of short period. In TX UMa,[42] which has a period of 3.06 days, Hiltner found a sharp *Ca* II line before and after mid-eclipse, superposed over a broad line of the F2 eclipsing star. This sharp line is stronger than the normal *Ca* II line of the B8 star, which is the one eclipsed at principal minimum. Hiltner considers it probable that the sharp line in the partial phases is caused by a ζ Aur effect.

b) Distinct satellite lines were found during the partial phases of principal eclipse in β Lyr.[22] These lines appear a short time before mid-eclipse and show large positive velocities; they disappear at mid-eclipse but reappear a few hours later, with large negative velocities. But the contours and the relative intensities of the violet and red satellites are not the same. It has been suggested that they originate in a stream of gas which passes from the B9 component, swings around the following side of the F5 component, and returns along the preceding side of the F5 component. Two related phenomena were observed in U Cep.[3] During the partial phases of eclipse the *H* absorption lines look as though they consist of a sharp line unsymmetrically superposed over a normal, rotationally broadened line. There are also observed distinct satellite lines of *Ca* II, whose velocities do not agree with either of the two components. These phenomena have been tentatively attributed to absorption in a gaseous stream similar to that observed in β Lyr, SX Cas, and RX Cas; but the information at present available is still insufficient. The unsymmetrical superposition of a narrow *H* line upon a broader line results in an apparent reduction of the range of the rotational disturbance, as measured in *H*, if compared to that measured in *He* I, *Ca* II, *Mg* II, and *Si* II. This phenomenon closely resembles that observed in RZ Sct (see Sec. I [*c*]) and attributed there to stratification in an extended atmosphere, because there was no suspicion of superposition. It is impossible to decide whether we have the same phenomenon, produced by one cause; and, if we do, whether we have stratification with diminishing rotational velocities at the greater heights or stream motion which is independent of the atmosphere of either component star. Perhaps there is less distinction between the two pictures than appears at first sight.

IX. LIMB SPECTRA

Redman[60] made the first attempt to measure the equivalent widths of several absorption lines in U Cep and U Sagittae during the partial phases of the eclipse. Both stars have deep total eclipses, and the secondary components are faint and belong to a different spectral class than the primaries. The Balmer lines behave as though they vanish at the limb of U Cep and become distinctly fainter at the limb of U Sge. Redman remarked that this could be due either to the effect of true absorption in the formation of the lines

[56] *Proc. Amer. Phil. Soc.*, **81**, 227, 1939.

[57] Wellmann, *Veröff. Berlin-Babelsberg*, Vol. **12**, No. 4, 1939.

[58] Goedicke, *Pub. U. Michigan Obs.*, Vol. **8**, No. 1, 1939.

[59] Frost, Struve, and Elvey, *Pub. Yerkes Obs.*, Vol. **7**, Part 2, 1932.

[60] *M.N.*, **96**, 488, 1936.

or to the reduction in Stark effect at the higher levels of the reversing layers. Some confirmatory evidence came from my own observations of U Cep.[3] In addition, I found that the lines of He I were also weaker at the limb, while Ca II and Si II did not change appreciably. My conclusion was that in all probability "the lines appear weaker at the limb because the pressures are lower and not because the mechanism of line formation is predominantly one of pure absorption." In RZ Sct[4] the lines of H, He I, and Mg II became much weaker at the limb; and in the case of H this is accompanied by a tendency for the lines to become sharper and narrower. The weakening of Mg II (which was also suspected in U Cep) may mean that this line is actually formed by the process of pure absorption. Because of the many complicated phenomena occurring during the eclipses, the observations and their interpretation are difficult.

X. SPECTROSCOPIC ASYMMETRY DURING ECLIPSE

Attention has already been drawn to the difference in appearance of the satellite lines in β Lyr before and after mid-eclipse—only a few hours apart. Completely unexplained differences were also noticed in U Cep before and after totality. Thus, "in the first half of the eclipse the lines of He I are systematically weaker than in the second half. . . . . There appears to be a real difference in the intensities of lines produced by the receding and approaching limbs of the B8 star, the former having the weaker lines. . . . ." It is possible that this is related to the asymmetry of the light-curve. It may well be that both these phenomena are caused by the streams discussed previously, or to some sort of tidal lag in a rapidly rotating star, as was suggested by Dugan.[61]

XI. UNUSUAL SPECTRA OF BINARY COMPONENTS

a) If the semiamplitudes of the velocity-curves of spectroscopic binaries of a single spectral class are plotted as ordinates against log P as abscissae, the stars fall within an area limited by the x-axis at the bottom, by a curve whose shape is given by[62]

$$K_1 = CP^{-1/3}$$

at the top, and by a vertical straight line at the left. For the B-type stars this straight line occurs approximately at

$$P = 1.3 \text{ days} .$$

The reason for this limitation on the diagram is that in a given spectral class most systems have approximately the same values of

$$r_1 + r_2 = \text{constant} ,$$

$$m_1 + m_2 = \text{constant} .$$

$$m_2 / m_1 = \text{constant}$$

$$e = \text{constant} ,$$

so that only the inclination produces a large scatter in the values of K. Stars which fall outside the occupied area are abnormal in some respect. The sharp limitation of the area by the straight line on the left must be caused by the average value of $r_1 + r_2$. Components cannot come closer to one another unless $r_1 + r_2$ is abnormally small. It is important to discover systems which fall outside the occupied area in order to find whether the spectra of these stars are also abnormal. Among the B's the two systems lying above the limiting curve $K_1 = CP^{-1/3}$, namely, β Lyr and BD+6°1309, are both abnormal in spectrum. The question arises whether there are stars which lie to the left of the line $P = 1.3$ days for the B's or $P = 1.0$ day for the A's. Several eclipsing variables are

[61] *Contr. Princeton U. Obs.*, No. 5, 1920   [62] *M.N.*, **86**, 63, 1925.

known which fall to the left of these lines—for example, S Antliae,[63] with $P = 0.64$ day and two spectra of classes A8 and A8, and UX Ursae Majoris,[64] with $P = 0.197$ day and a spectrum of class A. The masses of these stars are small, compared to those of normal A stars; and the systems behave as typical W UMa stars. It is clear that all A and B stars with $P < 1$ day deserve attention. Such stars must be small, and their spectra should be interesting. Thus, the spectrum of UX UMa suggests underluminous components.

b) Another procedure for picking out abnormally small stars is to select those systems which combine a large value of e with a small value of P. Among such systems RU Monocerotis should be especially interesting, since it has $P = 3.58$ days and $e = 0.38$. The latter value was determined by Krat from the light-curve and was confirmed by Shapley.

c) The results described in Section XI (a) have been confirmed by D. J. Martinoff,[65] who plotted the periods of eclipsing variables against their spectral types. The diagram shows that the points are scattered over a wide area, whose edge is sharply limited by a curve running from about $P = 5$ days, Sp. $= O5$, through $P = 1$ day, Sp. $= B0$; $P = 0.5$ day, Sp. $= A0$; $P = 0.3$ day, Sp. $= F0$; $P = 0.25$ day, Sp. $= G0$. This curve corresponds to the vertical lines in the diagrams described under Section XI (a). If we write Kepler's third law in the form

$$\frac{a^3}{P^2(m_1 + m_2)} = 74.5,$$

where m_1 and m_2 are in units of $\odot$ and P is in days, and if we assume for contact that $a = r_1 + r_2 = 2r_1$, then the minimum value of P is given by

$$\frac{r_1^3}{P^2_{\min}(m_1 + m_2)} = 9.3.$$

Martinoff uses typical values for r_1 and $(m_1 + m_2)$ and finds by computation:

Sp.	$P_{\min}$
O8.5.............	2.56 days
B5..............	0.80
A0..............	0.54
F0..............	0.33
G0..............	0.30

These values agree sufficiently well with the observed data and with the evidence from spectroscopic binaries. The conclusion arrived at again is that only stars which have abnormally small values of $r_1 + r_2$ can lie beyond the limits of the curve; and these stars, like UX UMa, are very rare. An interesting point in Martinoff's work is the pronounced tendency of eclipsing variables showing ellipticity variation (β Lyr and W UMa types) to crowd near the limiting curve, while ordinary Algol-type stars are located farther from the curve.

XII. RELATIVE INTENSITIES OF DOUBLE LINES IN SPECTROSCOPIC BINARIES

A considerable amount of work has been done by R. M. Petrie[66] on the measurement of equivalent widths of double lines. From these measurements information may be ob-

[63] Joy, *Ap. J.*, **64**, 287, 1926.

[64] Zverev and Kukarkin, *Ver. St. Nishni–Novgorod*, **5**, 125, 1937. A spectrogram taken by me with small dispersion shows only broad *H* absorption lines. Dr. G. P. Kuiper has informed me that he has obtained additional spectrograms of this variable in order to determine the range in velocity.

[65] *Astr. J. Soviet Union*, **14**, 306, 1937.

[66] *Pub. Dom. Ap. Obs., Victoria*, Vol. **7**, No. 12, 1939; *Pub. A.A.S.*, **10**, 333, 1944.

tained concerning the relative brightnesses of the component stars and their spectral types. It is important to determine whether the data thus determined are in accord with the mass ratio derived from the velocity-curves and with the mass-luminosity relation; also, whether the observed values of Δm agree with the results from the photometric orbits in eclipsing variables.

XIII. THE WOLF-RAYET SPECTROSCOPIC BINARIES

The problems connected with these stars are somewhat different from those encountered in other spectral classes, and they will not be discussed in this paper. Reference is made to recent papers by O. C. Wilson,[67] C. S. Beals,[68] and W. A. Hiltner.[69] There is a suspicion that all typical Wolf-Rayet stars may be binaries and that those which have little or no variation in velocity (of which there seem to be only a few) are binaries with small inclinations.

XIV. SPECIAL SYSTEMS

In addition to the more general problems discussed in the preceding sections, several stars present individual problems. Among these may be listed:

a) The expanding spiraling structure of the nebulosity around β Lyr [70]

b) The interpretation of the eclipse spectrum of ϵ Aur and the nature of the semitransparent infrared component of this system[71]

c) The question of the binary nature of such stars as ϕ Per[72] and ζ Tauri,[73] which have peculiar velocity-curves but whose periods are strictly uniform

d) The interpretation of the sharp absorption lines of VV Cep[58] and of the peculiar structure of the emission lines of this star

e) The suspected duplicity of the lines of Algol near mid-eclipse[74] and the asymmetry of their rotational disturbance

f) The spectroscopic consequences of intrinsic variations in light observed in the components of several binaries—for example, ϵ Aur[75] and RX Cas[76]

g) The reported differences in the spectroscopic phenomena observed in ζ Aur during two eclipses[57]

h) The system of υ Sagittarii,[77] which is unique because of its hydrogen-poor spectrum,[78] its stationary[79] emission lines of forbidden[80] and permitted[81] Ca II, and its strong He I absorption lines.[82] The period is 138 days, and the semiamplitude of the velocity-curve is 48 km/sec

i) The faint companions of several visual binaries, which have already been mentioned as possessing abnormal spectra (bright [Fe II] in α Sco, etc.). Among these objects, the companion to Mira Ceti is one of the most remarkable.[83]

[67] *Ap. J.*, **95**, 402, 1942.

[68] *M.N.*, **104**, 205, 1944.

[69] *Ap. J.*, **99**, 273, 1944; **101**, 356, 1945; see also O. Struve, *Ap. J.*, **100**, 384, 1944.

[70] Kuiper, *Ap. J.*, **93**, 133, 1941.

[71] Kuiper, Struve, and Strömgren, *Ap. J.*, **86**, 570, 1937.

[72] Hynek, *Ap. J.*, **100**, 151, 1944.

[73] Hynek and Struve, *Ap. J.*, **96**, 425, 1942.

[74] Morgan, *Ap. J.*, **81**, 348, 1935; Melnikov, *Poulkovo Obs. Circ.*, No. 30, p. 65, 1940.

[75] C. M. Huffer, *Ap. J.*, **76**, 1, 1932.

[76] S. Gaposchkin, *Ap. J.*, **100**, 230, 1944.

[77] R. E. Wilson, *Lick Obs. Bull.*, **8**, 134, 1914.

[78] Greenstein, *Ap. J.*, **91**, 438, 1940.

[79] Merrill, *Pub. A.S.P.*, **56**, 42, 1944.

[80] *Ibid.*, **55**, 242, 1943.   [81] Weaver, *Ap. J.*, **98**, 131, 1943.

[82] J. S. Plaskett, *Pub. Dom. Ap. Obs. Victoria*, **4**, 1, 1927.

[83] See Merrill, *Spectra of Long-Period Variable Stars*, p. 79, Chicago, 1940.

XV. MASS FUNCTIONS

A rather surprising result is the small average value of the mass function derived from our studies of eclipsing variables. Almost without exception, those systems which were chosen because they were known to have deep eclipses have turned out to have relatively small velocity amplitudes. In some cases the mass functions are so small as to impose severe restrictions upon the masses themselves, or upon the mass-ratios, or upon both. In BD Virginis,[84] for example, we have only one component, of spectral type A5. The mass function is

$$\frac{m_2^3}{(m_1 + m_2)^2} \sin^3 i = 0.02 \odot .$$

If we designate $m_1/m_2 = a$ and if we require that the mass of the fainter component, whose spectrum from the light-curve should be G, be $m_2 = 1 \odot$, we find $a = 6$ (since $\sin^3 i \sim 1$); or, more generally,

m_2	a	m_1
2 $\odot$........	9	18$\odot$
1.............	6	6
0.5..........	4	2

To obtain reasonable masses, a must be at least equal to 4. From the light-curve of this, and of other similar stars we infer that the faint, cool, companions are larger than the hot primary star and may well be considered as subgiants. We now infer that at least in some of these systems the subgiant secondaries have relatively small masses, probably of the order of from 0.5 $\odot$ to 1.0 $\odot$, and that the mass ratios are somewhere between $a = 4$ and $a = 6$. I am mentioning this here because a star of class G whose mass is less than the solar mass and whose radius is several times larger than that of the sun is not often encountered among the single stars, and its spectrum should be of interest. In several eclipsing variables, though not in BD Vir, these spectra can be observed at principal minimum.

The frequent occurrence of these systems among the eclipsing variables is a result of observational selection. Spectroscopic binaries are discovered as such when the brighter components have large values of K_1. This happens when $a = 1$. Ordinarily, a cannot be less than 1, except in Wolf-Rayet binaries. On the other hand, deep eclipses are discovered if $k = r_1/r_2$ is very small and if, at the same time, J_2/J_1 is also small. This favors systems of the kind of BD Vir.

XVI. THICK ATMOSPHERES

Some of the points discussed in the preceding sections must have a bearing upon the theory of thick atmospheres recently developed by S. Gaposchkin and C. Payne-Gaposchkin.[85] As these authors have stated, the photometric observations of SX Cas and RX Cas and a few other stars show striking differences between light-curves obtained in yellow light and those obtained in blue light. These differences can be formally explained by adopting very different radii for the two effective wave lengths; and this, in turn, leads to the suspicion that the stars have very extensive atmospheres, whose heights are comparable to the radii of the two components. A theoretical discussion of the limb darkening as a function of wave length by the Gaposchkins has given a series of predicted light-curves which strikingly resemble those observed in SX Cas. The spectrographic observations have given reliable evidence of the existence of tenuous gase-

[84] *Ap. J.*, **100**, 181, 1944.   [85] *Ap. J.*, **101**, 56, 1945.

ous streams in SX Cas and a number of other systems. The question arises whether these tenuous streams are identical with the thick atmospheres postulated by the Gaposchkins. The theory of the latter depends upon their strong continuous absorption. The streams produce emission lines and absorption lines. There has been no indication from the spectrograms that they also produce appreciable continuous absorption. On the whole, I had thought that these extra absorption lines are somewhat weaker than the corresponding lines in normal stellar atmospheres. Their contours are suggestive of turbulence, so that moderately strong lines can be produced by relatively small numbers of atoms. Hence I was inclined to disregard the continuous absorption in the streams, except in such peculiar objects as β Lyr. But these arguments are not sufficiently strong to carry any weight against the photometric evidence. In β Lyr and in a few other stars the absorption lines of the component which is supposedly involved in a nebulous ring appear strongly at all phases and are not noticeably cut down by the continuous absorption within the shell. I suppose that this is consistent with the hypothesis of a ring, but it is difficult to see how a distinct line of type B9 could be observed through a "thick atmosphere," surrounding the star completely. In SX Cas the spectrographic evidence is different. It is not possible to separate the absorption lines from the shell and from the star. Hence it is entirely possible that we have a thick atmosphere in the sense required by the theory of the Gaposchkins and that the outer layers of this atmosphere produce the spectroscopic shell phenomena. Such a view might help to reconcile the conflicting pictures of the rotational disturbance in such stars as U Cep and RZ Sct. But there is, as yet, no photometric evidence of a "thick atmosphere" in RZ Sct, while in U Cep it is present but is much less conspicuous than in SX Cas or RZ Ophiuchi. Perhaps all that can be stated with certainty at the present time is that those stars which best show the photometric phenomenon attributed to "thick atmospheres" also show spectroscopic features which have been attributed to absorbing streams of gas located outside the classical boundaries of the two components.

THE OBSERVATIONS

Nearly all eclipsing variables for which there are no spectrographic observations are fainter than magnitude 8 at maximum light, and a useful program should reach stars to magnitude 10 or even 10.5 at maximum. We have seen that the average rate of increase in the number of eclipsing binaries observed with slit spectrographs has been of the order of one or two systems per year. It is clear that in order to secure answers to our questions within a reasonable length of time we must greatly accelerate the observational work.

CONCLUSIONS

Out of the profusion of observational results secured in the past few years on the spectra of eclipsing variables there emerge several phenomena which appear to me to be of major importance. Most of the questions which I discussed in the beginning of this paper can be related to one of these major phenomena. It is possible that the first and the second may later be found to be produced by the same cause, but there is no such evidence available now.

1. The occurrence of rapidly moving streams of gas in close binary systems is not a rare phenomenon; it manifests itself in the form of emission lines, of absorption lines which are seen to be superposed over the normal stellar absorption lines, or of distortions in the velocity-curves produced by lines which cannot be seen as distinct features on top of the stellar lines. These streams show certain regularities in respect to direction of motion, intensity of lines, etc. There are few, if any, criteria to predict which stars are most likely to show these effects. Among eclipsing variables chosen at random, we may expect, perhaps, one star out of eight or ten to show some easily observable manifestation of these streams.

2. Periodic variations in the line intensities of binary components may be related to

such effects as reflection and gravity, and there is a strong suspicion that the spectral type may be slightly different at different phases. This type of variation is most easily observed in double-lined binaries, but this is perhaps due only to the greater ease with which comparisons can be made in these objects. It is difficult to estimate the frequency of such variable-line binaries.

3. Stars chosen either because they fall outside the area defined by the relation between K and P and by P_{min} = constant or because they have a large eccentricity with a short period are likely to have unusual spectra. For values of P which are smaller than the limiting value established for normal stars and for binaries of abnormally large eccentricity the spectra must correspond to the characteristics of underluminous stars.

4. There is some indication that the bright lines of Ca II do not originate in streams which are detached from the two stellar bodies but are produced in the atmosphere of one of the components. The fact that these components are sometimes seen with great intensity at principal mid-eclipse certainly[99] suggests that they are produced even in that hemisphere of the eclipsing component which is turned away from the eclipsed component. Lack of appreciable broadening or doubling precludes their coming from a nebulous layer of dimensions comparable to the radius of the star. It is more probable that we are here concerned with a reversal whose origin may not be very different from the origin of the reversals in the Ca II lines observed on the disk of the sun.

5. The origin of the bright lines of H must be entirely different from that of Ca II lines and must be related to the small optical depth for continuous absorption in the gaseous streams. It is instructive to find that in many systems the streams produce only absorption lines, while in others (RW Per) the emission lines make their appearance only during the eclipse.

6. The principal result which we have obtained is the recognition of the complexity of the absorption lines in several binaries. Theories of line formation must be applied with great caution to these peculiar spectra, and it is important that we should realize that it is often impossible to conclude from the appearance of a line that it represents the superposition of a normal stellar absorption line and of a line produced in a gaseous stream high above the reversing layer. In RW Per the emission lines appear only during the eclipse (except Ha). There must be other binaries in which the emission lines are never observed because the inclination may not be sufficient to produce a deep eclipse. It is possible that if we should apply to the lines of such a star the usual formulae of radiative transfer in a reversing layer the conclusions would be erroneous. It is tempting to try to speculate on the relative equivalent widths of the H absorption and emission lines. If we disregard cyclic processes and consider only a two-state problem, within a "thin" shell, then the absorption line will predominate so long as the radius of the shell is comparable to the radius of the star. But in large shells the two equivalent widths should be approximately the same: were it not for the rotation of the shell, the emission line would exactly fill in the absorption line, and no trace of the shell would appear in the spectrum. It is only because of the Doppler effect that we observe the emission lines. These considerations might suggest that our spectrograms would give us no indication of any *extended stationary atmospheres* whose projected diameters were smaller than the slit of the spectrograph and in which the assumptions we have made were realized.

7. It should be pointed out that the various spectroscopic anomalies which have been discussed in this paper are not the usual thing among spectroscopic binaries. Although these anomalies are not infrequent, the great majority of close systems have probably normal spectra.

SPECTROSCOPIC BINARIES

D. M. POPPER

University of California, Los Angeles

I. HISTORICAL INTRODUCTION

Spectrographic investigations of individual spectroscopic binary stars may be grouped loosely into two categories. In one the principal aim is the determination of orbital elements, from which masses, radii, apsidal rotation, and other fundamental properties may be derived. In the other, the purpose is to study physical phenomena such as the origin of emission lines, gas streams, effects of tidal interactions, etc., peculiar to the system under investigation. The first two published contributions by Struve, appearing as successive papers in the *Astrophysical Journal* (1923a, 1923b) consist of one paper of each category. It is somewhat prophetic that the very first of Struve's papers falls in the second class, for it is in this direction that his later interest developed and his major contributions to the study of close binary stars were made.

More than twenty years after these first contributions, Struve could still point out (in his 1945 paper reproduced just preceding this article [1945]) that little attention was paid by stellar spectroscopists to the physical characteristics of the stars in spectroscopic binary systems. After another twenty years, the outlook is now vastly different, very largely because of the work of Struve and his collaborators (of whom, I must add, I was not one). Although his first important contribution toward understanding physical phenomena in close binary systems was made in 1934 with the first of his many papers on β Lyr (Struve 1934), it was not until the period 1944–47 that Struve undertook with the 82-inch McDonald reflector his massive attack in this domain. The observational results by Struve and his co-workers were presented in a long series of papers in the *Astrophysical Journal*, and were summarized in his book (Struve 1950).

The impact of this work on investigations of close binaries has been profound. Studies of physical phenomena immediately became a major aspect of such investigations. The conclusion expressed in his 1945 paper that, "Most close systems have probably normal spectra," was almost certainly no longer held by Struve in later years. Indeed, because of possible effects of absorption or emission by matter not part of a stellar photosphere, it has become necessary to justify the assumption, when made, that the measured radial-velocity curves do, in fact, represent true orbital motion.

Discussions by Struve and others of physical phenomena in close binaries have emphasized two themes. On the one hand, the observations of a particular system are used to attempt to construct a physical model for the system that will account for as many of the observations as possible. On the other, some unifying concepts are sought which will serve to coordinate the variety of observational results found and models constructed. Struve's great contribution and strong interest were always centered in the former theme, and his acceptance of unifying concepts was contingent upon direct support afforded by observations. Thus in 1955 he surveyed progress in interpretation of close binaries (Struve 1955), pointing to the importance of the McDonald work on gaseous streams and envelopes and of Wood's recognition that in many eclipsing systems the larger, less massive component fills the critical zero-velocity surface of the system which passes through the inner Lagrangian point, a property correlated with gas streams and with period changes. In the 1955 survey, Struve also referred to a related unifying concept; namely, the expansion of one or both components as a consequence of post-main sequence evolution, proposed by Kopal (1954) and Crawford (1955). Although this theory is generally accepted as providing the framework for discussing the present state of a close binary system, in his later years Struve was unconvinced that observations unequivocally supported the scheme for most systems (Struve 1960, 1962). Mrs. Hack (1963) has reviewed arguments concerning various aspects of the theory.

After his most active period of observing eclipsing binaries at the McDonald Observatory, spectrographic studies of physical phenomena in close binaries were continued by Struve and his collaborators and students (including Hiltner, Sahade, Hardie, McNamara, Huang, Abhyankar) at a considerably slower pace. The later observations tended to consolidate and elaborate upon the ideas developed earlier. Relatively little observational work of this kind has been carried out by other astronomers, the major exception being the studies of novae, U Gem stars, and related objects by Kraft, Herbig, and others (Kraft 1963), following discoveries by Joy and by Walker of their binary nature. An event of major importance for interpreting physical conditions in extended binary envelopes is the discovery of variable polarization in β Lyr by Shakovskoy (1962; see also Appenzeller 1965 and Hiltner 1966). Another discovery of possible relevance is that by Walker and Popper (1964) of a strong ultraviolet continuum in the spectrum of KU Cyg.

In this paragraph I take the liberty of commenting in a personal vein. Although I was associated with Professor Struve for a considerable period, both at the McDonald and Yerkes Observatories and at the University of California, my work had relatively few points of contact with his. Two quotations are relevant in this regard. In the paper partially reproduced preceding this article (Struve 1945), the passage immediately following that reproduced reads as follows: "For this reason a fairly comprehensive program of spectrographic work was undertaken at the McDonald Observatory by D. M. Popper. . . . After Dr. Popper's departure on a leave of absence for war research, I revised the program and continued it with the efficient cooperation of Dr. W. A. Hiltner, Dr. Carlos U. Cesco, and Dr. Jorge Sahade. . . ." This was Struve's diplomatic way of indicating that he felt a major change in direction for attacking problems of eclipsing binaries was in order. Fifteen years later in an introductory survey of observational data on stellar evolution (Struve 1960), he said, "How important are these phenomena [*i.e.*, gaseous streams in close binary systems] in causing significant modifications in the evolutionary tracks of binary stars? Kopal has ably defended the view that major evolutionary changes must result. On the other hand, Popper has demonstrated that at least some close binaries are essentially 'normal' with respect to luminosity, mass, spectrum, etc. I am inclined to take an intermediate position."

In the following sections it is my aim to review some of the consequences of Struve's "revision" of the observing program of eclipsing binaries, with particular attention paid to the consequences in which I am most interested; namely those affecting derivation of masses and radii and their interpretation. A review of the program in which I am engaged, along with some of its results, has appeared recently (Popper 1967a).

II. THE STRUVE REVOLUTION

The list of "Astrophysical Questions Encountered in the Study of Spectroscopic Binaries" appearing in the 1945 contribution of Struve, reproduced above, was prepared early in the period of intensive observation, and not as a result of retrospective analysis. It is, nevertheless, instructive to consider the current status of items on the list from the observational standpoint. With the exceptions of the three items, XI (criteria for selecting underluminous stars), XIII (Wolf-Rayet binaries), and XIV (a miscellany of problems meriting analysis in its own right), the list may be grouped into three categories: problems of components as individual stars (questions I, VIIIa, IX, XII, XV, and XVI); problems arising because the stars are close together (II, III and IV); and problems involving effects of gaseous streams. Study of the questions of this last group—problems of emission lines (V), deformation of velocity curves (VI), shell absorption (VII), "satellite" lines (VIIIb), and spectroscopic asymmetry during eclipse (X)—constitute the mainstream of the contributions made by Struve and his co-workers over the years. A review of these effects has been given by Sahade (1960), and they will not be discussed further here. They provide conclusive evidence of the flow of matter in the systems, with each binary presenting its own special problems. From the standpoint of mass and radius determination, it is these effects that cause difficulty through the production of extraneous lines in the spectra and the distortion of photospheric lines. Workers in this field are now agreed that in a variety of systems, spectrum lines that were originally assumed to belong to the fainter component have their origin, at least in part, in matter having motion differing from that of either star.

Questions II, III and IV remain largely unanswered, in my opinion. II concerns variations with phase of the intensities of absorption lines in the spectra of binary components. In some systems the variations are clearly associated with other effects of gas streams, as in the distortion of the velocity curve of U Cep (Struve 1963a). In the case of double-lined binaries of types F and G, the variations may result from blending of lines of the two components. For earlier types, this explanation does not apply. The statement 20 years ago by Struve: "The existence of this effect should be checked, and an attempt should be made to explain it," is as relevant now as it was then, particularly for "detached" systems with no clear evidence of extraneous matter in the system. Question III concerns spectroscopic phenomena attributed to the reflection effect. The principal phenomenon under consideration (IIIb) is that the spectral types of the components in numerous cases appear to be more nearly equal than one would expect from the difference in depths of the two minima of the light curve, or in some cases (e.g., α Vir) from the mass and light ratios. In most of these cases it is now realized that the earlier observers were overly optimistic with respect to the presence of lines of the fainter component or with respect to spectral classification from lines of marginal appearance. There remains, however, a residue of cases for which the situation is still unclear. These cases fall into two small groups. First are early-type stars such as α Vir (Struve, Sahade, Huang, and Zebergs 1958) and SX Aur (Popper 1943), for which the spectroscopic observations of the secondary are obtained near quadrature, and there is no question as to the clear visibility of its lines. Huang (1959) has proposed corpuscular radiation as a possible explanation in the case of α Vir, whereas Struve et al. (1958) invoke reflection effects. The second group consists of systems in which lines of the secondary do not appear at quadrature but are reportedly seen at primary minimum. The most striking example is RX Gem, for which the only relevant observation is still that of Wyse (1934): spectral type A2 outside of minimum and A3 at total eclipse, the surface brightnesses of the two components being greatly different. The observation needs to be repeated, and colors at primary minimum would be helpful.

Another uncertain case is U CrB. The lines of a secondary of type B9, originally reported as seen near quadrature, are illusory, but strengthening of the K line during minimum (partial eclipse), reported by Sahade and Struve (1945) is confirmed by my observations and resembles an effect of blending an A-type spectrum, rather than the expected G-type, with that of the primary B star. In both RX Gem and U CrB the cause may be shell absorption. RX Gem shows strong Hα emission outside eclipse, with central reversal (Popper, unpublished), although no such evidence of a shell is present in U CrB. The W UMa stars, not under consideration in Struve's article, provide examples in which the differences in spectral type and in luminosity are very much less than expected from the mass ratios.

Struve's statement under question IV is still valid; namely, that there is no reliable information concerning differences in the spectra of components as seen at different phase angles that can be interpreted as due to differences in gravity. It would not be a simple observational matter to establish such differences unequivocally. Precise color observations might provide some evidence.

We turn now to the last group of questions, those concerning the individual stars themselves rather than effects produced by the star's existence in a close binary system. Question I, problems of axial rotation, is concerned principally with the problem of synchronism between orbital and axial angular motions. Struve's last article in *Sky and Telescope* (1963*b*) reviewed the subject. As he noted, mechanisms and time scales for establishing synchronism are still not well enough understood for conclusions to be drawn from the well-established lack of synchronism in a number of systems. On the other hand, if precise synchronism does exist in a variety of systems, one should be able to infer something about the magnitude of the forces of tidal dissipation. This topic is deserving of further observational study.

Question VIIIa concerns the ζ Aurigae effect. The use of this effect for studying the properties of extended atmospheres of supergiant stars is well established. Its presence in the systems 31 and 32 Cyg was shown after the 1945 paper was written. Evidence for similar absorption in the atmospheres of smaller stars, giants or subgiants, is scanty. Struve (1945) refers to Hiltner's work on TX UMa, and the case of the somewhat similar system, U CrB, has been referred to under question III above. Our lack of knowledge of the structure of the atmospheres of normal giants is complete.

The difficult observational problem of limb spectra, IX, has received little attention. In contrast to the results reported for U Cep, U Sge, and RZ Sct by Struve (1945), Glushneva (1964) reports an increase in the strength of the H lines of β Per at the limb, corresponding to a temperature decrease of 1000°. The problem of spectrophotometric determination of the magnitude difference between the components of spectroscopic binaries, XII, was mentioned by Struve in passing. I have been skeptical whether such observations can lead to definitive results except in the most favorable cases of equal or nearly equal components.

A result of basic importance that emerged from the observational program at McDonald, XV, was the set of very small mass functions for single-lined spectroscopic binaries having deep eclipses, the hotter component usually being a main-sequence star of relatively normal spectrum. Although the existence of a small mass function for R CMa had been known for many years, the McDonald work showed them to be of relatively frequent occurrence. It is recognized, as first pointed out by Wood (1946), that the subgiant secondary of such a system usually fills its lobe of the inner contact surface. Since the spectra of the more luminous smaller components appear to be those of normal main-sequence stars, the assumption has frequently been made that their masses are normal for their spectral types, an assumption leading to small masses, usually considerably less than 1 ○, for the subgiants. These small masses have recently been verified directly in a few cases by measuring the lines of the subgiant in the visual region of the spectrum: AW Peg, $\mathfrak{M} \approx 0.3$ ○ (Hilton and McNamara 1961); DN Ori, $\mathfrak{M} \approx 0.2$ ○ (Smak 1964); XZ Sgr, $\mathfrak{M} \approx 0.25$ ○ (Smak 1965); AS Eri, $\mathfrak{M} \approx 0.2$ ○ (Popper unpublished). The main sequence primaries of these systems have approximately normal masses. Small masses may also occur among late-type giant secondary components, as in the case of RZ Cnc, type K4 III, (Popper 1962). These stars are, of course, very much more luminous than main-sequence stars of similar mass and very much less massive than subgiants or giants with similar luminosity in detached systems. Clues for the existence of such stars may be furnished by the evolutionary histories of close binary systems, including effects of mass exchange, as discussed for example, by Morton (1960), Smak (1964), Kippenhahn, Kohl, and Weigert (1967), and Paczynski (1967).

Although the large difference of mass between components of a W UMa system nearly equal to each other in luminosity had been found earlier (Popper 1943), Struve's work on these systems (1948) had not commenced at the time he wrote the 1945 paper. The first rational attempt to explain the anomalous properties of the W UMa stars appears to be that of Lucy (1968).

There has been considerable discussion of another class of binaries, the so-called R CMa stars, in which not only the subgiant secondaries (as described above) but also the main-sequence primaries with normal-appearing spectra, are taken to be greatly overluminous for their masses (Kopal 1956; Smak 1961; Sahade 1963) on the basis of indirect arguments involving limiting equipotential surfaces. The current situation with respect to observations of these systems is summarized by Plavec and Grygar (1965), to which summary should be added the observation by Smak (1965) that the primary component of XZ Sgr has normal mass, not the small value that had been assumed previously, and that the masses of TU Mon are not abnormal (Popper 1967b). The conclusion is that both observational and theoretical uncertainties (including possible effects of nonsynchronous rotation on the size of the limiting contact surface) are such that there is inadequate evidence for the existence of such undermassive main-sequence primaries.

It has recently become realized that stars considerably underluminous for their masses also exist in close binary systems. The only direct, though not very precise, determination for such a case is the hotter, more massive F-type component of KU Cyg (Popper 1965), $\mathfrak{M} \approx 5 \odot$, $M_{bol} \approx +0.5$. Recent interpretations of β Lyr (e.g., Woolf 1965) lead to $\mathfrak{M} \approx 12 \odot$, $M_{bol} \approx -2$ for the cooler, more massive, less luminous, F-type component. Sahade has proposed (1966) that there may be a considerable group of spectroscopic binaries of large mass with one underluminous component. β Lyr is an outstanding example of a case in which the interpretation was influenced for many years by the assumption, no longer considered valid, that the more luminous component of a binary system must also be the more massive.

Question XVI, thick atmospheres, has to do with the possible effects of continuous opacity in extended atmospheres of eclipsing binaries. While there have been a few attempts (Gaposchkin and Gaposchkin 1945; Popper 1955; Lukatskaya and Rubashevsky 1961) to interpret the light curves of eclipsing binaries in terms of occultations of extended atmospheres, this very difficult subject has not received definitive treatment, and its status is uncertain. Another approach to this problem is to study the increase in duration of eclipse by an extended atmosphere with decreasing wavelength, an effect discovered by Roach and Wood (1952) for the supergiant component of ζ Aur. I have sought, without success, for a similar effect in the less favorable cases of the giant components of KU Cyg and TW Cnc. Wood and Richardson (1964) report its possible existence for the K-type giant AL Vel. In systems such as KU Cyg and SX Cas, where the absorption-line spectrum as a whole is suspected of being produced in a thick shell, the continuous opacity of the shell may be appreciable. Wolf-Rayet binaries such as V444 Cyg present special problems of a related kind (e.g., Underhill, 1960).

III. CONSEQUENCES OF THE REVOLUTION

With the exception of question XV (small mass functions), the list of problems of close binaries discussed here in Section II does not involve derivation of orbital elements, *per se*, nor the determination of fundamental properties of the stars. It would be unjust, of course, to infer that Struve ignored questions of this category.

His paper (Struve 1948) on departures from the mass-luminosity relation, for example, is a major contribution. But he was much more interested in the physical phenomena. Even though (and perhaps partly because) his first work at Yerkes was concerned mainly with the determination of orbital elements, as early as 1927 Struve was to refer to the need for studying the "physical" as well as the "mechanical" properties of spectroscopic binaries (Struve 1927). His work has, nevertheless, had a very great effect on studies of the "mechanical," or as one may prefer to term them, the "fundamental," properties of close binaries.

In the first place, the status of every close binary system, with respect to its suitability for providing reliable masses and radii of the components, has been changed by virtue of Struve's work, from "suitable unless proven unsuitable" to "unsuitable unless proven suitable." While each case must be considered in its own light, one may list a few relevant criteria of suitability. Only systems with the lines of both components measurable will be considered, since only they provide masses and radii directly. (1) The velocity curves of both components should have the same orbital elements except for the amplitudes of velocity variation. The elements most sensitive to the effects of gas streams are e, ω, and the velocity of the center of mass. As has been emphasized by Petrie (1962), a small eccentricity, perhaps too small to be well evaluated spectroscopically, can give rise to an apparent difference in the center-of-mass velocity if the orbits are assumed to be circular. (2) Lines of different atoms in the spectrum of a component should give the same velocity curve. One must be careful to avoid lines (particularly the hydrogen lines in many cases) for which blending of the two components may give the appearance of a smaller amplitude of velocity variation than is actually the case. (3) The elements from the velocity curve, in particular $e \cos \omega$ and the epoch of nodal passage, must agree with the elements from the light curve. (4) The difference in spectral type of the components, if well evaluated, should not contradict prediction from the light curve (including color variations).

Thus the impact of the work of Struve and his colleagues has been to decrease the fraction of systems considered suitable for fundamental work. But even more important, it has resulted in an increase in the reliability of our results by forcing us to look for and avoid effects that could cause incorrect masses and radii to be obtained.

It is well known that the assumption of a single mass-luminosity relation for all Population I stars except white dwarfs delayed progress in the development of theories of stellar interiors and evolution as well as in the interpretation of observations of close binary stars, β Lyr being an outstanding example of the latter. Although an important outcome of modern work is the realization that stars in close binary systems may depart greatly from a single mass-luminosity relation (see the discussion on question XV above), it is fair to say that observational data on the masses of stars above the main sequence have been so scarce as to have had very little influence on the development of current evolutionary theory (see $e.g.$, Oke and Schwarzschild, 1952). The discussion by Struve (1948) already referred to (see also Parenago and Masevich 1950; Struve and Gould 1954) might have been taken to indicate that stars overluminous for their mass do exist. But it was not until Struve's work on Capella (Struve and Kung 1953) showed that the mass of the G-type giant component is considerably less than had been accepted for a generation that an evolutionary interpretation ($i.e.$, an inhomogeneous structure) of giant stars found direct observational support. Among eclipsing binaries, the less extreme cases of detached subgiant stars overluminous for their masses, as well as the revised mass of the supergiant component of ζ Aur (Popper 1961) are reasonably in accord with theoretical expectations. But the existence of stars both greatly overluminous (subgiants in some semidetached systems) and underluminous (β Lyr, KU Cyg) for their masses, and the uncertain nature of interactions within the binary systems have made it questionable whether fundamental data derived for stars in close binary systems are applicable to the interpretation of the structure and evolution of single and wide double stars.

Thus, while we have concluded that the Struve revolution in studying the physical properties of close binaries has led to a much clearer understanding as to what data should be used for obtaining reliable fundamental properties of the stars, we are left in an uncertain position with respect to the interpretation of the properties. It is essential that the uncertainty be resolved, since all our direct determinations of masses of O- and B-type stars, and most of those of A- and F-type stars come from analyses of eclipsing binaries, which also provide some of the very small amount of information available on the masses of subgiant, giant, and supergiant stars.

The nature of the uncertainty involves two questions. The first is whether, in any given case, the star is an equilibrium configuration in the sense of stellar structure, with its radius and luminosity dependent only on its mass and the distribution of its atomic constituents. Departures from equilibrium, due either to a contraction phase in normal evolution or to exchange of mass in a close binary, may be assumed to have short durations relative to equilibrium phases (Morton 1960; Smak 1962; Kippenhahn et al. 1967). As noted before, Struve was skeptical at the end (1960, 1962) whether observations of gas streams and changes of period are adequate evidence of secular changes in binary systems. Without passing judgment on whether non-equilibrium stellar configurations are found among known systems, we are probably justified in assuming that most stars are equilibrium configurations and that the few exceptions, if any exist, are to be found among systems showing evidence of mass transfer. But these are just the systems that are unsuitable anyway for reliable fundamental results because of extraneous effects in their spectra. The underluminous stars KU Cyg and β Lyr both fall in this category. The greatly overluminous stars do not appear to do so in general.

There remains the question of judging for which classes of eclipsing binary systems, not currently involved with gas streams, the present equilibrium configurations of the components are likely to have their counterparts in the general stellar population. For the purposes of discussing this question, it is useful to classify systems into the groups proposed by Kopal (1955): detached, semidetached, and contact systems. Contact systems need not be considered further, since solutions of their light curves are too uncertain to furnish reliable results. The most favorable systems are detached systems with the components separated sufficiently far (radii less than 0.2, say, of the distance between centers) that tidal interaction is unimportant. The origin and pre-main sequence history of such systems is an outstanding problem. But during an earlier epoch, the chemical compositions of the two component stars were presumably homogeneous and the same, so that any interchange of matter that may have occurred could not cause a current equilibrium configuration differing from that of a single star with the same fundamental properties, including chemical composition and age since reaching the main sequence. Systems such as ζ Aurigae, with the components separated so far that, even though one is a supergiant, there are no effects of tidal interaction, may be assumed a fortiori comparable to single stars in their histories. Observational evidence has been given (Popper 1959a, 1959b, 1965) supporting the expectation that detached main-sequence eclipsing binaries of types A to G on the one hand, and single stars and visual binaries on the other, have very similar properties in the H-R and mass-luminosity diagrams. If further investigations maintain this conclusion, we can perhaps have confidence in the use of the components of detached eclipsing binaries to provide observational checks on the theory of the structure and evolution of single stars of all types for which adequate data are available. It may be pointed out, parenthetically, that the detached eclipsing binaries for which the best data are obtainable consist of two similar main-sequence components. If the components differ by as much as one magnitude in luminosity, the lines of the fainter component are not likely to be reliably measurable. It is assumed that this selection effect does not influence the conclusions drawn here.

In evaluating the observed positions of eclipsing binaries in the mass-luminosity diagram, it is essential to know, for comparison of observation with theory, whether the stars lie on the main sequence of the H-R diagram (only Population I stars are in question) or are above it by a determined amount. There are several observational approaches to obtaining this knowledge. The most direct one is to replace, for the binaries, the H-R diagram by a color-radius diagram, since the radii are obtained directly in the analysis. Then the lower envelope of radii in such a diagram corresponds to the lower envelope of the distribution (the "zero-age main sequence") of the H-R diagram, and departures from this sequence for a given color are also directly obtained. A second approach is to employ calibrated narrow-band photometry alone for determining positions in the equivalent of an H-R diagram, as in the work of Strömgren and others. Colors and hydrogen-line equivalent widths could be used similarly. There are serious difficulties in evaluating these observational criteria in a composite spectrum.

While the status of the components of detached systems appears favorable for the question under discussion, the situation with respect to semidetached systems is less clear. In the first place, the observational determination of the masses and radii is usually difficult because of the difference of luminosity of the components. But more fundamental is the question of interpretation of the results. Consider first the components filling their limiting zero-velocity surfaces. These are the stars for which a large range of masses, about 0.2 to 1.2 $\odot$, is found for stars of similar luminosity. Let us accept the usual view that the filling of the contact surface of such a star is a result of processes interrupting the normal evolutionary development of one or both components, accompanied by possible redistribution of mass and of the atomic species. Whatever the details of the processes may be, even though such a star may have a mean radius and luminosity governed by equilibrium considerations of stellar structure, we are probably not justified, without firm knowledge, in assuming the star to have its counterpart among the normal stellar population. Next consider the primary component of a semidetached system. In a number of cases such a star appears to have a spectrum and radius similar to those of main-sequence stars. But, not knowing certainly the past history of the system, we cannot be sure that it has a "normal" structure, nor, as a consequence, that it is suitable for our purposes. The answer to this question should be left until more observational evidence can be brought to bear.

We are thus left with the conclusion that, if one is to take a conservative view, only the masses of stars in detached eclipsing binary systems, in which post-main-sequence mass transfer has not occurred, can be assumed suitable for providing fundamental information relevant to problems of the structure and evolution of single and wide double stars, with the possibility that detached components in semidetached systems may eventually prove suitable also.

It is, of course, just the study of the properties of the components of "nonsuitable" semidetached and contact systems that may furnish clues as to the past histories of evolved close binaries. Possible examples of such clues are the high helium abundance in the atmosphere of the B8 component of β Lyr (Boyarchuk 1960; Struve and Zebergs 1961), the very small masses of some subgiants, the suggested pulsation of one component of UX Mon (Lynds 1957), and the discovery of what are reported to be W UMa systems in the clusters M67 and NGC 188 (Efremov, Kholopov, Kukarkin, and Sharov 1964; Kurochkin 1965).

Assistance from a contract with the Office of Naval Research is acknowledged.

REFERENCES

Appenzeller, I. 1965, *Ap. J.*, **141**, 1390.
Boyarchuk, A. A. 1960, *Soviet Astron.—A. J.*, **3**, 748.
Crawford, J. A. 1955, *Ap. J.*, **121**, 71.
Efremov, Y. N., Kholopov, P. N., Kukarkin, B. V., and Sharov, A. S. 1964, *Inform. Bull. Var. Stars*, No. 75.
Gaposchkin, S. and Gaposchkin, C. P. 1945, *Ap. J.*, **101**, 56.
Glusheva, I. N. 1964, *Soviet Astron.—A. J.*, **8**, 163.
Hack, M. 1963, *Internat. School. Phys. "Enrico Fermi,"* **28**, 452.
Hiltner, W. A. 1966, *Ap. J.*, **143**, 1008.
Hilton, W. B. and McNamara, D. H. 1961, *Ap. J.*, **134**, 839.
Huang, S.-S. 1959, *Ann. d'Ap.*, **22**, 527.
Kippenhahn, R., Kohl, K., and Weigert, A. 1967, *Zs. f. Ap.*, **66**, 58.
Kopal, Z. 1954, *Mém. Soc. Roy. Sci. Liège*, **15**, 84.
———. 1955, *Ann. d'Ap.*, **18**, 379.
———. 1956, *Ann. d'Ap.*, **19**, 299.
———. 1959, *Close Binary Systems* (London: Chapman and Hall), pp. 529 *et seq.*
Kraft, R. P. 1963, *Advances in Astronomy and Astrophysics* (New York and London: Academic Press) **2**, 43.
Kurochkin, N. E. 1965, *Inform. Bull. Var. Stars*, No. 79.
Lucy, L. 1968, *Ap. J.*, **153**, 877.
Lukatskaya, F. I. and Rubashevsky, A. A. 1961, *Peremennye Zvezdy*, **13**, 345.
Lynds, C. R. 1957, *Ap. J.*, **126**, 69.
Morton, D. C. 1960, *Ap. J.*, **132**, 146.
Oke, J. B. and Schwarzschild, M. 1952, *Ap. J.*, **116**, 317.
Paczynski, B. 1967, *Acta Astr.*, **17**, 193.
Parenago, P. P. and Masevich, A. G. 1950, *Proc. Shternberg Astr. Inst.*, No. 20.
Petrie, R. M. 1962, *Astronomical Techniques*, ed. W. A. Hiltner (Chicago: University of Chicago Press), Chap. 23.
Plavec, M. and Grygar, J. 1965, *Kl. Veröff. Bamberg*, **4**, Nr. 40, 213.
Popper, D. M. 1943, *Ap. J.*, **97**, 394.
———. 1955, *Ap. J.*, **121**, 56.
———. 1959, *Ap. J.*, **129**, 659.
———. 1960, *Mém. Soc. Roy. Sci. Liège*, 5th Ser., **3**, 96.
———. 1961, *Ap. J.*, **134**, 828.
———. 1962, *Pub. A.S.P.*, **74**, 129.
———. 1965a, *Ap. J.*, **141**, 126.
———. 1965b, *Ap. J.*, **141**, 314.
———. 1967a, *Ann. Rev. Astr. Ap.*, **5**, 85.
———. 1967b, *Pub. A.S.P.*, **79**, 493.
Roach, F. E. and Wood, F. B. 1952, *Ann. d'Ap.*, **15**, 21
Sahade, J. 1960, *Stellar Atmospheres*, ed. J. L. Greenstein (Chicago: University of Chicago Press), Chap. 12.
———. 1963, *Ann. d'Ap.*, **26**, 80.
———. 1966, *Stellar Evolution*, ed. R. F. Stein and A. G. W. Cameron (New York: Plenum Press), p. 449.
Sahade, J. and Struve, O. 1945, *Ap. J.*, **102**, 480.
Shakovskoy, N. M. 1962, *Soviet Astron.—A. J.*, **6**, 587.
Smak, J. 1961, *Acta Astr.*, **11**, 171.
———. 1964, *Pub. A.S.P.*, **76**, 210.
———. 1965, *Acta Astr.*, **15**, 327.
Struve, O. 1923a, *Ap. J.*, **58**, 138.
———. 1923b, *Ap. J.*, **58**, 141.
———. 1927, *Mirovedenie*, **16**, 53. (Abstracted in *Astr. Jahresbericht*, **29**, 177, 1928).
———. 1934, *Observatory*, **57**, 265.
———. 1945, *Ap. J.*, **102**, 74.
———. 1948, *Ann. d'Ap.*, **11**, 117.
———. 1950, *Stellar Evolution* (Princeton: Princeton University Press).
———. 1955, *Sky and Telescope*, **15**, 64.
———. 1960, *Mém. Soc. Roy. Sci. Liège*, 5th Ser., **3**, 36.
———. 1962, *Symposium on Stellar Evolution* (La Plata: National University of La Plata), p. 225 *et seq.*
———. 1963a, *Pub. A.S.P.*, **75**, 207.
———. 1963b, *Sky and Telescope*, **25**, 199.

Struve, O. and Gould, N. 1954, *Pub. A.S.P.*, **66**, 28.
Struve, O. and Kung, S. M. 1953, *Ap. J.*, **117**, 1.
Struve, O., Sahade, J., Huang, S.-S., and Zebergs, V. 1958, *Ap. J.*, **128**, 310.
Struve, O. and Zebergs, V. 1961, *Ap. J.*, **134**, 161.
Underhill, A. B. 1960, *Stellar Atmospheres*, ed. J. L. Greenstein (Chicago: University of Chicago Press), Chap. 10.
Walker, M. F. and Popper, D. M. 1964, *Ap. J.*, **139**, 168.
Wood, F. B. 1946, *Contrib. Princeton Univ. Obs.*, No. 21.
Wood, F. B. and Richardson, R. R. 1964, *Astr. J.*, **69**, 297.
Woolf, N. J. 1965, *Ap. J.*, **141**, 155.
Wyse, A. B. 1934, *Lick Obs. Bull.*, **17**, 37.

INDEXES

SUBJECT INDEX

AUTHOR INDEX

Only the first author of a paper with multiple authorship is referenced.

457

INDEX OF STARS, NEBULAE, AND CLUSTERS

461